VOLUME SIX HUNDRED AND SEVENTY SIX

METHODS IN ENZYMOLOGY

Biochemical Pathways and Environmental Responses in Plants: Part A

METHODS IN ENZYMOLOGY

VOLUME SIX HUNDRED AND SEVENTY SIX

METHODS IN ENZYMOLOGY

Biochemical Pathways and Environmental Responses in Plants: Part A

Edited by

JOSEPH JEZ

Department of Biology,
Washington University in St. Louis,
United States

ELSEVIER

ACADEMIC PRESS
An imprint of Elsevier

Academic Press is an imprint of Elsevier
50 Hampshire Street, 5th Floor, Cambridge, MA 02139, United States
525 B Street, Suite 1650, San Diego, CA 92101, United States
The Boulevard, Langford Lane, Kidlington, Oxford OX5 1GB, United Kingdom
125 London Wall, London, EC2Y 5AS, United Kingdom

First edition 2022

ISBN: 978-0-323-95561-4
ISSN: 0076-6879

For information on all Academic Press publications
visit our website at https://www.elsevier.com/books-and-journals

Publisher: Zoe Kruze
Developmental Editor: Federico Paulo S. Mendoza
Production Project Manager: Sudharshini Renganathan
Cover Designer: Greg Harris

Typeset by STRAIVE, India

Contents

Contributors

Ikuro Abe
Graduate School of Pharmaceutical Sciences; Collaborative Research Institute for Innovative Microbiology, The University of Tokyo, Tokyo, Japan

Angeles Aroca
Institute of Plant Biochemistry and Photosynthesis, CSIC-University of Seville, Seville, Spain

Naveen C. Bisht
National Institute of Plant Genome Research, New Delhi, India

Clint Chapple
Department of Biochemistry; Purdue Center for Plant Biology, Purdue University, West Lafayette, IN, United States

Sixue Chen
Plant Molecular and Cellular Biology Program, University of Florida; Department of Biology, Genetics Institute, University of Florida; Proteomics and Mass Spectrometry, Interdisciplinary Center for Biotechnology Research (ICBR), University of Florida; University of Florida Genetics Institute (UFGI), University of Florida, Gainesville, FL, United States

Swarup Roy Choudhury
Donald Danforth Plant Science Center, St. Louis, MO, United States; Department of Biology, Indian Institute of Science Education and Research (IISER) Tirupati, Tirupati, Andhra Pradesh, India

Jerry D. Cohen
Department of Horticultural Science and Microbial and Plant Genomics Institute, University of Minnesota, Saint Paul, MN, United States

Katrina Cornish
Department of Horticulture and Crop Science; Department of Food, Agricultural and Biological Engineering, College of Food Agricultural and Environmental Sciences, The Ohio State University, Wooster, OH, United States

Craig P. Dufresne
Thermo Fisher Scientific Training Institute, West Palm Beach, FL, United States

Parker Evans
Indiana Center for Regenerative Medicine & Engineering, Indiana University Health Comprehensive Wound Center, Department of Surgery, Indiana University School of Medicine, Indianapolis, IN, United States

Milos R. Filipovic
Leibniz Institute for Analytical Sciences, ISAS e.V., Dortmund, Germany

Cecilia Gotor
Institute of Plant Biochemistry and Photosynthesis, CSIC-University of Seville, Seville, Spain

Süleyman Günal
Institute for Plant Sciences, Cluster of Excellence on Plant Sciences (CEPLAS), University of Cologne, Cologne, Germany

Kate Harline
Donald Danforth Plant Science Center, St. Louis, MO, United States

Adrian D. Hegeman
Department of Horticultural Science and Microbial and Plant Genomics Institute; Department of Plant and Microbial Biology, University of Minnesota, Saint Paul, MN, United States

Cynthia K. Holland
Department of Biology, Williams College, Williamstown, MA, United States

Sarah Joe
Department of Biology, Genetics Institute, University of Florida, Gainesville, FL, United States

Ana Jurado-Flores
Institute of Plant Biochemistry and Photosynthesis, CSIC-University of Seville, Seville, Spain

Sunhyung Kim
Division of Plant Science and Technology, University of Missouri, Columbia, MO, United States

Wonseok Kim
Division of Plant Science and Technology, University of Missouri, Columbia, MO, United States

Stanislav Kopriva
Institute for Plant Sciences, Cluster of Excellence on Plant Sciences (CEPLAS), University of Cologne, Cologne, Germany

Hari B. Krishnan
Division of Plant Science and Technology, University of Missouri; Plant Genetics Research Unit, USDA, Agricultural Research Service, Columbia, MO, United States

Roshan Kumar
National Institute of Plant Genome Research, New Delhi, India; Institute of Science and Technology Austria, Klosterneuburg, Austria

Eric Lam
Department of Plant Biology, Rutgers, The State University of New Jersey, New Brunswick, NJ, United States

Chuwei Lin
Department of Biology, Genetics Institute; University of Florida Genetics Institute (UFGI), University of Florida, Gainesville, FL, United States

Chang-Jun Liu
Biology Department, Brookhaven National Laboratory, Upton, NY, United States

Deng Liu
Department of Biology, Washington University, St. Louis, MO, United States

Haijiao Liu
Department of Material Science and Chemical Engineering, Stony Brook University, Stony Brook, NY, United States

Qun Liu
Department of Material Science and Chemical Engineering, Stony Brook University, Stony Brook; Biology Department, Brookhaven National Laboratory, Upton; Department of Biochemistry and Cell Biology, Stony Brook University, Stony Brook, NY, United States

Takahiro Mori
Graduate School of Pharmaceutical Sciences; Collaborative Research Institute for Innovative Microbiology, The University of Tokyo, Tokyo; PRESTO, Japan Science and Technology Agency, Saitama, Japan

Hiroyuki Morita
Institute of Natural Medicine, University of Toyama, Toyama, Japan

Yu Nakashima
Institute of Natural Medicine, University of Toyama, Toyama, Japan

Ülo Niinemets
Crop Science and Plant Biology, Estonian University of Life Sciences, Tartu, Estonia

Joel Nott
The Protein Facility of the Iowa State University Office of Biotechnology, Ames, IA, United States

Himadri B. Pakrasi
Department of Biology, Washington University, St. Louis, MO, United States

Sona Pandey
Donald Danforth Plant Science Center, St. Louis, MO, United States

Noé Perron
Plant Molecular and Cellular Biology Program, University of Florida, Gainesville, FL, United States

Bahtijor Rasulov
Crop Science and Plant Biology, Estonian University of Life Sciences, Tartu, Estonia

Michael Reichelt
Department for Biochemistry, Max Planck Institute for Chemical Ecology, Jena, Germany

Luis C. Romero
Institute of Plant Biochemistry and Photosynthesis, CSIC-University of Seville, Seville, Spain

Abira Sahu
MSU-DOE Plant Research Laboratory, Plant Resilience Institute, and Department of Biochemistry and Molecular Biology, Michigan State University, East Lansing, MI, United States

Annesha Sengupta
Department of Biology, Washington University, St. Louis, MO, United States

Thomas D. Sharkey
MSU-DOE Plant Research Laboratory, Plant Resilience Institute, and Department of Biochemistry and Molecular Biology, Michigan State University, East Lansing, MI, United States

Jeffrey Simpson
Department of Biochemistry; Purdue Center for Plant Biology, Purdue University, West Lafayette, IN, United States

Hisham Tadfie
Department of Biology, Williams College, Williamstown, MA, United States

Bowen Tan
Department of Biology, Genetics Institute, University of Florida, Gainesville, FL, United States

Qian Tang
Department of Horticultural Science and Microbial and Plant Genomics Institute, University of Minnesota, Saint Paul, MN, United States

Sarathi M. Weraduwage
MSU-DOE Plant Research Laboratory, Plant Resilience Institute, and Department of Biochemistry and Molecular Biology, Michigan State University, East Lansing, MI, United States; Department of Biology/Biochemistry, Bishop's University, Sherbrooke, QC, Canada

Incheol Yeo
Department of Biochemistry & Biophysics, Texas A&M University, College Station, TX, United States

Qingfang Zhang
Department of Biochemistry and Cell Biology, Stony Brook University, Stony Brook, NY, United States

Guimei Zhao
Department of Biology, Genetics Institute, University of Florida, Gainesville, FL, United States

Xianhai Zhao
Biology Department, Brookhaven National Laboratory, Upton, NY, United States

Ping Zhu
Biology Department, Brookhaven National Laboratory, Upton, NY, United States

Preface

The diversity of plant biochemistry matches that of the plants studied. From understanding how individual proteins function and examining the vast array of metabolites to revealing how biotic and abiotic stresses impact pathways, proteins, and genes, the tools available to study biochemical pathways and environmental responses in plants continue to provide new insights into various species. The timing of this could not be more critical. Facing population growth and changing climate, understanding plants from genotype to phenotype is essential for future innovations that will create the next green revolution.

Originally, I imagined a single volume with chapters providing examples of new approaches to studying the proteins and metabolites in plant biochemical pathways and tools for monitoring plant environmental responses; however, 1 volume grew to 3 volumes and nearly 50 chapters. Each volume contains chapters on proteins, metabolites, and the environment.

The chapters on proteins cover a wide range of topics such as strategies on how to express and purify plant proteins, how to modulate reactions and catalysis, approaches to study different enzyme families (aromatic aminotransferases, formate-tetrahydrofolate ligases, methyltransferases, a metallolyase in glucosinolate biosynthesis, trypsin inhibitors, BAHD acyltransferases, dirigent proteins, and carboxysomes), structural biology from "classic" protein crystallography to microcrystals and small-angle X-ray scattering, computational methods for testing protein-ligand interactions and homology modeling, and applications of chemical cross-linking and mass spectrometry to study multiprotein complexes.

In parallel to new techniques for examining proteins, the tools for identifying plant metabolites, monitoring metabolic changes, and discovering new compounds are being rapidly developed. Irrespective of using isotope labeling and flux analysis, monitoring specialized metabolite synthesis, mapping the metabolic origins of molecules, characterizing hydrocarbon/isoprenoid emissions, measuring the indole auxin phytohormone, isoprenyl phosphates, or diacylglycerol derivatives, examining metabolism using general profiling or targeted analyses either *in vivo* or *ex vivo*, or combining multiomic analysis, mass spectrometry reigns supreme in versatility and power for studying plant metabolism.

The proteins and biochemical pathways in plants are dynamic and continually respond to environmental stresses and signals. Importantly, different plants (and other green organisms) are not the same and need specialized environments for their study. The chapters on environment cover G-protein signaling systems, how salinity stress impacts nitrogenase, use of CRISPR genome editing in cyanobacteria to engineer metabolic processes, approaches to controlling plant growth environments for cultivation, application of RNA-seq to examine pre-mRNA splicing in response to photoreceptor signals, proteomic and phosphoproteomic strategies to study the transition of C3 to CAM metabolism, kinase interactome labeling, and redox modifications, and new visual labeling strategies to monitor lipid-protein interactions, cellular calcium signal changes, responses to flooding, and environmentally triggered redox changes.

I thank all the colleagues who helped with this effort and provided snapshots of the work and methods used in their research groups. Although these three volumes offer only a small glimpse into the range of studies examining plant biochemical pathways and environmental responses, they highlight how the field is integrated from the molecular to the environmental scale. I hope this collection is a useful resource for the community and helps those entering this area of research appreciate the field's truly multidisciplinary perspective.

Joseph Jez
Department of Biology
Washington University in St. Louis

SECTION I

Protein

CHAPTER ONE

Structure, function, and engineering of plant polyketide synthases

Takahiro Mori[a,b,c], Yu Nakashima[d], Hiroyuki Morita[d,*], and Ikuro Abe[a,b,*]

[a]Graduate School of Pharmaceutical Sciences, The University of Tokyo, Tokyo, Japan
[b]Collaborative Research Institute for Innovative Microbiology, The University of Tokyo, Tokyo, Japan
[c]PRESTO, Japan Science and Technology Agency, Saitama, Japan
[d]Institute of Natural Medicine, University of Toyama, Toyama, Japan
[*]Corresponding authors: e-mail address: hmorita@inm.u-toyama.ac.jp; abei@mol.f.u-tokyo.ac.jp

Contents

Methods in Enzymology, Volume 676
ISSN 0076-6879
https://doi.org/10.1016/bs.mie.2022.06.003

Abstract

Type III polyketide synthases (PKSs) catalyze sequential condensations of acyl-CoA thioesters to generate a variety of polyketide scaffolds with remarkable structural diversity and significant biological activities. These enzymes share a similar thiolase fold and use a common catalytic triad for the polyketide chain elongation and cyclization reactions. Structural and biochemical analyses of type III PKSs revealed that the functional diversity of these highly homologous enzymes is attributable to subtle changes in their active site volumes and architectures. The accumulated knowledge of their detailed catalytic versatility and mechanisms provides a platform for enzyme engineering *via* structure-guided approaches for the generation of unnatural novel polyketides. In this chapter, the methods for identification, biochemical characterization, and structure-guided engineering of polyketide synthases will be described in detail, along with brief overviews of the structures and functions of these enzymes.

1. Introduction

The type III polyketide synthase (PKS) superfamily enzymes catalyze the formation of diverse core structures of polyketides, including medicinal plant polyketides such as chalcones, flavonoids, stilbenes, pyrones, chromones, resorcinols, phloroglucinols, diarylheptanoids, xanthones, quinolones, and acridones (Abe, 2020; Abe & Morita, 2010; Austin & Noel, 2003; Morita, Wong, & Abe, 2019; Yu & Jez, 2008). These enzymes catalyze iterative condensation reactions of starter and extender acyl-CoA thioesters. The typical plant-specific type III PKSs chalcone synthase (CHS) and stilbene synthase (STS) utilize *p*-coumaroyl-CoA as a starter substrate and three molecules of malonyl-CoA as extender substrates to generate naringenin chalcone and resveratrol, respectively (Fig. 1A and B) (Yamaguchi et al., 1999). In these reactions, chalcone and resveratrol are generated *via* C—C bond formations with Claisen-type and decarboxylative aldol-type cyclization reactions, respectively, from a common linear poly-β-keto intermediate. In contrast, 2-pyrone synthase (2-PS) cyclizes the linear poly-β-keto intermediate through a C—O bond formation reaction to form the 2-pyrone structure (Fig. 1C) (Jez et al., 2000). In addition to these C—C and C—O bond formation reactions, some PKSs catalyze C—N bond formation. Acridone synthase (ACS) and quinolone synthase (QNS) catalyze the decarboxylative condensation of three molecules and one molecule of malonyl-CoA with *N*-methylanthranyloyl-CoA as a starter substrate to generate acridone and quinolone scaffolds, respectively, through C—N bond formation between the enzyme-tethered thioester and the nitrogen atom

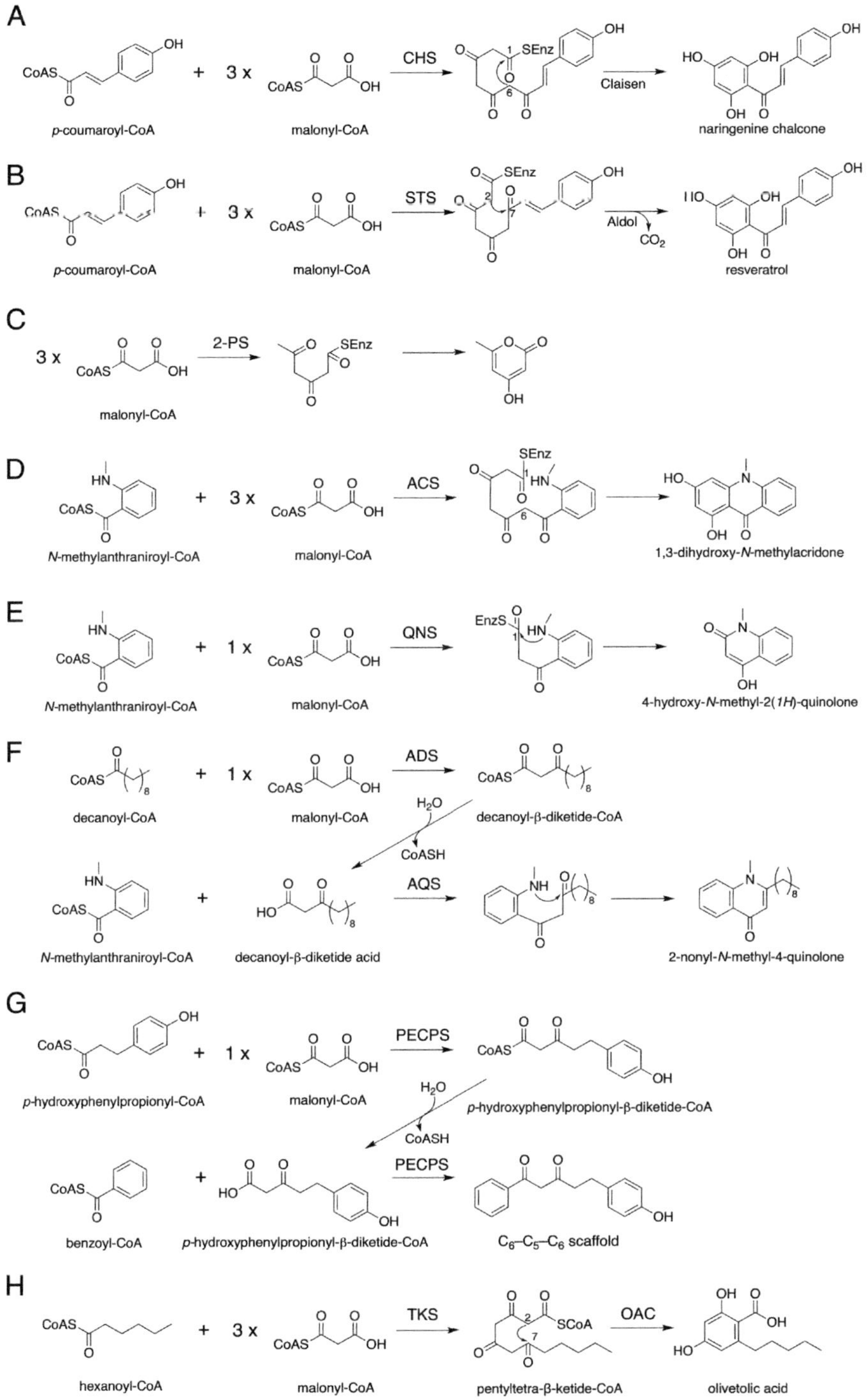

Fig. 1 Representative reactions of plant type III PKSs. Reactions of (A) CHS (Ferrer, Jez, Bowman, Dixon, & Noel, 1999), (B) STS (Schröder & Schröder, 1990), (C) 2-PS (Jez et al., 2000), (D) ACS (Springob, Lukacin, Ernwein, Gröning, & Matern, 2000), (E) QNS (Mori et al., 2013), (F) ADS and AQS (Matsui et al., 2017), (G) PECPS (Wang et al., 2022), and (H) TKS and OAC (Yang et al., 2016).

of the *N*-methylanthranyloyl-intermediate (Fig. 1D and E) (Jez, Bowman, & Noel, 2002; Lukacin et al., 1999; Mori et al., 2013; Resmi, Verma, Gokhale, & Soniya, 2013; Springob et al., 2000). Furthermore, the second generation of type III PKSs reportedly mediates the interesting chemistry of condensation reactions of distinct acyl-CoAs to produce various core skeletons of secondary metabolites in plants (Katsuyama, Kita, Funa, & Horinouchi, 2009; Katsuyama, Miyazono, Tanokura, Ohnishi, & Horinouchi, 2011; Matsui et al., 2017; Morita et al., 2010; J. Wang et al., 2016). For example, in the biosynthesis of 2-alkylquinolone alkaloids in Rutaceae plants, alkyldiketide-CoA synthase (ADS) and alkylquinolone synthase (AQS) cooperatively catalyze the condensation of a fatty acyl-CoA, *N*-methylanthranyloyl-CoA, and malonyl-CoA to yield 2-alkylquinolone alkaloids (Fig. 1F) (Matsui et al., 2017). In these reactions, ADS accepts a fatty acyl-CoA and a molecule of malonyl-CoA to generate an alkyl-β-diketide-CoA, which is non-enzymatically hydrolyzed to form an alkyl-β-diketide acid. This is accepted by AQS to catalyze the decarboxylative condensation of the alkyl-β-diketide acid with *N*-methylanthranyloyl-CoA to produce a 2-alkylquinolone compound. Similarly, in the biogenesis of 2-(2-phenylethyl) chromones in agarwood, the 2-(2-phenylethyl) chromone precursor synthase (PECPS) firstly accepts 4-hydroxyphenylpropionyl-CoA and one molecule of malonyl-CoA to yield the 4-hydroxyphenylpropionyl-β-diketide-CoA intermediate, which is released from the active site and hydrolyzed non-enzymatically to generate the corresponding β-diketide acid intermediate (Wang et al., 2022). The PECPS then accepts benzoyl-CoA and catalyzes the decarboxylated condensation of the 4-hydroxyphenylpropionyl-β-diketide acid intermediate to form the C_6–C_5–C_6 scaffold compound (Fig. 1G). Thus, in this case, a single enzyme catalyzes the condensations of three distinct substrates to produce the medicinally important plant secondary metabolite in a one-pot enzymatic synthesis.

Due to the high sequence identity among plant type III PKS-superfamily enzymes, it is quite difficult to predict their catalytic functions from the amino acid sequences (Jez et al., 2001). Biochemical and structural analyses of various type III PKSs have revealed that these enzymes possess highly homologous homodimeric αβαβα thiolase structures (Fig. 2A) (Ferrer et al., 1999; Morita et al., 2019). Type III PKSs have a characteristic CoA-binding tunnel to bind the CoA-linked starter substrate and employ a Cys-His-Asn catalytic triad to load the acyl moiety of the substrate on the active site Cys residue (Fig. 2B). Subsequent elongation of the

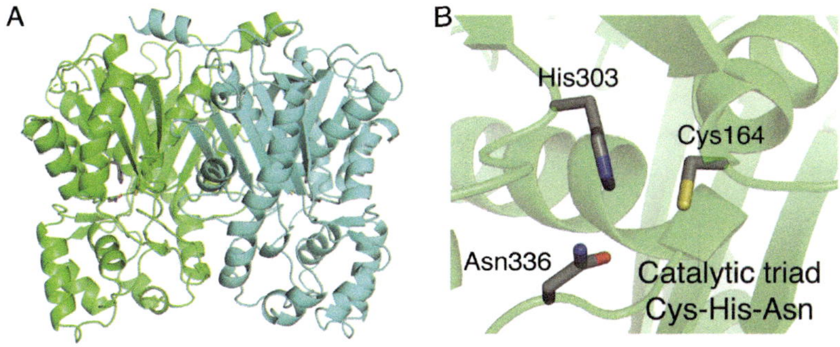

Fig. 2 Structure of a plant type III PKS (Ferrer et al., 1999; Morita et al., 2019). (A) Overall structure and (B) Cys-His-Asn catalytic triad of the plant type III PKS.

polyketide chain *via* decarboxylative condensation of the extender substrate generates the linear poly-β-ketide intermediate. Finally, the reaction is usually terminated by cyclization of the intermediate through an intramolecular Claisen, aldol condensation, lactonization, and/or C—N bond formation reaction.

The functional diversity of these highly homologous type III PKSs is attributable to the subtle changes in their active site volumes and architectures. Specifically, the active site forming residues 197, 256, 338, 215, and 265 (numbering is from *Medicago sativa* CHS, the first solved crystal structure of a type III PKS) determine these factors (Ferrer et al., 1999; Morita et al., 2019). These differences in the active site establish the starter and extender substrate preferences, the elongation reaction numbers, and the modes of cyclization reactions. These structural observations also suggest that structure-based mutagenesis can rationally alter the substrate and product specificities of enzymes. Indeed, the structure-based engineering of type III PKSs has been demonstrated to generate novel polyketide compounds (I. Abe, 2008; Jez et al., 2000).

In addition to polyketide chain formation reactions by PKSs, a structurally distinct type of the plant polyketide biosynthetic enzyme that catalyzes the cyclization reaction of the linear poly-β-ketide intermediate has been identified. Olivetolic acid cyclase (OAC), in the biosynthesis of cannabinoid from *Cannabis sativa*, is the only known plant polyketide cyclase (Gagne et al., 2012; Taura et al., 2009). OAC catalyzes the C2–C7 aldol cyclization of a linear pentyltetra-β-ketide intermediate, which is produced by a type III PKS tetraketide synthase (TKS), to generate olivetolic acid without any cofactors (Fig. 1F). The structural analysis of OAC revealed that it is composed of a dimer of an α + β fold with a 9 Å long pentyl intermediate-binding pocket

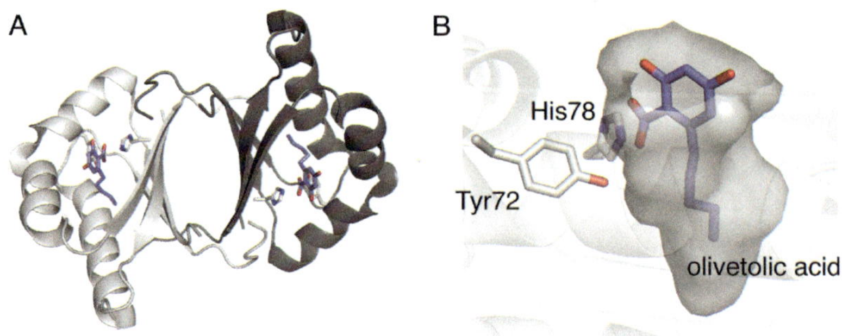

Fig. 3 Structure of OAC (Yang et al., 2016). (A) The overall structure and (B) the active site of OAC with olivetolic acid.

(Fig. 3A and B) (Yang et al., 2016). This pocket consists of hydrophobic residues that bind the pentyl moiety of the olivetolic acid. The Tyr72-His78 catalytic dyad, located close to the active site entrance, plays a catalytic role as an acid/base catalyst (Fig. 3B). The structure-based engineering of OAC has expanded the substrate specificity for the aliphatic moieties of substrates to generate alkylresorcylic acids with alkyl side chains ranging from C5 to C11 (Lee, Nakashima, Kodama, Chen, & Morita, 2022).

A summary of the engineering of plant type III PKSs was described previously (Wakimoto, Morita, & Abe, 2012). This chapter provides a follow-up for the methodologies currently used for the preparation, assay, crystallization, and structure-based mutagenesis of type III PKSs (ACS, QNS, ADS, AQS, PECPS, TKS, and OAC) in our laboratory during the past decade.

2. Acridone synthase (ACS) and quinolone synthase (QNS)

2.1 Overview

Citrus microcarpa acridone synthase (ACS) and quinolone synthase (QNS) catalyze the formation of the tetraketide 1,3-dihydroxy-*N*-methylacridone and the diketide 4-hydroxy-*N*-methylquinolone, respectively. *C. microcarpa* ACS (CmACS) accepts *N*-methylanthraniloyl-CoA as a starter substrate and catalyzes the decarboxylative condensation of three molecules of malonyl-CoA to generate the tetraketide intermediate, which is subsequently cyclized to produce 1,3-dihydroxy-*N*-methylacridone (Mori et al., 2013). While *C. microcarpa* QNS (CmQNS) also accepts *N*-methylanthraniloyl-CoA as a starter substrate, the enzyme catalyzes the condensation of one molecule of malonyl-CoA and the following cyclization reaction to produce 4-hydroxy-*N*-methylquinolone. CmACS shares 56–61% amino acid sequence identities with CmQNS (60%), *Medicago sativa* chalcone synthase (MsCHS: 61%) (Ferrer et al., 1999), *Ruta graveolens* ACS (56%) (Lukacin et al., 1999; Springob et al., 2000), and *Aegle marmelos* QNS (AmQNS: 59%) (Resmi et al., 2013). CmQNS shares ~60% amino acid sequence identities with MsCHS (60%) and AmQNS (59%). CmQNS generates the diketide 4-hydroxy-*N*-methylquinolone as a single product, in contrast to the previously reported quinolone- and acridone-producing AmQNS. The genes encoding CmACS and CmQNS have been identified in *C. microcarpa* leaves. Here, we report the cloning, functional characterization, and structural analysis of CmACS and CmQNS by heterologous expression of the enzymes in *E. coli*.

2.2 Cloning, overexpression, and purification of CmACS and CmQNS

2.2.1 Materials

- RNeasy plant mini kit (Qiagen)
- Superscript II RT (Invitrogen)
- 5′-RACE system (Invitrogen)
- TaKaRa PCR Thermal Cycler DiceTM
- pQE80L (QIAGEN)
- Restriction enzymes BamHI and XhoI (TaKaRa)
- Escherichia coli M15
- Luria Bertani (LB) media
- LB agar plates containing 100 μg/mL of ampicillin
- Ampicillin, 100 mg/mL
- 1 M isopropyl-1-thio-β-D-galactopyranoside (IPTG)
- Erlenmeyer flask (2 L)
- Temperature-controlled shaker
- Lysis buffer: 50 mM Tris-HCl buffer (pH 7.5), 0.2 M NaCl, 5% (v/v) glycerol, and 5 mM imidazole
- Wash buffer: 50 mM Tris-HCl buffer (pH 7.5), 0.2 M NaCl, 5% (v/v) glycerol, and 10 mM imidazole
- Elution buffer: 50 mM Tris-HCl buffer (pH 7.5), 0.2 M NaCl, 5% (v/v) glycerol, and 300 mM imidazole
- Ion exchange buffer A: 50 mM Tris-HCl buffer (pH 7.5), 5% (v/v) glycerol, and 2 mM DTT
- Ion exchange buffer B: 50 mM Tris-HCl buffer (pH 7.5), 1 M NaCl,5% (v/v) glycerol, and 2 mM DTT

- Gel filtration buffer: 20 mM Tris-HCl buffer (pH 7.5), 100 mM NaCl, and 2 mM DTT
- High speed refrigerated microcentrifuge
- Ni-NTA resin (Cytiva)
- Resource-Q column (Cytiva)
- HiLoad™ 16/600 Superdex™ 200 pg (Cytiva)
- ÄKTAexplorer (Cytiva)

2.2.2 Procedure for cloning of CmACS and CmQNS

1. Total RNA from *Citrus microcarpa* Bunge leaves is extracted with an RNeasy plant mini kit (Qiagen).
2. Reverse-transcription of total RNA is performed by using the oligo(dT) primer (RACE 32 = 5′-GGCCACGCGT-CGACTAGTACTTTTTT TTTTTTTTTTT-3′) and Superscript II RT (Invitrogen).
3. The polyketide synthase genes are amplified by PCR with the obtained cDNA and inosine-containing degenerate primers, designed based on the conserved sequences of CHSs: 112S: 5′-(A/G)A(A/G)GCIITI(A/C) A(A/G)GA(A/G)TGGGGICA-3′, 174S: 5′-GCIAA(A/G)GA(T/C)ITI GCIGA(A/G)AA(T/C)AA-3′, 368A: 5′-CCC(C/A)(A/T)ITCIA(A/G) ICCITCICCIGTIGT-3′, and 380A: 5′-TCIA(T/C)IGTIA(A/G)ICCI GGICC(A/G)AA-3′ (the primer's name indicates the amino acid number of MsCHS).
4. The nested PCR is performed with the 112S and 380A primers. Subsequently, the core fragments of CmACS and CmQNS are amplified by PCR with the 174S and 368A primers
5. The 3′-RACE method is used to amplify the 642-bp fragment of CmACS and the 472-bp fragment of CmQNS with two specific primers (207S: 5′-GATTCTCTAGTGGGTCAGGCTCTTT-3′ and 245S: 5′-GCTTCTGCCTGATTCTGATGGTGCAA-3′ for CmACS, and 209S: 5′-ACA-TCTTGGTTGGGCAAGCTGTGTTTG-3′ and 252S: 5′-ACCAACAAGTTTATCAGGGCACACGTG-3′ for CmQNS), and the RACE32 primer
6. The 681-bp fragment of CmACS and the 643-bp fragment of CmQNS are amplified using the 5′-RACE system (Invitrogen) and two specific primers (253A: 5′-ATTGCACCATCAGAATCAGGCAGAAGC-3′ and 215A: 5′-AAAGAGCCTGACCCACTAGAGAATC-3′ for CmACS, and 261A: 5′-TTTCACCTGTGTGATAAACTTGTTG GT-3′ and 218A: 5′-CCAAACACAGCTTGCCCAACCAAGA TG-3′ for CmQNS).

7. The full-length cDNAs encoding CmACS and CmQNS are amplified with specific primers (5′-GCGAGATCTATGGTAACCATGGAGGAGATTAGAAGGGCT-3′ and 5′-CGCCTCGAGTCATGCTTCTATAGGGAGACTGTGCAACA-3′ for CmACS, and 5′-GCGAGATCTATGGAATCAATGGCGAAAGTGAAAAATTTTCTTAATGCCAAG-3′ and 5′-CGCCTCGAGTCAACAGGCGGAATCAATAGGGACGCT-3′ for CmQNS, with the *Bgl*II and *Xho*I digestion sites underlined, respectively
8. The full-length CmACS and CmQNS cDNA fragments are sub-cloned into the BamHI/XhoI sites of the pQE80L expression plasmid (Novagen), respectively

Notes

- The typical yields of total RNA are 20–50 μg per 100 mg of leaves.
- Although we subcloned the genes into pQE80L, other plasmids such as pET vectors can also be used.

2.2.3 Procedure for expression and purification of CmACS and CmQNS

1. The full-length cDNAs encoding CmACS and CmQNS in pQE80L are individually transformed into *E. coli* M15 and selected by cultivation on Luria-Bertani (LB) agar plates with 100 μg/mL ampicillin, at 37 °C for 16 h
2. A single colony is selected and incubated overnight in 10 mL LB broth containing 100 μg/mL ampicillin at 37 °C
3. The seed culture is transferred into 1 L of LB broth containing 100 μg/mL ampicillin and incubated at 37 °C until the optical density at 600 nm (OD_{600}) becomes 0.6.
4. Protein expression is induced by the addition of IPTG (final concentration 1 mM).
5. The culture is incubated at 23 °C for another 16 h
6. Cultured cells are harvested by centrifugation at 5000 × *g* for 10 min, and the cell pellet can be stored at −20 °C until use
7. The harvested cells are resuspended in 10 mL of lysis buffer
8. The cells are disrupted by sonication on ice, and the cell debris is removed by centrifugation at 12,000 × *g* for 30 min
9. The supernatant is loaded onto a 5 mL column of Ni-NTA resin equilibrated with lysis buffer. This procedure is repeated at least two times
10. The resin is washed with 10 column volumes (CVs) of wash buffer
11. The recombinant protein is subsequently eluted with 5 CVs of elution buffer

12. The protein solution is concentrated to 2 mL with a 10 kDa MWCO Amicon ultrafiltration unit (Merck), and then diluted 5-fold with Ion exchange buffer A
13. The sample is injected onto a Resource-Q column (Cytiva). The column is washed with Ion exchange buffer A containing 50 mM NaCl, and the protein is eluted using a linear gradient of 50–300 mM NaCl
14. The fraction containing the target protein is concentrated to 5 mL with a 10 kDa MWCO Amicon ultrafiltration unit (Merck), and then purified to homogeneity by gel filtration chromatography on a HiLoad™ 16/600 Superdex™ 200 pg column (Cytiva). The fraction containing the target protein is concentrated to 10 mg/mL in gel filtration buffer
15. The purified protein is flash frozen in liquid nitrogen and can be stored at −80 °C until use

Notes
- The proteins are expressed with a His_6-tag at the N-terminus
- Parameters such as the concentration of IPTG, incubation time before induction, speed of shaking during expression can be optimized to improve the yields of proteins
- Other *E. coli* strains; *i.e.*, BL21 (DE3) CodonPlus and Rosetta2 (DE3), allow the use of rare codons of *E. coli*, which can enhance the expression levels
- All purification procedures are performed at 4 °C to prevent enzyme aggregation.
- The purification procedures are designed for a 4 L expression culture
- The typical yields for both enzymes are 5–10 mg per 1 L culture
- The purity of the isolated enzymes is examined by SDS-PAGE
- For crystallization of enzymes, the entire purification should be performed within 1 day

2.3 *In vitro* characterizations of CmACS and CmQNS

Since it is difficult to predict the enzyme substrates, various acyl-CoAs should be used as substrates to clarify the substrate specificity of the enzyme. CmACS shows broad substrate specificity and accepts various starter substrates, including *N*-methylanthranyloyl-CoA, *p*-coumaroyl-CoA, benzoyl-CoA, and hexanoyl-CoA as starter substrates to generate acridone, quinolone, phloroglucinol, benzophenone, and chalcone, respectively. Some of the acyl-CoA substrates are not commercially available, and thus must be synthesized (Ikuro Abe, Morita, Nomura, & Noguchi, 2000; I. Abe, Watanabe, & Noguchi, 2004; Mori et al., 2013). In contrast to

CmACS, CmQNS exhibits specific substrate and product tolerances. CmQNS can only accept *N*-methylanthranyloyl-CoA as the starter substrate to produce quinolone as the single product. The enzyme reaction products can be separated and analyzed by LC-MS.

2.3.1 Materials

- HPLC 1100 series (Agilent Technologies)
- Esquire 4000 ion trap mass spectrometer (Bruker Daltonics)
- COSMOSIL 5C18-MS-II column (4.6 × 150 mm) (nacalai)
- Malonyl-CoA
- *N*-methylanthranyloyl-CoA
- *p*-Coumaroyl-CoA
- Benzoyl-CoA
- Hexanoyl-CoA
- Buffer A: 100 mM potassium phosphate buffer (pH 7.0)

2.3.2 Procedure for enzyme reaction of CmACS and CmQNS

1. Enzyme Reactions of CmACS and CmQNS are performed on a 500 μL scale in a 1.5 mL tube. The reaction mixtures contain 54 μmol of starter CoA (*N*-methylanthranyloyl-CoA, *p*-coumaroyl-CoA, benzoyl-CoA, or hexanoyl-CoA), 108 μmol of malonyl-CoA, and 20 μg of enzyme in buffer A
2. Incubate the reaction solutions at 30 °C for 1 h
3. Add 50 μL of 20% HCl to quench the reactions
4. Extract the enzyme reaction products with 500 μL of ethyl acetate, three times
5. After concentration, products can be analyzed by LC-MS using a COSMOSIL $5C_{18}$-MS-II column. The elution is monitored by UV absorbance detection at 280 nm and MS. Gradient elution is performed with ddH_2O (phase A) and MeOH (phase B), both containing 0.1% trifluoroacetic acid: 0–5 min, 30% B; 5–17 min, 30–60% B; 17–25 min, 60% B; 25–27 min, 60–70% B; 27–35 min, 70% B; and 35–40 min, 70–100% B

Notes

- Parameters such as reaction temperature, buffer pH, and the ratio between enzyme and substrate can be optimized to improve the yields of enzyme reaction products
- Addition of HCl is important not only for quenching the reaction, but also for the efficient extraction with ethyl acetate from the water layer

2.4 Structural analysis and structure-based mutagenesis of CmACS and CmQNS

Residues 197, 256, 338, 215, and 265 (numbering in MsCHS) are important for the substrate specificity and product selectivity, by determining the shape of the active site cavity (Ferrer et al., 1999; Morita et al., 2019). It should be noted that the size of the active site entrance also determines the substrate specificity. Therefore, these active site residues, as well as the shape of the active site cavity and the size of the active site entrance, should be focused on in the structural analysis of type III PKSs. The crystal structures of CmACS and CmQNS were obtained at 2.35 and 2.47 Å resolutions, respectively. The comparison of their active sites indicated the structural basis for their substrate and product selectivities (Fig. 4) (Mori et al., 2013). In the structure of CmACS, the gatekeeper residue Phe265 in MsCHS is substituted with a Val residue. Furthermore, the conformational changes of the active site residues Phe215 and Leu267 increase the entrance area of the CmACS active site, as compared with that of MsCHS. Similarly, the substitution of the gatekeeper residue Phe265 with Leu, and the conformational changes of the loop region of 159–164 and the following α-helix between 165 and 179 also expand the active site entrance of CmQNS. The resulting estimated total areas of the active site entrances of CmACS and CmQNS are 33 and 47 $Å^2$, respectively, which are significantly larger than that of *M. sativa* CHS (17 $Å^2$) (Fig. 4A–C). The comparison of the active site cavities of CmACS and CmQNS indicated that the substitutions of the active site-forming residues Ser132, Thr194, and Thr197 in CmACS with the large Met132, Met194, and Tyr197 residues, respectively, significantly decrease the active site volume of CmQNS. The estimated active site cavity volume of CmQNS is 290 $Å^3$, which is 2.6 times smaller than that of CmACS (760 $Å^3$) (Fig. 4D–F). Thus, these Ser132, Thr194, and Thr197 residues could be targets for enzyme engineering. The site-directed mutagenesis of CmACS and CmQNS demonstrated the functional conversion of these enzymes. The S132M, T194M, and T197Y variants of CmACS lost the ability to form the tetraketide acridone and chalcone from *N*-methylanthranyloyl-CoA and *p*-coumaroyl-CoA, respectively. Instead, these variants generated the quinolone as the single product from *N*-methylanthranyloyl-CoA, as in the case of CmQNS. Thus, the function of CmACS can be converted into CmQNS by a single substitution. The Y197A variant of CmQNS expanded the substrate specificity and newly accepted *p*-coumaroyl-CoA to generate a triketide lactone and a diketide benzalacetone. In following section, we report the detailed

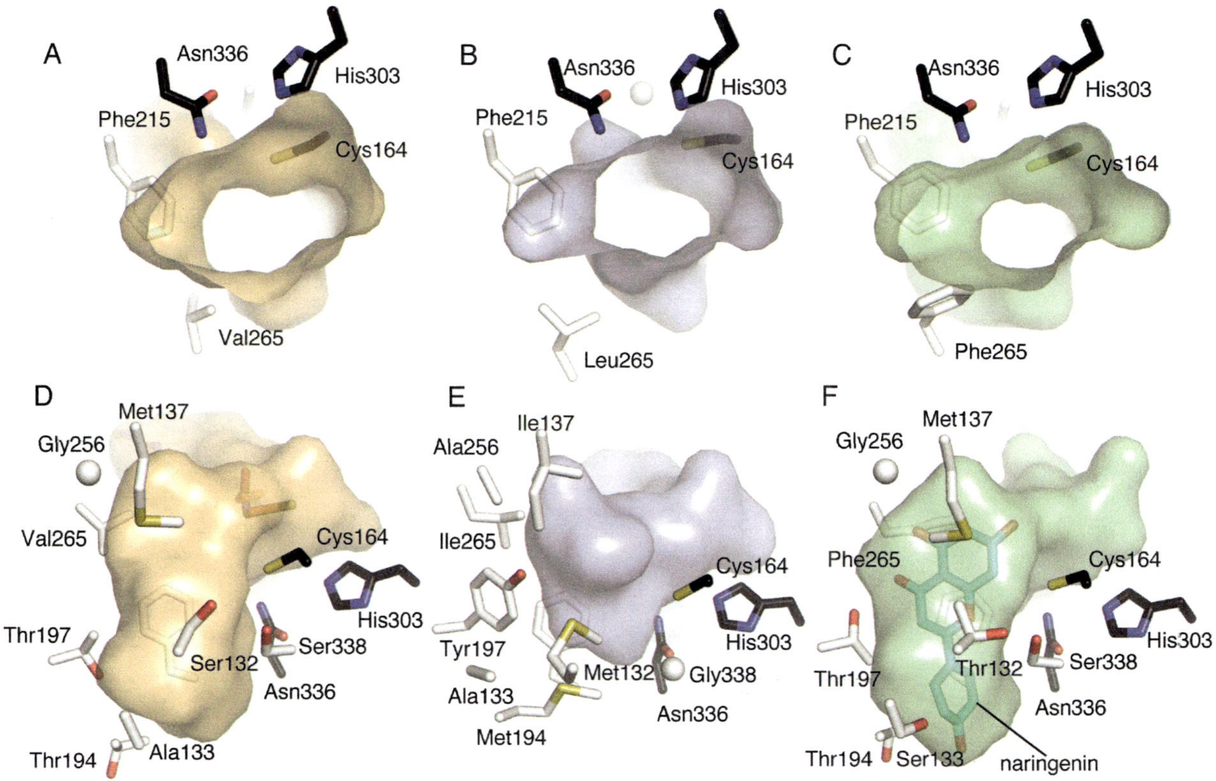

Fig. 4 Active sites of CmACS and CmQNS. (A–C) Active site entrances of (A) CmACS (Mori et al., 2013), (B) CmQNS (Mori et al., 2013), and (C) MsCHS (Ferrer et al., 1999). (D–F) Active site cavities of (D) CmACS, (E) CmQNS, and (F) MsCHS.

protocols for X-ray crystallographic analysis and structure-based mutagenesis experiments to clarify the catalytic mechanisms and alter the functions of CmACS and CmQNS.

2.4.1 Materials

- Sitting drop crystallization plates
- Sealing film
- 20 °C incubator
- Microscope
- Crystallization solutions:
  CmACS: 100 mM HEPES-NaOH (pH 7.5), 1400 mM ammonium sulfate, 2 mM $NiCl_2$, and 2 mM CoASH
  CmQNS: 50 mM Tris-HCl (pH 7.0), 18% (*w*/*v*) PEG3350, and 4% (*v*/v) 1-propanol
- QuikChange site-directed mutagenesis kit (Stratagene)
- TaKaRa PCR Thermal Cycler Dice

2.4.2 Procedure for X-ray crystallography of CmACS and CmQNS

1. Initial crystallization screenings of CmACS and CmQNS are performed by using the 96-condition crystallization screening kit originally designed by Mitsubishi Chemical Corp. Add 50 μL of the precipitant solution in the reservoir
2. Mix 0.5 μL CmACS or CmQNS (10 mg/mL) and 0.5 μL precipitant solution in the drop well, and store the plates at 20 °C
3. The hit conditions are further optimized by using an additive screen kit (Hampton Research). Add 5 μL of the additive screen solution to each crystal solution. Then, the protein solution and the precipitant solution containing the additive are mixed 1:1. Samples are incubated according to the sitting drop vapor diffusion method at 20 °C
   - Diffraction quality crystals of CmACS are obtained in 100 mM HEPES-NaOH (pH 7.5), 1400 mM ammonium sulfate, 2 mM $NiCl_2$, and 2 mM CoASH (Fig. 5).
   - Diffraction quality crystals of CmQNS were finally obtained in 50 mM Tris-HCl (pH 7.0), 18% PEG3350, and 4% (v/v) 1-propanol by the sitting drop vapor diffusion method at 20 °C (Fig. 5).
4. The crystals of CmACS and CmQNS are transferred into a cryoprotectant solution, consisting of the final crystallization buffer containing 18% glycerol
5. The crystals are captured by a cryoloop (Hampton Research) and flash-cooled on a goniometer under a nitrogen gas stream (−173 °C).

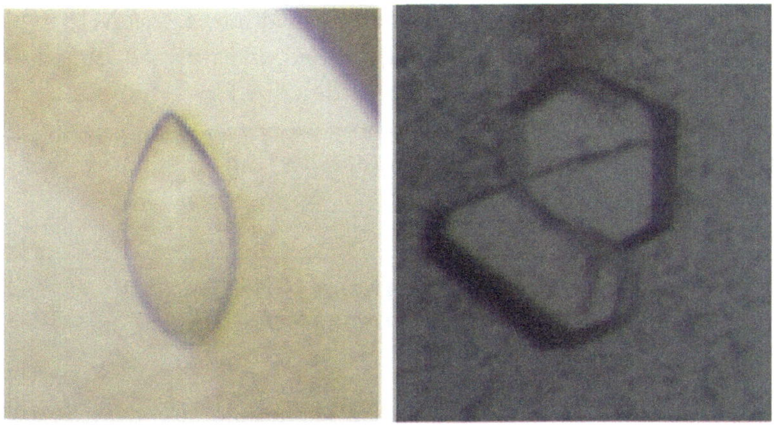

Fig. 5 Crystals of CmACS (left) and CmQNS (right).

6. X-ray diffraction datasets are collected at −173 °C at 0.9800 Å. A 1° oscillation angle and 1 s exposure time are used for the collection of 180 frames

Notes

- The rest of the protein solution remaining after crystallization cannot be refrozen
- Optimal crystallization results were obtained with the above precipitant conditions; however, we recommend optimizing with a 2-dimensional drop screen around these conditions. For example, start by changing the concentration of ammonium sulfate between 1200 and 1600 mM in one dimension and the concentration of $NiCl_2$ between 0.5 and 3 mM in the second dimension

2.4.3 Procedure for site-directed mutagenesis of CmACS and CmQNS

1. PCR amplifications of mutants are performed using the following pairs of primers:

S132M of CmACS: 5′-CATCTCATTTTCTGCACAATGGCAGGCGTCGACATGCC-3′ and 5′-GGCATGTCGACGCCTGCCATTGTGCAGAAAATGAGATG-3′), T194M of CmACS: 5′-GTTTGCTCTGAGAACATGATCCCCACTTTCCGTG-3′ and 5′-CACGGAAAGTGGGGATCATGTTCTCAGAGCAAAC-3′, and T197Y of CmACS: 5′-GAGAACACAATCCCCTATTTCCGTGGGCCG-3′ and 5′-CGGCCCACGGAAATAGGGGATTGTGTTCTC-3′. Y197A of CmQNS: 5′-GATATCATGAACATGGCTTTTCATGAGCCG-3′ and 5′-CGGCTCATGAAAAGCCATGTTCATGATATC-3′.

2. The plasmids are independently transformed into *E. coli* M15
3. The variant enzymes are expressed, purified, and analyzed using the same procedures as those for the wild type described in Sections 2.2 and 2.3

Notes

- The yields of the CmACS and CmQNS variants are similar to wild type enzymes

3. Alkyldiketide-CoA synthase (ADS) and alkylquinolone synthase (AQS)

3.1 Overview

Alkyldiketide-CoA synthase (ADS) and alkylquinolone synthase (AQS), from the Rutaceae medicinal plant *Evodia rutaecarpa*, are involved in the biosynthesis of 2-alkylquinolones. ErADS and ErAQS share 46–76% amino acid sequence identities with other characterized type III PKSs from plants (Matsui et al., 2017). ErADS and ErAQS share 61% amino acid sequence identity with each other. ErADS shares 62% amino acid sequence identity with MsCHS, 59% identity with CmQNS, and 64% identity with CmACS. ErAQS shares 76% amino acid sequence identity with CmQNS, 61% identity with CmACS, and 60% identity with MsCHS. The collaboration of ErADS and ErAQS produces 2-alkylquinolones *via* condensations of a fatty acyl-CoA, *N*-methylanthraniloyl-CoA, and malonyl-CoA. Firstly, ErADS catalyzes the condensation of malonyl-CoA with a fatty acyl-CoA to generate an alkyldiketide-CoA, and then ErAQS catalyzes the coupling of an alkyldiketide acid, non-enzymatically hydrolyzed from alkyldiketide-CoA, with *N*-methylanthraniloyl-CoA to yield the 2-alkylquinolone through the C–C/C–N bond formations. The genes encoding ErADS and ErAQS are identified from *E. rutaecarpa* buds and subcloned into pQE80L, and the plasmid is used to transform *E. coli* M15. In this section, we report the cloning, functional characterization, and structural analyses of ErADS and ErAQS by expression of the enzymes in *E. coli*.

3.2 Cloning, overexpression, and purification of ErADS and ErAQS

3.2.1 Materials

- RNeasy plant mini kit (Qiagen)
- Superscript II RT (Invitrogen)
- SMARTer RACE cDNA amplification kit (Clontech)
- TaKaRa PCR Thermal Cycler Dice

- pQE80L (QIAGEN)
- Restriction enzymes BamHI, *Xho*I, and *Sal*I (TaKaRa)
- Lysis Buffer: 50 mM Tris-HCl buffer (pH 8.0), 200 mM NaCl, 5% (*v*/v) glycerol
- Ion exchange buffer A: 50 mM Tris-HCl (pH 8.0) and 5% (v/v) glycerol
- Ion exchange buffer B: 50 mM Tris-HCl buffer (pH 8.0), 1 M NaCl, and 5% (v/v) glycerol
- Gel filtration buffer: 50 mM Tris-HCl buffer (pH 8.0), 200 mM NaCl, 5% (v/v) glycerol
- Lysozyme from egg white (Merck)
- Benzonase nuclease (Merck)
- High speed refrigerated microcentrifuge
- HisTrap HP column (Cytiva)
- Resource-Q column (Cytiva)
- HiLoad™ 16/600 Superdex™ 200 pg (Cytiva)
- ÄKTAexplorer (Cytiva)
- Microcep 10 k centrifugation device (PALL Life Sciences).

3.2.2 Procedure for cloning of ErADS and ErAQS

1. Total RNA is extracted from *E. rutaecarpa* buds, using an RNeasy plant mini kit (Qiagen).
2. Reverse-transcription of Total RNA is performed by using the oligo(dT) primer (5′-GACTCGAGTCGACATCGATTTTTTTTTT TTTT-3′) and Superscript II RT (Invitrogen).
3. The obtained cDNA is used for the PCR amplification of PKS genes with inosine-containing degenerate primers, 112S, 174S, 368A, and 380A (same primers as in 2.2.1).
4. The 555-bp core fragments of ErADS and ErAQS are amplified by nested PCR with the primer sets. Primers 112S and 380A are used for the first PCR and then 174S and 368A are used for the second PCR
5. The 5′- and 3′-RACEs are performed by using the SMARTer RACE cDNA amplification kit (Clontech) with the gene-specific primers (For 5′-RACE of ErADS: 5′-CCATCCTCTGAATCAGGGATTGTTG TTTGATTTGC-3′. For 3′-RACE of ErADS: 5′-AATGGCTTC AAAGGTGGAGAGTCGGCAGGCAGCAG-3′. For 5′-RACE of ErAQS: 5′-AAAACAGCTTGCCCATATAATAAATCAACGCGG TT-3′. For 3′-RACE of ErAQS: 5′-TGCTGGAGCTCGAGTTCT TTTAGTGTTTTGTGATC-3′).

6. The full-length cDNAs are obtained by PCR using the specific primers. ErADS: 5′-GCGGGATCCATGGCTTCAAAGGTGGAGAGTCGCA-3′ and 5′-CGCCTCGAGTCAATTGATAGCAACACTACGGAGGA-3′ (BamHI and XhoI sites are underlined). ErAQS: 5′-GCGGGATCCATGGCTTCATCTTTTTCAATGGAAAA-3′ and 5′-CGCCTCGAGATGGTTGGAATCTAGAGGGACACTGA-3′ (BamHI and XhoI sites are underlined).
7. The full-length cDNA fragments of ErADS and ErAQS are treated with BamHI/XhoI and individually sub-cloned into the BamHI/SalI sites of pQE80L

Note

- Although we subcloned the genes into pQE80L, other plasmids such as pET vectors can also be used

3.2.3 Procedure for expression and purification of ErADS and ErAQS

1. The ErADS/pQE80L and ErAQS/pQE80L plasmids are individually transformed into *E. coli* M15
2. The cells are cultured in LB medium containing 100 μg/mL ampicillin at 37 °C until the OD_{600} reaches 0.6
3. The protein expression is induced by the addition of IPTG (final concentrations: 1 mM for ErADS, 0.1 mM for ErAQS).
4. The cells are further incubated at 27 °C for 16 h
5. The *E. coli* cells are harvested by centrifugation at 6000 × *g* for 20 min. The cell pellets are resuspended in 20 mL of lysis buffer, containing 10 mM imidazole. To this suspension, 0.5 mg/mL lysozyme and 0.1 units/mL benzonase nuclease (Merck) are added
6. The cells are disrupted by sonication, and the solution is centrifuged at 8000 × *g* for 20 min
7. The supernatant is loaded onto a HisTrap HP column (Cytiva), which is pre-equilibrated with lysis buffer
8. The column is washed with lysis buffer and the bound protein is subsequently eluted with lysis buffer, containing 600 mM imidazole
9. The fraction containing the target protein is diluted 5-fold with ion exchange buffer A and injected onto a Resource Q column (Cytiva).
10. The column is washed with Ion exchange buffer A containing 50 mM NaCl, and the protein is eluted using a linear gradient of 50–300 mM NaCl (5–30% B concentration).

11. The fractions containing ErADS or ErAQS are collected and concentrated to 5 mL with a Microcep 10k centrifugation device (PALL Life Sciences).
12. The concentrated solution is further purified by gel filtration chromatography on a HiLoad 16/60 Superdex 200 column (Cytiva), and the fractions containing ErADS or ErAQS are concentrated in gel filtration buffer with a Microcep 10k centrifugation device (PALL Life Sciences).
13. The purified protein is flash frozen in liquid nitrogen and can be stored at −80 °C until use

Notes

- All purification procedures are performed at 4 °C to prevent enzyme aggregation
- The entire purification process should be performed within a day for enzyme crystallization
- The purity of the isolated enzymes is examined by SDS-PAGE
- The purification procedure is designed for a 4 L expression culture
- The typical yields for both enzymes are 5–10 mg per 1 L culture
- A Mono Q column (Cytiva), instead of a Resource Q column, can be used with the same gradient

3.3 *In vitro* characterization of ErADS and ErAQS

In vitro co-incubation assays of ErADS and ErAQS with substrates indicated that both enzymes are required to generate a 2-alkylquinolone (Matsui et al., 2017). The products of the combination enzyme reaction assay of ErADS and ErAQS can be analyzed by LC-MS. The single enzyme assays of ErADS and ErAQS revealed that ErADS accepts lauroyl-, myristoyl-, and palmitoyl-CoA as starter substrates to produce alkyldiketide-CoAs, while ErAQS accepts *N*-methylanthranyloyl-CoA as a starter substrate and C_6- to C_{16}-alkyldiketide acids as the extender substrates. To analyze the ErADS reaction, the products are detected by a TLC-based assay due to the low UV absorption of alkyldiketide acids. The enzyme reaction is performed in the presence of 2-^{14}C-labeled malonyl-CoA. The enzymatically produced alkyldiketide-CoAs are hydrolyzed to the corresponding alkyldiketide acids by an alkaline hydrolysis reaction. Subsequently, the alkyldiketide acids are detected by a TLC-based assay.

3.3.1 Materials

- HPLC 1100 series (Agilent Technologies)
- Esquire 4000 ion trap mass spectrometer (Bruker Daltonics)

- TSK-gel ODS-80Ts column (4.6 mm × 150 mm) (Tosoh)
- Malonyl-CoA
- *N*-methylanthranyloyl-CoA
- Fatty acyl-CoA (C_6, C_8, C_{10}, C_{12}, C_{14}, or C_{16}-CoAs)
- Buffer A: 100 mM potassium phosphate buffer (pH 7.5)
- Analytical TLC plates (Merck silica gel 60 F_{254}).
- Typhoon FLA-9500 biomolecular imager (Cytiva)
- ImageQuant TL software (Cytiva)
- [2-^{14}C] malonyl-CoA (52.4 mCi/mmol) (PerkinElmer Life Sciences)
- HPLC 1290 series (Agilent Technologies)
- Esquire 4000 ion trap mass spectrometer (Bruker Daltonics)
- TSK-gel ODS-80Ts column (4.6 mm × 150 mm) (Tosoh)
- Malonyl-CoA
- *N*-methylanthranyloyl-CoA

3.3.2 Procedure for co-incubation assays of ErADS and ErAQS

1. 20 μg ErADS and 20 μg ErAQS are incubated with 55 μM *N*-methylanthranyloyl-CoA, 60 μM fatty acyl-CoA (C_6, C_8, C_{10}, C_{12}, C_{14}, or C_{16}-CoAs), and 100 μM malonyl-CoA, in a final volume of 500 μL of buffer A, at 37 °C for 16 h
2. The reactions are quenched by adding 20 μL of 20% HCl, and the enzyme reaction products are extracted with 500 μL of EtOAc twice
3. The extracted solutions are evaporated under reduced pressure, and the residues are dissolved in 80 μL of 100% methanol
4. 20 μL of the sample is injected into the LC-ESI-MS system. The enzyme reaction products are separated on a TSK-gel ODS-80Ts column with a gradient of H_2O (Phase A) and CH_3CN (Phase B), both containing 0.1% trifluoroacetic acid: 0–5 min, 20% B; 5–45 min, 20–100% B at a flow rate of 0.6 mL/min

Notes

- Optimization of reaction conditions should be performed. Temperature, buffer pH, and the ratio between enzyme and substrate can be verified to improve the yield of enzyme reaction products
- Addition of HCl is important not only for quenching the reaction, but also for the efficient extraction with ethyl acetate from the water layer

3.3.3 Procedure for assay of the diketide formation activity of ADS

1. 4 μg of ErADS, 60 μM starter substrate (C_6, C_8, C_{10}, C_{12}, C_{14}, or C_{16}-CoAs), and 101.6 μM [2-^{14}C] malonyl-CoA (1.1 mCi/mmol) are mixed in buffer A (total 100 μL) and incubated at 37 °C for 16 h

2. The produced alkyldiketide-CoAs are hydrolyzed to the corresponding alkyldiketide acids by the addition of 5 μL of 20% NaOH and an incubation at 65 °C for 15 min
3. 10 μL of 20% HCl is added to the solution to quench the hydrolysis reaction, and the products are extracted with 100 μL of EtOAc
4. Compounds are loaded onto analytical TLC plates (Merck silica gel 60 F_{254}) and developed with a hexane/EtOAc/acetic acid solvent system (9:1:1, *v*/v/v).
5. Radioactivities of products are quantified by a Typhoon FLA-9500 biomolecular imager (Cytiva).
6. The radioactive compounds on the TLC plates are analyzed using the ImageQuant TL software (Cytiva).

Note

- The alkyldiketide-CoAs can also be detected by HPLC. In this case, 20% HCl is not added to the reaction and the reaction solution is injected into the HPLC after removing the protein. The enzyme reaction products are separated on a Octyl-80Ts column with a gradient of 25 mM KPB (pH 6.5) and CH_3CN: 0–5 min, 30% CH_3CN; 5–21 min, 30–42% CH_3CN; 21–26 min, 42–80% CH_3CN; 26–40 min, 80% CH_3CN

3.3.4 *Procedure for assay of the alkylquinolone formation activity of ErAQS*

1. 20 μg of ErAQS is mixed with 55 μM malonyl-CoA and 60 μM alkyldiketide acid (C_6, C_8, C_{10}, C_{12}, C_{14}, or C_{16}-diketide acids), in a final volume of 100 μL of buffer A and incubated at 37 °C for 16 h
2. The reactions are quenched by adding 4 μL of 20% HCl
3. 20 μL of the enzyme reaction mixture is directly injected into the HPLC. The enzyme reaction products are separated on a TSK-gel ODS-80Ts column (4.6 mm × 150 mm; Tosoh) with 25 mM potassium phosphate buffer (pH 6.5) (Phase A) and a CH_3CN (phase B) gradient system: 0–5 min, 15% B; 5–21 min, 15–42% B; 21–26 min, 42–80% B; 26–40 min, 80% B

Note

- Parameters such as reaction temperature, buffer pH, ratio between enzyme and substrate can be optimized to improve the yield of enzyme reaction products

3.4 Structural analysis and structure-based mutagenesis of ErADS and ErAQS

In addition to the cavity shape and active site entrance, the size of the CoA-binding tunnel should be considered in the analysis of the substrate selectivity toward CoA and a diketide acid. The crystal structures of ErADS and ErAQS were obtained at 1.80 and 2.20 Å resolutions, respectively. The structural analysis of ErADS revealed that the bulky side chains of Trp332 and Met257 decrease the space of the active site, as compared to those of MsCHS and the functionally similar benzalacetone synthase (BAS) (Morita et al., 2010; Shimokawa, Morita, & Abe, 2012), which catalyzes the formation of benzalacetone through the decarboxylative condensation of malonyl-CoA with *p*-coumaroyl-CoA (Fig. 6). Furthermore, one of the gatekeeper residues, Phe265 in MsCHS (Phe258 in BAS), is uniquely substituted with Tyr259 in the active site of ErADS, and this residue protrudes toward the opposite side of the active site cavity. As a result, the side chain of the neighboring residue Phe209 is positioned toward the active site and decreases the volume of the active site cavity. Thus, the total volume of the active site is decreased to 200 $Å^3$ in ErADS, which is significantly smaller than those of RpBAS (350 $Å^3$) and CmQNS (290 $Å^3$). A structure-based mutagenesis study of Trp332 significantly altered the substrate and product specificities of ErADS. The results of a mutagenesis study can be confirmed by X-ray crystallization, which provides insights for further engineering. The crystal structures of W332Q and W332G indicated that the estimated volumes of the active sites of the ErADS W332Q and W332G variants are 240 and 295 $Å^3$, which are larger than that of the wild type, indicating that this residue is important for the substrate specificity for the fatty acyl-CoA by controlling the cavity volume (Fig. 6).

The residues in the active site of ErAQS are well conserved, as compared to those in the active site of CmQNS (Figs. 4 and 7). The significant difference between ErAQS and CmQNS is the conformation of Met138. Thus, the size of the ErAQS active site cavity (330 $Å^3$) is comparable to that of CmQNS (290 $Å^3$). In contrast, the substitution of the gatekeeper residue Val215 in CmQNS with Tyr215 in AQS reduces the space of the CoA-binding tunnel. These observations suggest that the narrow CoA-binding tunnel and the wide entrance area of the active site in ErAQS determine the substrate specificity to accept *N*-methylanthranyloyl-CoA and alkyldiketide acids as the starter and extender substrates, respectively. Interestingly, the structure-based mutagenesis study of Tyr215 with Val indicated that the ErAQS Y215V variant newly accepts benzoyl-CoA and *p*-coumaroyl-CoA as starter substrates. The crystal structure of the

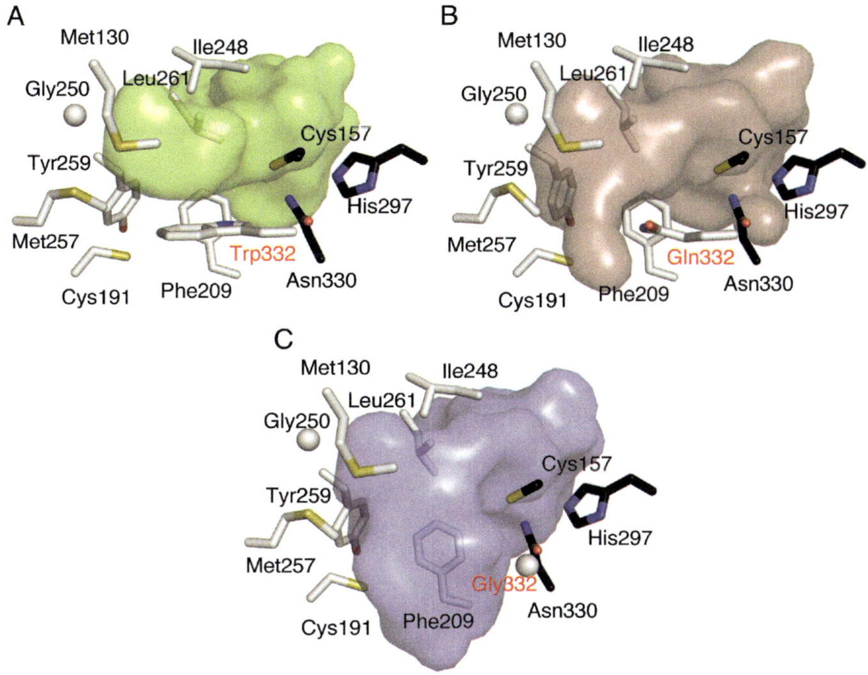

Fig. 6 The active site architectures of ErADS wild type and variants (Matsui et al., 2017). The active sites of (A) ErADS wild type, (B) ErADS W332Q variant, and (C) W332G variant.

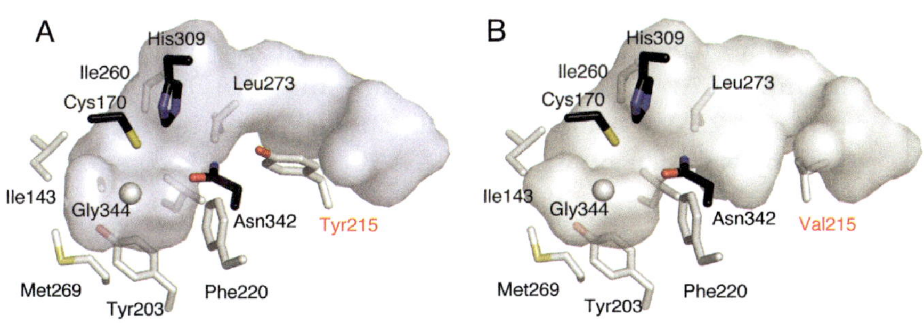

Fig. 7 Active sites of (A) ErAQS wild type and (B) Y215V variant (Matsui et al., 2017).

ErAQS Y215V variant confirmed the widened CoA-binding tunnel (Fig. 7). In these sections, we describe the detailed protocols for X-ray crystallographic analysis and structure-based mutagenesis experiments of ErADS and ErAQS.

3.4.1 Materials

- Sitting drop crystallization plates
- Sealing film

- 20 °C incubator
- Microscope
- Crystallization solutions:

ErADS wild type: 0.1 M Bis-Tris (pH 6.2), 0.2 M ammonium sulfate, and 25% (*w/v*) PEG3350.

ErADS W332Q variant: 0.1 M MES (pH 6.1), 0.2 M lithium sulfate, and 20% (w/v) PEG3350.

ErADS W332G variant: 0.1 M MES (pH 6.1), 0.2 M lithium sulfate, 25% (w/v) PEG3350, and 10 mM CoASH.

ErAQS wild type: 0.1 M MES, 0.2 M magnesium chloride, and 10% (w/v) PEG8000.

ErAQS Y215V variant: 0.1 M sodium citrate (pH5.6) and 1.99 M ammonium sulfate

- QuikChange site-directed mutagenesis kit (Stratagene)
- TaKaRa PCR Thermal Cycler DiceTM

3.4.2 Procedure for X-ray crystallography of ErADS, ErAQS, and their variants

1. Initial crystallization screenings of ErADS and ErAQS are performed by using the 96-condition crystallization screening kits originally designed by Mitsubishi Chemical Corp. Add 50 μL of the precipitant solution to the reservoir
2. Mix 0.5 μL of purified ErADS (10 mg/mL), ErADS W332Q variant (5 mg/mL), ErADS W332G variant (5 mg/mL), ErAQS (22 mg/mL), or ErAQS Y215V variant (10 mg/mL) and 0.5 μL precipitant solution in the drop well. The plates are stored at 20 °C and checked periodically for crystal formation (Fig. 8).

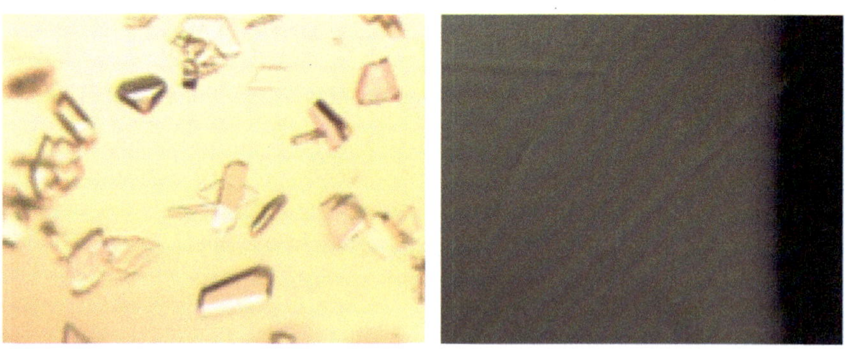

Fig. 8 Crystals of ErADS (left) and ErAQS (right).

3. The crystals are transferred into a cryoprotectant solution, consisting of the final crystallization buffer containing 12% ethylene glycerol
4. The crystals are captured by a cryoloop (Hampton Research, Inc.) and flash-cooled on a goniometer under a nitrogen gas stream (−173 °C).
5. X-ray diffraction datasets are collected at 100K at 1.000 Å. A 1° oscillation angle and 1 s exposure time are used for the collection of 180 frames

Notes

- The rest of the protein solution remaining after crystallization cannot be refrozen
- Optimal crystallization results were obtained with the above precipitant conditions; however, we recommend optimizing with a 2-dimensional drop screen around these conditions

3.4.3 Procedure for site-directed mutagenesis of ErADS and ErAQS

1. PCR amplification of mutants using the following pairs of primers: W332Q of ErADS: 5′-TTCTTAGTGAATATGGCAATATGCAGGGTCCAAGCGTGCTG-3′ and R: 5′-CAGCACGCTTGGACCCTGCATATTGCCATATTCACTAAGAA-3′, W332G of ErADS: 5′-TTAGTGAATATGGCAATATGGGGGTCCAAGCGTG-3′ and 5′-CACGCTTGGACCCCCATATTGCCATATTCACTAA-3′, Y215V of ErAQS: 5′-GCGCCTCAAAACCGCGTTGATTTATTAGTTGGGCAAGCTGTTT-3′ and 5′-AAACAGCTTGCCCAACTAATAAATCAACGCGGTTTTGAGGCGC-3′.
2. The plasmids are independently transformed into *E. coli* M15
3. The variants are expressed, purified, and analyzed using the same procedures as described in Sections 3.2 and 3.4. The procedures for the crystallization of these variants are shown in Section 3.3.2

Note

- The yields of the ErADS and ErAQS variants are similar to those of the wild-type enzymes

4. 2-(2-Phenylethyl)chromone precursor synthase (PECPS)

4.1 Overview

The diarylpentanoid-producing polyketide synthase PECPS, which catalyzes the formation of the C_6–C_5–C_6 scaffold of 5-(4-hydroxyphenyl)-1-phenylpentane-1,3-dione from 4-hydroxyphenylpropionyl-CoA, malonyl-CoA, and benzoyl-CoA, was identified from *Aquilaria sinensis* as a key enzyme in the biosynthesis of the agarwood fragrance, phenylethylchromones (Wang et al., 2022). PECPS shares 58% and 47% amino acid sequence identities with MsCHS and *Oryza sativa* curcuminoid synthase (OsCUS), respectively (Katsuyama, Matsuzawa, Funa, & Horinouchi, 2007; Morita, Wanibuchi, et al., 2010). The gene encoding PECPS is subcloned into pET28a, and the plasmid is used to transform *E. coli* BL21(DE3). Here, we report the cloning, functional characterization, and structural analysis of PECPS by heterologous expression of the enzyme in *E. coli*.

4.2 Cloning, overexpression, and purification of PECPS

4.2.1 Materials

- Total RNA Purification Reagent (Norgen)
- M-MLV transcriptase (Promega)
- TaKaRa PCR Thermal Cycler DiceTM
- pET28a (Novagen)
- Restriction enzymes BamHI and *Hin*dIII (TaKaRa)
- *Escherichia coli* BL21(DE3)
- Luria Bertani (LB) media
- LB agar plates containing 50 μg/mL of kanamycin
- Kanamycin, 50 mg/mL
- 1 M isopropyl-1-thio-β-D-galactopyranoside (IPTG)
- Erlenmeyer flask (2 L),
- Temperature-controlled shaker
- Lysis buffer: 20 mM Tris-HCl (pH 8.0), 100 mM NaCl, and 10% (*v*/v) glycerol
- Wash buffer: 20 mM Tris-HCl (pH 8.0), 100 mM NaCl, 10% (v/v) glycerol, and 50 mM imidazole
- Elution buffer: 20 mM Tris-HCl (pH 8.0), 100 mM NaCl, 10% (v/v) glycerol, and 500 mM imidazole
- Ion exchange buffer A: 50 mM HEPES-NaOH (pH 7.0), 20 mM NaCl, and 2 mM DTT
- Ion exchange buffer B: 50 mM HEPES-NaOH (pH 7.0), 1.0 M NaCl, and 2 mM DTT
- Gel filtration buffer: 20 mM HEPES-NaOH (pH 7.0), 100 mM NaCl, and 2 mM DTT
- High speed refrigerated microcentrifuge
- Ni-NTA resin (Cytiva)
- Resource-Q column (Cytiva)
- HiLoad™ 16/60 Superdex™ 200 pg (Cytiva)

- ÄKTAexplorer (Cytiva)
- Macrosep Advanced Centrifugal Device-30 K (PALL Corporation)

4.2.2 Procedure for cloning, expression, purification of PECPS

1. Total RNA from *A. sinensis* calli is extracted with the Total RNA Purification Reagent (Norgen).
2. The first cDNA strand is synthesized using an oligo $d(T)_{18}$ primer and M-MLV transcriptase (Promega).
3. The full-length cDNA of PECPS was obtained using gene-specific PCR primers: 5′-CGCGGATCCATGGCAGCCCAACCTGTGGA GTGGGTG-3′ and 5′-CCCAAGCTTCTATGCGTCTGTGAGCG TTGCAGTAG-3′ with the BamHI and HindIII digestion sites underlined, respectively
4. The full-length PECPS cDNA fragment is sub-cloned into the BamHI/HindIII sites of pET28a (Novagen).
5. The full-length cDNA encoding PECPS in pET28a is transformed into *E. coli* BL21(DE3) and selected by cultivation on Luria-Bertani (LB) agar plates with 50 μg/mL kanamycin, at 37 °C for 16 h
6. A single colony is selected and incubated in 2 mL LB broth containing 50 μg/mL kanamycin at 37 °C overnight
7. The seed culture is transferred into 1 L of LB broth containing 50 μg/mL kanamycin and incubated at 37 °C until the optical density at 600 nm (OD_{600}) becomes 0.6
8. Protein expression is induced by adding IPTG (final concentration 1 mM)
9. The culture is incubated at 20 °C for another 20 h
10. Cultured cells are harvested by centrifugation at 5000 × *g* for 10 min, and the cell pellet can be stored at −80 °C until use
11. The harvested cells are resuspended in 20 mL of lysis buffer
12. The cells are disrupted by sonication on ice, and the cell debris is removed by centrifugation at 12,000 × *g* for 20 min
13. The supernatant is loaded onto a 5 mL column of Ni-NTA resin equilibrated with lysis buffer. This procedure is repeated at least twice
14. The resin is washed with 30 column volumes (CVs) of wash buffer
15. The recombinant protein is subsequently eluted with 5 CVs of elution buffer
16. The protein solution is concentrated to 5 mL with a Macrosep Advanced Centrifugal Device-30K (PALL Corporation)

17. The sample is injected into a Resource-Q column (Cytiva). The column is washed with Ion exchange buffer A, and the protein is eluted using a linear gradient of 20–1000 mM NaCl
18. The fraction containing the target protein is concentrated to 5 mL with a Macrosep Advanced Centrifugal Device-30K (PALL Corporation) and then purified to homogeneity by gel-filtration chromatography on a HiLoad™ 16/60 Superdex™ 200 pg column (Cytiva). The fraction containing the target protein is concentrated to 15 mg/mL in gel filtration buffer
19. The purified protein is flash-frozen with liquid nitrogen and can be stored at −80 °C until use

Notes

- All purification procedures are performed at 4 °C using the indicated buffers
- The purification procedure is designed for an 8 L expression culture
- Four polymorph forms of PECPS are observed in the ion-exchange column purification step. Usually, it is better to separate the peaks for further assays, including enzyme reaction assays and crystallization
- The typical yield of wild type of PECPS is 1 mg per 1 L culture

4.3 *In vitro* characterization of PECPS

PECPS catalyzes two reaction steps in the same active site cavity (Wang et al., 2022). The first reaction is the decarboxylative condensation of 4-hydroxyphenylpropionyl-CoA as the starter substrate with one molecule of malonyl-CoA to form the 4-hydroxyphenyl-propionyl-β-diketide-CoA intermediate, which is released from the active site and nonenzymatically hydrolyzed for conversion into a 4-hydroxyphenyl-propionyl-β-diketide acid. The enzyme then accepts benzoyl-CoA, 3-hydroxybenzoyl-CoA, or 4-hydroxybenzoyl-CoA as a substrate to form the catalytic Cys-tethered benzoyl thioester intermediate. The subsequent tail-to-tail decarboxylative condensation between the 4-hydroxy-phenylpropionyl-β-diketide acid and the enzyme-bound benzoyl unit generates the corresponding C_6–C_5–C_6 scaffold. For the *in vitro* assays, by-products can be suppressed by adding the second starter-substrate, benzoyl-CoA, 1–2 h after the start of the enzyme reaction. The enzyme reaction products can be separated and analyzed by LC-MS.

4.3.1 *Materials*

- LCMS-IT-TOF system (Shimadzu)
- Agilent Eclipse XDB C_{18} column (250 × 4.6 mm) (Agilent Technologies)
- 4-Hydroxyphenylpropionyl-CoA
- Benzoyl-CoA
- 3-Hydroxybenzoyl-CoA
- 4-Hydroxybenzoyl-CoA
- Malonyl-CoA
- 4-Hydroxyphenylpropionyl-β-diketide acid
- Buffer A: 100 mM potassium phosphate buffer (pH 7.0)

4.3.2 *Procedure for* in vitro *analysis of PECPS*

1. Enzyme reactions of PECPS are performed on a 500 μL scale in a 1.5 mL tube. The reaction mixtures contain 26 μmol of starter CoA (4-hydroxyphenylpropionyl-CoA/benzoyl-CoA/3-hydroxybenzoyl-CoA/4-hydroxybenzoyl-CoA), 29 μmol of malonyl-CoA/4-hydroxyphenylpropionyl-β-diketide acid, and 0.4 nmol of enzyme in buffer A
2. The reaction solutions are incubated at 37 °C for 12 h
3. Add 20 μL of 20% HCl to quench the reactions
4. The enzyme reaction products are extracted with 800 μL of ethyl acetate, three times
5. After concentration, products can be analyzed by LC-MS using an Agilent Eclipse XDB C_{18} column. Gradient elution is performed with ddH_2O (phase A) and CH_3CN (phase B): 0–10 min, 30% B; 10–20 min, 30–60% B; 20–25 min, 60–70% B; 25–30 min, 70–80% B; 30–35 min, 80–100% B; and 35–40 min, 100% B

Notes

- Optimization of reaction conditions should be performed. Temperature, buffer pH, and ratio between enzyme and substrate can be verified to improve the yield of enzyme reaction products

4.4 Structural analysis and structure-based mutagenesis of PECPS

In some cases, the observed size of an active site cavity is smaller than the expected volume, because of the conformations of active site residues. In such cases, the site-directed mutagenesis study of cavity forming residues is critical for understanding the functions of the residues and the substrate binding site. The crystal structure of PECPS was obtained at 1.98 Å resolution (Wang et al., 2022). The comparison of the active sites among PECPS, MsCHS, and OsCUS indicated that the substitution of MsCHS's Ser338 with Phe340 decreases the volume of pocket A, which locks the aromatic moiety of the intermediates, in the active site of PECPS. Furthermore, the residues Asn199, Thr213, and Pro267 in PECPS separated pocket B, the binding site of the aromatic moiety of the diketide acid intermediate in OsCUS (Morita, Wanibuchi, et al., 2010), from the main active site cavity. As a result, PECPS possesses a small active site cavity (247 $Å^3$) as compared to those of MsCHS (742 $Å^3$) and OsCUS (642 $Å^3$) (Fig. 9). Interestingly, the catalytic activity of PECPS was maintained even when pocket A or pocket B was eliminated by site-directed mutagenesis (F340W for pocket A and A210E for pocket B), suggesting that both pockets are not required during the enzyme reaction. In contrast, the substitution of Asn199, a gatekeeper residue of pocket B, with Leu and Phe decreased the catalytic activity of PECPS by 35% and 56%, respectively, indicating the critical role of Asn199 for the interaction with ligands. These structural analyses and structure-based mutagenesis experiments suggested that PECPS employs a unique one-pot C_6–C_5–C_6 scaffold formation mechanism, including the diketide-CoA release step, in this small cavity pocket. In the following section, we provide the detailed protocol for X-ray crystallographic analyses and structure-based mutagenesis experiments to understand the catalytic mechanisms of PECPS.

4.4.1 Materials

- Sitting drop crystallization plates
- Sealing film
- 20 °C incubator
- Microscope
- Crystallization solutions: 100 mM Tris-HCl (pH 8.5), 0.12 M KF, 4% butanediol, and 24% (*w*/*v*) PEG 8000
- QuikChange site-directed mutagenesis kit (Stratagene)
- TaKaRa PCR Thermal Cycler Dice

4.4.2 Procedure for X-ray crystallography of PECPS

1. Initial crystallization screening of PECPS is performed by using the 96-condition crystallization screening kit originally designed by Mitsubishi Chemical Corp. Add 50 μL of the precipitant solution to the reservoir

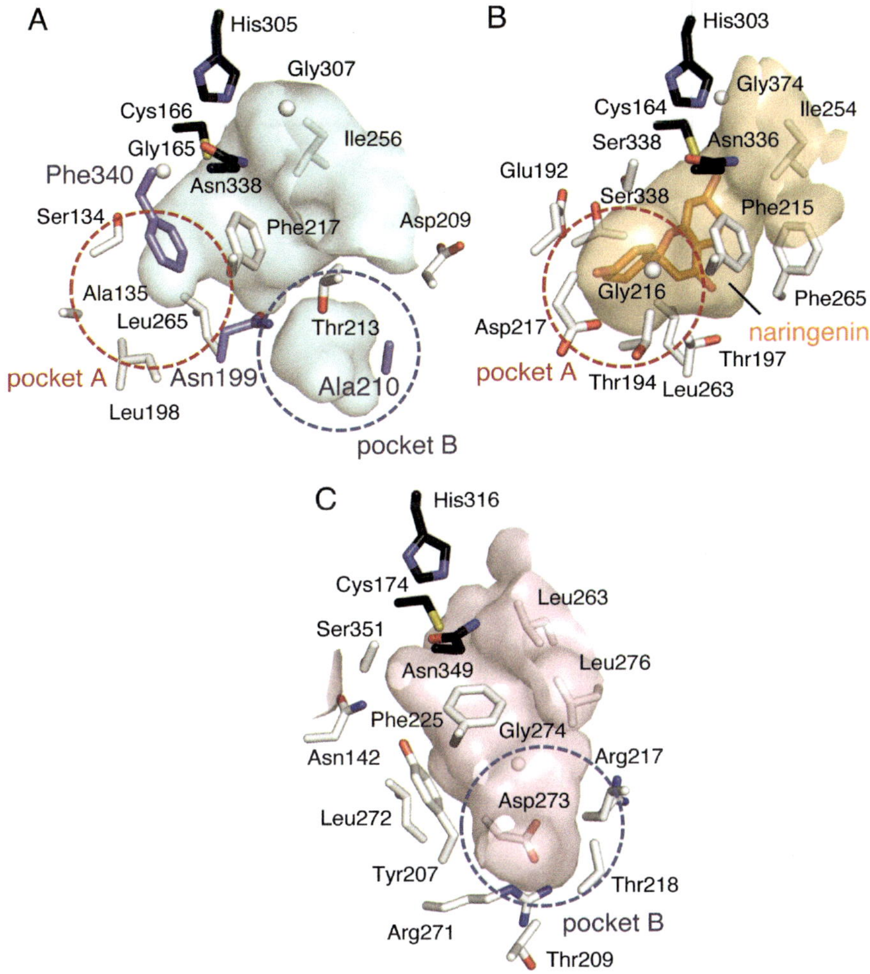

Fig. 9 Active sites of PECPS. Comparison of the active sites of (A) PECPS (Wang et al., 2022), (B) MsCHS with naringenin (Ferrer et al., 1999), and (C) OsCUS (Morita, Wanibuchi, et al., 2010).

2. Mix 0.5 μL PECPS (15 mg/mL) and 0.5 μL precipitant solution in the drop well. Store the plates at 20 °C
3. The hit conditions are further optimized by using an additive screen kit (Hampton Research). 5 μL of additive screen solution is added to each crystal solution. The protein solution and precipitant solution containing the additive are then mixed 1:1. Samples are incubated by the sitting drop vapor diffusion method at 20 °C

– Diffraction quality crystals of PECPS are obtained in 100 mM Tris-HCl (pH 8.5), 0.12 M KF, 4% butanediol, and 24% (w/v) PEG 8000 (Fig. 10).

Fig. 10 Crystal of PECPS.

4. The crystals of PECPS are transferred into a cryoprotectant solution, consisting of the final crystallization buffer containing 20% (*v*/v) glycerol
5. The crystals are captured by a cryoloop (Hampton Research) and flash-cooled on a goniometer under a nitrogen gas stream (−173 °C).

Note

- As described in Section 4.2, four types of polymorphs are observed in the ion-exchange column purification step. These four purified polymorphs of PECPS are independently crystallized, and well-diffracting crystals are obtained from one of the polymorphs

4.4.3 Procedure for site-directed mutagenesis of PECPS

1. PCR amplifications of mutants are performed using the following pairs of primers: A210E: 5′-TCGGAGACCCACATCGACGAGCTC ATAACTCAATCT-3′ and 5′-AGATTGAGTTATGA GCTCGTC GATGTGGGTCTCCGA-3′, F340W: 5′-GTGAGTTTGGCAATA TGTGGAGTGCCACCG-3′ and 5′-CCACATATTGCCAAACTCA CTAAGCACGTG-3′, N199L: 5′-CGGAGGCCAACCTACTCCT CTTCCGGGGCCCG-3′ and 5′-GAGGAGTAGGTTGGCCTCCG AGCAGACGACAAG-3′, and N199F: 5′-CGGAGGCCAACCTA CTCTTCTTCCGGGGCCCG-3′ and 5′-GAAGAGTAGGTTGGC CTCCGAGCAGACGACAAG-3′.
2. The plasmids are independently transformed into *E. coli* BL21(DE3).
3. The variant enzymes are expressed, purified, and analyzed according to the same procedures used for the wild type, described in Sections 4.2 and 4.3

Notes
- The yields of the PECPS variants, A210E, F340W, N199L, and N199F, are comparable to wild type PECPS
- Since these mutants are somewhat more unstable than the wild type, aggregated enzymes should be removed by centrifugation, and then the accurate enzyme concentrations are measured before the *in vitro* reactions

5. Tetraketide synthase (TKS) and olivetolic acid cyclase (OAC)

5.1 Overview

Tetraketide synthase (TKS), a type III PKS, and olivetolic acid cyclase (OAC), a polyketide cyclase, are biosynthetic enzymes involved in olivetolic acid formation in *Cannabis sativa* (Gagne et al., 2012; Taura et al., 2009). CsTKS accepts hexanoyl-CoA as a starter substrate and catalyzes the sequential condensations of three molecules of malonyl-CoA to generate a linear pentyltetra-β-ketide-CoA. Subsequently, the pentyltetra-β-ketide-CoA intermediate is cyclized by OAC to form the resorcylic acid skeleton. *In vivo* analyses and *in vitro* assays of wild-type CsTKS and OAC pairs revealed that the resorcylic acid products with alkyl moieties longer than C_5 are not generated by this combination system (Luo et al., 2019). The genes encoding CsTKS and OAC have been identified from *C. sativa* leaves and subcloned into the pCold II and pQE80L vectors as the *N*-terminal GST tag fusion proteins linked to the *N*-termini of CsTKS and OAC, respectively, *via* an 8-residue (LEVLFQGP) linker that can be cleaved by PreScission Protease. The plasmids are used to transform *E. coli* C43(DE3) and *E. coli* M15(pREP$_4$), respectively. Here, we report the cloning, functional characterization, and structural analyses of TKS and OAC by heterologous expression of the enzymes in *E. coli*.

5.2 Cloning, overexpression, and purification of CsTKS and OAC

5.2.1 Materials

- ReverTra Ace (Toyobo)
- TaKaRa PCR Thermal Cycler DiceTM
- pCold II (TaKaRa)
- pQE80L (QIAGEN)
- Restriction enzymes *Sma*I, *Sal*I, and *Xba*I (TaKaRa)

- *Escherichia coli* C43(DE3)
- Luria Bertani (LB) media
- LB agar plates containing 100 μg/mL ampicillin
- Ampicillin, 100 mg/mL
- 1 M isopropyl-1-thio-β-D-galactopyranoside (IPTG)
- Erlenmeyer flask (2 L)
- Temperature-controlled shaker
- *E. coli* M15(pREP$_4$)
- LB agar plates containing 100 μg/mL ampicillin and 25 μg/mL kanamycin
- Kanamycin, 25 mg/mL
- Lysis buffer: 50 mM Tris-HCl buffer (pH 8.0), 200 mM NaCl, and 5% (*v*/v) glycerol
- Wash buffer: 50 mM Tris-HCl buffer (pH 8.0), 50 mM NaCl, and 5% (v/v) glycerol
- Gel filtration buffer: 50 mM Tris-HCl buffer (pH 8.0), 25 mM NaCl, and 5% (v/v) glycerol
- High speed refrigerated microcentrifuge
- COSMOGEL GST-Accept resin (nacalai)
- Pre-Scission Protease (GE Healthcare)
- 3 kDa MWCO Amicon ultra centrifugal filtration unit (Merck)
- HiLoad™ 16/60 Superdex™ 200 pg (Cytiva)
- ÄKTA start (Cytiva)

5.2.2 Procedure for cloning of CsTKS

1. Total RNA from *C. sativa* leaves is extracted by the acid-guanidinium-phenol-chloroform method
2. Reverse-transcription of total RNA is performed by using an oligo-dT primer (5′-(T)$_{17}$–3′) and oligo-dT primer possessing an adapter (5′-GACTCGTCTAGAGGATCCCG(T)$_{17}$-3′) with Moloney murine leukemia virus reverse transcriptase (ReverTra Ace, Toyobo).
3. PCR amplify the core genes (240-bp) of polyketide synthases with the obtained cDNA and inosine-containing degenerate primers, designed based on the conserved sequences of CHSs: 5′-GGGGTACCAT (A/C/T)TT(C/T)TGGAT(A/C/T)GCICA(C/T)C-3′ and 5′-GCTC TAGAA(A/G)ICC(A/C/G/T)GGICC(A/G)AAICC(A/G)A-3′.
4. Amplify the cDNA fragment for the 5′-upstream region close to the protein *N*-terminus (900-bp) with degenerate and gene-specific primer

pairs: 5′-CGGGATCC(A/G)(A/C)ITA(C/T)CCIGA(C/T)TA(C/T)TA(C/T)T-3′ and 5′-GGGGTACCAGCACATGACGTGAATCCA-3′.

5. Use the 3′-RACE method to amplify the 370-bp fragment of CsTKS with a gene-specific primer and an adapter primer: 5′-GGGGTACCTGAGTTGAGGAAGAGGTCG-3′ and 5′-GACTCGTCTAGAGGATCCCG-3′.
6. The 5′-RACE product (200-bp) is amplified as follows. The first round of PCR is performed with gene-specific primer and adapter primer pairs: 5′-CAGTTACGTTTCCTTATCATAC-3′ and 5′-GACTCGTCTAGAGGATCCCG$(T)_{17}$–3′, in the presence of poly(dA)-tailed cDNA. The nested PCR is performed with gene-specific primer pairs: 5′-GGGGTACCCGAAACTTTTCTTTGAGTTGAG-3′ and 5′-GACTCGTCTAGAGGATCCCG-3′.
7. The full-length cDNA encoding CsTKS is amplified with specific primer pairs: 5′-TTCCAGGGCCCGGGAATGAACCACCTTCGTGCAGAAGGTCCAGCA-3′ and 5′-GATTACCTATCTAGATCAATATTTGATCGGTACGGAGCGGACCAC-3′, with SmaI and XbaI digestion sites underlined, respectively
8. The full-length CsTKS cDNA fragments are sub-cloned into the SmaI/XbaI sites of the pCold II vector (TaKaRa) for expression as *N*-terminal GST-tag fusion proteins

Note

- We extracted the total RNA from *C. sativa* leaves *via* the reported procedure (acid-guanidinium-phenol-chloroform method) (Chomczynski & Sacchi, 1987), without a commercial kit. This technique is suitable for extracting RNA from plants with high pigmentation

5.2.3 Procedure for cloning of OAC

1. The cDNA encoding full-length OAC from *C. sativa*, with SmaI/SalI sites, is purchased from Eurofins Genomics
2. The full-length CsTKS cDNA fragment is sub-cloned into the SmaI/SalI sites of the pQE80L vector (QIAGEN), for expression as an *N*-terminal GST-tag fusion protein

5.2.4 Procedure for CsTKS expression

1. The full-length cDNA encoding CsTKS in pCold II is transformed into *E. coli* C43(DE3) and selected by cultivation on Luria-Bertani (LB) agar plates with 100 μg/mL ampicillin at 37 °C for 16 h

2. A single colony is selected and incubated in 10 mL LB broth containing 100 μg/mL ampicillin at 37 °C overnight
3. The seed culture is transferred into 1 L of LB broth containing 100 μg/mL ampicillin and incubated at 37 °C until the optical density at 600 nm (OD_{600}) becomes 0.6
4. Protein expression is induced by adding IPTG (final concentration 0.5 mM).
5. The culture is incubated at 15 °C for another 16 h
6. Cultured cells are harvested by centrifugation at 6500 rpm for 15 min, and the cell pellet can be stored at −80 °C until use.

Notes

- The proteins are expressed with a His_6-tag at the N-terminus
- Parameters such as the IPTG concentration, incubation time before induction, and shaking speed during expression can be optimized to improve the yield of proteins

5.2.5 Procedure for OAC expression

1. The full-length cDNA encoding OAC in pQE80L is transformed into *E. coli* M15($pREP_4$) and selected by cultivation on Luria-Bertani (LB) agar plates with 100 μg/mL ampicillin and 25 μg/mL kanamycin, at 37 °C for 16 h
2. A single colony is selected and incubated in 10 mL LB broth containing 100 μg/mL ampicillin and 25 μg/mL kanamycin, at 37 °C overnight
3. The seed culture is transferred into 1 L of LB broth containing 100 μg/mL ampicillin and 25 μg/mL kanamycin and incubated at 37 °C until the optical density at 600 nm (OD_{600}) becomes 0.6
4. Protein expression is induced by adding IPTG (final concentration 1 mM).
5. The culture is incubated at 17 °C for another 16 h
6. Cultured cells are harvested by centrifugation at 6500 rpm for 15 min, and the cell pellet can be stored at −80 °C until use

Notes

- The proteins are expressed with a His_6-tag at the N-terminus
- Parameters such as the IPTG concentration, incubation time before induction, and shaking speed during expression can be optimized to improve the yield of proteins

5.2.6 Procedure for purification of CsTKS and OAC

1. The harvested cells are resuspended in 10 mL of lysis buffer

2. The cells are disrupted by sonication on ice, and the cell debris is removed by centrifugation at 12,000 × *g* for 15 min
3. The supernatant is loaded onto a 5 mL COSMOGEL GST-Accept resin column (nacalai) equilibrated with lysis buffer. This procedure is repeated at least twice
4. The resin is washed with 30 column volumes (CVs) of wash buffer
5. The GST-tag is cleaved on the column by Pre-Scission Protease (GE Healthcare) for 16 h
6. The recombinant protein is subsequently eluted with 5 CVs of wash buffer
7. The protein solution is concentrated to 5 mL with a 3 kDa MWCO Amicon ultrafiltration unit (Merck) and then purified to homogeneity by gel-filtration chromatography on a HiLoad™ 16/60 Superdex™ 200 pg column (Cytiva). The target protein fractions are concentrated to 5.0 mg/mL (CsTKS) and 4.2 mg/mL (OAC) in gel filtration buffer
8. Purified proteins are flash-frozen with liquid nitrogen and can be stored at −80 °C until use

Notes

- All purification procedures are performed at 4 °C using the indicated buffers
- For the heterologous expression of CsTKS and OAC in *E. coli*, the *N*-terminal GST tag is essential for high yields of the purified enzymes
- The purification procedure is designed for an 8 L expression culture
- The typical yields of the purified wild-type CsTKS and OAC enzymes are ∼0.5 mg and ∼1.0 mg from 1 L culture, respectively

5.3 *In vitro* characterization of CsTKS and OAC

In vitro co-incubation assays of CsTKS and OAC with substrates indicated that both enzymes are required to generate alkylresorcylic acids. The enzyme reaction products of the combination assay of CsTKS and OAC can be analyzed by LC-MS. The *in vitro* assays of CsTKS and OAC revealed that CsTKS accepts pentyl- and heptyl-CoA as starter substrates to produce linear tetra-β-ketide-CoAs. Subsequently, OAC accepts these linear tetra-β-ketide-CoAs as starter substrates and produces alkylresorcylic acids. Due to the instability of the linear tetra-β-ketide-CoAs, the non-enzymatic products such as alkyltetraketide pyrone and alkylresorcinol are generated. Therefore, CsTKS and OAC must be mixed before starting the reaction for the co-incubation assays of CsTKS and OAC.

5.3.1 Materials

- 1260 Infinity LC system (Agilent Technologies)
- 6420 Triple Quad LC/MS spectrometer (Agilent Technologies)
- TSK-gel ODS-80Ts column (4.6mm × 150mm) (Tosoh)
- Malonyl-CoA
- Fatty acyl-CoA (C_6, C_8, C_{10}, or C_{12}-CoAs)
- Buffer A: 100mM potassium phosphate buffer (pH 7.0)

5.3.2 Procedure for co-incubation assays of CsTKS and OAC

1. 10 μg CsTKS and 30 μg OAC are incubated with 60 μM fatty acyl-CoA (C_6, C_8, C_{10}, or C_{12}-CoAs), and 100 μM malonyl-CoA, in a final volume of 500 μL of buffer A, at 20 °C for 16h
2. The reactions are quenched by adding 50 μL of 20% (*v*/v) HCl, and the enzyme reaction products are extracted with 1.0mL of EtOAc, three times
3. The extracted solutions are evaporated under reduced pressure, and the residues are dissolved in 100 μL of 100% methanol
4. 20 μL of the sample is injected into LC-ESI-MS. The enzyme reaction products are separated on a TSK-gel ODS-80Ts column with a gradient of H_2O (Phase A) and methanol (Phase B), both containing 0.1% formic acid: 0–5min, 60% B; 5–7min, 60–70% B; 7–15min, 70% B; 15–70min, 70–80% B, at a flow rate of 0.6 mL/min

Note

- Temperature, buffer pH, and ratio between enzyme and substrate can be optimized to improve the yields of the enzyme reaction products
- The efficiency of the enzyme reaction can be improved when the ratio between the starter-CoA and malonyl-CoA is 1:2
- Although the enzyme products can be detected in 2h, 16h are required for the reaction to consume all substrates

5.4 Structural analysis and structure-based mutagenesis of CsTKS and OAC

To rationally engineer the substrate specificity, the analysis of the hidden pockets and the substitutions of pocket forming residues with amino acids bearing smaller side chains are critical. The crystal structures of wild type CsTKS and OAC were obtained at 1.39 and 1.70 Å resolutions, respectively (Yang et al., 2016). In the CsTKS structure, the Met187 and Leu190 residues separate the main active-site cavity from a pocket behind these residues, which determines the specificity for the alkyl chain length of the substrate (Fig. 11). Based on these observations, Leu190 in CsTKS was

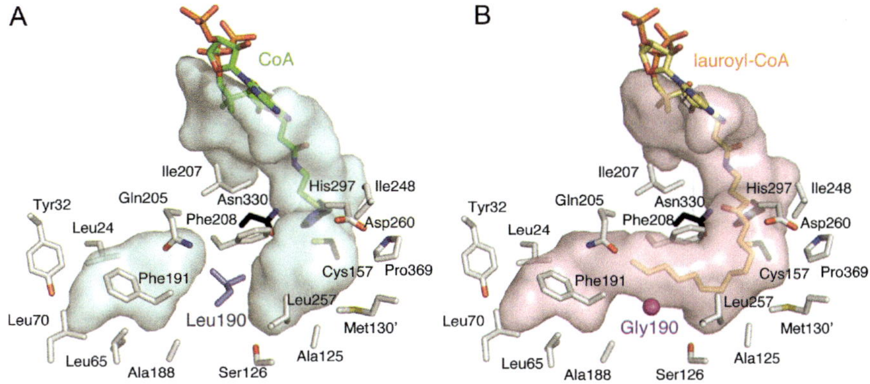

Fig. 11 Active sites of (A) CsTKS wild type with CoASH and (B) CsTKS L190G variant with lauroyl-CoA (Lee et al., 2022).

substituted with glycine to open the gate of the hidden pocket (Lee et al., 2022). *In vitro* analyses of this variant with different lengths of fatty acyl-CoAs indicated that the enzyme accepts up to C_{12} of fatty acyl-CoA as a substrate to generate the corresponding tetraketide intermediate. The crystal structure of CsTKS L190G in complex with lauroyl-CoA confirmed the increased cavity size of CsTKS, from 417 to 471 Å^3. Furthermore, the terminal alkyl chain of lauroyl-CoA is located in the elongated pocket.

The crystal structure of wild type OAC with C_5-alkylresorcylic acid revealed that the bulky Phe24 residue constructs the bottom surface of the substrate-binding pocket. To increase the active site volume of OAC, the Phe24 was substituted. These structure-based engineerings of CsTKS and OAC demonstrated that the enzyme pair of CsTKS L190G and OAC F24I has expanded catalytic ability to produce C_5- to C_{11}-alkylresorcylic acids from hexanoyl-, octanoyl-, decanoyl-, and dodecanoyl-CoAs as starter substrates. The crystal structures of OAC F24I with C_7-alkylresorcylic acid and C_9-alkylresorcylic acid indicated that the F24I substitution expanded the active site and formed a longer alkyl-binding pocket length (12.3 Å) than that of the wild-type pocket (9.0 Å) to accommodate the undecyltetra-β-ketide-CoA as a substrate (Fig. 12).

In the following section, we report the detailed protocol for X-ray crystallographic analysis and structure-based mutagenesis experiments for expanding the substrate tolerances of CsTKS and OAC.

5.4.1 Materials

- Sitting drop crystallization plates
- PEG/Ion screening kit (Hampton)

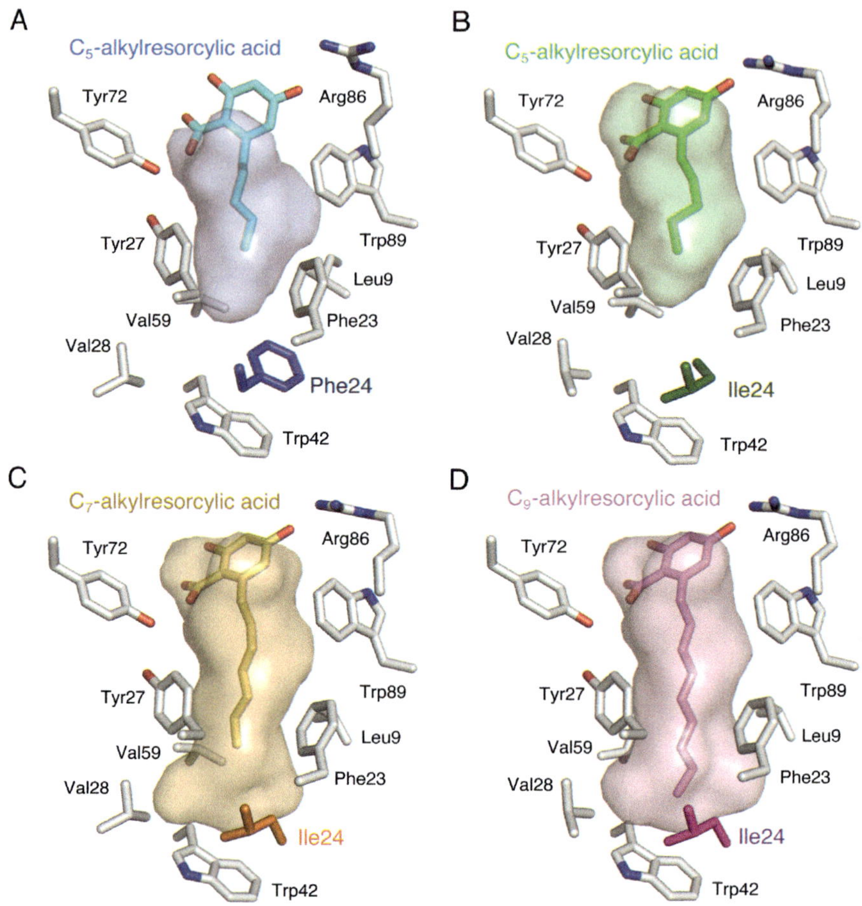

Fig. 12 Comparison of the active sites of OAC wild type and F24I variant (Lee et al., 2022). The active sites of (A) OAC wild type in complex with C_5-alkylresorcylic acid, (B) OAC F24I in complex with C_5-alkylresorcylic acid, (C) C_7-alkylresorcylic acid, and (D) C_9-alkylresorcylic acid.

- Sealing film
- 4 °C incubator
- Microscope
- Crystallization solutions:

CsTKS/CoA: 200 mM potassium nitrate, 20% (*w*/*v*) PEG3350, and 2 mM CoA

CsTKS L190G/lauroyl-CoA: 200 mM ammonium acetate, 20% (w/v) PEG3350, and 2 mM lauroyl-CoA

OAC/C_5-alkylresorcylic acid: 100 mM Tris-HCl (pH 8.8), 25% (w/v) PEG6000, and 50 mM C_5-alkylresorcylic acid

OAC F24I/C_5-alkylresorcylic acid: 100 mM MES-NaOH (pH 6.5), 30% (w/v) PEG3350, 1.2% (*v*/v) (±)-1,3-butanediol, and 2 mM C_5-alkylresorcylic acid

OAC F24I/C_7-alkylresorcylic acid: 100 mM HEPES (pH 7.5), 35% (w/v) PEG3350, and 2 mM C_7-alkylresorcylic acid

OAC F24I/C_9-alkylresorcylic acid: 100 mM MES-NaOH (pH 6.5), 25% (w/v) PEG3350, and 2 mM C_9-alkylresorcylic acid

- QuikChange site-directed mutagenesis kit (Stratagene)
- TaKaRa PCR Thermal Cycler Dice

5.4.2 Procedure for X-ray crystallography of CsTKS and OAC

1. Initial crystallization screening of CsTKS and OAC is performed by using the 96-condition crystallization screening kit designed initially by Mitsubishi Chemical Corp or PEG/Ion developed by Hampton Research. Add 50 μL of the precipitant solution to the reservoir
2. Mix 0.5 μL CsTKS (10 mg/mL), CsTKS L190G (20 mg/mL), OAC (25 mg/mL), or OAC F24I (10 mg/mL) and 0.5 μL precipitant solution in the drop well. Store the plates at 4 °C
   - Diffraction quality crystals of CsTKS are obtained in 200 mM potassium nitrate, 20% (w/v) PEG3350, and 2 mM CoA (Fig. 13).
   - Diffraction quality crystals of the CsTKS L190G variant are obtained in 200 mM ammonium acetate, 20% (w/v) PEG3350, and 2 mM lauroyl-CoA

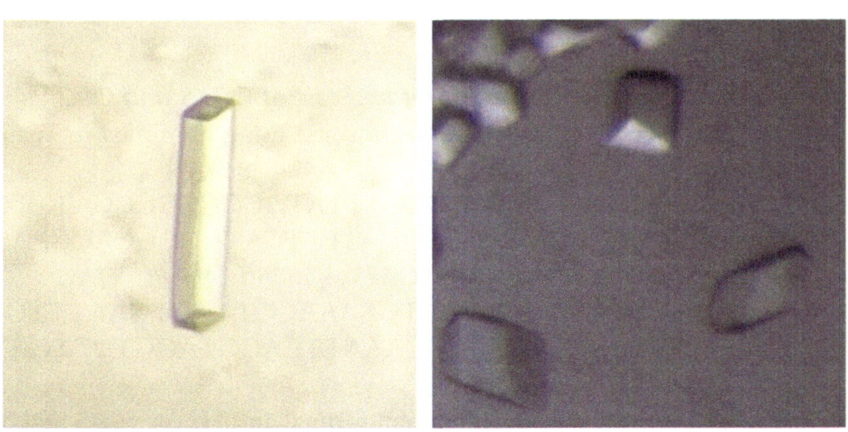

Fig. 13 crystals of CsTKS (left) and OAC (right).

- Diffraction quality crystals of OAC with C_6-alkylresorcylic acid are obtained in 100 mM Tris-HCl (pH 8.8), 25% (w/v) PEG6000, and 50 mM C_6-alkylresorcylic acid (Fig. 13).
- Diffraction quality crystals of the OAC F24I variant with C_6-alkylresorcylic acid are obtained in 100 mM MES-NaOH (pH 6.5), 30% (w/v) PEG3350, 1.2% (v/v) (±)-1,3-butanediol, and 2 mM C_6-alkylresorcylic acid
- Diffraction quality crystals of the OAC F24I variant with C_8-alkylresorcylic acid are obtained in 100 mM HEPES (pH 7.5), 35% (w/v) PEG3350, and 2 mM C_8-alkylresorcylic acid
- Diffraction quality crystals of the OAC F24I variant with C_{10}-alkylresorcylic acid are obtained in 100 mM MES-NaOH (pH 6.5), 25% (w/v) PEG3350, and 2 mM C_{10}-alkylresorcylic acid

3. The CsTKS and OAC crystals are transferred into cryoprotectant solutions, consisting of the final crystallization buffer containing 20% (v/v) glycerol (CsTKS and CsTKS L190G), 10% (v/v) glycerol (OAC), or 30% (v/v) glycerol (OAC F24I).
4. The crystals are captured by a cryoloop (Hampton Research) and flash-cooled on a goniometer under a nitrogen gas stream (−173 °C).

Notes

- For the co-crystallization of CsTKS with lauroyl-CoA, the buffer conditions for the CsTKS purification are critical. We could not obtain a ligand-binding structure under conditions with reducing agents such as DTT or 2-mercaptoethanol
- The soaking method does not work to obtain complex structures of OAC and ligands

5.4.3 Procedure for site-directed mutagenesis of CsTKS and OAC

1. PCR amplification of mutants is performed using the following pairs of primers:

   L190G of CsTKS: 5′-TGCGACATTATGGCCTGCGGATTTC GTGGCCCATCAGAG-3′ and 5′-CTCTGATGGGCCACGAAA TCCGCAGGCCATAATGTCGCA-3′, and F24I of OAC: 5′-GCACAGAAAGAGGAGTTCATCAAAACCTACGTGAATTTG-GTGAAC-3′ and 5′-GTTCACCAAATTCACGTAGGTTTTGAT GAACTCCTCTTTCTGTGC-3′.
2. The plasmids are independently transformed into *E. coli* C43 (DE3) (CsTKS) and *E. coli* M15 ($pREP_4$) (OAC).

3. The variant enzymes are expressed, purified, and analyzed using the same wild-type procedures described in Sections 5.2 and 5.3

Notes

- The yields of the CsTKS and OAC variants are similar to the wild-type enzymes

6. Summary

Due to their catalytic potential and broad substrate promiscuity to generate chemically different and structurally diverse molecular scaffolds, the plant type III PKS-superfamily enzymes have attracted keen attention as a source of powerful and unique biocatalysts. Recent functional and structural investigations of type III PKSs have increased our knowledge toward understanding their detailed catalytic versatility and mechanisms. In this chapter, we have described our approaches (the tools and methods) for gene identifications, enzyme preparations, and enzyme assays of plant type III PKSs and polyketide cyclase to explore novel enzymes from the plant kingdom, and the structural analyses of the enzymes to understand their catalytic mechanisms and to alter their functions by protein engineering. These investigations will provide important insights into the manipulation of the enzyme reactions for the further production of medicinally important natural and unnatural molecules.

Acknowledgments

The work was supported in part by a Grant-in-Aid for Scientific Research from the Ministry of Education, Culture, Sports, Science and Technology, Japan (JSPS KAKENHI Grant Number JP16H06443, JP19K15703, JP20H00490, JP20KK0173, and JP20K22700), the New Energy and Industrial Technology Development Organization (NEDO, Grant Number JPNP20011), AMED (Grant Number JP21ak0101164), the PRESTO program from Japan Science and Technology Agency (Grant Number JPMJPR20DA).

References

Abe, I. (2008). Engineering of plant polyketide biosynthesis. *Chemical & Pharmaceutical Bulletin (Tokyo)*, *56*(11), 1505–1514. https://doi.org/10.1248/cpb.56.1505.

Abe, I. (2020). Biosynthesis of medicinally important plant metabolites by unusual type III polyketide synthases. *Journal of Natural Medicines*, *74*(4), 639–646. https://doi.org/10.1007/s11418-020-01414-9.

Abe, I., & Morita, H. (2010). Structure and function of the chalcone synthase superfamily of plant type III polyketide synthases. *Natural Product Reports*, *27*(6), 809–838. https://doi.org/10.1039/b909988n.

Abe, I., Morita, H., Nomura, A., & Noguchi, H. (2000). Substrate specificity of chalcone synthase: Enzymatic formation of unnatural polyketides from synthetic cinnamoyl-CoA analogues. *Journal of the American Chemical Society*, *122*(45), 11242–11243. https://doi.org/10.1021/ja0027113.

Abe, I., Watanabe, T., & Noguchi, H. (2004). Enzymatic formation of long-chain polyketide pyrones by plant type III polyketide synthases. *Phytochemistry*, *65*(17), 2447–2453. https://doi.org/10.1016/j.phytochem.2004.08.005.

Austin, M. B., & Noel, J. P. (2003). The chalcone synthase superfamily of type III polyketide synthases. *Natural Product Reports*, *20*(1), 79–110. https://doi.org/10.1039/b100917f.

Chomczynski, P., & Sacchi, N. (1987). Single-step method of RNA isolation by acid guanidinium thiocyanate-phenol-chloroform extraction. *Analytical Biochemistry*, *162*(1), 156–159. https://doi.org/10.1006/abio.1987.9999.

Ferrer, J. L., Jez, J. M., Bowman, M. E., Dixon, R. A., & Noel, J. P. (1999). Structure of chalcone synthase and the molecular basis of plant polyketide biosynthesis. *Nature Structural Biology*, *6*(8), 775–784. https://doi.org/10.1038/11553.

Gagne, S. J., Stout, J. M., Liu, E., Boubakir, Z., Clark, S. M., & Page, J. E. (2012). Identification of olivetolic acid cyclase from Cannabis sativa reveals a unique catalytic route to plant polyketides. *Proceedings of the National Academy of Sciences of the United States of America*, *109*(31), 12811–12816. https://doi.org/10.1073/pnas.1200330109.

Jez, J. M., Austin, M. B., Ferrer, J., Bowman, M. E., Schröder, J., & Noel, J. P. (2000). Structural control of polyketide formation in plant-specific polyketide synthases. *Chemistry & Biology*, 7(12), 919–930. https://doi.org/10.1016/s1074-5521(00)00041-7.

Jez, J. M., Bowman, M. E., & Noel, J. P. (2002). Expanding the biosynthetic repertoire of plant type III polyketide synthases by altering starter molecule specificity. *Proceedings of the National Academy of Sciences of the United States of America*, *99*(8), 5319–5324. https://doi.org/10.1073/pnas.082590499.

Jez, J. M., Ferrer, J. L., Bowman, M. E., Austin, M. B., Schröder, J., Dixon, R. A., et al. (2001). Structure and mechanism of chalcone synthase-like polyketide synthases. *Journal of Industrial Microbiology & Biotechnology*, *27*(6), 393–398. https://doi.org/10.1038/sj.jim.7000188.

Katsuyama, Y., Kita, T., Funa, N., & Horinouchi, S. (2009). Curcuminoid biosynthesis by two type III polyketide synthases in the herb Curcuma longa. *The Journal of Biological Chemistry*, *284*(17), 11160–11170. https://doi.org/10.1074/jbc.M900070200.

Katsuyama, Y., Matsuzawa, M., Funa, N., & Horinouchi, S. (2007). In vitro synthesis of curcuminoids by type III polyketide synthase from Oryza sativa. *The Journal of Biological Chemistry*, *282*(52), 37702–37709. https://doi.org/10.1074/jbc.M707569200.

Katsuyama, Y., Miyazono, K., Tanokura, M., Ohnishi, Y., & Horinouchi, S. (2011). Structural and biochemical elucidation of mechanism for decarboxylative condensation of beta-keto acid by curcumin synthase. *The Journal of Biological Chemistry*, *286*(8), 6659–6668. https://doi.org/10.1074/jbc.M110.196279.

Lee, Y. E., Nakashima, Y., Kodama, T., Chen, X., & Morita, H. (2022). Dual engineering of olivetolic acid cyclase and tetraketide synthase to generate longer alkyl-chain olivetolic acid analogs. *Organic Letters*, *24*(1), 410–414. https://doi.org/10.1021/acs.orglett.1c04089.

Lukacin, R., Springob, K., Urbanke, C., Ernwein, C., Schröder, G., Schröder, J., et al. (1999). Native acridone synthases I and II from Ruta graveolens L. form homodimers. *FEBS Letters*, *448*(1), 135–140. https://doi.org/10.1016/s0014-5793(99)00355-5.

Luo, X., Reiter, M. A., d'Espaux, L., Wong, J., Denby, C. M., Lechner, A., et al. (2019). Complete biosynthesis of cannabinoids and their unnatural analogues in yeast. *Nature*, *567*(7746), 123–126. https://doi.org/10.1038/s41586-019-0978-9.

Matsui, T., Kodama, T., Mori, T., Tadakoshi, T., Noguchi, H., Abe, I., et al. (2017). 2-Alkylquinolone alkaloid biosynthesis in the medicinal plant Evodia rutaecarpa involves collaboration of two novel type III polyketide synthases. *The Journal of Biological Chemistry*, *292*(22), 9117–9135. https://doi.org/10.1074/jbc.M117.778977.

Mori, T., Shimokawa, Y., Matsui, T., Kinjo, K., Kato, R., Noguchi, H., et al. (2013). Cloning and structure-function analyses of quinolone- and acridone-producing novel type III polyketide synthases from Citrus microcarpa. *The Journal of Biological Chemistry*, *288*(40), 28845–28858. https://doi.org/10.1074/jbc.M113.493155.

Morita, H., Shimokawa, Y., Tanio, M., Kato, R., Noguchi, H., Sugio, S., et al. (2010). A structure-based mechanism for benzalacetone synthase from Rheum palmatum. *Proceedings of the National Academy of Sciences of the United States of America*, *107*(2), 669–673. https://doi.org/10.1073/pnas.0909982107.

Morita, H., Wanibuchi, K., Nii, H., Kato, R., Sugio, S., & Abe, I. (2010). Structural basis for the one-pot formation of the diarylheptanoid scaffold by curcuminoid synthase from Oryza sativa. *Proceedings of the National Academy of Sciences of the United States of America*, *107*(46), 19778–19783. https://doi.org/10.1073/pnas.1011499107.

Morita, H., Wong, C. P., & Abe, I. (2019). How structural subtleties lead to molecular diversity for the type III polyketide synthases. *The Journal of Biological Chemistry*, *294*(41), 15121–15136. https://doi.org/10.1074/jbc.REV119.006129.

Resmi, M. S., Verma, P., Gokhale, R. S., & Soniya, E. V. (2013). Identification and characterization of a type III polyketide synthase involved in quinolone alkaloid biosynthesis from Aegle marmelos Correa. *The Journal of Biological Chemistry*, *288*(10), 7271–7281. https://doi.org/10.1074/jbc.M112.429886.

Schröder, J., & Schröder, G. (1990). Stilbene and chalcone synthases: Related enzymes with key functions in plant-specific pathways. *Zeitschrift fuer Naturforschung, C: Journal of Biosciences*, *45*(1–2), 1–8. https://doi.org/10.1515/znc-1990-1-202.

Shimokawa, Y., Morita, H., & Abe, I. (2012). Benzalacetone synthase. *Frontiers in Plant Science*, *3*, 57. https://doi.org/10.3389/fpls.2012.00057.

Springob, K., Lukacin, R., Ernwein, C., Gröning, I., & Matern, U. (2000). Specificities of functionally expressed chalcone and acridone synthases from Ruta graveolens. *European Journal of Biochemistry*, *267*(22), 6552–6559. https://doi.org/10.1046/j.1432-1327.2000.01746.x.

Taura, F., Tanaka, S., Taguchi, C., Fukamizu, T., Tanaka, H., Shoyama, Y., et al. (2009). Characterization of olivetol synthase, a polyketide synthase putatively involved in cannabinoid biosynthetic pathway. *FEBS Letters*, *583*(12), 2061–2066. https://doi.org/10.1016/j.febslet.2009.05.024.

Wakimoto, T., Morita, H., & Abe, I. (2012). Engineering of plant type III polyketide synthases. *Methods in Enzymology*, *515*, 337–358. https://doi.org/10.1016/b978-0-12-394290-6.00016-1.

Wang, X. H., Gao, B. W., Nakashima, Y., Mori, T., Zhang, Z. X., Kodama, T., et al. (2022). Identification of a diarylpentanoid-producing polyketide synthase revealing an unusual biosynthetic pathway of 2-(2-phenylethyl)chromones in agarwood. *Nature Communications*, *13*(1), 348. https://doi.org/10.1038/s41467-022-27971-z.

Wang, J., Wang, X. H., Liu, X., Li, J., Shi, X. P., Song, Y. L., et al. (2016). Synthesis of unnatural 2-substituted quinolones and 1,3-diketones by a member of type III polyketide synthases from Huperzia serrata. *Organic Letters*, *18*(15), 3550–3553. https://doi.org/10.1021/acs.orglett.6b01501.

Yamaguchi, T., Kurosaki, F., Suh, D. Y., Sankawa, U., Nishioka, M., Akiyama, T., et al. (1999). Cross-reaction of chalcone synthase and stilbene synthase overexpressed in Escherichia coli. *FEBS Letters*, *460*(3), 457–461. https://doi.org/10.1016/s0014-5793(99)01403-9.

Yang, X., Matsui, T., Kodama, T., Mori, T., Zhou, X., Taura, F., et al. (2016). Structural basis for olivetolic acid formation by a polyketide cyclase from Cannabis sativa. *The FEBS Journal*, *283*(6), 1088–1106. https://doi.org/10.1111/febs.13654.

Yu, O., & Jez, J. M. (2008). Nature's assembly line: Biosynthesis of simple phenylpropanoids and polyketides. *The Plant Journal*, *54*(4), 750–762. https://doi.org/10.1111/j.1365-313X.2008.03436.x.

CHAPTER TWO

An LC-MS/MS assay for enzymatic characterization of methylthioalkylmalate synthase (MAMS) involved in glucosinolate biosynthesis

Roshan Kumar[a,b], Michael Reichelt[c], and Naveen C. Bisht[a,*]
[a]National Institute of Plant Genome Research, New Delhi, India
[b]Institute of Science and Technology Austria, Klosterneuburg, Austria
[c]Department for Biochemistry, Max Planck Institute for Chemical Ecology, Jena, Germany
*Corresponding author: e-mail address: ncbisht@nipgr.ac.in

Contents

Abstract

Brassicaceae are blessed with specialized metabolites called glucosinolates (GSLs), which along with their degradation products, are beneficial in agriculture and human health. To date, more than 130 GSL structures have been identified, mostly derived from the amino acid methionine. The biosynthesis of methionine-derived aliphatic GSLs starts with a side-chain elongation step involving a recursive three-step cyclic process that incorporates a new methylene group into the 2-oxo acid to form a series of elongated

Methods in Enzymology, Volume 676
ISSN 0076-6879
https://doi.org/10.1016/bs.mie.2022.07.019

2-oxo acids. Methylthioalkylmalate synthase (MAMS) catalyzes the first committed step in the side-chain elongation of methionine-derived GSLs. The substrate specificity of MAMS with different 2-oxo acids determines whether reaction products of a given cycle enter for an additional round of chain elongation or enter into core GSLs structure formation. Multiple MAMS encoding genes are present in the *Brassicaceae* species and are known to play a central role in shaping the diverse profile of aliphatic GSLs. We recently established a highly sensitive LC-MS/MS-based methodology that quantifies the MAMS activity by estimating the amount of the next intermediate of the pathway, the 2-malate derivatives. Overall, this chapter describes the protocol for the expression, purification, and steady-state kinetic analysis of the recombinant MAMS protein.

1. Introduction

Plants possess an enormous diversity of structurally diverse specialized metabolites formed by diverse biosynthetic enzymes. Numerous studies demonstrate that specialized metabolism evolved by incorporating enzymes and pathways derived from primary metabolism (Khersonsky & Tawfik, 2010; Schwab, 2003). A notable example connecting the evolution of specialized metabolism and primary metabolism can be seen in members of the *Brassicaceae* family, where the evolution of methylthioalkylmalate synthase (MAMS) from isopropylmalate synthase (IPMS) of leucine biosynthesis, catalyzes the committed step in the biosynthesis of precursors to glucosinolates (GSLs) (de Kraker & Gershenzon, 2011; de Kraker, Luck, Textor, Tokuhisa, & Gershenzon, 2007). The basic reaction mechanism of MAMS and IPMS is conserved, where both the enzymes catalyze the condensation reaction between acetyl-CoA with 2-oxo acids derived from methionine. However, the plant MAMS and IPMS have significant structural differences. The former lack the C-terminal regulatory domain interacting with leucine, along with a few amino acid residues different in its active sites, thus changing its affinity for 2-oxo acids (de Kraker et al., 2007; Kumar et al., 2019; Textor et al., 2004). This difference appears to be critical to the molecular evolution of MAMS from IPMS, possibly leading to the deployment of GSL biosynthesis parallel to leucine biosynthesis in the *Brassicaceae* family.

GSLs are the well-defined group of plant secondary metabolites abundantly present in members of the Brassicaceae family, which includes globally cultivated *Brassica* species and the model plant *Arabidopsis thaliana*

(Augustine & Bisht, 2017; Halkier & Gershenzon, 2006). GSLs are thioglycosides and are structurally defined by the presence of a sulfonated oxime, a glucose moiety, and a variable aglycone side-chain (R) derived from α-amino acids (Fig. 1A) (Blažević et al., 2020; Halkier & Gershenzon, 2006). Under normal conditions, GSLs are generally stable inside plant cells. However, tissue injury caused by herbivory or chewing leads to the hydrolysis of thioglycosidic linkage of GSL by myrosinase, which results in formation of β-D-glucose and unstable aglycone intermediate (Angelino et al., 2015; Halkier & Gershenzon, 2006). Further, aglycone molecules can spontaneously rearrange to various active compounds, including isothiocyanates, thiocyanates, and nitriles, depending on the side-chain (R) of the parent GSL and physical and chemical conditions of the medium (Halkier & Gershenzon, 2006; Nguyen, Stewart, Lopez, Ioannou, & Allais, 2020). The GSL and their hydrolyzed products exhibit different biological, pharmacological and dietary properties, thereby playing an important role in plant defense against pathogens and insect herbivores, and also contributing to animal and human health (Fahey, Zhang, & Talalay, 1997; Halkier & Gershenzon, 2006; Hopkins, van Dam, & van Loon, 2009).

Plants contain more than 130 GSL structures, which are grouped into three categories depending on the chemical structure of precursor amino acids: (i) aliphatic, (ii) indole, and (iii) aromatic GSL (Blažević et al., 2020; Halkier & Gershenzon, 2006). The model plant *Arabidopsis* and related *Brassica* species possess huge diversity of aliphatic GSL, which are predominantly derived from methionine (Augustine & Bisht, 2017; Sønderby, Geu-Flores, & Halkier, 2010). The basic biosynthesis of aliphatic GSLs derived from methionine comprises three steps viz., side-chain elongation, core structure formation, and side-chain modification (Halkier & Gershenzon, 2006; Hopkins et al., 2009; Sønderby et al., 2010). The side-chain elongation is the first committed step providing the chemical diversity to aliphatic GSL and catalyzed by MAMS (EC: 2.3.3.17) (Halkier & Gershenzon, 2006; Kumar et al., 2019). This step defines the side-chain length (R) of aliphatic GSL (Fig. 1B). The side-chain elongation proceeds through a three-step cyclic reaction, where MAMS catalyzed the initial aldol condensation of acetyl-CoA with 2-oxo acid derived from precursor methionine to form a substituted 2-alkylmalic acid, a 2-malate derivative (Halkier & Gershenzon, 2006). After that, it undergoes

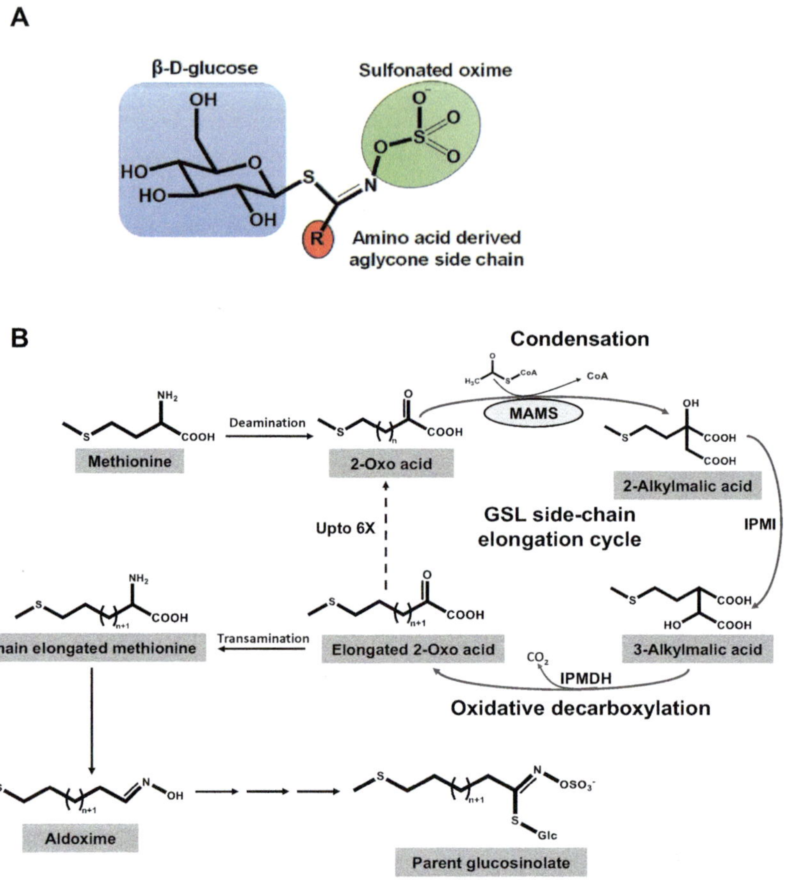

Fig. 1 Glucosinolate structure and side-chain elongation of methionine-derived aliphatic GSLs. (A) The core molecular structure of GSL consists of a β-D-glucose, a sulfonated oxime and a variable aglycone side chain (R). (B) Outline of side-chain elongation of methionine-derived GSL. Initially, deamination of methionine produces 2-oxo acids, which enter into a three-step chain elongation cycle. By catalyzing the condensation reaction between 2-oxo acid and acetyl-CoA, MAMS produces the substituted 2-alkyl malate derivative (2-alkylmalate), which upon isomerization and oxidative decarboxylation reactions forms the 2-oxo acid that is elongated by one methylene (CH_2) moiety. Eventually, the elongated 2-oxo acid is transaminated into elongated methionine or is re-entered into a subsequent round of chain elongation. Up to six rounds of side-chain elongation have been reported from *A. thaliana*.

isomerization via a 1,2-shift of a hydroxyl group by isopropylmalate isomerase (IPMI) to give the substituted 3-alkylmalic acid, a 3-malate derivative (He, Chen, Pang, Strul, & Chen, 2010). Finally, oxidative decarboxylation by isopropylmalate dehydrogenase (IPMDH) results in new 2-oxo acid having one more additional methylene group (He et al., 2009, 2011). After that, the substrate specificity of MAMS with the elongated 2-oxo acid determines whether reaction products of a given cycle enter for an additional round of chain elongation or enter into core GSL structure formation (Halkier & Gershenzon, 2006; Kumar et al., 2019).

The MAMS introductory chemistry is the same as other DRE-TIM family metallolyases in that it orients the substrate for catalysis based on its metal site and forms carbon-carbon bonds between acetyl-CoA and α-ketoacids (Forouhar et al., 2006; Kumar et al., 2019). MAMS contains Arg-89 and Asp-90, which are the key features of the DRE-TIM metallolyase active site signature, where the aspartate coordinates the divalent cation, and the arginine is responsible for catalysis (Forouhar et al., 2006; Kumar et al., 2019). Structure determination of *Brassica juncea* MAMS with acetyl-CoA and a 2-oxo substrate 4-methylthio-2-oxobutanoic acid (4MTOB) reveals a dimeric protein, with each monomer folded into N-terminal α/β-barrel domain and C-terminal α-helical domain (Kumar et al., 2019). The active site residues (V182, E233, A253 and P255) interact with the 4MTOB, the Mn^{2+} and acetyl CoA in BjMAM1.1.

The plant species belonging to the *Brassicaceae* family are known to generate a wide range of GSLs with variable side-chain lengths, but only limited information on the MAMS chemistry in these plants is available (Gohain et al., 2021; Wang et al., 2011). Depending on the side-chain length, aliphatic GSL can be divided into two groups, short-chain (C3-C5) and long-chain (C6-C9) GSLs. The *Arabidopsis* genome encodes three MAMS proteins, wherein MAMS1 and MAMS2 are involved in the catalysis of C3-C5 GSLs, whereas MAMS3 also produces the C6-C9 GSL (Benderoth, Pfalz, & Kroymann, 2008; de Kraker et al., 2007; Field et al., 2004; Textor et al., 2004). Beyond *Arabidopsis*, the MAMS chemistry has only been reported from a polyploid oilseed crop, *B. juncea* which encodes four BjMAMS isoforms (Kumar et al., 2019). Biochemical studies show that BjMAM1-A and BjMAM1-B catalyze the formation of C3-C5 GSLs similar to the Arabidopsis MAMS1, whereas BjMAM2-A and BjMAM2-B catalyze the synthesis of C3-GSLs only. Since MAMS activity primarily dictates the diversity of aliphatic GSLs in the *Brassicaceae*, understanding the MAMS function is a necessary first step for the manipulation of beneficial GSL compounds for animal health and plant protection.

In this context, the characterization of the activity of MAMS and their interaction with 2-oxo acid substrates is of critical importance. Previously, radiolabeled-based MAMS activity assays were developed to study the 2-oxo acid product profiles obtained from the condensation reaction between acetyl-CoA and 2-oxo acids (Kroymann et al., 2001; Textor, de Kraker, Hause, Gershenzon, & Tokuhisa, 2007). Several drawbacks associated with these methods include being relatively insensitive, requiring higher amounts of protein and substrates, are time-consuming, and of all above using the radiolabeled substrate, thereby making the protocol cumbersome. Moreover, artificial or tagged substrate use in the radiolabeled, fluorescence and calorimetric assays might also alter enzyme activity. Recently, LC-MS/MS has become increasingly popular in measuring accurate detection of a plethora of compounds that has its potential use in enzyme activity and kinetic assays (Bothner et al., 2000). Enzymatic reactions analyzed by mass spectroscopic techniques such as liquid-chromatography (LC) coupled with tandem mass spectrometry (MS/MS) have apparent advantages like the possibility of using a physiological substrate and a reduced amount of enzyme or protein. Moreover, the ability to detect substrate and reaction products at a very lower concentration also makes this technique more sensitive and reproducible.

Here we describe the highly sensitive LC-MS/MS-based detection method to detect and quantify the intermediate 2-malate product obtained from the MAMS catalyzed condensation reaction between acetyl-CoA and 2-oxo acids (Table 1), which eventually gets transformed into the different lengths of aliphatic GSLs by a different group of enzymes. We first describe the expression and purification of recombinant MAMS in the bacterial system; second, the endpoint and kinetics of MAMS; third, the standard curve preparation of 2-hexyl malate (a 2-malate derivative) to calculate the unknown concentrations of intermediate 2-malate derivatives; and fourth, detection and quantification of intermediate 2-malate derivatives by LC-MS/MS. These steps will be described using the *A. thaliana* MAMS1 (aka AtMAMS1, AtMAM1) as a representative MAMS candidate.

Table 1 Summary of various 2-oxo substrates and 2-malate intermediates involved in side-chain elongation steps of the Met-derived aliphatic GSLs.

2-oxo substrate	Catalog no. and source	Structure	2-malate intermediates	Structure	Chain elongation step
4-Methyl-thio-2-oxo-butanoate (4MTOB)	K6000 (Sigma-Aldrich, Germany)	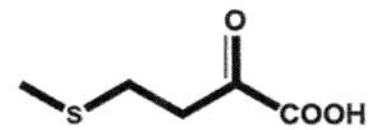	2-[2-(Methylthio) ethyl]-malate (2MTEM)	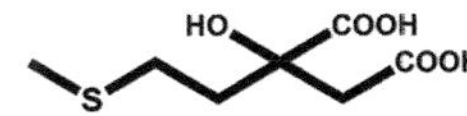	C2 ➔ C3
5-Methyl-thio-2-oxo-pentanoate (5MTOP)	#A4705,0001 (AppliChem, Germany)	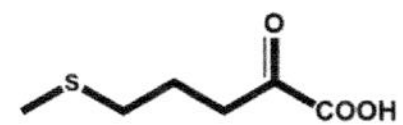	2-[3-(Methylthio) propyl]-malate (2MTPM)	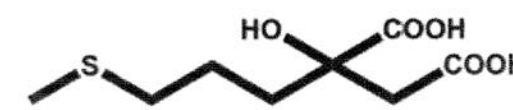	C3 ➔ C4
6-Methyl-thio-2-oxo-hexanoate (6MTOH)	A4851,0001 (AppliChem, Germany)	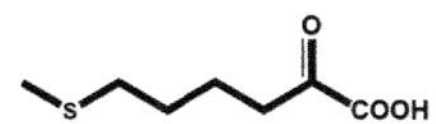	2-[4-(Methylthio) butyl]-malate (2MTBM)	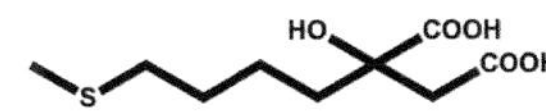	C4 ➔ C5
2-Oxo-nonanoate (2ON)	Synthethized by de Kraker as described in Textor et al., 2007	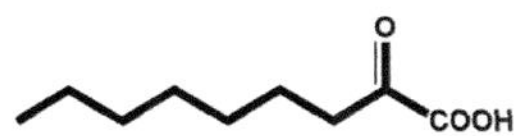	2-Heptyl-malate (2HptM)	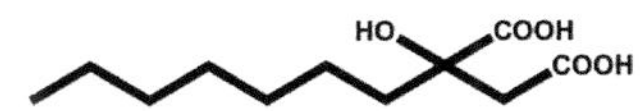	C5 ➔ C6
8-Methyl-thio-2-oxo-octanoate (8MTOO)	Synthethized by de Kraker as described in Textor et al., 2007	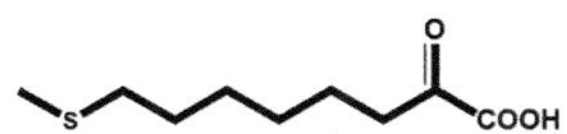	2-[6-(Methylthio) hexyl]-malate (2MTHM)	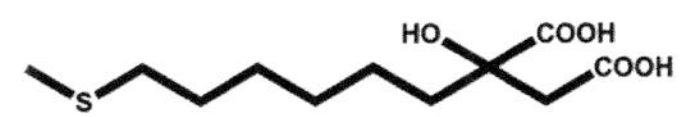	C6 ➔ C7
9-Methyl-thio-2-oxo-nonoate (9MTON)	#A4852,02000 (AppliChem, Germany)	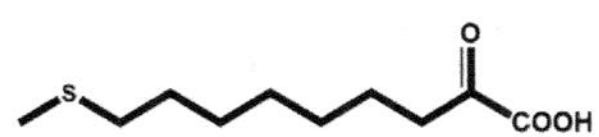	2-[7-(Methylthio) heptyl]-malate (2MTHptM)		C7 ➔ C8
2-Oxo-dodecanoate (2ODD)	Synthethized by de Kraker as described in Textor et al., 2007		2-Decyl malate (2DM)	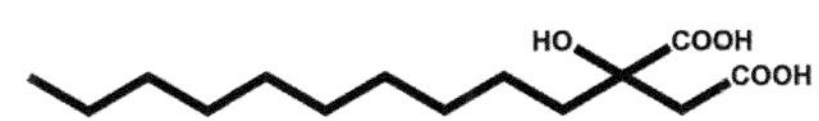	C8 ➔ C9

2. Cloning, expression and purification of methylthioalkylmalate synthase (MAMS)

2.1 Cloning and expression of MAMS

2.1.1 Materials, reagents and equipment

- Restriction enzymes (NEB)
- pET28a expression vector (Novagen)
- DH5α competent cells
- BL21 (DE3) competent cells
- Luria-Bertani (LB) media
- Terrific Broth (TB) media
- Kanamycin
- Isopropyl-1-thio-β-D-galactopyranoside (IPTG)
- Incubator shaker at 18 °C and 37 °C
- Centrifuge at 4 °C

2.1.2 Procedure

1. Amplify the coding DNA sequence of MAMS genes, lacking the N-terminal Chloroplast Transit Peptide (CTP) sequence, using gene-specific primers. The forward and reverse primers should contain appropriate restriction sites that are present in pET28a expression vector (e.g., *NheI* and *NotI*)
2. Ligate the amplified MAMS sequence within *NheI* and *NotI* sites of the pET-28a expression vector and transformed it into DH5α competent cells using standard bacterial transformation procedures
3. Transform 100 ng of the recombinant pET28a-MAMS plasmid into BL21 (DE3) competent cells and select the transformants on LB agar plates containing kanamycin (50 μg/mL)
4. Pick three to five colonies to inoculate 5 mL of LB media containing kanamycin (50 μg/mL) and incubate for 16 h at 37 °C to grow the starter culture with shaking (200 rpm). Additionally, sterile the 500 mL flask containing the 100 mL of TB media
5. Inoculate 100 mL of TB media with 500 μL starter culture with kanamycin (50 μg/mL) and incubate at 37 °C with shaking (200 rpm) for approximately 4–5 h or until OD_{600} reaches 0.6–0.8
6. Cool down the culture at 18 °C with shaking (200 rpm) and add isopropyl-1-thio-β-D-galactopyranoside (IPTG) to a final concentration of 1 mM to induce the recombinant protein synthesis

7. Further, allow the culture to grow overnight (16–17 h) at 18 °C with shaking (200 rpm)
8. Harvest the cells by centrifugation at 2100 (×g) for 30 min at 4 °C and store the cell pellet at −80 °C freezer till further use

2.2 Purification of the recombinant MAMS protein

2.2.1 Materials, reagents and equipment

- Lysis buffer: 50 mM Tris-HCl pH 8.0, 20 mM imidazole, 500 mM NaCl, 1% [v/v] Tween-20, and 10% [v/v] glycerol
- Wash buffer: 50 mM Tris-HCl pH 8.0, 20 mM imidazole, 500 mM NaCl, and 10% [v/v] glycerol
- Elution buffer: 50 mM Tris-HCl pH 8.0, 250 mM imidazole, 500 mM NaCl, and 10% [v/v] glycerol
- Phenylmethylsulfonyl fluoride (PMSF; Sigma, Cat. No. P7626)
- Ni-NTA Agarose (Qiagen, Cat. No. 30230)
- Column PD-10 (Cytiva, Cat. No. GE17-0435-01)
- Protein Assay Dye Reagent Concentrate (Bio-Rad, Cat. No.5000006)
- SDS-PAGE electrophoresis system
- Sonicator (UP200S, Hielscher Ultrasound Technology, Germany)

2.2.2 Procedure

1. Thaw and suspend the cell pellet into 5 mL of lysis buffer containing 1 mM PMSF with continuous agitation at 4 °C for 1–2 h
2. Sonicate the resuspended cells 4 to 5 times (0.5 s ON/OFF cycle for 1 min) with 60–70% amplitude on ice
3. Centrifuge the sonicated cell lysates at 54000 (×g) at 4 °C for 45 min
4. Filter the supernatant containing soluble fraction using a 0.22 μm filter and keep it on ice
5. Add 750 μL of Ni-NTA agarose resin into the PD-10 gravity-flow column and equilibrate with 5 × resin volume of lysis buffer at 4 °C
6. Transfer the filtered supernatant onto the Ni-NTA agarose resin previously equilibrated with lysis buffer at 4 °C
7. Wash the column three times with 5 mL of wash buffer
8. Elute the His-tagged MAMS protein two times with 500 μL of elution buffer
9. Quantify the MAMS protein using 20 μL of each fraction and 980 μL of protein assay dye reagent and measure the absorbance at 595 nm with bovine serum albumin as standard

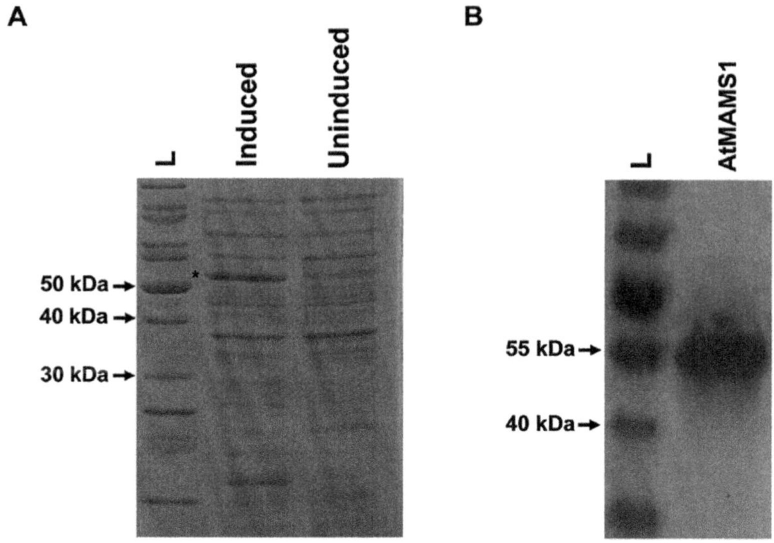

Fig. 2 Expression and purification of recombinant AtMAMS1 protein from *A. thaliana*. (A) Induction of expression of recombinant protein by 1 mM IPTG indicated as the asterisk. (B) SDS-PAGE analysis of the purified fraction of AtMAMS1 protein.

10. Keep the purified protein at 4 °C for enzymatic assays or stored it at −80 °C

11. Check the expression, purity and integrity of purified MAMS protein on 12% SDS-PAGE gel (Fig. 2)

Note: The above procedure can be used to purify the recombinant MAMS proteins from other species of the *Brassicaceae* family. The online tools (e.g., TargetP, ChloroP) can be used to predict the N-terminal Chloroplast Transit Peptide sequence in MAMS, which needs to be excluded for cloning and expression of the recombinant MAMS protein.

3. Determination of MAMS activity

3.1 Assays for MAMS protein

3.1.1 Materials, reagents and equipment

- Eppendorf tube (2 mL)
- 2 mL 96-well round polypropylene microplates

- Water bath at 30 °C
- Mili-Q water
- Tris buffer (pH-8.0)
- $MnCl_2$ (Sigma, Cat. No. 244589)
- Acetyl-CoA (Sigma, Cat. No. A2056)
- 2-Oxo acid substrates (source and catalog numbers are provided in Table 1)
- 2-Hexyl malate (synthesized according to Textor et al., 2007)
- ATP (Sigma, Cat. No. A1852)
- DDT (Sigma, Cat. No. 43815)
- Purified MAMS protein
- 100% Ethanol
- 1.5 mL glass vials (Macherey-Nagel, Düren, Germany, Catalog# 702282)
- Centrifuge

3.1.2 Procedure

3.1.2.1 End-point MAMS assay

The assay can be performed in 2 mL Eppendorf tubes or 96-well plates in a final volume of 50 μL. At a minimum, reactions are performed in duplicates.

1) Prepare the reaction mixture as shown in the table below and add the components in the order listed, where 2-oxo acid substrate should be added last onto the side wall of the tube

Components	Stock concentration	Final concentration	Volume/ reaction (μL)
Milli-Q	–	–	Variable
Tris-buffer (pH -8)	1 M	100 mM	5
$MnCl_2$	40 mM	4 mM	5
Acetyl CoA	10 mM	1 mM	5
ATP	20 mM	2 mM	5
DDT	10 mM	1 mM	5
Purified MAMS	–	1 μg	Variable
2-Oxo acid	30 mM	3 mM	5

2) In case of multiple reactions, prepare the master mix by adding all the components except the 2-oxo acid, gently mix the reaction mixture by flicking the tubes
3) Centrifuge the mixture at 130 (×g) for 15–30 s at room temperature
4) Carefully, add the 2-oxo acid to the side wall of the tubes, and mix gently by flicking to avoid the air bubble
5) Centrifuge the tubes at 130 (×g) for 15–30 s at room temperature
6) Incubate the reaction mixture at 30 °C for 1 h in the water bath
7) Add 150 μL of absolute ethanol to stop the reaction
8) Centrifuge the tubes at 18,000 (× g) for 10 min at room temperature
9) Carefully aliquot 150 μL of the supernatant into HPLC/LC-MS certified 1.5 mL clear glass vials and kept at −20 °C prior to LC-MS/MS analysis as described in Section 3.2
10) Calculate the amount of 2-malate product formed by MAMS catalysis using a standard 2-hexyl malate curve (Sections 3.2 and 3.3)

    Note: The reaction can also be performed in 1 mL volume by scaling up the components, however, no significant difference in the specific activity of MAMS was found between low and high-volume reactions.

3.1.2.2 Steady-state MAMS kinetic assay

After determination of the enzyme activity of MAMS by end-point assay, a full steady-state kinetic experiment can be performed. An established approach based on the Michaelis-Menten equation (MM) is used to understand the MAMS kinetics. The following protocol containing the fixed MAMS (enzyme) concentration against variable 2-oxo acid concentration (substrate) use to derive the MM equation containing the parameters like maximum reaction velocity (V_{max}) and Michaelis constant (K_m).

The kinetic assay is performed in 2 mL 96-well round polypropylene microplates in a 50 μL reaction.

1. At a minimum, reactions are performed in duplicates
2. Make an eight dilution of 2-oxo acid substrate solutions between 0.05 to 8 mM
3. In this case, one can use eight different concentrations of the 2-oxo acid substrate in duplicates and set up enough master mix by mixing all the components according to the Table given below, except for the substrate
4. Centrifuge the master mix at 130 (×g) for 15–30 s at room temperature

Components	Stock concentration	Final concentration	Volume/ reaction (μL)
Milli-Q	–	–	Variable
Tris-buffer (pH -8)	1 M	100 mM	5
$MnCl_2$	40 mM	4 mM	5
Acetyl CoA	10 mM	1 mM	5
ATP	20 mM	2 mM	5
DDT	10 mM	1 mM	5
Purified MAMS	–	250 ng	Variable

5. Aliquot 45 μL of master mix to the bottom of the 96-well plate
6. Individually or using a multichannel pipette add the different concentrations of the 2-oxo acid substrate onto the side wall of the designated well
7. Centrifuge the plate at 500 (×g) for 1 min at room temperature
8. Incubate the plate at 30 °C for 10 min in the water bath
9. Add 150 μL of absolute ethanol to stop the reaction
10. Centrifuge the tubes at 18000 (×g) for 10 min at room temperature
11. Carefully aliquot 150 μL of the supernatant into HPLC/LC-MS certified 2 mL clear glass vials and kept at −20 °C prior to LC-MS/MS analysis as described in Section 3.2
12. Calculate the amount of 2-malate product formed by MAMS catalysis using the standard 2-hexyl malate curve (Sections 3.2 and 3.3)

    Note: The final reaction product should be quantified on LC-MS/MS on the same or next day or maximum within 1 week after performing the reactions. Perform the reaction using the same batch of MAMS protein for the different 2-oxo acid substrates.

3.2 Standard curve preparation of 2-malate derivative

3.2.1 Materials, reagents and equipment

- Milli-Q water (Milli-Q-Synthesis A10, MerckMillipore, Darmstadt, Germany)
- Formic acid (for LC-MS; VWR, Rosny-spus-Bois-cedex, France; Cat. No. 84865.180)

- Acetonitrile (for HPLC LC-MS-grade, VWR, Rosny-spus-Bois-cedex, France; Cat. No. 83640.320)
- 2-hexyl malate (synthesized according to Textor et al. 2007, Textor et al. 2007: Plant Physiol. 144: 60–71)
- LC-MS/MS (API3200, AB Sciex, Darmstadt, Germany coupled to Agilent 1200 HPLC system (Agilent Technologies, Waldbronn, Germany))
- HPLC vials (1.5 mL, Macherey-Nagel, Düren, Germany, Cat. No. 702282)

3.2.2 Procedure

1. To optimize MS detection of 2-malate derivatives, prepare the standard stock solution of 2-hexyl malate (500 μg/mL) by dissolving 0.5 mg of powder in 1 mL of methanol
2. Dilute the above-mentioned stock solutions in ultrapure water in different concentrations between 0.024 and 25 μM
3. Store the stock and working standard solutions of 2-hexyl malate at −20 °C
4. Carry out the chromatographic separation on a Zorbax Eclipse XDB-C18 column
5. The mobile phase A (0.05% Formic acid in water) and mobile phase B (Acetonitrile) are programmed to mix in a chromatographic linear gradient as follows:

Time (minutes)	%A	%B
0.0	95.0	5
0.5	95.0	5
3.0	42.2	57.8
3.1	0.0	100
4.0	0.0	100
4.1	95.0	5
6.5	95.0	5

6. Maintain the flow rate of 1.1 mL/min and a column temperature of 25 °C
7. Set the mass spectrometer in a negative Turbo Ion spray ionization mode with the following parameters: ion spray voltage—4500 eV; turbo gas temperature 700 °C; nebulizer gas at 60 psi, curtain gas at 25 psi, heating gas at 60 psi and collision gas at 7 psi
8. Use multiple reaction monitoring (MRM) to detect analyte parent ion-to-product ion transition (Table 2)
9. Maintain Q1 and Q3 quadrupoles at unit resolution
10. Inject the 1 μl of working standard solutions into LC-ESI TripleQuad-MS and calculate the peak area (Fig. 3A)
11. Data acquisition and processing was carried out using Analyst 1.5 software (Applied Biosystems)
12. Prepare the calibration curve of 2-hexyl malate by plotting the peak areas (Y-axis) versus 2-hexyl malate concentrations (X-axis)
13. Determine the regression parameters including slope, intercept, and correlation coefficient
14. The peak intensity of 2-hexyl malate standard should follow the linear trend with regression squared (R^2) value >0.98 over 1000-fold dilution (Fig. 3B)

Table 2 MRM table of the analysis of 2-malate derivatives by LC-MS/MS (HPLC 1200 (Agilent Technologies)-API 3200 (Applied Biosystems)) in negative ionization mode.

2-Malate compound	Q1	Q3	RT (min)	DP	CE
2-Isopropyl-malate (2IPM)	175	115	2.68	−30	−30
2-(2-Methylthioethyl)-malate (2MTEM)	207	147	2.67	−30	−20
2-(3-Methylthiopropyl)-malate (2MTPM)	221	161	2.91	−30	−20
2-(4-Methylthiobutyl)-malate (2MTBM)	235	175	3.18	−30	−20
2-Hexyl-malate (2HM)	217	157	3.76	−30	−20
2-Heptyl-malate (2HptM)	231	171	4.05	−30	−20
2-(6-Methylthiohexyl)-malate (2MTHM)	263	203	n.d.	−30	−20
2-(7-Methylthioheptyl)-malate (2MTHptM)	277	217	n.d.	−30	−20
2-Decyl-malate (2DM)	273	213	n.d.	−30	−20

Q1: *m/z* quadrupole 1; Q3: m/z quadrupole 3; RT: retention time; DP: declustering potential, CE: collision energy; n.d.: not determined.

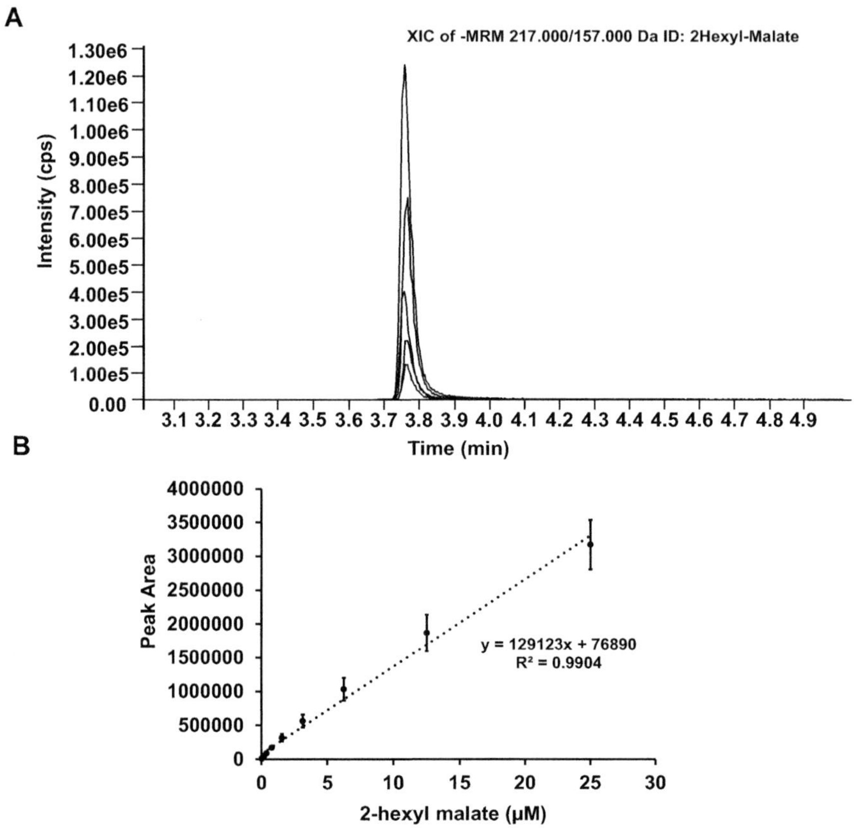

Fig. 3 Standard curve preparation of 2-hexyl malate, for quantification of various 2-malate products obtained by MAMS reaction. The response of the malate standard, 2-hexyl-malate was analyzed over a range of concentrations (24.4 nM to 25 μM), injected in ESI-TripleQuad-MS (ABI-3200), each in five replicates. (A) An overlay chromatogram showing the peak intensity of 1 μL of the standard mix over a range of concentrations. (B) The peak intensity of both malate standards overlapped and followed the linear trend with regression squared (R^2) value >0.98 over 1000-fold dilution.

3.3 Detection and quantification of various 2-malate derivatives using the liquid chromatography-tandem mass spectrometer (LC-MS/MS)

3.3.1 Materials, reagents and equipment

- Milli-Q water (Milli-Q-Synthesis A10, MerckMillipore, Darmstadt, Germany)
- Acetonitrile (for HPLC LC-MS-grade, VWR, Rosny-spus-Bois-cedex, France; Cat. No. 83640.320)

- 2-Hexyl malate (synthesized according to Textor et al. 2007, Plant Physiol. 144: 60-71)
- LC-MS/MS (API3200, AB Sciex, Darmstadt, Germany coupled to Agilent 1200 HPLC system (Agilent Technologies, Waldbronn, Germany))
- HPLC vials (1.5 mL, Macherey-Nagel, Düren, Germany, Cat. No. 702282)

3.3.2 Procedure

1. Program the LC-MS/MS instrument as performed earlier for the running of 2-hexyl malate (please follow steps 4–9 in Section 3.2.2).
2. Inject a 1 μL aliquot of the final MAMS reaction product
3. Data acquisition and processing were carried out using Analyst 1.5 software (Applied Biosystems)
4. Quantify the 2-malate derivative by the MAMS enzyme reaction based on an external standard curve of 2-hexyl-malate with a theoretical response factor of 1
5. Calculating the malate product formed in the MAMS assays using the equation derived from the external curve prepared for 2-hexyl malate, as follows:

$$y \text{ (peak area)} = 129123 \times + 76890$$

which implies;

if, $x = 1\ \mu M$ (i.e. 1 μmole/L or 1 pmol/μL)

then, $y \text{ (peak area)} = 206013$

or

1 pmol of malate standard is equivalent to 206013 AU of peak area (as only 1 μL of the reaction product was injected).

Note: The mass spectrometer transition from Q1 to Q3 is a loss of 60 Da (loss of $C_2H_4O_2$) for all 2-malate derivatives. The 2-nonyl malate can also be used to create an external standard curve. The comparison of standard curves of 2-hexyl malate and 2-nonyl malate on a molar concentration basis showed that they have a similar slope. It was deduced that the response of all malate derivatives on a molar basis is equal. AtMAMS1 catalyzes the conversion of 2-oxo acid substrates to 2-malate products involving the formation of short-chain aliphatic GSL. Peak intensity showing the MRM of 2-malate intermediates catalyzed by AtMAMS1 is shown in Fig. 4. AtMAM3 catalyzes the formation of all aliphatic GSL chain lengths in Arabidopsis.

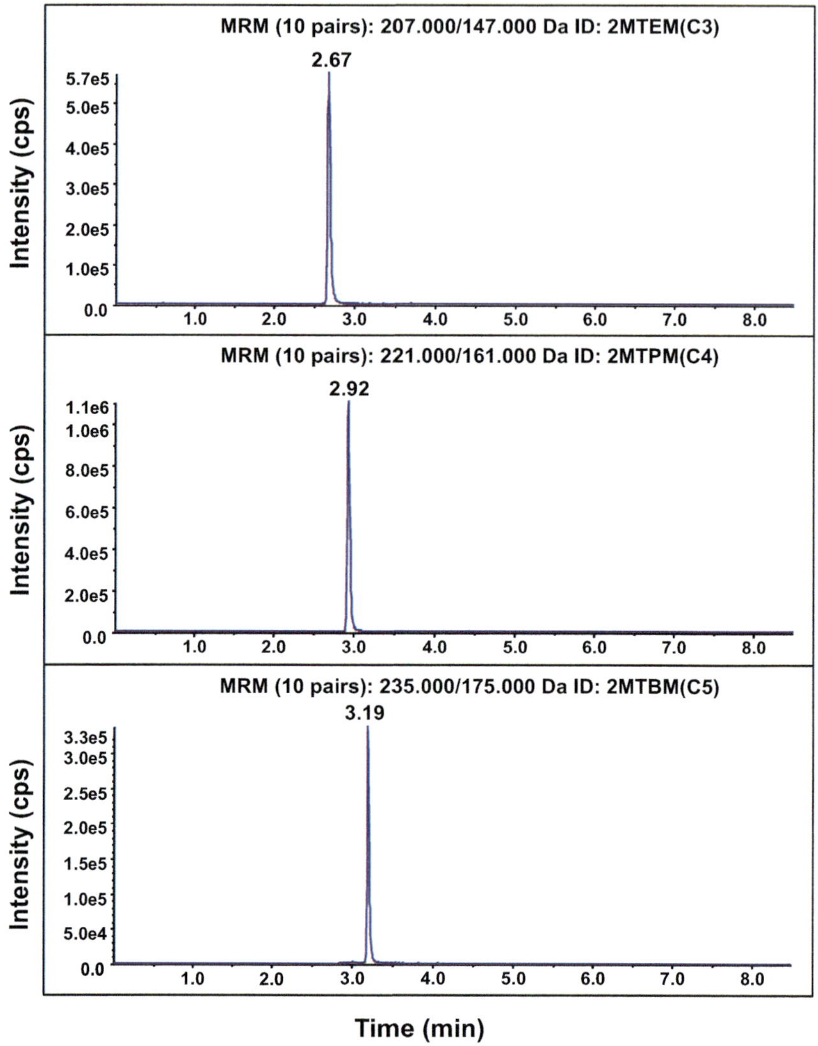

Fig. 4 Chromatograms showing the MRM of 2-malate intermediates involved in the formation of short-chain (C3-C5) aliphatic GSL. The purified AtMAMS1 (200 ng) protein was used to catalyze the condensation reaction of acetyl-coA (1 mM) with 4-Methyl-thio-2-oxo-butanoate (4MTOB), 5-Methyl-thio-2-oxo-pentanoate (5MTOP), and 6-Methyl-thio-2-oxo-hexanoate (6MTOH) (3 mM) to form the 2-malate products namely 2-(2-methylthioethyl)-malate (2MTEM), 2-(3-methylthiopropyl)-malate (2MTPM), and 2-(4-methylthiobutyl)-malate (2MTBM), respectively (please see Table 1).

4. Safety considerations and standards

Protein purification and assay components are not particularly hazardous to health or safety. However, the implementation of regular lab safety procedures is imperative.

5. Conclusions

We have presented a highly sensitive MS-based MAMS enzymatic assay, which enables the direct biochemical characterization of MAMS protein. The end-point and the steady-state kinetic assay can, in principle, be adapted to examine the MAMS activity against a variety of 2-oxo acid substrates, thereby understanding the regulation of side-chain elongation of aliphatic GSL. Overall, the protocol provided in this chapter allows researchers interested to study the GSL biochemistry across the species belonging to the *Brassicaceae* family. In addition, the protocol will also find its future application to study the biochemistry of the members of the Claisen condensation-like subgroup enzymes include isopropylmalate synthase, citramalate synthase, and homocitrate synthase.

Acknowledgments

We thank the Central Instrumentation Facility at the National Institute for Plant Genome Research (NIPGR). We are grateful to Prof. Jonathan Gershenzon (Max Planck Institute for Chemical Ecology, Jena, Germany) and Prof. Joseph M. Jez (Washington University, St. Louis, Missouri, USA) who made contributions to our glucosinolate studies over the years. The work was supported by the Department of Biotechnology, India grants (BT/06/IYBA/2012 and BT/HRD-NBA-NWB/38/2019-20(10)) and a Max Planck India Fellowship to N.C.B.

References

Angelino, D., Dosz, E. B., Sun, J., Hoeflinger, J. L., Van Tassell, M. L., Chen, P., et al. (2015). Myrosinase-dependent and -independent formation and control of isothiocyanate products of glucosinolate hydrolysis. *Frontiers in Plant Science*, *6*, 831. https://doi.org/10.3389/fpls.2015.00831.

Augustine, R., & Bisht, N. C. (2017). Regulation of glucosinolate metabolism: From model plant Arabidopsis thaliana to Brassica crops. In J.-M. Mérillon, & K. G. Ramawat (Eds.), *Glucosinolates* (pp. 163–199). Springer International Publishing. https://doi.org/10.1007/978-3-319-25462-3_3.

Benderoth, M., Pfalz, M., & Kroymann, J. (2008). Methylthioalkylmalate synthases: Genetics, ecology and evolution. *Phytochemistry Reviews*, *8*(1), 255. https://doi.org/10.1007/s11101-008-9097-1.

Blažević, I., Montaut, S., Burčul, F., Olsen, C. E., Burow, M., Rollin, P., et al. (2020). Glucosinolate structural diversity, identification, chemical synthesis and metabolism in plants. *Phytochemistry*, *169*, 112100. https://doi.org/10.1016/j.phytochem.2019.112100.

Bothner, B., Chavez, R., Wei, J., Strupp, C., Phung, Q., Schneemann, A., et al. (2000). Monitoring enzyme catalysis with mass spectrometry. *The Journal of Biological Chemistry*, *275*(18), 13455–13459. https://doi.org/10.1074/jbc.275.18.13455.

de Kraker, J.-W., & Gershenzon, J. (2011). From amino acid to Glucosinolate biosynthesis: Protein sequence changes in the evolution of Methylthioalkylmalate synthase in Arabidopsis[W][OA]. *The Plant Cell*, *23*(1), 38–53. https://doi.org/10.1105/tpc.110.079269.

de Kraker, J.-W., Luck, K., Textor, S., Tokuhisa, J. G., & Gershenzon, J. (2007). Two Arabidopsis genes (IPMS1 and IPMS2) encode isopropylmalate synthase, the branchpoint step in the biosynthesis of leucine. *Plant Physiology*, *143*(2), 970–986. https://doi.org/10.1104/pp.106.085555.

Fahey, J. W., Zhang, Y., & Talalay, P. (1997). Broccoli sprouts: An exceptionally rich source of inducers of enzymes that protect against chemical carcinogens. *Proceedings of the National Academy of Sciences of the United States of America*, *94*(19), 10367–10372. https://doi.org/10.1073/pnas.94.19.10367.

Field, B., Cardon, G., Traka, M., Botterman, J., Vancanneyt, G., & Mithen, R. (2004). Glucosinolate and amino acid biosynthesis in Arabidopsis. *Plant Physiology*, *135*(2), 828–839. https://doi.org/10.1104/pp.104.039347.

Forouhar, F., Hussain, M., Farid, R., Benach, J., Abashidze, M., Edstrom, W. C., et al. (2006). Crystal structures of two bacterial 3-hydroxy-3-methylglutaryl-CoA lyases suggest a common catalytic mechanism among a family of TIM barrel metalloenzymes cleaving carbon-carbon bonds. *The Journal of Biological Chemistry*, *281*(11), 7533–7545. https://doi.org/10.1074/jbc.M507996200.

Gohain, B., Kumar, P., Malhotra, B., Augustine, R., Pradhan, A. K., & Bisht, N. C. (2021). A comprehensive vis-NIRS equation for rapid quantification of seed glucosinolate content and composition across diverse Brassica oilseed chemotypes. *Food Chemistry*, *354*, 129527. https://doi.org/10.1016/j.foodchem.2021.129527.

Halkier, B. A., & Gershenzon, J. (2006). Biology and biochemistry of glucosinolates. *Annual Review of Plant Biology*, *57*, 303–333. https://doi.org/10.1146/annurev.arplant.57.032905.105228.

He, Y., Chen, B., Pang, Q., Strul, J. M., & Chen, S. (2010). Functional specification of Arabidopsis isopropylmalate isomerases in glucosinolate and leucine biosynthesis. *Plant and Cell Physiology*, *51*(9), 1480–1487. https://doi.org/10.1093/pcp/pcq113.

He, Y., Galant, A., Pang, Q., Strul, J. M., Balogun, S. F., Jez, J. M., et al. (2011). Structural and functional evolution of isopropylmalate dehydrogenases in the leucine and glucosinolate pathways of Arabidopsis thaliana. *The Journal of Biological Chemistry*, *286*(33), 28794–28801. https://doi.org/10.1074/jbc.M111.262519.

He, Y., Mawhinney, T. P., Preuss, M. L., Schroeder, A. C., Chen, B., Abraham, L., et al. (2009). A redox-active isopropylmalate dehydrogenase functions in the biosynthesis of glucosinolates and leucine in Arabidopsis. *The Plant Journal*, *60*(4), 679–690. https://doi.org/10.1111/j.1365-313X.2009.03990.x.

Hopkins, R. J., van Dam, N. M., & van Loon, J. J. A. (2009). Role of glucosinolates in insect-plant relationships and multitrophic interactions. *Annual Review of Entomology*, *54*, 57–83. https://doi.org/10.1146/annurev.ento.54.110807.090623.

Khersonsky, O., & Tawfik, D. S. (2010). Enzyme promiscuity: A mechanistic and evolutionary perspective. *Annual Review of Biochemistry*, *79*, 471–505. https://doi.org/10.1146/annurev-biochem-030409-143718.

Kroymann, J., Textor, S., Tokuhisa, J. G., Falk, K. L., Bartram, S., Gershenzon, J., et al. (2001). A gene controlling variation in Arabidopsis glucosinolate composition is part of the methionine chain elongation pathway. *Plant Physiology*, *127*(3), 1077–1088. https://doi.org/10.1104/pp.010416.

Kumar, R., Lee, S. G., Augustine, R., Reichelt, M., Vassão, D. G., Palavalli, M. H., et al. (2019). Molecular basis of the evolution of methylthioalkylmalate synthase and the diversity of methionine-derived glucosinolates. *The Plant Cell*, *31*(7), 1633–1647. https://doi.org/10.1105/tpc.19.00046.

Nguyen, V. P. T., Stewart, J., Lopez, M., Ioannou, I., & Allais, F. (2020). Glucosinolates: Natural occurrence, biosynthesis, accessibility, isolation, structures, and biological activities. *Molecules*, *25*(19), 4537. https://doi.org/10.3390/molecules25194537.

Schwab, W. (2003). Metabolome diversity: Too few genes, too many metabolites? *Phytochemistry*, *62*(6), 837–849. https://doi.org/10.1016/S0031-9422(02)00723-9.

Sønderby, I. E., Geu-Flores, F., & Halkier, B. A. (2010). Biosynthesis of glucosinolates—Gene discovery and beyond. *Trends in Plant Science*, *15*(5), 283–290. https://doi.org/10.1016/j.tplants.2010.02.005.

Textor, S., Bartram, S., Kroymann, J., Falk, K. L., Hick, A., Pickett, J. A., et al. (2004). Biosynthesis of methionine-derived glucosinolates in Arabidopsis thaliana: Recombinant expression and characterization of methylthioalkylmalate synthase, the condensing enzyme of the chain-elongation cycle. *Planta*, *218*(6), 1026–1035. https://doi.org/10.1007/s00425-003-1184-3.

Textor, S., de Kraker, J.-W., Hause, B., Gershenzon, J., & Tokuhisa, J. G. (2007). MAM3 catalyzes the formation of all aliphatic glucosinolate chain lengths in Arabidopsis. *Plant Physiology*, *144*(1), 60–71. https://doi.org/10.1104/pp.106.091579.

Wang, H., Wu, J., Sun, S., Liu, B., Cheng, F., Sun, R., et al. (2011). Glucosinolate biosynthetic genes in Brassica rapa. *Gene*, *487*(2), 135–142. https://doi.org/10.1016/j.gene.2011.07.021.

CHAPTER THREE

A structure-guided computational screening approach for predicting plant enzyme–metabolite interactions

Cynthia K. Holland* and Hisham Tadfie
Department of Biology, Williams College, Williamstown, MA, United States
*Corresponding author: e-mail address: ckh2@williams.edu

Contents

Abstract

Plants are molecular factories that have spent millions of years evolving the enzymes needed to synthesize diverse primary and specialized metabolites. Despite the wealth of metabolites that plants produce, many of the enzymes responsible for generating these molecules have yet to be identified. For enzymes with known substrates, the extent of substrate promiscuity and small-molecule regulation remains unexplored. Many computational methods for identifying metabolic enzymes focus on gene-based approaches that rely on transcriptomics, metabolomics, and comparative genomics. With new AI-based tools for accurate protein structure prediction, protein-based

Methods in Enzymology, Volume 676
ISSN 0076-6879
https://doi.org/10.1016/bs.mie.2022.07.025

strategies that screen a library of small molecules against a high-quality protein model can facilitate the identification of substrates, products, or inhibitors. Virtual screening has been used for structure-based drug design in the pharmaceutical industry for decades and easily translates to investigating plant metabolic enzymes. Here, we present a method for rapid, user-friendly, and open-source virtual screening using the *Arabidopsis thaliana* UGT74F2 with a curated library of specialized metabolites and herbicides and AutoDock Vina as an example. This method may be applied broadly to metabolic enzymes, and compound libraries can be easily adapted. Compounds are ranked based on their relative binding affinities and the resulting binding modes are evaluated using a molecular visualization program, like PyMOL. Because this is a computational approach, results from the virtual screen will need to be validated using in vitro or in vivo activity, binding, or inhibition assays. Virtual screening may aid in identifying substrates for enzymes of unknown function, revisiting substrate selectivity, or identifying natural or synthetic inhibitors.

1. Introduction

Plants produce as many as a million compounds with diverse roles in growth and development, defense against pathogens and herbivores, protection from abiotic stresses, pollinator attraction, and reproduction (Fang, Fernie, & Luo, 2019). Over millions of years, plants have evolved enzymes that synthesize these specialized metabolites, many of which are the result of gene duplication followed by neofunctionalization of genes from primary metabolism (Maeda & Fernie, 2021). Because many of these compounds are useful to humans for their roles in medicine, nutrition, or crop improvement, researchers have spent the past few decades searching for enzymes that produce these specialized metabolites. As the cost of whole-genome sequencing and transcriptome sequencing has decreased, identification and functional characterization of plant metabolic enzymes has advanced rapidly in recent years. Computational identification of candidate genes coupled to functional assays has led to the identification of entire biosynthetic pathways. While many important specialized metabolic pathways have been elucidated, including the pathways for pharmaceuticals such as morphine, etoposide, and vinblastine, the enzymes involved in synthesizing many pharmaceutically and agriculturally relevant metabolites remain to be identified (Caputi et al., 2018; Schultz, Kim, Lau, & Sattely, 2019; Singh, Menéndez-Perdomo, & Facchini, 2019). To build on existing computational approaches that have been successful in identifying candidate genes, a protein-based approach to predict the function of enzymes is now possible due to recent advances in AI-based structural protein modeling (Jumper et al., 2021).

1.1 Identifying candidate genes using computational and systems biology approaches

In the two decades since the *Arabidopsis thaliana* genome was completed, over a thousand plant genomes have been sequenced, due in great part to advances in next-generation sequencing such as short- and long-read sequencing and the decreasing costs of whole-genome sequencing (Sun, Shang, Zhu, Fan, & Guo, 2022). Despite the availability of this data, only a fraction of plant genes have been experimentally characterized and many enzymes with unknown functions remain (Rhee & Mutwil, 2014). The establishment of model plants such as *A. thaliana* has played a substantial role in the development of comparative genomics, a field which uses genome sequence homology as a means of identifying gene function in previously uncharacterized organisms (Smith et al., 2019). Although comparative genomics is a useful tool for identifying gene homologs and syntenic regions of genomes, it becomes less practical when genes of interest are part of genus-specific specialized metabolic pathways that lack clear homologs in other plants. Specialized metabolites have become a key area of study in pharmacology, agriculture, and cosmetics, and so there exists a significant need for new techniques that would better allow for the identification and characterization of these divergently evolved pathways.

Sequencing plant genomes has revealed new information about the genomic organization of metabolic genes. While it was previously thought that eukaryotic biosynthetic genes were distributed non-continuously across distinct chromosomes, recent findings suggest that plant genes associated with specialized metabolic pathways may physically aggregate within the genome in biosynthetic gene clusters similar to a bacterial operon (Nützmann, Huang, & Osbourn, 2016). In plants, biosynthetic gene clustering is hypothesized to allow for the coinheritance of entire specialized pathways, thereby decreasing the likelihood of incomplete pathway inheritance that could lead to toxic intermediate build-up (Kim & Buell, 2015). Mining biosynthetic gene clusters can be useful for identifying specialized metabolic genes that contribute to the same pathway (Nützmann et al., 2016). The online platform plantiSMASH can be used to identify genomic loci that resemble biosynthetic gene clusters and can also use transcriptomics data to prioritize candidates based on coexpression (Kautsar, Suarez Duran, & Medema, 2018). In the absence of biosynthetic gene clusters, candidate gene identification relies on transcriptomic or metabolic approaches that require gene expression induction, which can lead to larger pools of candidate genes. While there have been many recent advances in biosynthetic gene cluster identification, the products of many of these metabolic clusters have not yet been elucidated (Polturak, Liu, & Osbourn, 2022).

One additional technique for identifying candidate metabolic genes is by correlating tissue-level expression with metabolite presence and abundance. This method is especially useful in plants, which often synthesize and compartmentalize defense metabolites to ensure that they reach their target efficiently without inducing autotoxicity (Delli-Ponti, Shivhare, & Mutwil, 2021). In order to correlate gene expression with metabolite presence, it is important to first identify the tissue, cell type, or developmental stage where the specialized metabolite is synthesized. This can be done using mass spectrometry, where metabolite(s) of interest are detected using mass spectrometry and their relative abundance is compared across samples (Saito & Matsuda, 2010). After generating metabolomics data, transcriptomics can be used to identify genes that are differentially expressed across the same tissues, cell types, or developmental stages that were used for collecting metabolomics data. Correlating gene expression with metabolite presence and abundance can be very effective in identifying genes in biosynthetic pathways. This approach is less effective when the specialized metabolites are not synthesized in the tissue in which they are localized. For example, the plant defensive compound nicotine is synthesized in the roots of *Nicotiana tabacum* (tobacco) and transported to the leaves, where it serves as an insecticide (Baldwin, 1989). Although this approach has its limitations, it is a useful technique for identifying candidate primary and secondary metabolic genes.

In addition to forming biosynthetic gene clusters, specialized metabolic genes are often coregulated by a common set of transcription factors and consequently, coexpressed. One method for identifying these expression patterns is to use global coexpression network analysis, where genes, which are represented by nodes, are linked together based on overlapping expression profiles to form modules (Wisecaver et al., 2017). Though this approach offers a high-throughput method for identifying candidate specialized metabolic pathways, it is important to recognize that many specialized pathway genes within a pathway are not necessarily coexpressed. Additionally, it can be difficult to delineate separate metabolic pathways that may be expressed in response to the same elicitors. Despite these limitations, global coexpression network analysis remains a powerful tool for identifying candidate secondary metabolic genes in plants.

While the above-mentioned computational techniques have been instrumental in predicting gene function and for biosynthetic pathway discovery, these approaches may still yield long lists of candidate genes that all need to be screened using in vivo or in vitro enzyme assays. To narrow down lists of candidate enzymes further, computational substrate docking using virtual screens is a protein-based approach that may be used to predict enzyme function. Additionally, virtual screening may be applied broadly to studying metabolic enzymes that are targets for herbicides, promiscuous enzymes, or enzymes of unknown function.

1.2 Docking and virtual screening as a computational tool for functional prediction of plant metabolic enzyme activity

Docking is a commonly used computational tool used to model interactions between a three-dimensional protein structure and a small molecule. Similarly, virtual screening is a docking approach that iteratively docks a library of molecules with a target protein. For decades, protein biochemists, molecular biologists, and pharmaceutical industries have been using protein structure determination or protein homology modeling coupled with small-molecule docking to investigate molecular interactions. This line of research has been made possible by the availability of open-source docking software that does not require advanced programming knowledge (Villoutreix et al., 2007). One of the most widely used docking programs, AutoDock Vina, uses a simple scoring function to efficiently evaluate intermolecular interactions within a given protein–ligand complex and outputs a prediction of protein–ligand binding affinities and binding conformations, which can be modeled in three-dimensional visualization programs such as PyMol (Trott & Olson, 2010). Docking is an excellent tool for identifying candidate substrates, cofactors, and regulators for an enzyme of interest, and in plant biology research, it has primarily been used to model interactions between a single protein and one metabolite. Examples of instances where docking has been used to study plant enzymes include: the *Arabidopsis* GH3.15 that conjugates amino acids to the auxinic hormone indole-3-butyric acid; a noroxomaritidine reductase involved in alkaloid biosynthesis in daffodils (*Narcissus* spp.); and a rice naringenin *O*-methyltransferase involved in phytoalexin synthesis (Kilgore, Holland, Jez, & Kutchan, 2016; Murata et al., 2020; Sherp, Westfall, Alvarez, & Jez, 2018).

In the past, protein-based methods for narrowing down lists of candidate enzymes in biosynthetic pathways have been limited by the availability of reliable protein structures or structural models. Though cryo-electron microscopy, nuclear magnetic resonance (NMR), and X-ray crystallography have been instrumental for solving protein structures, each of these approaches has their challenges. The development of web-based programs with simple interfaces for homology-based modeling, such as SWISS-MODEL and Phyre2, has allowed for the generation of highly accurate three-dimensional protein structures (Kelley, Mezulis, Yates, Wass, & Sternberg, 2015; Waterhouse et al., 2018). However, proteins that lack known homologs, as in the case of many secondary metabolic enzymes, may be prone to structural inaccuracy. Recent advances in AI-based structure prediction with the release of AlphaFold has substantially improved our ability to predict the three-dimensional structure of enzymes from any species (Jumper et al., 2021). AlphaFold is a machine learning program that integrates knowledge of biophysical dynamics with protein evolutionary history analysis to generate highly accurate structural predictions, even if no structural homolog is known. Being able to generate a structural model of any protein of interest has the potential to improve our ability to study specialized metabolism and decipher the molecular underpinnings of plant metabolism broadly.

In medicinal chemistry and pharmacology, it is common practice to screen compound–enzyme interactions computationally before proceeding with empirical experiments, especially when there are hundreds or thousands of compounds and several target proteins (Rester, 2008). This same approach can be used for investigating plant metabolism and aid in identifying candidate enzymes in biosynthetic pathways. Virtual screens can be implemented for all metabolic enzymes, including glycosyltransferases, oxygenases (i.e., cytochrome P450s), oxidoreductases, ligases, hydrolases, or terpene synthases. Aside from identifying enzyme substrates, virtual screening may be used to investigate many open questions in plant metabolism, including understanding the substrate promiscuity of enzymes or identifying competitive and allosteric inhibitors of an enzyme. The structure-based virtual screening methods described here use open access programs that only require a local PC or Mac computer. Because AutoDock Vina is open-source, fast, and has a wealth of online tutorials, user manuals, and discussion forums available to support new users, the protocol will focus on the use of this program (Eberhardt, Santos-Martins, Tillack, & Forli, 2021; Trott & Olson, 2010). While having a working understanding of protein structural visualization programs such as PyMOL or Chimera would be helpful, it is not necessary, and resources such as PyMOL Wiki are available online.

2. Designing a virtual screen experiment

Before conducting a virtual screen, the enzyme of interest and the library of compounds that will be screened will need to be prepared. While there are several programs available that will run a virtual screen, the information below will focus on preparing compounds and enzymes for docking using the widely used and freely available program AutoDock Vina (Trott & Olson, 2010), and the files will be prepared for docking using AutoDock Tools, a graphical user interface that is part of the MGLTools software suite. Because these programs are popular for docking, many online resources and published protocols are readily available (Forli et al., 2016).

To demonstrate how virtual screens are executed, the protocols below will use a glycosyltransferase from the model plant *A. thaliana,* UGT74F2 (AT2G43820), that has been functionally characterized as a UDP-dependent glycosyltransferase (UGT) that glycosylates the carboxylate of the plant hormone salicylic acid (SA), forming an SA glucose ester (SGE) (Lim et al., 2002). This enzyme is known to be promiscuous and can also glycosylate the hydroxyl of SA (forming SA 2-O-beta-D-glucose; SAG), as well as other benzoate substrates, including the tryptophan pathway intermediate anthranilate (2-aminobenzoate), benzoic acid, and 3-hydroxybenzoic acid (George Thompson, Iancu, Neet, Dean, & Choe, 2017; Lim et al., 2002; Quiel & Bender, 2003). UGTs use nucleotide-activated sugars as sugar donors in the transferase reaction, and UGT74F2, as well as other plant UGTs, uses UDP-glucose as the sugar donor (Akere et al., 2020). These UGTs have a variable N-terminal domain and a C-terminal nucleotide-sugar binding domain that contains a conserved 44 amino acid motif known as a Plant Secondary Product GT box, and substrates bind UGTs in a cleft between these two domains.

To prepare ligands and protein input files for virtual screening, you will need to begin by downloading and installing MGLTools from: http://mgltools.scripps.edu/downloads. On a Mac, users will also need to download X11 from http://xquartz.org in order to run AutoDock Tools.

2.1 Generating a compound library

The first step in conducting a virtual screen is to decide which compounds to include in the screen given the enzyme of interest. While several online databases of ligand files exist, the compounds in these databases are primarily targeted for human health and medicine, including ZINC Docking (Irwin et al., 2020; Sterling & Irwin, 2015), PubChem (Kim et al., 2016), and ChEMBL (Bento et al., 2014). However, because numerous plant metabolites are used as pharmaceuticals or nutraceuticals, ready-to-dock molecules for many commonly investigated plant-produced compounds are available for download in a ".mol2" file format from ZINC Docking, including intermediates in primary metabolism, plant hormones, specialized metabolites, and herbicides (Irwin et al., 2020; Sterling & Irwin, 2015). While by no means comprehensive, a list of plant metabolites that are available for download from ZINC Docking has been included in Tables 1 and 2. Aside from the hormones, metabolites, and herbicides listed in Tables 1–3, primary

Table 1 List of plant hormones and hormone-related metabolites included in the compound library ranked by molecular mass.

Metabolite	Formula	Molecular mass (g/mol)
Phenylacetic acid	$C_8H_8O_2$	136.15
Salicylic acid	$C_7H_6O_3$	138.12
Indole-3-acetic acid (IAA)	$C_{10}H_9NO_2$	175.18
Indole-3-butyric acid	$C_{12}H_{13}NO_2$	203.24
4-Chloroindole-3-acetic acid	$C_{10}H_8ClNO_2$	209.63
Jasmonic acid	$C_{12}H_{18}O_3$	210.27
Kinetin	$C_{10}H_9N_5O$	215.21
cis-Zeatin	$C_{10}H_{13}N_5O$	219.24
6-Benzylaminopurine	$C_{12}H_{11}N_5$	225.25
Abscisic acid	$C_{15}H_{20}O_4$	264.32
IAA-glutamine	$C_{15}H_{17}N_3O_4$	303.31
Sorgolactone	$C_{18}H_{20}O_5$	316.30
Gibberellic acid	$C_{19}H_{22}O_6$	346.37
Strigol	$C_{19}H_{22}O_6$	346.40
Orobanchol	$C_{19}H_{22}O_6$	346.40
Kinetin riboside	$C_{15}H_{17}N_5O_5$	347.33
Kinetin-9-N-glucoside	$C_{16}H_{19}N_5O_6$	377.35

Table 2 List of plant metabolites included in the compound library ranked by general classification and molecular mass.

Specialized metabolite	Description	Formula	Molecular mass (g/mol)
Nicotine	Alkaloid	$C_{10}H_{14}N_2$	162.23
Caffeine	Alkaloid	$C_8H_{10}N_4O_2$	194.19
Camalexin	Alkaloid	$C_{11}H_8N_2S$	200.26
Morphine	Alkaloid	$C_{17}H_{19}NO_3$	285.34
Galantamine	Alkaloid	$C_{17}H_{21}NO_3$	287.35
Capsaicin	Alkaloid	$C_{18}H_{27}NO_3$	305.41
Quinine	Alkaloid	$C_{20}H_{24}N_2O_2$	324.4
Berberine	Alkaloid	$C_{20}H_{18}NO_4^+$	336.4
Strictosidine	Alkaloid	$C_{27}H_{34}N_2O_9$	530.57
Vinblastine	Alkaloid	$C_{46}H_{58}N_4O_9$	811
Benzaldehyde	Aromatic	C_7H_6O	106.12
Benzoate	Aromatic	$C_7H_5O_2^-$	121.11
Cinnamaldehyde	Aromatic	C_9H_8O	132.16
p-Coumaryl alcohol	Aromatic	$C_9H_{10}O_2$	150.17
Methyl anthranilate	Aromatic	$C_8H_9NO_2$	151.17
Vanillin	Aromatic	$C_8H_8O_3$	152.15
Methyl salicylate	Aromatic	$C_8H_8O_3$	152.15
Gallic acid	Aromatic	$C_7H_6O_5$	170.12
Ferulic acid	Aromatic	$C_{10}H_{10}O_4$	194.18
Anthraquinone	Aromatic	$C_{14}H_8O_2$	208.21
DIMBOA	Aromatic	$C_9H_9NO_5$	211.17
Resveratrol	Aromatic	$C_{14}H_{12}O_3$	228.24
Catechin	Aromatic	$C_{15}H_{14}O_6$	290.26
Quercetin	Aromatic	$C_{15}H_{10}O_7$	302.23
Dhurrin	Aromatic	$C_{14}H_{17}NO_7$	311.29
Tetrahydrocannabinol	Aromatic	$C_{21}H_{30}O_2$	314.45
Bergamottin	Aromatic	$C_{21}H_{22}O_4$	338.4

Continued

Table 2 List of plant metabolites included in the compound library ranked by general classification and molecular mass.—cont'd

Specialized metabolite	Description	Formula	Molecular mass (g/mol)
Rosmarinic acid	Aromatic	$C_{18}H_{16}O_8$	360.3
Podophyllotoxin	Aromatic	$C_{22}H_{22}O_8$	414.41
Allicin	Sulfur-containing	$C_6H_{10}OS_2$	162.28
Glucobrassicin	Sulfur-containing	$C_{16}H_{19}N_2O_9S_2^-$	447.46
Limonene	Terpene	$C_{10}H_{16}$	136.24
Linalool	Terpene	$C_{10}H_{18}O$	154.25
Campesterol	Terpene	$C_{28}H_{48}O$	400.68
Betulinic acid	Terpene	$C_{30}H_{48}O_3$	456.7
Lycopene	Terpene	$C_{40}H_{56}$	536.87
Beta-carotene	Terpene	$C_{40}H_{56}$	536.87

Aromatics refers to metabolites that contain an aromatic ring (i.e., flavonoids, polyphenols, coumarins, etc.), and terpenes include mono-, di-, and triterpenes.

Table 3 List of herbicides included in the compound library listed alphabetically.

Herbicide	Formula	Molecular mass (g/mol)
Aatrex	$C_8H_{14}ClN_5$	215.69
Bentazon	$C_{10}H_{12}N_2O_3S$	240.28
Caprylic acid	$C_8H_{16}O_2$	144.21
Clethodim	$C_{17}H_{26}ClNO_3S$	359.9
Clomazone	$C_{12}H_{14}ClNO_2$	239.7
Clopyralid	$C_6H_3Cl_2NO_2$	192
Cycloate	$C_{11}H_{21}NOS$	215.36
Ethalfluralin	$C_{13}H_{14}F_3N_3O_4$	333.26
Ethofumesate	$C_{13}H_{18}O_5S$	286.34
Fluazifop	$C_{19}H_{20}F_3NO_4$	327.25
Flumioxazin	$C_{19}H_{15}FN_2O_4$	354.1

Table 3 List of herbicides included in the compound library listed alphabetically.—cont'd

Herbicide	Formula	Molecular mass (g/mol)
Fomesafen	$C_{15}H_{10}ClF_3N_2O_6S$	460.7
Glyphosate	$C_3H_8NO_5P$	169.07
Halosulfuron-methyl	$C_{13}H_{15}ClN_6O_7S$	434.81
Imazethapyr	$C_{15}H_{19}N_3O_3$	289.33
Linuron	$C_9H_{10}Cl_2N_2O_2$	249.1
Metam sodium	$C_2H_4NNaS_2$	129.18
Mesotrione	$C_{14}H_{13}NO_7S$	339.32
Metribuzin	$C_8H_{14}N_4OS$	214.29
Napropamide	$C_{17}H_{21}NO_2$	271.16
Nicosulfuron	$C_{15}H_{18}N_6O_6S$	410.4
Norflurazon	$C_{12}H_9ClF_3N_3O$	303.04
Oxyfluorfen	$C_{15}H_{11}ClF_3NO_4$	361.7
Paraquat	$C_{12}H_{14}Cl_2N_2$	257.16
Pelargonic acid	$C_9H_{18}O_2$	158.23
Pendimethalin	$C_{13}H_{19}N_3O_4$	281.31
Phenmedipham	$C_{16}H_{16}N_2O_4$	300.31
Prometryn	$C_{10}H_{19}N_5S$	241.36
Pronamide	$C_{12}H_{11}Cl_2NO$	256.12
Pyraflufen-ethyl	$C_{15}H_{13}Cl_2F_3N_2O_4$	413.2
Pyroxasulfone	$C_{12}H_{14}F_5N_3O_4S$	391.06
Rimsulfuron	$C_{14}H_{17}N_5O_7S_2$	431.4
Saflufenacil	$C_{17}H_{17}ClF_4N_4O_5S$	500.9
Sethoxydim	$C_{17}H_{29}NO_3S$	327.5
Simazine	$C_7H_{12}ClN_5$	201.66
Tembotrione	$C_{17}H_{16}ClF_3O_6S$	440.8
Terbacil	$C_9H_{13}ClN_2O_2$	216.67
Trifluralin	$C_{13}H_{16}F_3N_3O_4$	335.28

metabolites like amino acids, nucleotides, and intermediates in carbohydrate and lipid metabolism that are conserved across domains of life are also available from ZINC Docking and may be useful to include in a virtual screening experiment.

In addition to the compounds included in Tables 1–3, the example virtual screen involving the Arabidopsis UGT74F2 will also include: 3-hydroxybenzoic acid, 4-hydroxybenzoic acid, 2,3-dihydroxybenzoic acid, 2,4-dihydroxybenzoic acid, 2,5-dihydroxybenzoic acid, 2,6-dihydroxybenzoic acid, 3,4-dihydroxybenzoic acid, 3-hydroxyanthranilate, chorismate, and tryptophan. The ".mol2" files for each of these ligands were downloaded from ZINC Docking.

While online databases of metabolites have many of the common plant metabolites, intermediates in specialized metabolism are not likely to be included. If a compound of interest is not available to download, as a flexible 3D formatted file, structural files can be interconverted into acceptable file formats, like ".mol2", using the open-source tool Open Babel (O'Boyle et al., 2011).

2.2 Preparing molecules for a virtual screen

1. The coordinate files for the ligands from Section 2.1 will now need to be converted to PDBQT (Protein Data Bank, Partial Charge (Q), & Atom Type (T)) files for downstream applications. The files can be opened as ligands in AutoDock Tools and converted to the ".pdbqt" file extension (Forli et al., 2016). To do this, click on "Ligand" in the toolbar and select "input" to open each of the ".mol2" files. To convert them to ".pdbqt" files, click on "Ligand" and select "output" and save the file in a directory where it can be easily located. The name for each ligand file should not contain spaces
2. Once the ligand files have been generated, save them in a folder named "bin" inside a "Vina" folder on your computer's desktop. Alternatively, the files may be renamed with an abbreviation or a number, and a spreadsheet could be used to connect the full name of the compound to the abbreviated file name
3. Using a text editor, create a file named Ligands.txt. Type the name of each ligand file (i.e., benzoate.pdbqt). Each line in this file should contain only a single ligand file name. This file should be saved in Desktop/Vina/bin. NOTE: If the spelling of the ligand file name does not exactly match the spelling in the Ligands.txt file, downstream codes will not run

4. To prepare the enzyme of interest, you will need to obtain structural coordinate of the protein. Several options exist for finding or generating a coordinate file for the enzyme:
   a. If a solved structure of the enzyme of interest is available, the ".pdb" file can be downloaded from the Protein Data Bank (rcsb.org). If the protein structure was solved in complex with a ligand that occupies the binding site (i.e., active site or allosteric site), then the ".pdb" file can be opened using a text editor, and the three-dimensional coordinates of the ligand can be manually deleted from the file. To determine that the ligand has been successfully removed, the modified ".pdb" file can be opened using a structural visualization program like PyMOL and visually inspected. Extraneous ions, water molecules, ligands, and cofactors can also be deleted using a text editor. Solved structures may also have multiple biological assemblies (e.g., dimers) within one asymmetric unit. While not necessary, additional biological assemblies may be deleted manually using a text editor and inspected using PyMOL
   b. As of this writing, the proteomes for the model plants *A. thaliana, Zea mays, Glycine* max, and *Oryza sativa* are available for download online from the AlphaFold Protein Structure Database (www.alpha fold.ebi.ac.uk; Jumper et al., 2021). For example, searching for "*Arabidopsis thaliana* UGT74F2" returns an entry for this protein, and a structural model can be downloaded as a ".pdb" file. Note that these models do not contain cofactors, metal ions, water molecules, or ligands. It may be important to add cofactors to the structure using a single docking run (see the note at the end of this section).
   c. For non-model plants or for proteins that are not available from the AlphaFold database, protein models can be generated from online servers, such as SWISS-MODEL (swissmodel.expasy.org) or Phyre2 (www.sbg.bio.ic.ac.uk/~phyre2), using an amino acid sequence for the enzyme of interest
5. Once a coordinate file for the enzyme(s) of interest has been obtained, the file can be prepared using AutoDock Tools. The protein used here, UGT74F2 from *A. thaliana*, has been previously crystalized (PDB ID: 5U6M; George Thompson et al., 2017), and the file was first prepared by removing the salicylic acid ligands from the dimer. The UDP that was cocrystalized with the enzyme was left in the file because it is the product of the reaction after the activated nucleotide sugar (i.e., UDP-glucose) transfers the sugar onto a nucleophilic acceptor substrate. Alternatively,

the AlphaFold-generated structural coordinates may have been used, but the protein structure would have been apo, meaning that UDP would have to be docked into the structure. To open the ".pdb" coordinate file for the enzyme, use the topmost toolbar and select "File" and then "Read Molecule." Protein structures typically do not include hydrogens, so these may be added to the enzyme using AutoDock Tools by clicking "Edit" in the toolbar and then "Hydrogens" (Forli et al., 2016). Hydrogens should appear on the enzyme in the protein viewing window. To export the protein as a ".pdbqt" file, click "Grid," then "Macromolecule" and "Choose" and select the protein. A window will pop up so the file can be saved in the Desktop/Vina/bin folder with a ". pdbqt" extension

6. To define the docking area that encompasses the active site or binding pocket of interest, click "Grid" and then "Grid Box." This is the area in which Vina will search for binding interactions between the enzyme and each ligand, so it is important that the space is large enough to cover the entire area of interest and allow the ligands to rotate freely but not so large that the molecules are predicted to bind the protein superficially. The size and center of the search space in x, y, and z dimensions of the grid box, can be altered so the entire search space is encompassed by the coordinates. It is also important to rotate the protein in the AutoDock Tools viewer to see that the binding pocket of interest is covered in all dimensions
7. Once the grid box dimensions have been determined, these values will be recorded in a text file saved in the Desktop/Vina/bin subfolder. This plain text file can be made using a plain text editor, like TextEdit or Notepad

```
conf_vs.txt

receptor = UGT74F2.pdbqt

center_x = 46.7
center_y = 76.1
center_z = 81.6

size_x = 40
size_y = 40
size_z = 40

exhaustiveness = 8
num_modes = 10
```

The name of the enzyme coordinate file is listed as "receptor =" and the center x, y, and z coordinates were determined from AutoDock

Tools, as was the size of the box in x, y, and z dimensions. The exhaustiveness of 8 is default and is the number of Monte-Carlo iterated searches. The number of modes (num_modes) is the maximum number of ligand conformations that will be generated per ligand. Increasing this number may increase the total time that it takes to run Vina.

8. This process was repeated to generate a structural model of UGT74F2 that had the activated nucleotide sugar donor UDP-glucose bound in the active site. To do this, the coordinates for UDP were removed from the 5U6M PDB file above using a text editor. The apo enzyme structure was used for a single docking run using AutoDock Vina, and the coordinates of the UDP-glucose conformation that had the highest binding affinity and bound in a logical conformation was exported as a PDB file and pasted into the text file of the apo enzyme model and saved as a new UDP-glucose bound model. This new combined file was opened using PyMOL to ensure that the structural model had UDP-glucose in the active site. This file was then opened in AutoDock Tools for conversion to a ".pdbqt" file and to determine the grid box dimensions

Note: If an important cofactor for the protein of interest—such as heme in cytochrome P450 monooxygenases, NADH or NADPH in dehydrogenases, or S-adenosyl methionine in SAM-dependent methyltransferases—is not part of the structural file, a single docking run using AutoDock Vina could be used to place the cofactor in the active site. The coordinate file for cofactors may be obtained from the Protein Data Bank, converted to a ".pdbqt" file, and docked into the active site. Coordinates for the best docking confirmation of the cofactor could be pasted into the ".pdb" file of the protein above "END" using a text editor.

3. Virtual screening using AutoDock Vina

The virtual screen presented here uses AutoDock Vina, which uses a rapid gradient-optimization conformational search (Forli et al., 2016). The scoring function estimates the force of non-covalent interactions between compounds and the target protein using mathematical models, and the user does not need to have working knowledge of these models in order to use the program. This straightforward approach is easily implemented by molecular biologists, and the length of time needed for the screen to run depends on the factors such as the number of ligands included in the virtual screen, as well as the number of modes for each ligand.

3.1 Software needed

Windows users will need to download Perl to run the virtual screen script from https://www.perl.org/get.html. Perl is installed on MacOS by default. AutoDock Vina can be downloaded online at http://vina.scripps.edu. The following protocol uses version 1.1.2 for its ease of use. Move the two Unix executable files (titled vina and vina_split) from the Autodock Vina folder that was downloaded into the Desktop/Vina/bin subfolder. These are needed to ensure the following steps proceed properly.

3.2 Script for running the virtual screen

A Perl script enables iterative docking runs with AutoDock Vina using a Mac or Windows system. This file should be added to the Desktop/Vina/bin folder. The file can be created using a text editor and should contain the following information:

Vina_mac.pl

```
#!/usr/bin/perl
print"Ligand_file:\t";
$ligfile=<STDIN>;
chomp $ligfile;
open (FH,$ligfile)||die "Cannot open file\n";
@arr_file=<FH>;

for($i=0;$i<@arr_file;$i++)
{
print"@arr_file[$i]\n";
@name=split(/\./,@arr_file[$i]);
}
for($i=0;$i<@arr_file;$i++)
{
    chomp @arr_file[$i];
    print"@arr_file[$i]\n";
    system("./vina –config conf_vs.txt –ligand @arr_file[$i] –log @arr_file[$i]_log.log");
}
```

Modification to Perl script for Microsoft OS:

```
system("vina.exe –config conf_vs.txt –ligand @arr_file[$i] –log @arr_file[$i]_log.log");
```

3.3 Procedure on a Mac OS

1. Before starting the run, check that the following files are in the Desktop/Vina/bin folder:

a. conf_vs.txt
b. Ligands.txt
c. Vina_mac.pl (or Vina_windows.pl)
d. vina
e. vina_split
f. UGT74F2.pdbqt
g. ".pdbqt" files for each ligand

2. Open Terminal (Mac) or Command Prompt (Windows), and type:
cd Desktop/Vina/bin
perl Vina_mac.pl (OR **perl Vina_windows.pl**)
When you hit "Enter," you will be prompted to type in the name of the ligand file. Type: **Ligands.txt** and hit "Enter." After this, Vina will start running. Here is an example of the Terminal log during one compound docking run:

```
cinnamaldehyde.pdbqt
WARNING: The search space volume > 27000 Angstrom^3 (See FAQ)
Output will be cinnamaldehyde_out.pdbqt
Detected 4 CPUs
Reading input ... done.
Setting up the scoring function ... done.
Analyzing the binding site ... done.
Using random seed: 1584341810
Performing search ...
0%   10   20   30   40   50   60   70   80   90   100%
|----|----|----|----|----|----|----|----|----|----|
***************************************************
done.
Refining results ... done.
mode |    affinity | dist from best mode
     | (kcal/mol) | rmsd l.b.| rmsd u.b.
-----+------------+----------+----------
1          -5.9      0.000      0.000
2          -5.6      9.406     10.829
3          -5.4      2.469      4.623
4          -5.1      8.247     10.127
5          -5.1     12.839     14.177
6          -5.0      9.418     11.085
7          -4.9      9.755     10.414
8          -4.9     12.924     14.300
9          -4.9      3.682      5.598
10         -4.8     14.032     15.374
Writing output ... done.
```

The length of the run will vary based on the number of ligand files and the number of modes. Do not close the Terminal window during this time.

3. When your run is complete, all of the coordinate files for each of the docked ligands (i.e., ligand_out.pdbqt), as well as the output log files that include the binding affinities for each conformation, will be in the Desktop/Vina/bin folder

Note: Before running a new virtual screen, it is important to move the ".log" and "..._out.pdbqt" files from the previous run to a new folder so they are not overwritten.

4. Interpreting and visualizing screening results

To interpret the results, the protein can be opened in a protein visualization program, like PyMOL or UCSF Chimera. PyMOL can be downloaded from https://pymol.org/2/. In the same window, each of the ligand files can be opened and visualized in relation to the active site or a specified cavity on the enzyme of interest. Because the conf_vs.txt file used a num_modes setting of 10, a maximum of 10 conformations were generated for each compound. These conformations should be manually inspected to determine the conformation that is most logical (i.e., oriented within the active site and not on the surface of the protein) and has the lowest computational binding affinity (i.e., a large negative number) in kcal/mol. The conformations are shown in increasing order of binding affinity, and while the first conformation has the highest binding affinity, it does not always bind within the active site or in an orientation that would make sense for catalysis. If many of the molecules are binding on the surface of the protein, the search space dimensions can be redrawn, and the screen can be repeated with a more refined search window. If the grid box dimensions are stringent and the compound does not bind within those constraints, the results may have fewer than 10 conformations for some compounds.

For metabolic enzymes, one way to assess the accuracy of ligand binding is to use the measurement feature in PyMOL to determine the distance between the compound and the cofactor (Fig. 1). In the toolbar of PyMOL, click on "Wizard" and select "Measurement." This allows for the distance between individual atoms to be determined. Click on an atom on the ligand that is acted upon by the enzyme (e.g., the predicted ligand hydroxyl that is glycosylated by the enzyme), and then click on a second atom elsewhere in the enzyme like a catalytic residue (if known) or an atom

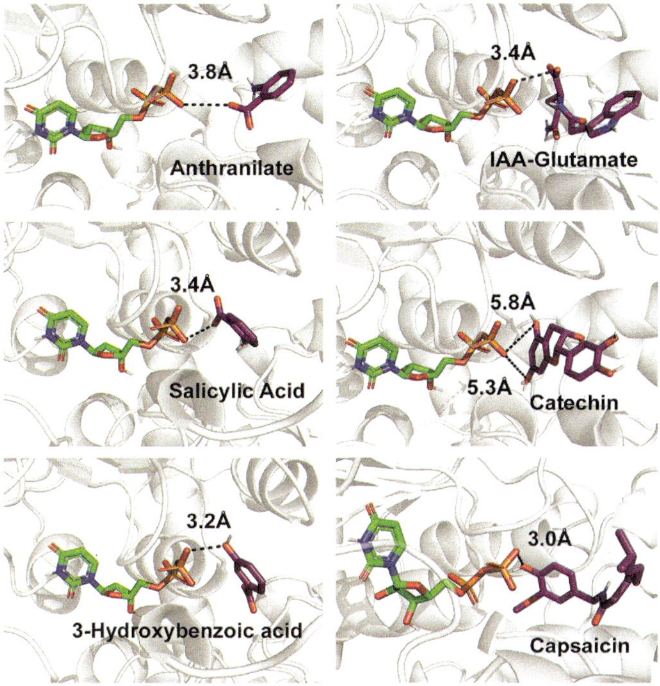

Fig. 1 Measured distances between functional groups on docked ligands and the nearest phosphate oxygen on UDP in the AtUGT74F2 active site. In the three panels on the left, three compounds (anthranilate, salicylic acid, and 3-hydroxybenzoic acid) that are confirmed substrates for the enzyme are shown, while compounds that have a high binding affinity but unknown activity (IAA-glutamate, catechin, and capsaicin) are shown in the three panels on the right.

on the cofactor. The distance will be displayed as a label in Angstroms (Å) above a dashed line that connects the two atoms that were selected. Once you are finished measuring distances between atoms, click "Done" under "Measurement" in the right-hand panel.

While inspecting each ligand, note the conformation(s) that have a low binding affinity and accurately bind within the active site. Vina has calculated the binding affinity for each mode, and the values can be found in the output ".log" files that were generated for each ligand. Taken together, the binding affinities for the ligand conformations that were noted by visual inspection may be added to a spreadsheet (Table 4 and 5). When the results have been tabulated, the compounds can be ranked by their binding affinity values from lowest to highest. The results from the virtual screens of UGT74F2 that had either UDP-glucose or UDP bound in the active site are included in Tables 4 and 5.

Table 4 Ranking of binding affinities from the *Arabidopsis thaliana* UGT74F2 virtual screen with UDP-glucose bound in the active site.

Compound	Binding mode	Binding affinity (kcal/mol)
Capsaicin	1	−7.2
2,4-Dihydroxybenzoic acid[a]	1	−6.5
2,3-Dihydroxybenzoic acid[a]	1	−6.5
Gallic acid	3	−6.1
3-Hydroxyanthranilate	3	−6.1
3,4-Dihydroxybenzoic acid[a]	5	−6.1
Salicylic acid[b]	3	−6
Nicotine	1	−6
Anthranilate[b]	2	−5.9
3-Hydroxybenzoic acid[b]	3	−5.9
4-Hydroxybenzoic acid[a]	3	−5.9
2,5-Dihydroxybenzoic acid[a]	5	−5.8
Phenylacetic acid	2	−5.7
p-Coumaryl alcohol	2	−5.6
Indole-3-butyric acid	7	−5.6
Benzoic acid[b]	3	−5.6
Clopyralid	4	−5.4
2,6-Dihydroxybenzoic acid[a]	3	−5
Glyphosate	4	−4.4
Metam	2	−2.8

[a]No activity detected in an in vitro assay.
[b]Activity has been confirmed in vitro.
Binding affinities are listed for the binding mode that was closest to the UDP-glucose cosubstrate (measured in angstroms). The distance represents the approximate distance between the nearest atom that could participate in the glycosyl transfer reaction (e.g., hydroxyl oxygen) and the hydroxyl on carbon-6 of glucose. Compounds with a predicted distance of >5 Å from the oxygen atoms of the gamma phosphate were not included in the table. Compounds are ranked from highest binding affinity to lowest binding affinity as determined by AutoDock Vina, and molecules that bound in the active site but did not have a functional group that could be glycosylated were not included in the table.

Table 5 Ranking of binding affinities from the *Arabidopsis thaliana* UGT74F2 virtual screen with UDP bound in the active site.

Compound	Binding mode	Binding affinity (kcal/mol)
IAA-glutamate	1	−8.8
Catechin	1	−8.8
Tryptophan	1	−7.8
Resveratrol	1	−7.4
6-Benzylaminopurine	1	−7.2
DIMBOA	1	−7.1
Linuron	2	−7.1
Kinetin-riboside	8	−7
Phenmedipham	3	−7
Dhurrin	6	−6.9
cis-Zeatin	3	−6.8
IAA	2	−6.8
Indole-3-butyric acid	2	−6.8
4-Chloroindole-3-acetic acid	1	−6.8
Capsaicin	4	−6.4
Ferulic acid	2	−6.4
Bentazon	5	−6.3
2,3-Dihydroxybenzoic acid[a]	4	−6.2
2,5-Dihydroxybenzoic acid[a]	4	−6.2
Caffeine	4	−6.2
2,6-Dihydroxybenzoic acid[a]	3	−6.1
2,4-Dihydroxybenzoic acid[a]	1	−6.1
3-Hydroxyanthranilate	6	−5.9
Anthranilate[b]	2	−5.9
Nicotine	3	−5.8
Terbacil	5	−5.8
Clopyralid	2	−5.8

Continued

Table 5 Ranking of binding affinities from the *Arabidopsis thaliana* UGT74F2 virtual screen with UDP bound in the active site.—cont'd

Compound	Binding mode	Binding affinity (kcal/mol)
p-Coumaryl alcohol	2	−5.7
Salicylic acid[b]	5	−5.6
Phenylacetic acid	4	−5.6
3-Hydroxybenzoic acid[b]	8	−5.5
Paraquat	3	−5.5
Simazine	3	−5.4
Methyl anthranilate	8	−5.3
4-Hydroxybenzoic acid[a]	8	−5.3
Methyl salicylate	10	−5
Glyphosate	5	−5
Caprylic acid	1	−4.9
Nonoate	2	−4.8
Metam	1	−3.1

[a]No activity detected in an in vitro assay.
[b]Activity has been confirmed in vitro.
Binding affinities are listed for the binding mode that was closest to the UDP product (measured in angstroms). The distance represents the approximate distance between the nearest atom that could participate in the glycosyl transfer reaction (e.g., hydroxyl oxygen) and the oxygen atoms on the beta-phosphate of UDP. Compounds with a predicted distance of greater than 6.0 Å from the terminal phosphate oxygen atoms were not included in the table. Compounds are ranked from highest binding affinity to lowest binding affinity as determined by AutoDock Vina, and molecules that bound in the active site but did not have a functional group that could be glycosylated were not included in the table.

For UGT enzymes, analyzing the results of a virtual screen with both the UDP-glucose cosubstrate as well as the UDP product bound is informative since the glucose moiety is large and occupies a portion of the active site (Figs. 1 and 2). When UDP-glucose is bound, larger molecules may be precluded from binding the active site pocket, especially if the pocket is small. However, a virtual screen that has UDP bound in the active site may allow larger or glycosylated molecules to bind. Taken together, the UDP-glucose bound structure is useful for identifying putative substrates for this enzyme, while the UDP-bound structure may inform putative products of the reactions. In either docking simulation, putative small-molecule regulators (i.e., competitive inhibitors) may also be identified.

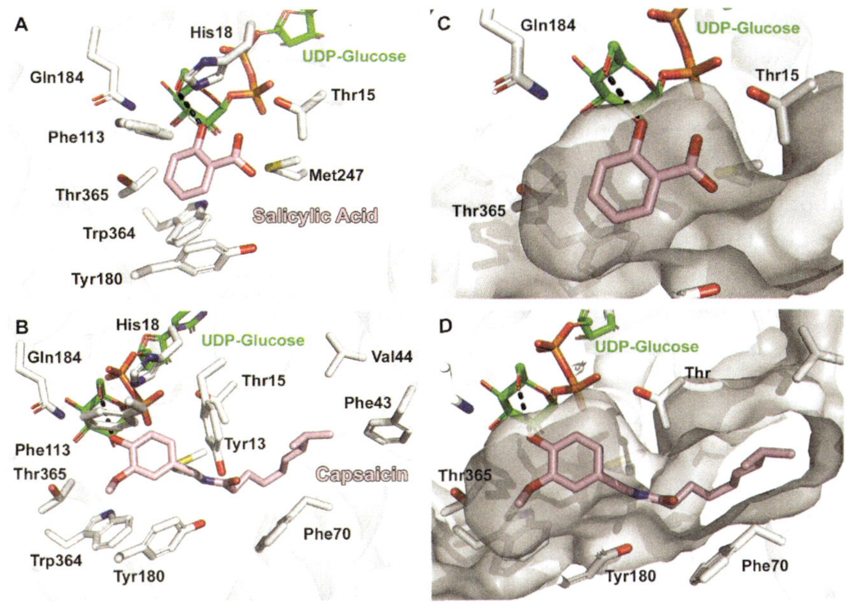

Fig. 2 Docking results for AtUGT74F2 with UDP-glucose in the active site. A. View of the active site with salicylic acid docked. Dashed lines are used to show how the distance between the hydroxyl of salicylic acid and the hydroxyl on carbon-6 of glucose was measured. Active site residues are displayed as sticks. B. View of the active site with capsaicin docked, including hydrophobic residues distal from the site of glycosyl transfer. Dashed lines are used to show how the distance between the hydroxyl of salicylic acid and the hydroxyl on carbon-6 of glucose was measured. Active site residues are displayed as sticks. C. Surface view of the active site with salicylic acid docked to show the space of the active site cavity. D. Surface view of the active site with capsaicin docked to show the full extent of capsaicin binding both close to the site of glycosyl transfer and in the distal hydrophobic space within the pocket. Note that while the distances were drawn between the ligand and the hydroxyl on carbon-6 of glucose, the hydroxyl on carbon-1 that is covalently attached to the beta-phosphate of UDP is the hydroxyl on which the ligand is transferred.

In the virtual screen of the *A. thaliana* UGT74F2 with UDP-glucose bound, we found an overall trend in our inspection of each compound (Table 4). Notably, small, aromatic compounds, such as 2,4-dihydroxybenzoic acid and salicylic acid, have a high binding affinity (−6.5 and −6 kcal/mol, respectively) and a hydroxyl or carboxyl oxygen on these molecules is a short distance from a hydroxyl oxygen on the glucose molecule (approximately 2.8 Å and 2.7 Å, respectively). Aromatic amino acids in the active site (Tyr13, Phe70, Phe113, Tyr180, and Trp364) likely stabilize intramolecular interactions between substrates and the protein through π-π stacking interactions (Fig. 2). His18 is a conserved catalytic residue that hydrogen bonds with substrates to position them for the glycosyl transfer reaction (George Thompson et al., 2017). When salicylic acid is the substrate, the protonation state of the basic active site amino acid His18 determines whether the carboxyl or hydroxyl are glucosylated by the enzyme. The carboxylate of SA likely forms hydrogen bonding interactions with the side-chain hydroxyl of Thr15, while the carboxylate of 2,4-dihydroxybenzoic acid is on the opposite side of the active site and is predicted to form hydrogen bonds with the side-chain hydroxyl of Thr365. These two threonine residues have been found to contribute to substrate binding conformation.

These results are consistent with previous functional data that found that UGT74F2 can glycosylate the carboxylate and the hydroxyl of SA (2-hydroxybenzoic acid), anthranilate, benzoic acid, and 3-hydroxybenzoic acid (George Thompson et al., 2017; Lim et al., 2002; Quiel & Bender, 2003). However, activity was not seen with 4-hydroxybenzoic acid, 2,4-dihydroxybenzoic acid or any other dihydroxybenzoates that were tested (Lim et al., 2002). This discrepancy highlights the importance of validating the results from a virtual screen using functional assays.

Interestingly, the molecule with the highest binding affinity that was within 5 Å of UDP-glucose was the alkaloid capsaicin (−7.2 kcal/mol). Capsaicin is an alkaloid produced in the *Capsicum* genus (Solanaceae) and is responsible for the pungent and spicy sensation associated with eating hot peppers.

Capsaicin bound the active site with a hydroxyl on the aromatic ring positioned 2.8 Å away from the hydroxyl on carbon-6 of glucose, and the hydrophobic 8-methyl-6-nonenoyl moiety oriented near hydrophobic residues in the active site pocket (Phe 43, Val 44, and Phe 70) (Fig. 2). Capsaicin glucosides have been detected in *Capsicum annuum* peppers, and one glucoside, capsaicin-β-D-glucopyranoside, was found to have 1/100th of the pungency of capsaicin (Higashiguchi, Nakamura, Hayashi, & Kometani, 2006; Kometani, Tanimoto, Nishimura, Kanbara, & Okada., S., 1993). While insect UGTs that glycosylate capsaicin as a detoxification mechanism have been characterized, a plant enzyme that glycosylated capsaicin remains to be identified (Ahn et al., 2011).

Other compounds in the table that have not been screened for in vitro activity but bind in the active site in the virtual screen include the trihydroxybenzoate gallic acid, the tryptophan catabolism intermediate 3-hydroxyanthranilate, the auxinic hormone phenylacetic acid, and the phenylpropanoid biosynthetic intermediate *p*-coumaryl alcohol. Additionally, three herbicides were predicted to bind the active site of the enzyme: clopyralid (3,6-dichloro-2-pyridinecarboxylic acid), glyphosate (*N*-(phosphonomethyl)glycine), and metam (methylcarbamodithioic acid). The three herbicides have lower binding affinities than the known substrates, but because this is a computational screen, the binding of these molecules to UGT74F2 should be confirmed experimentally. It may be that enzyme activity is unaffected by these herbicides, or the herbicides could serve as competitive inhibitors by binding and blocking substrates. Assays would also be needed to determine whether the carboxylate of clopyralid, the carboxylate of glyphosate, or the thiol of metam can be glycosylated, which may serve as a mechanism for herbicide resistance if the glycosylated herbicide has reduced toxicity (Gaines et al., 2020).

The virtual screen that included UDP bound in the active site generated more compounds that were able to bind in the active site than in the UDP-glucose structure, which is likely because the glucose moiety occupies a large space in the active site and excludes larger molecules (Table 5). For the UDP-bound enzyme, a cutoff of 6.0 Å between hydroxyl on carbon-6 of glucose and an atom on the compound that could be glycosylated was selected (Fig. 1). Known substrates, including anthranilate, salicylic acid, and 3-hydroxybenzoic acid, had predicted binding affinities of −5.9, −5.6, and −5.5 kcal/mol, respectively (Table 5). Unsurprisingly the results of docking with UDP and UDP-glucose bound in the active site varied, which highlights the importance of testing multiple cosubstrates, coproducts, or cofactors.

5. Limitations of virtual screens

While docking and virtual screens are an excellent way to computationally predict how a small molecule will interact with an enzyme of interest, this technique is not without its limitations. One of the limitations of the virtual screening method provided is that the protein or enzyme of interest will be screened for ligand binding based on a single, rigid protein conformation. Regardless of the source of the protein structure (i.e., X-ray crystal structure, homology model, etc.), flexibility of the ligand will not be taken into account during the virtual screen. Virtual screen results may not accurately depict the molecular interactions that would occur between the protein and ligand in another conformation. If protein flexibility is a key consideration in generating reliable docking results for the protein of interest, this AutoDock Vina-based virtual screen could be followed with a second screen using DOCK 6 with the AMBER scoring function, which performs molecular dynamics simulations (Allen et al., 2015; Maia, Assis, de Oliveira, da Silva, & Taranto, 2020). However, there are limitations to this approach as well, namely, the input preparation and the amount of time needed to run a more computationally intensive program.

Another limitation inherent to virtual screening is that there are false positives, and perhaps false negatives, within the data (Maia et al., 2020). For example, metabolites that are known to not exhibit activity with the *A. thaliana* UGT74F2, such as 3,4-dihydroxybenzoic acid, have a higher predicted binding affinity than known substrates, such as salicylic acid and anthranilate (Table 4). Like any computational screen, in vitro or in vivo experiments are needed to complement and confirm the results. For example, if the metabolite is a putative substrate, then the enzyme could be purified and assayed for activity with the metabolite(s) of interest. Gene knockout or knockdown experiments coupled with metabolomics could be used to determine the chemical phenotype in the absence of the enzyme. Depending on the number of ligands included in the screen, one may end up with a list of tens of metabolites to screen for activity, many of which may not serve as substrates. On the other hand, without knowing that anthranilate and salicylic acid are substrates for UGT74F2, one may have made an arbitrary cutoff for binding affinities that excluded these compounds from in vitro analyses, which would have meant that the substrates of the enzyme may not have been identified.

A key consideration for virtual screening is the space that cofactors and cosubstrates occupy in the active site. In the case presented here, docking was performed with both UDP-glucose, which is a cosubstrate, and the product UDP. While there were similarities between the results (Tables 4 and 5), there are also notable differences. The UGT74F2 active site was able to accommodate larger molecules when only UDP was bound, whereas smaller aromatic compounds predominantly bound the active site when UDP-glucose was present. Because of the speed of conducting a virtual screen, re-screening enzymes with multiple cofactors or cosubstrates is feasible.

Many simulation-based softwares are limited in their ability to model complex systems to perfect accuracy. Furthermore, the ability of protein-ligand docking programs such as AutoDock to provide accurate results is highly dependent upon the accuracy of the binding site of the structural model (Cross et al., 2009). Most protein modeling programs use homology-based techniques, where an unknown protein of interest is modeled using a known protein with a solved crystal structure as a template. In some previous pharmacological studies, unknown proteins that shared >50% sequence identity with the template structure were deemed eligible for docking (Hillisch, Pineda, & Hilgenfeld, 2004). More recent findings suggest that the 50% homology cutoff is not an effective method for assessing docking eligibility, but with recent advances in predictive modeling programs like Alpha Fold, highly accurate protein structures have never been more accessible (Bordogna, Pandini, & Bonati, 2011; Jumper et al., 2021). As these protein structures continue to increase in accuracy, so do the capabilities of docking programs to accurately evaluate and model protein–ligand interactions.

Protein-ligand docking programs often suffer from an inability to predict the influence of water molecules on protein–ligand interactions (Verdonk et al., 2005). This is especially problematic for enzymes that contain water molecules in their active site, as these molecules may play a vital role in facilitating catalysis. During docking, optimization of water molecule orientation has been found to significantly increase docking accuracy (Roberts & Mancera, 2008). In addition to increasing docking accuracy, inclusion of water molecules can also be a useful tool for discovering substrates that are able to displace water molecules from the active site (Verdonk et al., 2005). Some docking programs, such as GOLD, can be modified to factor water mediation and displacement during protein-ligand docking. More recently, AutoDock Vina version 1.2.0 has been updated to support the modeling of explicit water molecules via the hydrated docking protocol (Eberhardt et al., 2021). This method produces a more accurate portrayal of water–protein–ligand interactions; however, it comes at a high computational cost compared to the default solvent models.

6. Conclusions

While virtual screening is commonly used in the pharmaceutical and biotechnology industries for drug and small-molecule inhibitor design, there is an untapped potential for this computational method for exploring the function of plant enzymes. Now with advances in protein structure prediction and the availability of ready-to-dock ligands, conducting a virtual screen can be free, easy, relatively fast and complementary to gene-expression-based methods that have been used extensively to identify candidate enzymes in metabolic pathways.

Before starting in vitro or in vivo experiments to confirm the function of a metabolic enzyme, conducting a virtual screen of the putative substrates and/or products of the enzymes of interest may provide an additional mechanism to narrow down lists of candidate genes. Additionally, virtual screens may be applied to enzymes with known functions to identify alternative substrates, identify natural inhibitors, or for designing inhibitors, like herbicides. However, like any computational tool, the results of virtual screens need to be confirmed experimentally.

Specialized metabolites have revolutionized the medical, cosmetic, and agricultural industries, however, enzyme promiscuity in specialized metabolic pathways also poses significant threats, such as the evolution of herbicide resistance (Abdollahi et al., 2021). While the evolution of enzyme specificity played a major role in increasing metabolic efficiency, enzyme promiscuity continues to drive the emergence of secondary metabolic pathways, which often provide a selective advantage to the host organism (Leong & Last, 2017). Understanding the mechanisms that drive enzyme promiscuity is key to navigating these challenges and may even allow for improved manipulation of valuable specialized metabolic pathways. Protein-based computational techniques such as virtual screening are excellent tools for probing individual metabolic enzymes for large pools of potential substrates. High-throughput analysis of structurally similar substrates will reveal key details about enzyme activity and may enable the development of novel pathways for natural product biosynthesis.

Acknowledgments

Research in the Holland lab is supported by the U.S. National Science Foundation (MCB-2214883) and by Williams College. The Perl script in the methods was made publicly available by Dr. Babajan Banaganapalli. The authors thank current and former lab members for testing the virtual screening protocols.

References

Abdollahi, F., Alebrahim, M. T., Ngov, C., Lallemand, E., Zheng, Y., Villette, C., et al. (2021). Innate promiscuity of the CYP706 family of P450 enzymes provides a suitable context for the evolution of dinitroaniline resistance in weed. *The New Phytologist*, *229*(6), 3253–3268.

Ahn, S. J., Badenes-Pérez, F. R., Reichelt, M., Svatoš, A., Schneider, B., Gershenzon, J., et al. (2011). Metabolic detoxification of capsaicin by UDP-glycosyltransferase in three Helicoverpa species. *Archives of Insect Biochemistry and Physiology*, *78*(2), 104–118.

Akere, A., Chen, S. H., Liu, X., Chen, Y., Dantu, S. C., Pandini, A., et al. (2020). Structure-based enzyme engineering improves donor-substrate recognition of Arabidopsis thaliana glycosyltransferases. *The Biochemical Journal*, *477*(15), 2791–2805.

Allen, W. J., Balius, T. E., Mukherjee, S., Brozell, S. R., Moustakas, D. T., Lang, P. T., et al. (2015). DOCK 6: Impact of new features and current docking performance. *Journal of Computational Chemistry*, *36*(15), 1132–1156.

Baldwin, I. T. (1989). Mechanism of damage-induced alkaloid production in wild tobacco. *Journal of Chemical Ecology*, *15*(5), 1661–1680.

Bento, A. P., Gaulton, A., Hersey, A., Bellis, L. J., Chambers, J., Davies, M., et al. (2014). The ChEMBL bioactivity database: An update. *Nucleic Acids Research*, *42*(D1), D1083–D1090.

Bordogna, A., Pandini, A., & Bonati, L. (2011). Predicting the accuracy of protein-ligand docking on homology models. *Journal of Computational Chemistry*, *32*(1), 81–98.

Caputi, L., Franke, J., Farrow, S. C., Chung, K., Payne, R., Nguyen, T. D., et al. (2018). Missing enzymes in the biosynthesis of the anticancer drug vinblastine in Madagascar periwinkle. *Science (New York, N.Y.)*, *360*(6394), 1235–1239.

Cross, J. B., Thompson, D. C., Rai, B. K., Baber, J. C., Fan, K. Y., Hu, Y., et al. (2009). Comparison of several molecular docking programs: Pose prediction and virtual screening accuracy. *Journal of Chemical Information and Modeling*, *49*(6), 1455–1474.

Delli-Ponti, R., Shivhare, D., & Mutwil, M. (2021). Using gene expression to study specialized metabolism—A practical guide. *Frontiers in Plant Science*, *11*, 2074.

Eberhardt, J., Santos-Martins, D., Tillack, A. F., & Forli, S. (2021). AutoDock Vina 1.2. 0: New docking methods, expanded force field, and python bindings. *Journal of Chemical Information and Modeling*, *61*(8), 3891–3898.

Fang, C., Fernie, A. R., & Luo, J. (2019). Exploring the diversity of plant metabolism. *Trends in Plant Science*, *24*(1), 83–98.

Forli, S., Huey, R., Pique, M. E., Sanner, M. F., Goodsell, D. S., & Olson, A. J. (2016). Computational protein–ligand docking and virtual drug screening with the AutoDock suite. *Nature Protocols*, *11*(5), 905–919.

Gaines, T. A., Duke, S. O., Morran, S., Rigon, C., Tranel, P. J., Küpper, A., et al. (2020). Mechanisms of evolved herbicide resistance. *The Journal of Biological Chemistry*, *295*(30), 10307–10330.

George Thompson, A. M., Iancu, C. V., Neet, K. E., Dean, J. V., & Choe, J. Y. (2017). Differences in salicylic acid glucose conjugations by UGT74F1 and UGT74F2 from Arabidopsis thaliana. *Scientific Reports*, 7, 46629.

Higashiguchi, F., Nakamura, H., Hayashi, H., & Kometani, T. (2006). Purification and structure determination of glucosides of capsaicin and dihydrocapsaicin from various capsicum fruits. *Journal of Agricultural and Food Chemistry*, *54*(16), 5948–5953.

Hillisch, A., Pineda, L. F., & Hilgenfeld, R. (2004). Utility of homology models in the drug discovery process. *Drug Discovery Today*, *9*(15), 659–669.

Irwin, J. J., Tang, K. G., Young, J., Dandarchuluun, C., Wong, B. R., Khurelbaatar, M., et al. (2020). ZINC20—A free ultralarge-scale chemical database for ligand discovery. *Journal of Chemical Information and Modeling*, *60*(12), 6065–6073.

Jumper, J., Evans, R., Pritzel, A., Green, T., Figurnov, M., Ronneberger, O., et al. (2021). Highly accurate protein structure prediction with AlphaFold. *Nature*, *596*(7873), 583–589.

Kautsar, S. A., Suarez Duran, H. G., & Medema, M. H. (2018). Genomic identification and analysis of specialized metabolite biosynthetic gene clusters in plants using PlantiSMASH. *Methods in molecular biology (Clifton, N.J.)*, *1795*, 173–188.

Kelley, L. A., Mezulis, S., Yates, C. M., Wass, M. N., & Sternberg, M. J. (2015). The Phyre2 web portal for protein modeling, prediction and analysis. *Nature Protocols*, *10*(6), 845–858.

Kilgore, M. B., Holland, C. K., Jez, J. M., & Kutchan, T. M. (2016). Identification of a noroxomaritidine reductase with Amaryllidaceae alkaloid biosynthesis related activities. *Journal of Biological Chemistry*, *291*(32), 16740–16752.

Kim, J., & Buell, C. R. (2015). A revolution in plant metabolism: Genome-enabled pathway discovery. *Plant Physiology*, *169*(3), 1532–1539.

Kim, S., Thiessen, P. A., Bolton, E. E., Chen, J., Fu, G., Gindulyte, A., et al. (2016). PubChem substance and compound databases. *Nucleic Acids Research*, *44*(D1), D1202–D1213.

Kometani, T., Tanimoto, H., Nishimura, T., Kanbara, I., & Okada, S. (1993). Glucosylation of capsaicin by cell suspension cultures of *Coffea arabica*. *Bioscience, Biotechnology, and Biochemistry*, *57*(12), 2192–2193.

Leong, B. J., & Last, R. L. (2017). Promiscuity, impersonation and accommodation: Evolution of plant specialized metabolism. *Current Opinion in Structural Biology*, *47*, 105–112.

Lim, E. K., Doucet, C. J., Li, Y., Elias, L., Worrall, D., Spencer, S. P., et al. (2002). The activity of Arabidopsis glycosyltransferases toward salicylic acid, 4-hydroxybenzoic acid, and other benzoates. *The Journal of Biological Chemistry*, *277*(1), 586–592.

Maeda, H. A., & Fernie, A. R. (2021). Evolutionary history of plant metabolism. *Annual Review of Plant Biology*, *72*, 185–216.

Maia, E., Assis, L. C., de Oliveira, T. A., da Silva, A. M., & Taranto, A. G. (2020). Structure-based virtual screening: From classical to artificial intelligence. *Frontiers in Chemistry*, *8*, 343.

Murata, K., Kitano, T., Yoshimoto, R., Takata, R., Ube, N., Ueno, K., et al. (2020). Natural variation in the expression and catalytic activity of a naringenin 7-O-methyltransferase influences antifungal defenses in diverse rice cultivars. *The Plant Journal*, *101*(5), 1103–1117.

Nützmann, H. W., Huang, A., & Osbourn, A. (2016). Plant metabolic clusters - from genetics to genomics. *The New Phytologist*, *211*(3), 771–789.

O'Boyle, N. M., Banck, M., James, C. A., Morley, C., Vandermeersch, T., & Hutchison, G. R. (2011). Open babel: An open chemical toolbox. *Journal of Cheminformatics*, *3*(1), 1–14.

Polturak, G., Liu, Z., & Osbourn, A. (2022). New and emerging concepts in the evolution and function of plant biosynthetic gene clusters. *Current Opinion in Green and Sustainable Chemistry*, *33*, 100568.

Quiel, J. A., & Bender, J. (2003). Glucose conjugation of anthranilate by theArabidopsis UGT74F2 glucosyltransferase is required for tryptophan mutant blue fluorescence. *Journal of Biological Chemistry*, *278*(8), 6275–6281.

Rester, U. (2008). From virtuality to reality - virtual screening in lead discovery and lead optimization: A medicinal chemistry perspective. *Current Opinion in Drug Discovery & Development*, *11*(4), 559–568.

Rhee, S. Y., & Mutwil, M. (2014). Towards revealing the functions of all genes in plants. *Trends in Plant Science*, *19*(4), 212–221.

Roberts, B. C., & Mancera, R. L. (2008). Ligand-protein docking with water molecules. *Journal of Chemical Information and Modeling*, *48*(2), 397–408.

Saito, K., & Matsuda, F. (2010). Metabolomics for functional genomics, systems biology, and biotechnology. *Annual Review of Plant Biology*, *61*, 463–489.

Schultz, B. J., Kim, S. Y., Lau, W., & Sattely, E. S. (2019). Total biosynthesis for milligram-scale production of etoposide intermediates in a plant chassis. *Journal of the American Chemical Society*, *141*(49), 19231–19235.

Sherp, A. M., Westfall, C. S., Alvarez, S., & Jez, J. M. (2018). Arabidopsis thaliana GH3.15 acyl acid amido synthetase has a highly specific substrate preference for the auxin precursor indole-3-butyric acid. *Journal of Biological Chemistry*, *293*(12), 4277–4288.

Singh, A., Menéndez-Perdomo, I. M., & Facchini, P. J. (2019). Benzylisoquinoline alkaloid biosynthesis in opium poppy: An update. *Phytochemistry Reviews*, *18*(6), 1457–1482.

Smith, S. D., Angelovici, R., Heyduk, K., Maeda, H. A., Moghe, G. D., Pires, J. C., et al. (2019). The renaissance of comparative biochemistry. *American Journal of Botany*, *106*(1), 3–13.

Sterling, T., & Irwin, J. J. (2015). ZINC 15- -ligand discovery for everyone. *Journal of Chemical Information and Modeling*, *55*(11), 2324–2337.

Sun, Y., Shang, L., Zhu, Q. H., Fan, L., & Guo, L. (2022). Twenty years of plant genome sequencing: Achievements and challenges. *Trends in Plant Science*, *27*(4), 391–401.

Trott, O., & Olson, A. J. (2010). AutoDock Vina: Improving the speed and accuracy of docking with a new scoring function, efficient optimization, and multithreading. *Journal of Computational Chemistry*, *31*(2), 455–461.

Verdonk, M. L., Chessari, G., Cole, J. C., Hartshorn, M. J., Murray, C. W., Nissink, J. W., et al. (2005). Modeling water molecules in protein-ligand docking using GOLD. *Journal of Medicinal Chemistry*, *48*(20), 6504–6515.

Villoutreix, B. O., Renault, N., Lagorce, D., Sperandio, O., Montes, M., & Miteva, M. A. (2007). Free resources to assist structure-based virtual ligand screening experiments. *Current Protein & Peptide Science*, *8*(4), 381–411.

Waterhouse, A., Bertoni, M., Bienert, S., Studer, G., Tauriello, G., Gumienny, R., et al. (2018). SWISS-MODEL: Homology modelling of protein structures and complexes. *Nucleic Acids Research*, *46*(W1), W296–W303.

Wisecaver, J. H., Borowsky, A. T., Tzin, V., Jander, G., Kliebenstein, D. J., & Rokas, A. (2017). A global Coexpression network approach for connecting genes to specialized metabolic pathways in plants. *The Plant Cell*, *29*(5), 944–959.

CHAPTER FOUR

Plant metacaspase: A case study of microcrystal structure determination and analysis

Haijiao Liu[a,†], Ping Zhu[b,†], Qingfang Zhang[c], Eric Lam[d], and Qun Liu[a,b,c,*]

[a]Department of Material Science and Chemical Engineering, Stony Brook University, Stony Brook, NY, United States
[b]Biology Department, Brookhaven National Laboratory, Upton, NY, United States
[c]Department of Biochemistry and Cell Biology, Stony Brook University, Stony Brook, NY, United States
[d]Department of Plant Biology, Rutgers, The State University of New Jersey, New Brunswick, NJ, United States
*Corresponding author: e-mail address: qunliu@bnl.gov

Contents

† These authors contributed equally.

Methods in Enzymology, Volume 676
ISSN 0076-6879
https://doi.org/10.1016/bs.mie.2022.07.026

Abstract

Metacaspases are highly conserved in plants and play essential roles in mediating programmed cell death, biotic and abiotic stress responses, and damage-induced innate immunity. Ca^{2+} signaling induced by plant damage leads to activation of metacaspase 4 from *Arabidopsis thaliana* (*At*MC4), which subsequently processes a plant elicitor peptide to trigger downstream immuno-response. To understand the structural basis of *At*MC4 activation by Ca^{2+}, we previously determined its crystal structure and performed in-crystal Ca^{2+} treatment to probe activation-associated conformational changes. To enable structure determination and in-crystal Ca^{2+} activation analysis, we used microcrystals and related methods which were essential for our successful approach. Here we describe in detail the methods that we used for determination of *At*MC4 structure using single-wavelength isomorphous replacement with anomalous signals assembled from 22 microcrystals. We also describe the method for in-crystal Ca^{2+} soaking, microcrystal data collection, data assembly and analysis to obtain the activated structure of *At*MC4 from 91 micro-sized crystals. The described methods may be useful to study other plant metacaspases and more broadly other plant enzymes for their structure determination and in-crystal functional characterization.

1. Introduction

Caspases (cysteine-dependent aspartate-specific proteases) belong to clan CD family 14 proteases and play essential roles in mediating metazoan cell death mechanisms that activate apoptosis and pyrotosis (Ramirez & Salvesen, 2018). Family 14 cysteine proteases have diverse domain organizations with a core caspase domain consisting of a large p20 domain and very often a smaller p10 domain. In addition to the core caspase domain, most family 14 proteases have a regulatory domain or domains before, after, or in the middle of the core caspase domain. In addition, caspases can assist to induce inflammation which is a protective mechanism in response to tissue damage or pathogen infection. Since identification of the first Caspase-1 in 1993, structure and function of caspases have been well studied (Salvesen & Dixit, 1999).

Caspases can be classified as executioners and initiators where executioner caspases are activated by proteolytic cleavage by initiator caspases that are upstream in the signaling cascade (Shi, 2004). Activation of caspases requires dimerization in initiator caspases, which first undergo autolytic cleavage followed by processing of executioner caspases for their activation. Activated executioner caspases process numerous cellular components to trigger cellular morphology changes and molecular responses. Programmed cell death is a tightly controlled process for development as well as for response to biotic and abiotic stresses in multi-cellular organisms. In animals, caspases, upon activation by various upstream signaling events, process an array of protein substrates that lead to the execution of cell death (Jorgensen, Rayamajhi, & Miao, 2017). Upon activation through dimerization, it specifically cleaves substrates after an aspartate at the P1 position. Under resting conditions, caspases are called zymogens or pro-caspases. Upon activation, a cleavage between the p20 and p10 domains enables reorganization of loops around the active site for caspase activation and substrate processing.

It is intriguing that caspases exist ubiquitously in metazoans, but not in prokaryotes, protozoa, algae, and plants. In 2000, 7 years after the identification of caspases, structure-based searches in assembled genomes across multiple phyla identified metacaspases in plants, fungi, and protozoa; and paracaspases in metazoans and the slime mold Dictyostelium (Uren et al., 2000). Like caspases, both metacaspases and paracaspases contain a p20-like domain with the conserved active site cysteine, histidine dyad. However, metacaspases have two unique features compared to caspases. Firstly, metacaspases recognize either a lysine or arginine in the P1 position of the target substrate and thus prefer to cleave after a basic residue instead of an aspartate (Vercammen et al., 2004). Secondly, metacaspases have either a proline/glutamine rich motif or zinc finger-like domain in the type I, or a long linker domain between p20 and p10 domains in the type II enzymes. Whereas in the more recently discovered type III metacaspases, the positions for the p10 and p20 domains are swapped, along with a shorter linker than that found in type II separating these two domains (Klemencic & Funk, 2018). Recently, a new type of metacaspase, orthocaspases, was identified in prokaryotes and shown to be enzymatically active (Klemenčič, Novinec, & Dolinar, 2015). Orthocaspases contain only a recognizable p20 domain and are also Arg/Lys-specific for cleavage site specificity. Evolutionarily, it is postulated that type I metacaspases may be more ancestral and both caspases and metacaspases could share a common origin.

Functional roles of metacaspases in analogy to their caspase counterparts were recently reviewed in detail (Basak & Kundu, 2022). Plant metacaspases were identified to display equivalent roles in mediating cell deaths in plant development as well as in resistance to insect and pathogen attacks (Coll et al., 2010; Hander et al., 2019; Suarez et al., 2004). As an example, the model plant *Arabidopsis thaliana* has nine metacaspases (*At*MC1–9). *At*MC1 and *At*MC2 are type I metacaspases and their loss-of-function mutants have shown quantitative effects in programmed cell death activation by oxidative stress (Coll et al., 2010). *At*MC1 could also interact with Sm-like4 (LSM4) to regulate pre-mRNA splicing and modulate immunity (Wang et al., 2021). *At*MC4–9 are type II metacaspases. *At*MC4 and its orthologs have reported to mediate cell death activation during plant embryogenesis (Suarez et al., 2004), response to biotic and abiotic stresses (Watanabe & Lam, 2011a), damaged-induced immune response (Hander et al., 2019), and stress-induced remodeling of nuclear stress granules (Sundström et al., 2009; Vandana et al., 2018); while *At*MC9 is involved in post-mortem clearance of xylem elements (Bollhöner et al., 2018) and ozone induced cell death through cleavage of GRIM REAPER and PRK5 (Pollen-specific receptor-like kinase 5) activation (Wrzaczek et al., 2015).

Differently from caspases that are activated by initiator caspases (Salvesen & Dixit, 1999; Shi, 2004), there are no known proteases responsible for metacaspase activation. Instead, most metacaspases require Ca^{2+} for autolytic zymogen activation through cleavage followed by activation and maturation to process substrates (McLuskey et al., 2012; Wen et al., 2013; Wong, Yan, & Shi, 2012). In Arabidopsis, except *At*MC9 whose activation is pH dependent, *At*MC1–8 requires calcium for their activation (Klemencic & Funk, 2018; Minina, Stael, Van Breusegem, & Bozhkov, 2014). In *At*MC4, a lysine residue at position 225 was identified to be critically important for its Ca^{2+} activation to mediate oxidative stress-induced cell death (Watanabe and Lam, 2011b). Compared to well-studied caspases, activation mechanism remains elusive for Ca^{2+}-dependent metacaspases (Minina, Coll, Tuominen, & Bozhkov, 2017; Vercammen et al., 2004;

Watanabe & Lam, 2011b). Although caspases and metacaspases share common ancestors, they have evolved divergently and do not co-exist in the same organisms (Klemenčič & Funk, 2019).

The structures of two type I metacaspases from *Trypanosoma brucei* and yeast (McLuskey et al., 2012; Wong et al., 2012) have been previously determined. However, the two structures do not contain the linker domain which is specific to type II metacaspase and has regulatory roles in mediating metacaspase activation (Fortin & Lam, 2018). In addition, there was no structural information to understand the autoinhibitory and Ca^{2+}-dependent activation mechanism. We therefore grew *At*MC4 crystals of its wild-type and its C139A catalytically inactive mutant and determined their structures using X-ray crystallography (Zhu et al., 2020). The C139A variant crystals form thin (~5–10 μm) plates stacking on each other with multiple layers. So, we used microdiffraction beamlines at NSLS-II to separate signals from different crystal plates by diffraction and determined its structure using anomalous signals enhanced from multiple crystals. In addition, to study conformational changes associated with Ca^{2+} activation, we developed an in-crystal Ca^{2+} soaking protocol using wild-type *At*MC4 crystals of less than 10–20 μm in thickness. Both structure determination and in-crystal activation methods used tiny crystals and micro-sized beam at the synchrotron beamlines, which may be of general interest to study other plant enzymes.

In this chapter, we describe our experimental procedure for using microcrystals and micro-crystallography to determine crystal structure of *At*MC4 and to study Ca^{2+}-triggered conformational changes in microcrystals. The procedure may be adapted to solve other challenging plant protein structures and to study conformational changes associated with protein function.

2. *At*MC4 expression and purification

This section introduces the expression and purification of wide-type *At*MC4 and its C139A inactive mutant. To facilitate protein purification, both genes were cloned into pET23a with a C-terminal 6xHis-tag. There are two main purification steps. The first step is using His-tag resin to perform affinity purification. The second step is using a Superdex-200 10/300 GL column to perform size-exclusion chromatography (SEC) which further purify proteins based on their molecular weights and shapes.

According to the SEC profile and the SDS-PAGE, fractions containing purified protein can be pooled together, concentrated, and store at −80 °C for further use.

2.1 Equipment

- Shaker (Innova 44, New Brunswick Scientific).
- Homogenizer (EmulsiFlex-C3, Avestin).
- Refrigerated centrifuge (Centrifuge 5810R, Eppendorf).
- Ultracentrifuge (Optima L-100K, Beckman).
- Nanodrop spectrophotometer (NanoDrop 1000, Thermo Scientific).
- HisTrap FF column (Cytiva, Marlborough, MA).
- Superdex-200 10/300 GL column.
- Amicon Ultra-30 centrifugal filter (Milipore, Inc.).
- Water Bath.
- Incubator.
- UV–visible spectrometer.

2.2 Reagents

- Sterile LB medium.
- Ampicillin stock at 100 mg/mL.
- Chloramphenicol stock at 34 mg/mL.
- Agar-LB plate with 100 μg/ml ampicillin and 34 μg/ml chloramphenicol.
- *Escherichia coli* BL21 (DE3) pLysS (ThermoFisher Scientific Cat# C606010).
- Isopropyl β-D-1-thiogalactopyranoside (IPTG).
- 4–20% gradient precast SDS-PAGE gels (Genscript Cat# M42015).
- Lysis buffer (25 mM Tris, pH 7.6, 250 mM NaCl, 0.5 mM TCEP (tris 2-carboxyethyl phosphine), 5% (*v*/v) glycerol and protease inhibitors).
- Buffer A (25 mM Tris, pH 7.6, 250 mM NaCl, 0.5 mM TCEP, 5% (v/v) glycerol, protease inhibitors and 10 mM imidazole).
- Buffer B (25 mM Tris, pH 7.6, 250 mM NaCl, 0.5 mM TCEP, 5% (v/v) glycerol, protease inhibitors and 500 mM imidazole).
- Buffer C (20 mM HEPES, pH 8.0, 300 mM NaCl and 0.1 mM TCEP).

2.3 Procedure

1. Prepare LB agar plates with appropriate concentration of antibiotics.
2. Add 1 μL *At*MC4 plasmid (75 –150 ng/μL) to 6 μL *E. coli* BL21 (DE3) pLysS competent cells, incubate on ice for 10 min, heat in 42 °C water

bath for 50 s, and immediately incubate on ice for 2 min. Add 40 μL SOC medium and incubate in 37 °C incubator for 1 h, then spread all solution onto an agar plate and incubate it in a 37 °C incubator overnight.

3. Transfer and culture 3–5 single colonies from the overnight agar plate to LB medium tubes contain 100 μg/mL ampicillin and 34 μg/mL chloramphenicol antibiotics.
4. Shake 37 °C overnight, dispense 10 mL cell solution to inoculate 1 L LB medium in a 2.8 L flask, shake at 250 rpm at 37 °C until OD_{600} reaches 0.4–0.6.
5. Add 0.4 mM IPTG (final concentration) to induce protein overexpression at 22 °C for 3–6 h.
6. Centrifuge cells at 5000 × *g* for 10 min and resuspend pelleted cells in lysis buffer.
7. Lyse cells using EmulsiFlex-C3 Homogenizer at 15000 psi with 4–5 passages. Then centrifuge at 18,000 × *g* at 4 °C for 1 h and collect supernatant.
8. Purification by nickel-nitrilotriacetic acid affinity (Ni^{2+}-NTA) chromatography. Add Ni^{2+}-NTA resin into supernatant and incubate for 1–2 h with rotation or stir at 4 °C, pass the supernatant with resin through a disposable column, and wash 3 bed volume each of imidazole concentrations from 10 to 150 mM as listed in Table 1
9. Analyze the Ni^{2+}-NTA purified protein using SDS-PAGE to determine the level of protein expression and purity (Fig. 1A).
10. Using an Amicon Ultra-30 centrifugal filter to concentrate purified protein to about 300 μL.

Table 1 Imidazole concentration sets.

Imidazole conc (mM)	Buffer A (mL)	Buffer B (mL)
10	4	0
30	3.837	0.163
60	3.592	0.408
90	3.347	0.653
120	3.102	0.897
150	2.857	1.143

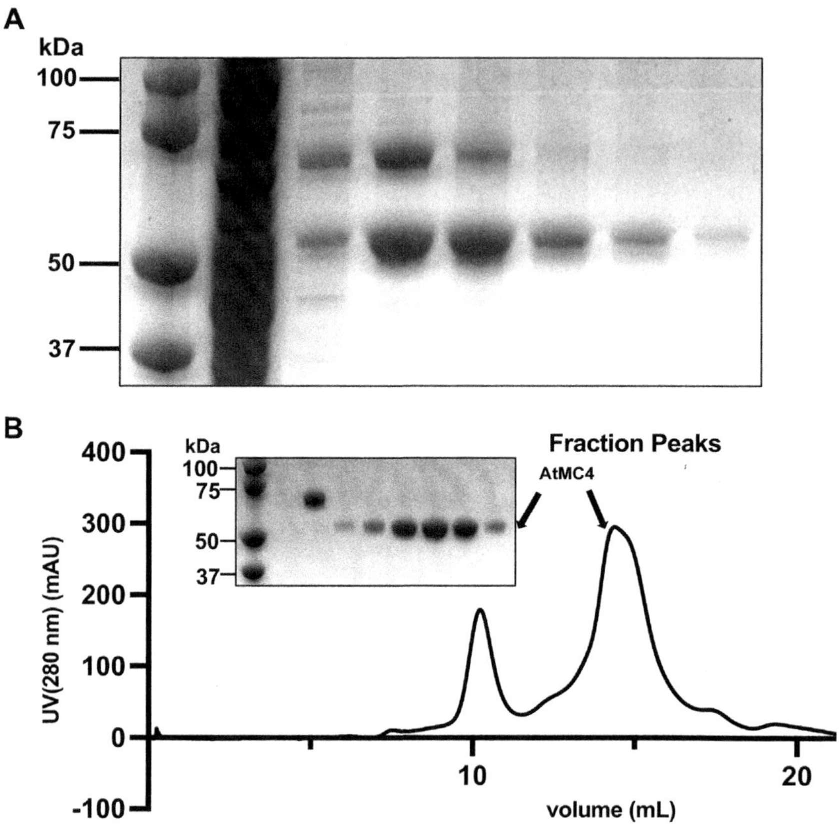

Fig. 1 Purification of *At*MC4. (A) SDS-PAGE analysis of Ni^{2+}-NTA purified protein fractions. (B) Purification of *At*MC4 by size exclusion chromatography (SEC). Inset is the SDS-PAGE gel of SEC profile peaks. Purified *At*MC4 is indicated by arrows.

11. Load the sample to Superdex-200 10/300 GL column preequilibrated in buffer C. Collect fractions containing the protein.
12. To determine the level of protein purity, take peak fractions for SDS-PAGE analysis (Fig. 1B).
13. Combine fractions containing *At*MC4, concentrate to 20 mg/ml and storage at −80 °C for future use.

2.4 Notes

1. At room temperature, the pH of Tris buffer is 7.6. However, in the cold room, its pH is around 8.0.
2. LB-agar plates are stored at 4 °C to prevent bacterial contaminants. Prior to use, it is suggested to warm the LB-agar plates in a 37 °C incubator for 30 mins.

3. *At*MC4 may be activated and cleaved during its expression in *E. coli*. To slow down the basal cleavage, the temperate is lowered to 22 °C prior to addition of IPTG for induction of protein expression. In addition, the expression time was limited to 3–4 h. Otherwise, *At*MC4 may be self-processed to small fragments.
4. Wild-type *At*MC4 may co-purify with a contaminant protein of a molecular weight at about 70 kDa (Fig. 1A). This contaminant protein can be removed from the *At*MC4 preparation by SEC (Fig. 1B).

3. Selenomethionine protein expression and purification

For de novo protein structure determination, anomalous diffraction from scatterers in protein crystals are typically used (Rice, Earnest, & Brunger, 2000). Selenomethionine substitution (SeMet) has been routinely used for anomalous phasing (Doublie, 2007; Walden, 2010). In this section, we describe the expression and purification of SeMet-labeled *At*MC4-C139A for crystallization and structure determination. The plasmid for expression is the same as unlabeled *At*MC4-C139A vector with a C-terminal 6xHis-tag for purification. In general, cells in the presence of SeMet grow slowly. Therefore, the expression temperature, time, and IPTG concentration should be optimally performed for maximizing protein yield. Then, a three-step purification approach by nickel-nitrilotriacetic acid affinity, ion exchange, and size-exclusion chromatographies can be applied for achieving homogeneous protein fractions with high purity. SDS-PAGE analysis is typically used to check protein purity.

3.1 Equipment

- Shaker (Innova 44, New Brunswick Scientific).
- Homogenizer (EmulsiFlex-C3, Avestin).
- Refrigerated centrifuge (Centrifuge 5810R, Eppendorf).
- Ultracentrifuge (Optima L-100K, Beckman).
- Nanodrop spectrophotometer (NanoDrop 1000, ThermoFisher Scientific).
- FPLC system (NGC, Bio-Rad).

3.2 Reagents

- Sterile LB medium (10 g/L tryptone, 10 g/L NaCl, and 5 g/L yeast extract).

- Sterile M9 medium (33.7 mM Na_2HPO_4, 22.0 mM KH_2PO_4, 8.55 mM NaCl, 9.35 mM NH_4Cl, 0.4% glucose, 1 mM $MgSO_4$, 0.3 mM $CaCl_2$, 1 μg biotin, 1 μg thiamin, 1 × trace elements).
- Lysis buffer (25 mM Tris, pH 7.6, 250 mM NaCl, 0.5 mM TCEP, 5% glycerol, and protease inhibitors).
- Ampicillin stock at 100 mg/mL.
- Chloramphenicol stock at 34 mg/mL.
- Isopropyl β-D-1-thiogalactopyranoside (IPTG).
- *E. coli* BL21 (DE3) pLysS (ThermoFisher Scientific Cat# C606010).

3.3 Procedure

1. Culture *E. coli* BL21 (DE3) pLysS cells carrying plasmids in M9 minimum medium at 37 °C for 3–6 h to reach OD_{600} of 0.4–0.6.
2. Add amino acid powder mix (100 mg/L of lysine, threonine, phenylalanine, 50 mg/L of leucine, isoleucine, valine, selenomethionine) to cell culture and incubate the culture at 16 °C for 1 h.
3. Add IPTG to a final concentration of 0.2 mM and culture the cells at 16 °C for 20 h.
4. Harvest cells by centrifugation at 5000 × *g* at 4 °C for 10 mins.
5. Resuspend cells in lysis buffer and lyse cells through 4–5 times passage with a homogenizer at 15,000 psi.
6. Centrifuge cell lysates at 18,000 × *g* at 4 °C for 1 h and collect supernatants.
7. Transfer the supernatants onto a Ni^{2+}-NTA affinity chromatography column, wash with 10 column volume of lysis buffer, and elute the protein using lysis buffer supplemented with 300 mM imidazole
8. Transfer the collected eluate onto an ion-exchange chromatography column and collect the eluate as fractions.
9. Check the fractions using SDS-PAGE to identify fractions containing *At*MC4-C139A protein.
10. Concentrate the fractions of target protein to 300 μL and load it onto a Superdex 200 column, collect protein fractions, and concentrate to ~20 mg/ml for crystallization. Fig. 2 is representative gel-filtration profile and SDS-PAGE for the purified SeMet *At*MC4-C139A protein.

3.4 Notes

- The amino acid powder should be mixed thoroughly prior to adding the mixture to the cell culture.

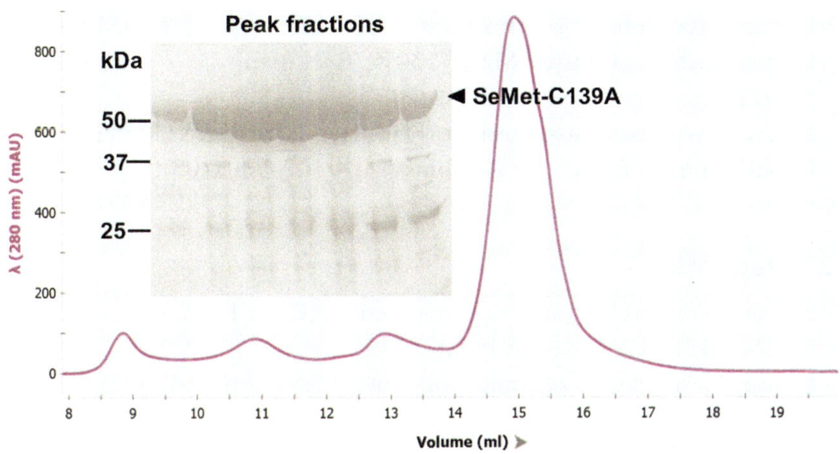

Fig. 2 Purification of *At*MC4-C139A. The SEC profile is shown as a magenta line. Peak fractions were used for SDS-PAGE analysis (inset).

- In addition to using the three-step purification procedure, one may use step-wise washes for Ni^{2+}-NTA purification and skip the ion-exchange step

4. Crystallization of *At*MC4 proteins

Protein crystallization is a sensitive process and is mediated by complicated physicochemical-parameters such as pH, temperature, pressure, humidity, and microenvironment (Chayen & Saridakis, 2008). Among these, protein concentration, precipitant concentration, pH, and temperature need to be primarily screened. This section describes the crystallization of *At*MC4 wild-type (WT), C139A mutant, and SeMet-C139A proteins. Initial crystallization screenings are performed by using well-defined commercial screen kits and a high-throughput crystallization robot. Primary small crystals are identified by using a light microscope. In later steps of crystallization optimization, a combined approach by setting various precipitant concentrations, buffer pH, detergents, and additives is applied for achieving diffraction-quality crystals. Finally, crystals with largest sizes are obtained and used for diffraction data collection and structure determination.

4.1 Equipment

- Crystallization robot (Gryphon, Art Robbins).

- Refrigerated benchtop centrifuge (Centrifuge 5810R, Eppendorf).
- Ultracentrifuge (Optima MAX-130 K, Beckman).
- Light microscope (Leica MZ75).
- VDX 24 well plate with sealant (Hampton Research Cat# HR3-171).
- Intelli-Plate 96 LVR (Art Robbins Cat# 102-0001-03).

4.2 Reagents

- Hampton research crystallization screen kit (Cat# HR2-086, HR2-130, HR2-134, HR2-136, and HR2-139).
- Hampton research optimization kit (Cat# HR2-138, HR2-241, HR2-407, and HR2-541).

4.3 Procedure

1. Prepare fresh WT, C139A, or SeMet-C139A proteins at various concentrations of 10, 20, 30 mg/mL and keep on ice until use.
2. Ultracentrifuge protein of interest at 50,000 × *g*, 4 °C, for 30 min before setting up crystallization trays.
3. Set up sitting drop crystallization screenings by mixing 0.2 μL protein with an equal volume of precipitant by a crystallization robot and incubate the plates at 4 °C and 18 °C for crystal growth.
4. Check for crystal growth from initial crystallization hits and optimize promising conditions using hanging drop method. Fig. 3 shows typical crystals of *At*MC4 C139A, wild-type and SeMet-C139A crystals.
5. Transfer crystals to new drops of mother liquor supplemented with 10% glycerol as cryoprotectant and then plunge freeze into liquid nitrogen for cryogenic data collection.

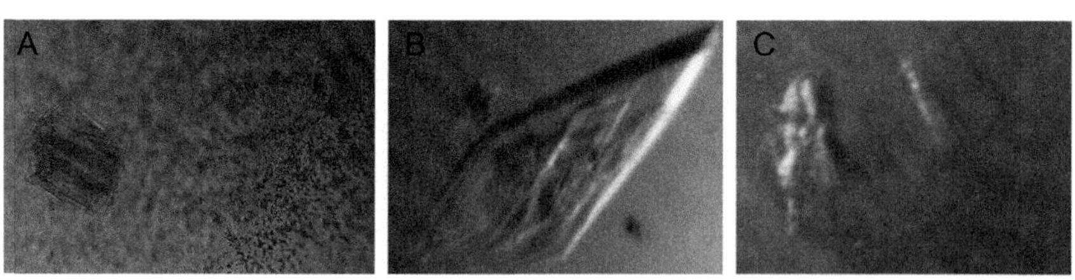

Fig. 3 Typical *At*MC4 crystals. (A) C139A mutant. (B) Wild-type. (C) SeMet C139A.

4.4 Notes

Since optimal crystallization reagents are prepared in-house, a 2-dimensional grid screen around initial crystallization conditions is recommended for crystal optimization.

In our experience, wild-type *At*MC4 crystals have poorer diffraction with a long-term growth at 18 °C. Only fresh crystals (1–2 days after reaching full size) are recommended for data collection.

5. Diffraction data collection and processing of *At*MC4-C139A

With the third generation synchrotrons and associated micro-diffraction beamlines, synchrotron micro-crystallography using high intensity X-rays of a few microns has been routine (Yamamoto et al., 2017). Together with photon-counting pixel CMOS detectors such as EIGER 9M and 16M from DECTRIS (Casanas et al., 2016), shutterless data collection is possible with exposure time down to a few microseconds. The AMX beamline at National Synchrotron Light Source II (NSLS-II) is a microdiffraction beamline with a beam size of $7 \times 5\,\mu m^2$ and ultra-brightness of $\sim 4 \times 10^{12}$ ph/s. *At*MC4 C139A mutant preferentially grows as stacking thin plates with a thickness of 5–10 μm (Fig. 3A). Even though we attempted to dissect those plates, many times we found that small pieces still contain multiple layers of crystals. Using the micro-sized beam at AMX, we collected multiple-lattice data sets from the layered crystals. Due to a small angle between the layers, these lattices do not overlap and can be indexed and integrated as individual data sets. DIALS can process multiple-lattice data through refinement of multiple orientation matrices and unit cell parameters (Waterman et al., 2016). Therefore, we used DIALS to process C139A data. With the processed data, we performed rejection of outlier crystals and frames following the procedure that we have developed for enhancing anomalous signals for structure determination which is implemented in PyMDA (Takemaru et al., 2020). PyMDA assembles diffraction data from radiation damaged and incomplete data to produce statistically-compatible scaled and merged data using DIALS (Winter et al., 2018) and CCP4 program AIMLESS (Winn et al., 2011). In this section we describe the procedure to collect and process micro-diffraction data from layered plate crystals.

5.1 Software

- lsdcGUI: Data collection program at NSLS-II AMX beamline.
- DIALS: Diffraction Integration for Advanced Light Sources (https://dials.github.io).
- CCP4: Collaborative Computational Project No.4. (https://www.ccp4.ac.uk).
- PyMDA: A python script for assembling diffraction data from multiple crystals. (https://github.com/qun-liu/PyMDA).

5.2 Procedure

1. SeMet crystals are mounted onto the AMX beamline using a sample loading robot at the beamline.
2. Do X-ray energy scan for fluorescence signals of selenium in the crystal. Set the X-ray energy above the peak on the X-ray fluorescence spectrum, e.g., 0.94985 Å or 13.053 keV.
3. Set detector to sample distance at 300 mm, exposure time 0.02 s, and rotation range of 0.3° per frame.
4. Using raster scanning with a step size of 5 μm to cover the entire crystal(s) to generate a heat map. Two orthogonal rasters are used to accurately position the crystal in the beam.
5. Based on the heat map (greener color means stronger diffraction), define a vector for data acquisition.
6. Collect rotational vector data of 180° from each crystal (600 frames each crystal).
7. Use one data set for index and integration using DIALS to obtain space group and unit-cell parameters.
8. Batch process the data in PyMDA with space group P222, derived unit cell dimension "109 215 40 90 90 90", and accumulative wedges of 50-frame increasement, i.e., 50, 100, 150, 200, ..., 600.
9. Reindex and integration using multi-lattice in DIALS. Use three lattices for each multi-layered data.
10. Perform crystal and frame rejection in PyMDA with resolution cutoff at 3.0, 3.5, and 4.0 Å, crystal rejection at 1, and without frame rejection.
11. Select the assembled data with the highest anomalous signals for phasing, reported as "DelAnom correlation between half-sets" in the AIMLESS log files.

5.3 Notes

1. In general, the X-ray wavelength used for SAD data collection is performed at the peak of the absorption edge from the X-ray fluorescence

spectrum scanning. To accommodate inaccuracy of the monochromator at a synchrotron beamline, the X-ray energy can be slightly above the absorption edge.

2. Selection of exposure time and number of frames to be collected is empirical. The rationale is to collect > 200 frames (60° of data) without significant radiation damage based on a pilot data collection. That is, we should see diffraction spots to a resolution of about 4.0 Å from the last data frame of the pilot data. With the limited number of crystals, it's better to collect as much data as possible from each crystal.
3. Raster-step size should be comparable to beam size. Otherwise, we may miss the region of high-resolution diffraction due to insufficient sampling.
4. The number of crystals needed may not be known ahead of time. For SeMet C139A, we collected data from 22 crystals that we obtained.
5. To get crystallographic parameters such as unit cell parameters and space group, one can also use programs such as HKL3000 (Minor, Cymborowski, Otwinowski, & Chruszcz, 2006), XDS (Kabsch, 2014), and MLSFLM (Leslie & Powell, 2007) for indexing. We choose DIALS as it can process multiple lattice data in one run.
6. Due to formation of layered plate crystals, it's challenging to collect a single-crystal dataset without diffraction from other layers of crystals. So, we collected diffraction data with the microdiffraction beam. Diffraction spots were well-separated and can thus be processed as separated data sets using DIALS.
7. Data assembly may be optimized using program PyMDA. The resolution limit, the steps of crystal rejection, and the frame rejection ratio may be optimized to improve anomalous signals.
8. When crystals are too small for vector-data collection, standard rotational data collection can be used to collect 30–50° of data from each crystal. These data may be processed and combined with vector data using PyMDA.

6. Structure determination of *At*MC4-C139A by SIRAS phasing

Crystal structures may be determined by molecular replacement when there is a structure homolog that has been previously determined. Otherwise, a structure will be determined with experimental phases. SeMet-substitution (Hendrickson, Horton, & Lemaster, 1990) and the use of anomalous signals collected at the Se—K edge have been dominant methods for multiple-wavelength anomalous diffraction (MAD) (Hendrickson, 1999) and single-wavelength anomalous diffraction (SAD) phasing. When such anomalous signals are not strong enough, enhanced anomalous signals may be obtained by combining data from multiple crystals (Liu, Zhang, & Hendrickson, 2011) or by combining phasing information by treating SeMet data as a heavy atom derivative of native data for SIRAS (Single Isomorphous Replacement with Anomalous Scattering) phasing (Nakane et al., 2016). For SIRAS phasing, a substructure is first determined by direct method that is implemented in SHELXD (Schneider & Sheldrick, 2002). Then, the substructure is refined and used for phasing in SHAPR/auotSHAPR (Vonrhein, Blanc, Roversi, & Bricogne, 2007). To improve the phase accuracy and map quality, isomorphous signals between the native and derivative data may be combined as implemented in SHARP/autoSHARP. In this section we describe the Se substructure determination using program SHELXD (Schneider & Sheldrick, 2002) and use of the determined substructure for phasing in SHARP/autoSHARP (Vonrhein et al., 2007) by SIRAS phasing, structure model building in COOT (Emsley, Lohkamp, Scott, & Cowtan, 2010) and refinement in PHENIX (Adams et al., 2010).

6.1 Software

- SHELXD: Use direct method for substructure determination (https://shelx.uni-goettingen.de).
- PHASER: Use SAD data in phaser for phasing (Read & McCoy, 2011).
- SHARP/autoSHARP (https://www.globalphasing.com/sharp).
- PHENIX: A comprehensive software package for macromolecular structure determination (https://phenix-online.org).
- COOT: macromolecular model building, model completion and validation (https://www2.mrc-lmb.cam.ac.uk/personal/pemsley/coot).

6.2 Procedure

1. Select the assembled data with the highest anomalous signals with a resolution cutoff at 4.0 Å.
2. Perform 2000 cycles of SHELXD searches with resolution cutoff at 4.5–6.5 Å with a step of 0.1, Emin cutoff 1.3–1.7 with a step size of 0.1, and search for 12 expected Se sites.
3. Check SHELXD search results using CCall/CCweak to identify the ones with CCall >30 and CCweak >12.

4. Refine substrate coordinates in PHASER using SAD phasing to complete substructure with options of "LLG COMPETE SIGMA 5.0" and "NCYC 100".
5. Perform SIRAS phasing in SHARP/autoSHARP with model building. The refined substructure from PHASER is used.
6. Check for output logfile LISTautoSHARP.html and look for the record of the number of sequenced residues.
7. Visualization of the model and maps to verify the correctness of automatically built model.
8. Perform manual model building in COOT to compete the model, fix misassigned segments and missing loops, and add water molecules.
9. Refine the model using PHENIX with automated search for NCS restraints. Refinement resolution is at 2.8 Å using the high-resolution native data.
10. Repeat 8 and 9 until the refinement is converged.
11. Validate the structure with MOLPROBITY (Williams et al., 2018) and correct the geometry outliers in COOT (Casañal, Lohkamp, & Emsley, 2020).
12. Final iteration of refinement with appropriate weights and validate the structure again for stereochemistry.

6.3 Notes

1. During the data assembly stage, more than one assembled dataset may have good anomalous signals. It is thus worth trying all datasets for substructure determination by SHELXD searches. In SHELXD searches, it is advantageous to use an assembled data at a lower resolution, for example 4.0 Å for substructure determination. With a lower-resolution cut-off, more frames will be combined for improved multiplicity and low-resolution anomalous signals which are critical for substructure determination.
2. The use of PHASER to refine and complete substructure is optional. One can also use SHELXD substructure directly for SIRAS phasing in SHARP/autoSHARP.
3. SHARP/autoSHARP integrates SHELXD and CCP4 programs for phasing and model building. In some cases, the build-in SHELXD strategy may work well to determine substructure. In our case, we separated the substructure determination from phasing and model building because we had to perform SHELXD searches in a more customized fashion.

4. When performing SHELXD and SHARP/autoSHARP searches against different datasets, using a multi-core computer or even a computer cluster will help to speed up the search process. The parallelization can be performed by a batch-queue job management system such as SLURM, SGE, or PBS which is dependent on a specific computer cluster. Minimally, one should make use of multiple cores for parallel SHELXD searches and perform SHARP/autoSHARP searches for all promising substructures found by SHELXD.
5. SHARP/autoSHARP has the built-in model building module using BUCCANEER (Cowtan, 2006) which worked quite well with *At*MC4 at 2.8 Å resolution. If the resolution is low and/or the phase error is high, the experimental density map may not allow for automated model building. Then, the electron density map from SHARP/autoSHARP may be used for manual model building in COOT.
6. In our structure determination, we used different datasets for substructure determination and phasing. Specifically, we used data processed to 4.0 Å for substructure determination where low-resolution anomalous signals are critical. For phasing, we used the high-resolution data processed at 3.0 Å using the previously determined and refined substructure. In addition, we used SIRAS phasing which included a native data processed at 2.8 Å. The high-resolution data of 2.8 Å helped to get the isomorphous difference of Se positions and benefited the high-resolution refinement. The refinement was performed using the 2.8-Å native data.

7. In-crystal Ca^{2+}-dependent *At*MC4 activation

Protein crystals contain about 30–80% water which maintains protein structure and likely also function in crystalline states. When crystals are large (> 100 μm), very often we observe cracking crystals and loss of diffraction associated with substrate or ligand soaking. It's also common to observe deteriorated diffractions when crystals are treated with cryoprotectants. Microcrystals of ~10 μm in size have unique properties for studying in-crystal conformational changes. Seeing conformational changes associated with protein function may be enabled using micro-sized crystals (Chang et al., 2014; Ishigami et al., 2019; Srajer & Schmidt, 2017; Stagno et al., 2017). When protein crystals are small enough, the diffusion time of substrate/ligand is expected to be in the millisecond range. It is then possible for substrate/ligand-triggered conformational changes to be synchronized without destroying the crystal lattice for structural analysis (Schmidt, 2020). The reduced diffusion time greatly reduces the stress during substrate/ligand soaking and may allow for large conformational changes to happen without destroying the lattice. Moreover, the synchronized structural movements in microcrystals may permit the capture of conformational states that are needed for a protein to function.

To optimize the conditions for Ca^{2+}-dependent *At*MC4 activation, we mixed *At*MC4 proteins with different $CaCl_2$ concentrations at room temperature and used SDS-PAGE to determine the best Ca^{2+}-dependent *At*MC4 cleavage. Besides, to study the Ca^{2+}-induced activation of *At*MC4, we crystallized wild-type *At*MC4. WT *At*MC4 crystals are thin and large plates. WT crystals exhibit serious anisotropy and diffracted X-rays to about 3.5–4.0 Å resolution. Ca^{2+} is a small divalent metal. To see Ca^{2+}-dependent *At*MC4 activation, we soaked pre-formed *At*MC4 crystals with $CaCl_2$. The Ca^{2+} treatment resulted in cracked crystals and cleavage after the inhibitory residue Lys225 (Zhu et al., 2020). Based on the solved crystal structure of *At*MC4, we hypothesize that after cleavage at the Lys225 site, strand β9 and loop L6 should move away from the active site so that a Lys/Arg residue from a substrate can gain access to the site for processing.

In this section, we use SDS-PAGE to determine optimal Ca^{2+} concentration to process *At*MC4 and describe the preparation and utilization of microcrystals for in-crystal activation by CCa^{2+}.To manipulate microcrystals for microdiffraction data collection, we used a pipette to transfer microcrystal slurries through gentle suction and put them on the custom-made MiTeGen well-mounts (Guo et al., 2019).

7.1 Equipment

- VDX 24 well plate with sealant (Hampton Research Cat# HR3-171).
- 22 mm siliconized glass cover slides (Hampton Research Cat# HR3-231).
- 4–20% gradient precast SDS-PAGE gels (Genscript Cat# M42015).
- Custom-made MiTeGen well-mounts for manipulation of micro-sized crystals (Guo et al., 2019).
- MiTeGen dual-thickness microloops with an aperture size of 50 or 100 μm.
- CrystalWand Magnetic (Hampton Research Cat# HR4-729).

- Cryogenic Foam Dewar (800 mL) (Hampton Research Cat# HR4-673).
- Universal V1-Puck (Uni-Puck, MiTeGen Cat# M-CP-111-021).
- Light microscope (LEICA MZ APO).
- 0.2–10 μL pipette.

7.2 Reagents

- Reaction solution (25 mM HEPES, pH 7.6, 250 mM NaCl, 0–50 mM $CaCl_2$, and 0.5 mM TCEP (tris(2-carboxyethyl)phosphine)).
- 50 mM EDTA.
- Soaking solution containing 100 mM sodium cacodylate, pH 6.4, 2.1 M ammonium sulfate, 10 mM $CaCl_2$, and 10% glycerol.

7.3 Procedure

1. Prepare reaction solutions with different $CaCl_2$ concentrations and soaking solution.
2. Add 5 μM of purified *At*MC4 into each 25 μL reaction solution tubes for 10 min at room temperature.
3. After 10 min, add 50 mM EDTA to stop reactions.
4. Pipette 20 μL solution from each tube for SDS-PAGE analysis, find optimal concentration of Ca^{2+} for self-cleavage
5. Transfer freshly grown wild-type *At*MC4 crystals to cold room. Use crystals in a day or two before they start to decay and lose diffraction.
6. In cold room, visualize crystal drops under a light microscope and select drops containing crystals.
7. Use a pipette to transfer 3–5 crystal-containing drops to a 22-mm glass slide.
8. Add three drops of soaking solution, 1 μL each on the same 22-mm slide.
9. Under a microscope, use Crystalwand and a 50 or 100 μm microloop to transfer crystals sequentially from crystallization drops to soaking drops with two times wash in the soaking solution.
10. Repeat step 5 and transfer 20–40 crystals to the soaking drop. Make sure crystals fill up the 1 μL soaking drop.
11. Smash crystals using the microloop by tapping crystals repeatedly to produce fragmented tiny crystals.

12. Invert the glass slide on one of the 24-well crystallization plate filled with crystallization precipitate, seal it, and move the plate out of the cold room.
13. Wait for 10 min for reaction to occur and move the plate back to cold room.
14. Wait for 5 min to let the plate cool down. Remove the glass slide from the crystallization well and put it under the microscope.
15. Put a Uni-Puck in the 800 mL foam dewar, fill the dewar slowly with liquid nitrogen so that the puck is 2–3 cm below the surface. Wait for the puck to cool down.
16. Withdraw 0.3–0.5 μL of microcrystal slurry using a pipette.
17. Using CrystalWand to hold a well-mount, dispense crystals on top of the well-mount, and use a filter paper to remove solvent from the bottom of the well-mount.
18. Rapid freeze micro-sized crystals by manually plunging the well-mount into liquid nitrogen.
19. Store the sample in liquid nitrogen for diffraction data acquisition.

7.4 Notes

1. Procedure 2–4 is for the optimization of Ca^{2+} concentration determination, 5–19 is for preparation and utilization of microcrystals for Ca^{2+} soaking.
2. When handling liquid nitrogen, wear gloves and safety glasses to prevent accidental spill-over and splashes of liquid nitrogen.
3. Wild-type *At*MC4 crystals will lose diffraction in about 3 days after their growth into full size. For best diffraction, it is suggested to prepare microcrystals for soaking with freshly produced crystals within a day.
4. We suggest using custom-made well-mounts by MiTeGen for convenient manipulation of microcrystals. These mounts were patterned with wells of ~10 μm, each with a 1–2 μm hole in the bottom which allow the removal of solvents from the bottom using the tip of a filter paper. Alternatively, one can use MiTeGen microloops with an aperture of 100 μm. Pay attention to dispense a thin layer of crystal slurries to the loop. Plunge freeze the crystal-containing microloop without paper blotting.
5. When there are large amounts of crystals to be smashed. One can transfer crystals to a microcentrifuge tube and use a seeding bead to create

microcrystals by vortexing. Care should be taken to not vortex too much which may cause extra damage to crystals. Smashed crystals may be concentrated by centrifuge at 500 × *g* for 5 min at 4 °C.
6. Concentration (10 mM of $CaCl_2$) and soaking time (10 min at room temperature) are empirical based on a pilot in-solution Ca^{2+}-activated *At*MC4 cleavage test.

8. Data assembly to reveal in-crystal enzyme activation

A Python program for microcrystal data assembly (PyMDA) is an effective program which can improve data quality for assembling multi-crystal datasets (Takemaru et al., 2020). There are three main steps to assemble multi-crystal data. The first step is using DIALS to process each single crystal datasets and using POINTLESS and AIMLESS in CCP4 (Winn et al., 2011) to scale these sets as wedges. The second step is using unit-cell parameters to classify these single crystal datasets to several classes. Step three is assembling datasets by running crystal and frame rejections. The assembled data will be used for further structure determination and refinement. For *At*MC4 microcrystals treated with Ca^{2+}, we used one dataset for indexing and integration using DIALS to obtain space group and unit cell parameters, processed the data in PyMDA with the obtained space group and performed crystal and frame rejection in PyMDA. From the comparison of *At*MC4 structures after data analysis and structure refinement, Ca^{2+} activated structure, i.e., cleavage after Lys225 and release/disordering of loop L6 and β9 from the active site, can be deduced. In this section, we describe the data collection, data assembly, and structure analysis.

8.1 Software

- lsdcGUI: Data collection program at NSLS-II FMX beamline.
- DIALS: Diffraction Integration for Advanced Light Sources (https://dials.github.io).
- CCP4: Collaborative Computational Project No. 4 (https://www.ccp4.ac.uk).
- PyMDA: A python script for assembly diffraction data from multiple crystals (https://github.com/qun-liu/PyMDA).

8.2 Procedure

1. Load crystals to the FMX beamline using the autoloader equipped with the beamline.

2. Set the X-ray energy at 6 keV, e.g., 2.0664 Å, detector to sample distance at 135 mm, beam size $4 \times 5\,\mu m^2$, exposure time 0.02 s, and rotation range of 0.3° per frame.
3. Align the well-mount so that its surface is perpendicular to incident X-ray beam. Write down the omega angle.
4. Using a raster scanning with a step size of 5 μm to cover the entire well-mount and produce a heat map.
5. Based on the heat map (greener color means stronger diffraction), select multiple spots, save their positions, and queue them for batch data acquisition.
6. Collect rotational data of 20° from each of selected crystals (67 frames per crystal). Set the rotation start angle as the recorded omega value subtracted by 10. Therefore, 20° of data will be collected from plus and minus 10° away from the aligned well-mount.
7. Process each of the single-crystal dataset using DIALS in PyMDA with 10 wedges and space group $p222_1$.
8. Run classification for 20 classes based on unit-cell variations.
9. According to clustering plot, delete outlier classes manually, and retain the classed in green region (Fig. 4A).
10. Run crystal rejection with a rejection of 10 crystals at a time, try different number of desired classes with a resolution cutoff at 3.2 Å.
11. Run frame rejection and try different number of desired classes with a decay factor of 3.0 and resolution cutoff at 3.2 Å.
12. Select the assembled data with the highest $CC_{1/2}$ values, reported as "Mn(I) half-set correlation CC(1/2)" in the aimless log files.
13. Use PHENIX program for refinement. The map shows a cleavage after Lys225 and the release of loop L6 and β9 in the Ca^{2+} activated *At*MC4 structure (Fig. 4B and C).

8.3 Notes

1. We collected the data at 6 keV with an expectation to see the anomalous signals from Ca^{2+} that could help to identify its binding site within *At*MC4. However, due to weak interactions between Ca^{2+} and *At*MC4, we did not observe any specific site associated with calcium binding. Therefore, the data collection may be performed at a higher energy.
2. With microcrystals, it is not possible to collect complete data from a single crystal. In addition, due to small crystal size and small beam size,

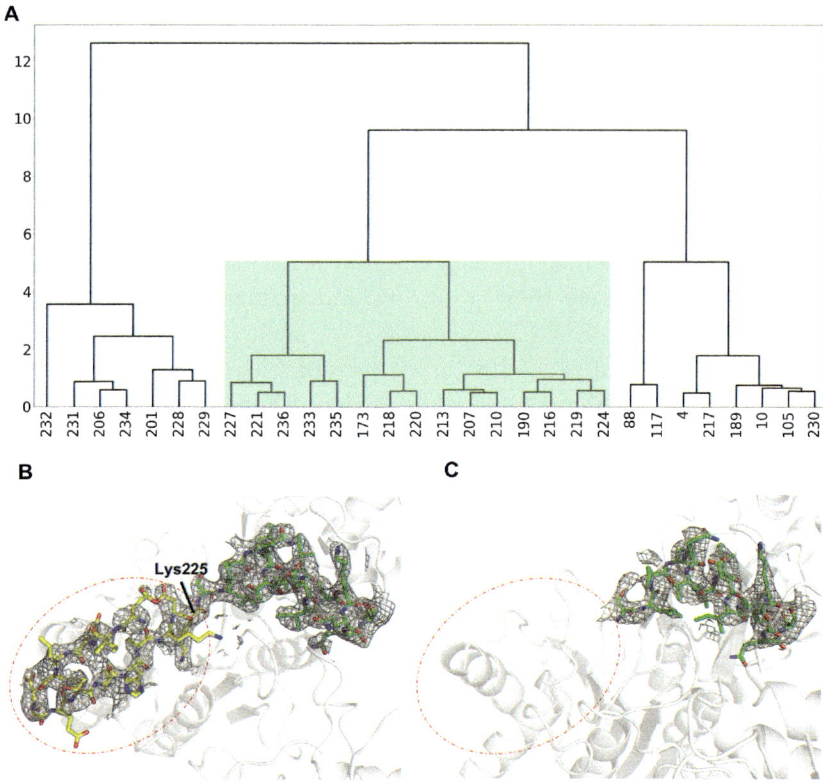

Fig. 4 Data analysis and comparison of *At*MC4 structure with its Ca^{2+}-activated structure. (A) Unit-cell variation analysis. (B) Inactivated *At*MC4 structure with Loop L6 and β9 (red circle). (C) Ca^{2+}-activated *At*MC4 structure without loop L6 and β9 (red circle).

X-rays may miss the crystal if a large rotation range is used for data collection. We tested the rotation range at the FMX beamline and found that collection of 20° of data (±10° around the aligned position) was accurate enough without missing crystals during rotation data collection. As a comparison, we found that collecting 30° of data will cause crystals to come in and out of the X-ray beam path.

3. Before Ca^{2+} treatment, the best diffraction from wild-type *At*MC4 crystals is about 3.5–4.0 Å. After Ca^{2+} treatment, crystals diffracted X-rays to higher resolution at about 3.0 Å associated with the change of unit cell dimensions from 126, 284, 58 Å to 50, 58, 300 Å and the change of space group from $P2_12_12$ to $P222_1$. The release of loop L6 and β9 from the active site may induce rearrangement of molecules in crystalline lattice with an improved packing.

9. Summary and conclusions

This chapter presents the application of micro-crystallography for structure determination of a challenging plant metacaspase *At*MC4 and in-crystal Ca^{2+}-activation to explore the associated conformational changes which was enabled by using micro-sized crystals. For structure determination, we expressed and purified an inactive C139A mutant protein with and without SeMet labeling. SeMet labeling provides phasing power through the combination of anomalous scattering and isomorphous signals. The anomalous signals from a single SeMet crystal were insufficient for structure determination. We thus combined data from 22 crystals which enhanced anomalous signals for structure determination by SIRAS phasing. Also for structure determination, we used three different datasets: low-resolution and highly redundant data for substructure determination at 4.0 Å; SIRAS phasing data at 3.0 Å; and structure refinement with a highest resolution native dataset at 2.8 Å. The strategy may be broadly applicable for structure determination of other challenging plant proteins.

The activation of *At*MC4 is strictly Ca^{2+} dependent. However, we couldn't obtain a *At*MC4/Ca^{2+} complex structure due to their weak interactions and Ca^{2+}-induced activation and self-cleavage in *At*MC4. To explore the activation process and associated conformational changes, we produced microcrystals of wild-type *At*MC4 which allowed in-crystal conformational changes without destroying their crystalline lattices. Through using microcrystals smashed from large crystals, we describe an in-crystal Ca^{2+} soaking procedure that enabled us to capture an activated conformation of *At*MC4 after cleavage of the inhibitory residue Lys225 and the release of the inhibitory L6 loop and β9 segments from the active site pocket. Subsequently, one can imagine that a substrate may be able to approach the active site and then be processed to trigger downstream signaling events (Hander et al., 2019). To observe conformational changes in their crystalline states, using microcrystals may be the key. Therefore, the described methods on microcrystal production, ligand/Ca^{2+} soaking, manipulation, micro-diffraction data collection, data assembly and analysis may be of broad applications to study a wide range of conformational changes in microcrystals of many plant proteins that could be crystallized.

Acknowledgments

We thank staff at the NSLS-II beamlines FMX and AMX for their assistance in data collection. P.Z. and Q.L. were supported by the U.S. Department of Energy (DOE), Office of Biological and Environmental Research, as part of the Quantitative Plant

Science Initiative at Brookhaven National Laboratory. H.L., Q.Z. and E.L. were supported by NSF grant IOS-2052997. The development of the structure determination method was supported in part by NIH grant RM107462. The work used National Synchrotron Light Source II (NSLS-II) which is supported in part by the U.S. DOE Office of Basic Energy Sciences under contract number DE-SC0012704. Beamlines FMX and AMX are supported by NIH P30GM133893 and by the DOE Office of Biological and Environmental Research.

References

Adams, P. D., Afonine, P. V., Bunkóczi, G., Chen, V. B., Davis, I. W., Echols, N., et al. (2010). PHENIX: A comprehensive Python-based system for macromolecular structure solution. *Acta Crystallographica Section D: Biological Crystallography*, *66*(2), 213–221.

Basak, S., & Kundu, P. (2022). Plant metacaspases: Decoding their dynamics in development and disease. *Plant Physiology and Biochemistry*, *180*, 50–63.

Bollhöner, B., Jokipii-Lukkari, S., Bygdell, J., Stael, S., Adriasola, M., Muñiz, L., et al. (2018). The function of two type II metacaspases in woody tissues of Populus trees. *New Phytologist*, *217*(4), 1551–1565.

Casañal, A., Lohkamp, B., & Emsley, P. (2020). Current developments in coot for macromolecular model building of electron cryo-microscopy and crystallographic data. *Protein Science*, *29*(4), 1055–1064.

Casanas, A., Warshamanage, R., Finke, A. D., Panepucci, E., Olieric, V., Noll, A., et al. (2016). EIGER detector: Application in macromolecular crystallography. *Acta Crystallographica Section D-Structural Biology*, *72*, 1036–1048.

Chang, Y. Q., Bruni, R., Kloss, B., Assur, Z., Kloppmann, E., Rost, B., et al. (2014). Structural basis for a pH-sensitive calcium leak across membranes. *Science*, *344*(6188), 1131–1135.

Chayen, N. E., & Saridakis, E. (2008). Protein crystallization: From purified protein to diffraction-quality crystal. *Nature Methods*, *5*(2), 147–153.

Coll, N. S., Vercammen, D., Smidler, A., Clover, C., Breusegem, F. V., Dangl, J. L., et al. (2010). Arabidopsis type I metacaspases control cell death. *Science*, *330*(6009), 1393–1397.

Cowtan, K. (2006). The Buccaneer software for automated model building. 1. Tracing protein chains. *Acta Crystallographica. Section D, Biological Crystallography*, *62*(Pt 9), 1002–1011.

Doublie, S. (2007). Production of selenomethionyl proteins in prokaryotic and eukaryotic expression systems. *Methods in Molecular Biology*, *363*, 91–108.

Emsley, P., Lohkamp, B., Scott, W. G., & Cowtan, K. (2010). Features and development of coot. *Acta Crystallographica. Section D, Biological Crystallography*, *66*(Pt 4), 486–501.

Fortin, J., & Lam, E. (2018). Domain swap between two type-II metacaspases defines key elements for their biochemical properties. *The Plant Journal*, *96*(5), 921–936.

Guo, G. R., Zhu, P., Fuchs, M. R., Shi, W. X., Andi, B., Gao, Y., et al. (2019). Synchrotron microcrystal native-SAD phasing at a low energy. *IUCrJ*, *6*, 532–542.

Hander, T., Fernandez-Fernandez, A. D., Kumpf, R. P., Willems, P., Schatowitz, H., Rombaut, D., et al. (2019). Damage on plants activates Ca^{2+}-dependent metacaspases for release of immunomodulatory peptides. *Science*, *363*(6433), eaar7486.

Hendrickson, W. A. (1999). Maturation of MAD phasing for the determination of macromolecular structures. *Journal of Synchrotron Radiation*, *6*, 845–851.

Hendrickson, W. A., Horton, J. R., & Lemaster, D. M. (1990). Selenomethionyl proteins produced for analysis by multiwavelength anomalous diffraction (MAD)—A vehicle for direct determination of three-dimensional structure. *EMBO Journal*, *9*(5), 1665–1672.

Ishigami, I., Lewis-Ballester, A., Echelmeier, A., Brehm, G., Zatsepin, N. A., Grant, T. D., et al. (2019). Snapshot of an oxygen intermediate in the catalytic reaction of cytochrome c oxidase. *Proceedings of the National Academy of Sciences of the United States of America, 116*(9), 3572–3577.

Jorgensen, I., Rayamajhi, M., & Miao, E. A. (2017). Programmed cell death as a defence against infection. *Nature Reviews. Immunology, 17*(3), 151–164.

Kabsch, W. (2014). Processing of X-ray snapshots from crystals in random orientations. *Acta Crystallographica, Section D: Biological Crystallography, 70*, 2204–2216.

Klemencic, M., & Funk, C. (2018). Type III metacaspases: Calcium-dependent activity proposes new function for the p10 domain. *New Phytologist, 218*(3), 1179–1191.

Klemenčič, M., & Funk, C. (2019). Evolution and structural diversity of metacaspases. *Journal of Experimental Botany, 70*(7), 2039–2047.

Klemenčič, M., Novinec, M., & Dolinar, M. (2015). Orthocaspases are proteolytically active prokaryotic caspase homologues: The case of Microcystis aeruginosa. *Molecular Microbiology, 98*(1), 142–150.

Leslie, A. G. W., & Powell, H. R. (2007). Processing diffraction data with MOSFLM. *Evolving Methods for Macromolecular Crystallography, 245*, 41–51.

Liu, Q., Zhang, Z., & Hendrickson, W. A. (2011). Multi-crystal anomalous diffraction for low-resolution macromolecular phasing. *Acta Crystallographica. Section D, Biological Crystallography, 67*(Pt 1), 45–59.

McLuskey, K., Rudolf, J., Proto, W. R., Isaacs, N. W., Coombs, G. H., Moss, C. X., et al. (2012). Crystal structure of a Trypanosoma brucei metacaspase. *Proceedings of the National Academy of Sciences of the United States of America, 109*(19), 7469–7474.

Minina, E., Coll, N., Tuominen, H., & Bozhkov, P. (2017). Metacaspases versus caspases in development and cell fate regulation. *Cell Death and Differentiation, 24*(8), 1314.

Minina, E. A., Stael, S., Van Breusegem, F., & Bozhkov, P. V. (2014). Plant metacaspase activation and activity. In P. V. Bozhkov, & G. Salvesen (Eds.), *Caspases, paracaspases, and metacaspases: Methods and protocols* (pp. 237–253). New York, NY: Springer New York.

Minor, W., Cymborowski, M., Otwinowski, Z., & Chruszcz, M. (2006). HKL-3000: The integration of data reduction and structure solution—From diffraction images to an initial model in minutes. *Acta Crystallographica. Section D, Biological Crystallography, 62*(Pt 8), 859–866.

Nakane, T., Hanashima, S., Suzuki, M., Saiki, H., Hayashi, T., Kakinouchi, K., et al. (2016). Membrane protein structure determination by SAD, SIR, or SIRAS phasing in serial femtosecond crystallography using an iododetergent. *Proceedings of the National Academy of Sciences of the United States of America, 113*(46), 13039–13044.

Ramirez, M. L. G., & Salvesen, G. S. (2018). A primer on caspase mechanisms. *Seminars in Cell & Developmental Biology, 82*, 79–85.

Read, R. J., & McCoy, A. J. (2011). Using SAD data in phaser. *Acta Crystallographica. Section D, Biological Crystallography, 67*(Pt 4), 338–344.

Rice, L. M., Earnest, T. N., & Brunger, A. T. (2000). Single-wavelength anomalous diffraction phasing revisited. *Acta Crystallographica, Section D: Biological Crystallography, 56*, 1413–1420.

Salvesen, G. S., & Dixit, V. M. (1999). Caspase activation: The induced-proximity model. *Proceedings of the National Academy of Sciences of the United States of America, 96*(20), 10964–10967.

Schmidt, M. (2020). Reaction initiation in enzyme crystals by diffusion of substrate. *Crystals, 10*(2), 116. https://doi.org/10.3390/cryst10020116.

Schneider, T. R., & Sheldrick, G. M. (2002). Substructure solution with SHELXD. *Acta Crystallographica, Section D: Biological Crystallography, 58*, 1772–1779.

Shi, Y. (2004). Caspase activation: Revisiting the induced proximity model. *Cell*, *117*(7), 855–858.
Srajer, V., & Schmidt, M. (2017). Watching proteins function with time-resolved X-ray crystallography. *Journal of Physics D: Applied Physics*, *50*(37). https://doi.org/10.1088/1361-6463/aa1087d1032.
Stagno, J. R., Liu, Y., Bhandari, Y. R., Conrad, C. E., Panja, S., Swain, M., et al. (2017). Structures of riboswitch RNA reaction states by mix-and-inject XFEL serial crystallography. *Nature*, *541*(7636), 242–246.
Suarez, M. F., Filonova, L. H., Smertenko, A., Savenkov, E. I., Clapham, D. H., von Arnold, S., et al. (2004). Metacaspase-dependent programmed cell death is essential for plant embryogenesis. *Current Biology*, *14*(9), R339–R340.
Sundström, J. F., Vaculova, A., Smertenko, A. P., Savenkov, E. I., Golovko, A., Minina, E., et al. (2009). Tudor staphylococcal nuclease is an evolutionarily conserved component of the programmed cell death degradome. *Nature Cell Biology*, *11*(11), 1347–1354.
Takemaru, L., Guo, G. R., Zhu, P., Hendrickson, W. A., McSweeney, S., & Liu, Q. (2020). PyMDA: Microcrystal data assembly using Python. *Journal of Applied Crystallography*, *53*, 277–281.
Uren, A. G., O'Rourke, K., Aravind, L., Pisabarro, M. T., Seshagiri, S., Koonin, E. V., et al. (2000). Identification of Paracaspases and Metacaspases: Two ancient families of caspase-like proteins, one of which plays a key role in MALT lymphoma. *Molecular Cell*, *6*(4), 961–967.
Vandana, Singh, A. P., Singh, J., Sharma, R., Akhter, M., Mishra, P. K., et al. (2018). Biochemical characterization of unusual cysteine protease of P. falciparum, metacaspase-2 (MCA-2). *Molecular and Biochemical Parasitology*, *220*, 28–41.
Vercammen, D., Cotte, B. V. D., Jaeger, G. D., Eeckhout, D., Casteels, P., Vandepoele, K., et al. (2004). Type II Metacaspases Atmc4 and Atmc9 of Arabidopsis thaliana cleave substrates after arginine and lysine. *The Journal of Biological Chemistry*, *279*(44), 45329–45336.
Vonrhein, C., Blanc, E., Roversi, P., & Bricogne, G. (2007). Automated structure solution with autoSHARP. *Methods in Molecular Biology*, *364*, 215–230.
Walden, H. (2010). Selenium incorporation using recombinant techniques. *Acta Crystallographica. Section D, Biological Crystallography*, *66*(Pt 4), 352–357.
Wang, S. N., Xue, M., He, C., Shen, D. Y., Jiang, C. H., Zhao, H. W., et al. (2021). AtMC1 associates with LSM4 to regulate plant immunity through modulating pre-mRNA splicing. *Molecular Plant-Microbe Interactions*, *34*(12), 1423–1432.
Watanabe, N., & Lam, E. (2011a). Arabidopsis metacaspase 2d is a positive mediator of cell death induced during biotic and abiotic stresses. *The Plant Journal*, *66*(6), 969–982.
Watanabe, N., & Lam, E. (2011b). Calcium-dependent activation and autolysis of Arabidopsis metacaspase 2d. *The Journal of Biological Chemistry*, *286*(12), 10027–10040.
Waterman, D. G., Winter, G., Gildea, R. J., Parkhurst, J. M., Brewster, A. S., Sauter, N. K., et al. (2016). Diffraction-geometry refinement in the DIALS framework. *Acta Crystallographica Section D-Structural Biology*, *72*, 558–575.
Wen, S., Ma, Q.-M., Zhang, Y.-L., Yang, J.-P., Zhao, G.-H., Fu, D.-Q., et al. (2013). Biochemical evidence of key residues for the activation and autoprocessing of tomato type II metacaspase. *FEBS Letters*, *587*(16), 2517–2522.
Williams, C. J., Headd, J. J., Moriarty, N. W., Prisant, M. G., Videau, L. L., Deis, L. N., et al. (2018). MolProbity: More and better reference data for improved all-atom structure validation. *Protein Science*, *27*(1), 293–315.
Winn, M. D., Ballard, C. C., Cowtan, K. D., Dodson, E. J., Emsley, P., Evans, P. R., et al. (2011). Overview of the CCP4 suite and current developments. *Acta Crystallographica Section D: Biological Crystallography*, *67*(4), 235–242.

Winter, G., Waterman, D. G., Parkhurst, J. M., Brewster, A. S., Gildea, R. J., Gerstel, M., et al. (2018). DIALS: Implementation and evaluation of a new integration package. *Acta Crystallographica Section D-Structural Biology*, *74*, 85–97.

Wong, A. H.-H., Yan, C., & Shi, Y. (2012). Crystal structure of the yeast Metacaspase Yca1. *The Journal of Biological Chemistry*, *287*(35), 29251–29259.

Wrzaczek, M., Vainonen, J. P., Stael, S., Tsiatsiani, L., Help-Rinta-Rahko, H., Gauthier, A., et al. (2015). GRIM REAPER peptide binds to receptor kinase PRK5 to trigger cell death in Arabidopsis. *EMBO Journal*, *34*(1), 55–66.

Yamamoto, M., Hirata, K., Yamashita, K., Hasegawa, K., Ueno, G., Ago, H., et al. (2017). Protein microcrystallography using synchrotron radiation. *IUCrJ*, *4*(5), 529–539.

Zhu, P., Yu, X. H., Wang, C., Zhang, Q. F., Liu, W., McSweeney, S., et al. (2020). Structural basis for Ca^{2+}-dependent activation of a plant metacaspase. *Nature Communications*, *11*(1), 2249.

CHAPTER FIVE

Biocatalytic system for comparatively assessing the functional association of monolignol cytochrome P450 monooxygenases with their redox partners

Xianhai Zhao and Chang-Jun Liu*
Biology Department, Brookhaven National Laboratory, Upton, NY, United States
*Corresponding author: e-mail address: cliu@bnl.gov

Contents

Methods in Enzymology, Volume 676
ISSN 0076-6879
https://doi.org/10.1016/bs.mie.2022.07.027

Abstract

Lignin is a complex heterogenous polymer derived from oxidative radical polymerization of three monolignols, i.e., *p*-coumaryl alcohol, coniferyl alcohol and sinapyl alcohol. These lignin monomeric precursors structurally differ in their methoxy groups of the benzene rings. In phenylpropanoid-monolignol biosynthetic pathway, the endoplasmic reticulum (ER)-resident cytochrome P450 monooxygenases, cinnamate 4-hydroxylase, coumaroyl ester 3′-hydroxylase and ferulate 5-hydroxylase, establish the key structural characteristics of monolignols. The catalysis of cytochrome P450 monooxygenase requires reducing power, which is supplied by the ER electron transfer chains, composed of cytochrome P450 oxidoreductase (CPR), cytochrome b_5 reductase (CBR) and/or cytochrome b_5 protein (CB5), from cofactor NADPH or NADH. While NADPH-dependent CPR serves as the typical electron donor for most P450 enzymes, in some cases, the CBR-CB5 or CPR-CB5 electron transfer system also transfers electrons to the terminal P450 enzymes. There are tremendous studies focusing on the discovery and characterization of cytochrome P450 monooxygenases. However, very limited attention has been paid to the versatility and the roles of electron transfer components in the P450 catalytic system. Due to the membrane-residence property of both P450 enzymes and electron transfer components, it is challenging to establish an effective experimental system to evaluate the functional association of P450s with their redox partners. This chapter describes a yeast cell biocatalytic system and the related experimental procedures for comparatively assessing the functional relationship of monolignol biosynthetic P450 enzymes and different redox partners in their catalysis.

Abbreviations

AtCB5D Arabidopsis cytochrome b_5 family member D
C3'H *p*-coumaroyl ester 3′-hydroxylase
C4H cinnamic acid 4-hydroxylase
CB5 cytochrome b_5
CBR NADH-dependent cytochrome b_5 reductase
CPR NADPH-dependent cytochrome P450 oxidoreductase
F3′5′H flavonoid 3′,5′-hydroxylase
F3′H flavonoid 3′-hydroxylase
F5H ferulate/coniferyl alcohol/coniferaldehyde 5-hydroxylase
NADH nicotinamide adenine dinucleotide
NADPH nicotinamide adenine dinucleotide phosphate
P450 cytochrome P450 monooxygenase

1. Introduction

Cytochrome P450 monooxygenases are heme-thiolate enzymes, constituting one of the largest and oldest protein superfamilies found in main life kingdoms, including bacteria, yeasts, mammals, and plants. In plants, these enzymes catalyze extremely diverse reactions leading to the precursors of structural macromolecules such as lignin, cutin, suberin, and sporopollenin, or are involved in biosynthesis or catabolism of small molecule natural products such as pigments, odorants, flavors, antioxidants, allelochemicals and defense compounds, plant hormone and signaling molecules, and in the xenobiotics metabolism (Bak et al., 2011; Bernhardt, 2006; Degtyarenko & Archakov, 1993; Nelson, 2009).

The P450-catalyzed reactions insert one of the oxygen atoms from active O_2 into the substrate and reduce the other to water. Therefore, the enzymes are classified as monooxygenases. In each catalytic cycle, a P450 enzyme coupled with its redox partner(s) acquires two electrons from reduced pyridine nucleotides (NADPH or NADH) to generate iron-oxygen intermediates that in turn oxidizes the substrates (Hannemann, Bichet, Ewen, & Bernhardt, 2007; Krest et al., 2013; Porter & Coon, 1991). In eukaryotic cells, there are two primary electron transfer systems that interact with microsomal P450s and deliver electrons from reductant cofactors to the prosthetic heme group of the P450 enzyme: NADPH-cytochrome P450 oxidoreductase (CPR) chain, and NADH-cytochrome b_5 reductase (CBR)-cytochrome b_5 (CB5) chain (Hannemann et al., 2007; Iyanagi, 2019) (Fig. 1). Since 1971, the involvement of CB5 in the CPR electron transfer system has been evidenced (Hildebrandt & Estabrook, 1971). Whether CB5 is obligatory for the catalysis of P450s and how CB5 participates in the CPR electron transfer chain remain controversial.

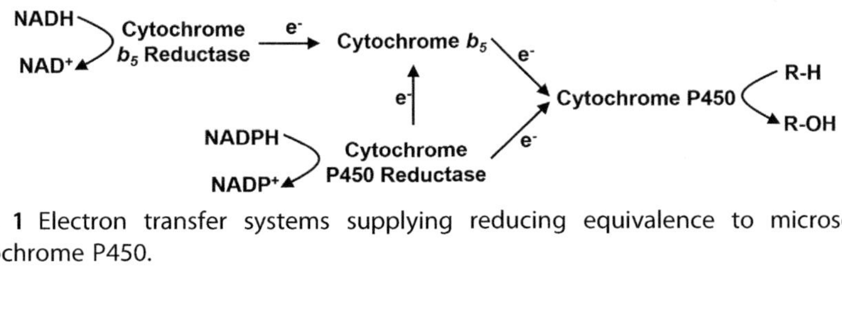

Fig. 1 Electron transfer systems supplying reducing equivalence to microsomal cytochrome P450.

Most P450 activities could be reconstituted with only CPR and P450 itself, which argues against an obligatory role for cytochrome b_5 in this pathway (Hannemann et al., 2007; Jensen & Moller, 2010); whereas many studies also reveal that a subset of P540-catalyzed reactions can be stimulated by CB5 protein (Porter, 2002). Moreover, CB5 displays multifaceted effects on P450 catalysis. The protein shows stimulation, no effect, or inhibition effect to the P450 monooxygenase activity in drug metabolism (Porter, 2002; Schenkman & Jansson, 2003). Collectively, two hypothetic roles are proposed to the CB5 protein in P450 catalysis: allosteric modulator or electron donor (Iyanagi, 2019; Porter, 2002; Schenkman & Jansson, 2003). CB5 is readily reduced by NADH-dependent CBR and NADPH-dependent CPR, suggesting CB5 can act as an electron carrier to shuttle reducing equivalent from either reductases to the terminal acceptors (Enoch & Strittmatter, 1979; Fukuchi-Mizutani, Mizutani, Tanaka, Kusumi, & Ohta, 1999). Cases from mice, human and fungi evidence that the CBR-CB5 electron transfer chain can donate entire electrons to some P450s (Estrada, Laurence, & Scott, 2016; Henderson, McLaughlin, & Wolf, 2013; Ichinose & Wariishi, 2012; Stiborová et al., 2016); whereas in CPR-CB5 pathway, CB5 was proposed to donate the second electron to oxyferrous P450 (Duggal et al., 2016; Porter, 2002). Furthermore, although CB5 can accept electron from CPR, the turnover efficiency is lower than that of CBR (Bhatt, Khatri, Rodgers, & Martin, 2017; Enoch & Strittmatter, 1979; Niu et al., 2017). Therefore, CB5 as an electron carrier to augment P450-catalyzed reactions probably have two different modes: (1) direct electron transfer of both required electrons from NADH-CBR to P450 in a pathway separate and independent of NADPH-CPR and (2) transfer of the second electron to oxyferrous P450 from either CPR or CBR (Porter, 2002) (Fig. 1).

In addition to the documented electron shuttle role, some studies also suggest the allosteric stimulation effect of CB5 protein to cytochrome P450 activity. The activities of some P450s could be stimulated with incorporation of an apo-CB5, devoid of the heme group (Yamazaki et al., 2002; Yamazaki, Johnson, Ueng, Shimada, & Guengerich, 1996); in P450 17A1-catalyzed reaction, the P450 conformation is changed with the binding of CB5 protein, which either alters the selective substrate affinity or increases the catalytic efficiency of the enzyme (Estrada et al., 2016). These data suggest an allosteric stimulation of CB5 to P450 enzyme without electron transfer.

While a subset of mammalian P450s involved in xenobiotics and drug metabolism were found to be affected by CB5 protein in their catalyzes (Porter, 2002), CB5 proteins have been recognized to functionally associate with plant P450-catalyzed reactions in just a couple cases so far (de Vetten et al., 1999; Gou et al., 2019). Nevertheless, those *in planta* evidence unambiguously demonstrate the indispensable role of CB5 as an electron donor in stimulating P450-catalzyed reactions in plant phenolic biosynthesis. An early genetic study discovered a petunia *cb5* locus, *Dif F,* that predominantly expresses in the flower of petunia and is required for the full activity of flavonoid 3′,5′-hydroxylase (F3′5′H) for the synthesis of 3′,5′ substitution of anthocyanins and the purple/blue coloration of flora. Disruption of *Dif F* gene by targeted transposon mutagenesis reduced F3′5′H enzymatic activity and the accumulation of 5′-substituted anthocyanins, resulting in an altered petunia flower color (de Vetten et al., 1999). Interestingly the loss of *cb5* locus does not influence the activities of other cytochrome P450s, including the functionally close homolog flavonoid 3′-hydroxylase in the anthocyanin biosynthetic pathway (de Vetten et al., 1999). These data suggest that cytochrome b_5 regulates the activity of specific P450s in plant metabolism.

Recently, among four ER-localized CB5 isoforms in Arabidopsis, AtCB5D was discovered from the immunoprecipitated proteins with monolignol biosynthetic P450s as bait (Gou et al., 2019). The protein was then found to specifically functionally associate with ferulate 5-hydroxylase (F5H) for syringyl lignin synthesis (Gou et al., 2019). The heterogenous biopolymer lignin plays a crucial role in plant growth and development and in the resistance to environmental stress. It confers hydrophobicity of vasculature for water conductance, mechanical strength for upright growth, and physical barrier against microbial and herbivory attacks (Boerjan, Ralph, & Baucher, 2003). *P*-coumaryl alcohol, coniferyl alcohol and sinapyl alcohol are three major lignin monomeric precursors (i.e., monolignols) that undergo oxidative radical polymerization to form *p*-hydroxyphenyl (H), guaiacyl (G) and syringyl (S) lignin subunits, respectively (Boerjan et al., 2003). Three cytochrome P450 monooxygenases catalyze the hydroxylation of the benzene ring of phenylpropanoid derivatives to yield structurally characteristic monolignols. They are *trans*-cinnamic acid 4-hydroxylase (C4H; CYP73A5), *p*-coumaroyl ester 3′-hydroxylase (C3′H; CYP98A3), and ferulate/coniferyl alcohol/coniferaldehyde 5-hydroxylase (F5H; CYP84A1). With *t*-cinnamic acid as starting precursor,

C4H catalyzes the 4-position hydroxylation of the phenyl ring yielding *p*-coumaric acid that is required for the formation of all three types of lignin monomeric precursors. C3′H and F5H catalyze the hydroxylation reactions at C3- and C5-positions of benzene ring, sequentially, leading to the biosynthesis of G and/or S lignin monomers.

Arabidopsis genome encodes two ER-localized CPRs (ATR1 and ATR2) and two CBR proteins with only one located to the ER (Shockey et al., 2005; Wayne, Wallis, Kumar, Markham, & Browse, 2013). ATR2 protein has been demonstrated to physically interact with monolignol biosynthetic C4H and C3′H (Guo, Ran, Martin, & Liu, 2018) and be involved in lignin biosynthesis (Gou et al., 2019; Sundin et al., 2014). The composition of Arabidopsis stem lignin and phenolic metabolites changed significantly in *atr2* mutants, indicating the involvement of ATR2 in lignin formation (Sundin et al., 2014). Notably, disruption of *ATR2* resulted in the accumulation of the biosynthetic precursors of all three monolignol biosynthetic P450 enzymes (Sundin et al., 2014), implicating that ATR2 likely functionally associates with C4H, C3′H and F5H-catalyzed reactions without discrimination. In addition to the genes encoding both reductases, six *CB5* and *CB5-like* genes are annotated in *Arabidopsis* genome which encode five canonical CB5 isoforms (AtCB5A ∼ AtCB5D) and a CB5-like protein (AtCB5LP). Four of them localize to the ER membrane (Hwang et al., 2004; Maggio, Barbante, Ferro, Frigerio, & Pedrazzini, 2007). CB5 is conventionally regarded to couple with CBR and is implicated in the fatty acid desaturation, hydroxylation, or elongation (Kumar, Tran, Neelakandan, & Nguyen, 2012; Wayne et al., 2013). Nevertheless, our recent study unveiled that disruption of AtCB5D, one of the ER-localized CB5 family members, significantly impaired the biosynthesis of S-lignin subunits but had no detrimental effect on G-lignin formation. Consistently, the loss of CB5D reduced microsomal F5H activity but not the C4H activity, suggesting that AtCB5D is an indispensable electron donor for F5H-catalyed 5-hydroxylation of lignin monomeric precursor (Gou et al., 2019). It is still not fully resolved which electron transfer chain(s) that AtCB5D might be involved in for delivering reducing equivalence to augment F5H-catalyzed reaction. Due to the complexity of plant genetics and the long-life cycle of plant, it is difficult to assess P450 monooxygenase-electron transfer system with an *in planta* system, a relatively clean and effective heterologous system is much desired.

Among heterologous expression systems, for example, *Escherichia coli*, mammalian cell lines, and insects cells, yeast is a preferential choice for expressing microsomal P450s (Pompon, Louerat, Bronine, & Urban, 1996). The insufficiency of endogenous CPR present in yeast cells and the low sequence similarities of yeast CPR and CB5 with mammalian or plant counterparts limit the capacity of yeast cells to be directly used in determining catalytic efficiency of P450s (Pompon et al., 1996). Nevertheless, these unique features of yeast cells offer opportunities for assessing the involvement of different electron transfer systems in a specific P450 monooxygenase-catalyzed reaction through heterologous expression of exogenous redox components. Furthermore, the availability of yeast mutants (partially) deficient in its endogenous electron transfer component(s) provides robust host strains with clean genetic background for heterologously determining the dependence of a particular P450 enzyme on electron transfer systems. Therefore, we adopt yeast cells deficient in their endogenous *CPR*, *CBR*, or *CB5* gene to establish a cell culture-based biocatalytic system, in which the plant-origin electron supply chains are selectively co-expressed with monolignol biosynthetic P450 enzymes, then biotransformation rate of the fed phenolic substrate is monitored and used as read-out to evaluate the functional association of the monolignol P450s with their electron donors. This established experimental system can be applied to evaluating the functional relationship of other cytochrome P450 monooxygenases from different metabolic processes with their electron donors.

This chapter describes the experimental procedure to establish and utilize the yeast biocatalytic system for comparatively assessing Arabidopsis C4H and F5H1 in the requirements for their redox partners in vivo and in vitro. C3′H was not included in the study due to the unavailability of its substrate. To determine the preferred electron transfer chains of C4H and F5H1, we co-expressed both enzymes with different combinations of *Arabidopsis thaliana* ATR1, ATR2, CBR1 and CB5D in yeast cells. A mutant version of AtCB5D, 2muCB5D, was adopted as a control. 2muCB5D has it's His-40 and His-64 substituted with alanines; both histidines are key axial ligands for heme binding and are critical for transferring electrons to the iron center of the heme (Cowley et al., 2002; Schenkman & Jansson, 2003; Vergeres & Waskell, 1995). Therefore, H40A and H64A mutations abolish the electron donor property of the mutant variant (Gou et al., 2019). Aided with the self-cleaving 2A peptide, polycistronic gene expression strategy was employed to integrate both C4H and F5H1 with the selected redox components into a same operon (Fig. 2). This strategy therefore ensures the co-expression of the targeted P450s and electron transfer components in the same yeast cells under the control of a same

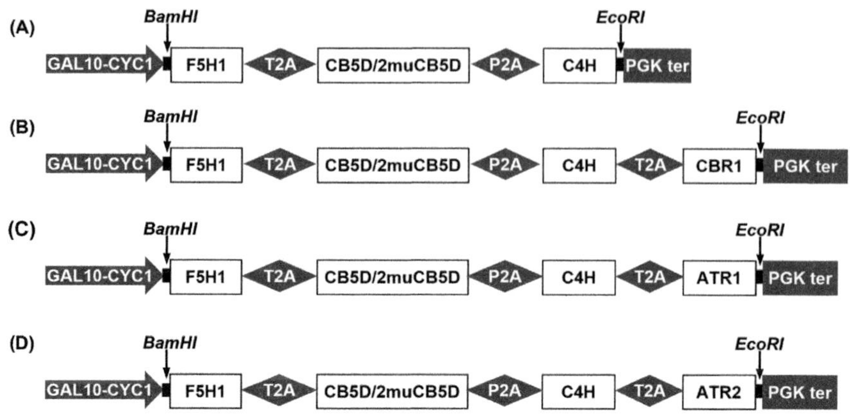

Fig. 2 The schema of T/P2A-based polycistronic expression cassettes for heterologous yeast protein production. (A) F5H1-CB5D/2muCB5D-C4H. (B) F5H1-CB5D/2muCB5D-C4H-CBR1. (C) F5H1-CB5D/2muCB5D-C4H-ATR1. (D) F5H1-CB5D/2muCB5D-C4H-ATR2. GAL10-CYC1 promoter and PGK terminator were located at the pYeDP60 vector. *BamHI* and *EcoRI* recognition sites were used to insert PCR fragments of expressing cassettes.

promoter. With this design, C4H and F5H1 operate in the exact same redox condition, thus facilitating the precise and reliable comparation of the needs of two P450 enzymes for particular electron transfer chains.

2. Plasmid construction

pYeDP60 vector, developed for the effective P450 protein expression under the control of inducible *GAL10-CYC1* promoter (Pompon et al., 1996), was adopted in the whole yeast cell biocatalytic system. The vector was digested with *BamHI* and *EcoRI* restriction enzymes for subcloning. The coding sequences of *Arabidopsis* P450 genes, *C4H* and *F5H1*, and the electron transfer components *ATR1*, *ATR2*, *CBR1* and *CB5D* were amplified by polymerase chain reaction (PCR) using Phusion High-Fidelity DNA Polymerase (NEB) with gene specific primers and fusion primers where T2A or P2A sequence was incorporated into gene specific primers (Table 1). The purified PCR products were subcloned to pYeDP60 vector with Gibson assembly method (Gibson et al., 2009), which generated a set of expression cassettes, including pYeDP60-F5H1-2muCB5D-C4H (as control), pYeDP60-F5H1-CB5D-C4H, pYeDP60-F5H1-CBR1-2muCB5D-C4H, pYeDP60-F5H1-CBR1-CB5D-C4H, pYeDP60-F5H1-ATR1-2muCB5D-C4H, pYeDP60-F5H1-ATR1-CB5D-C4H, pYeDP60-F5H1-ATR2-2muCB5D-C4H and pYeDP60-F5H1-ATR2-CB5D-C4H plasmids (Fig. 2).

Table 1 The primers used in this study.

Primer name	Sequence (5′ to 3′)
F5H1-F	aatacacacactaaattacc**ggatcc**atggagtcttctatatcacaaac
F5H1-R	ctcctccacgtcaccgcatgttagaagact tcctctgccctcaagagcacagatgaggcg
CB5D-F	ctaacatgcggtgacgtggaggagaatccc ggccctatgggcggagacggaaaag
CB5D-R	ctccagcttgcttaagaagagaaaaattag tagctccagatccagaagaaggagccttgg
C4H-F	tcttcttaagcaagctggagatgttgaaga aaatcctggacccatggacctcctcttgct
C4H-R	ctccacgtcaccgcatgttagaagacttcc tctgccctcacagttccttggtttcataa
C4H-Rs[a]	agacatgggagatcccccgc**gaattc**ttaacagttccttggtttcataa
CBR1-F	ctaacatgcggtgacgtggaggagaatccc ggccctatggataccgagtttctc
CBR1-R	agacatgggagatcccccgc**gaattc**tcagaactggaattgcatc
ATR1-F	ctaacatgcggtgacgtggaggagaatccc ggccctatgacttctgctttgtatgcttc
ATR1-R	agacatgggagatcccccgc**gaattc**tcaccagacatctctgaggtatc
ATR2-F	ctaacatgcggtgacgtggaggagaatccc ggccctatgtcctcttcttcttcttc
ATR2-R	agacatgggagatcccccgc**gaattc**ttaccatacatctctaagatatc

[a]C4H-Rs contains a stop codon which is designed for F5H1-CB5D-C4H and F5H1-2muCB5D-C4H tricistrons. The bold letters indicated *BamHI* and *EcoRI* restrictive enzyme sites.

2.1 Materials

1. Bio-Rad S1000 thermal cycler
2. New Brunswick Innova 42 incubator shaker (Eppendorf)
3. Thermomixer or water bath
4. LB liquid medium: tryptone 10 g/L, yeast extract 5 g/L, sodium chloride 10 g/L
5. LB agar plate: LB liquid medium supplemented with 2% agar

6. Milli-Q water
7. Phusion high fidelity DNA polymerase (NEB, #M0530)
8. OneTaq 2 × Master Mix with Standard Buffer (NEB, #M0482)
9. Wizard SV Gel and PCR Clean-Up System (Promega, #A9282)
10. GeneJET Plasmid Miniprep Kit (Thermo Fisher Scientific, #K0503)
11. Agarose (Thomas Scientific, #C748D75)
12. *BamHI* and *EcoRI* restriction enzymes (Bio-Rad)

2.2 Procedure

1. Design primers according to the requirement of Gibson assembly method. To facilitate this procedure, *pYeDP60-genes* final construct maps are created first with Snapgene (Dotmatics, San Diego, US), then proper primer sequences were chosen (see note 1).
2. Solubilize primers to 100 μM in milli-Q water for long-term stocks and 20 μM working solutions are prepared
3. Amplify the DNA fragments with Phusion high fidelity DNA polymerase (NEB), resolve PCR products on 1% agarose gel and recover target DNA fragments
4. Digest 2 μg *pYeDP60* vector plasmid with *BamHI* and *EcoRI* restriction enzymes, resolve vector fragment on 1% agarose gel and recover vector backbone
5. Prepare 2 X Gibson assembly master mix according to the describe procedure (Gibson et al., 2009).
6. Set up Gibson assembly reaction on ice with 100 ng of vector backbone, 1–2 fold molar excess of each insert DNAs, 5 μL of 2 X Gibson assembly master mix and bring up the total volume to 10 μL with milli-Q water (see note 2).
7. Incubate the assembly reactions in a thermocycler at 50 °C for 40 min
8. Place samples on ice or keep at −20 °C for subsequent transformation
9. Transform DH5α competent cells with heat shock method. Mix 50 μL competent cells with 5 μL of the assembly reaction, incubate on ice for 30 min, heat shock at 42 °C for 60 s, incubate on ice for 3 min, then add 1 mL LB medium, and incubate the cells at 37 °C for 30 min
10. Spin down *E. coli* cells at 5000 × *g* (Eppendorf 5424R centrifuge) for 5 min and discard supernatant
11. Resuspend cells in 100 μL water and plate them on LB agar plates containing 100 μg/mL ampicillin or carbenicillin

12. Incubate the plates at 37 °C, overnight
13. Conduct colony PCR screening of positive clones. Mix 10 μL OneTaq® 2× Master Mix (NEB), 9 μL water, 0.5 μL forward and reverse primers to prepare PCR reaction (see note 3). Pick a single colony with a sterile toothpick or pipette tip, stripe on a LB agar plate with grid marking to save clones for later culture, swirl the rest strain in PCR reaction mixture to introduce PCR template. For each transformation, 4 colonies are generally enough to find positive clones
14. Analyze PCR product size in an agarose gel
15. Inoculate one or two positive clones to LB liquid medium and culture at 37 °C for overnight
16. Isolate plasmid DNA from overnight *E. coli* culture and verify the insert sequence with Sanger sequencing

2.3 Notes

1. Most suppliers will increase the rate for oligonucleotides longer than 60 bp. When designing the gene-T2 fusion primers, the primer length and Tm value should be considered simultaneously
2. The NEB suggests 1: 1 M ratio of the each insert DNA over the vector DNA fragment. In the practice, 1 μL of the prepared insert DNA from the described procedures works efficiently
3. Since multi-genes are assembled into one vector backbone, each of them should be verified with PCR. But for a time-saving purpose, the method described here is to use backbone primers to amplify the entire insert and then the products are resolved on agarose gel to estimate the PCR product size

3. Yeast transformation

Yeast cells harbor endogenous electron transfer components, i.e., CB5, CBR1 and CPR (NCP1, NADPH-Cytochrome P450 reductase) on the ER membrane. To eliminate the potential disturbance of those endogenous redox components, *Saccharomyces cerevisiae* yeast mutant strains deficient in their endogenous *CB5*, *CBR1* or *NCP1* gene were adopted. Among them, the strain *ncp1* defective in *CPR* is a heterozygous diploid; while strains *cyb5* and *cbr1* deficient in *CB5* and *CBR1* are the homozygous.

3.1 Materials

1. Yeast strains: *cyb5* (MATa, BY4741 background, his3Δ1 leu2Δ0 met15Δ0 ura3Δ0 cyb5Δ), *cbr1* (MATa, BY4741 background, his3Δ1 leu2Δ0 met15Δ0 ura3Δ0 cbr1Δ) and *ncp1* (heterozygous diploids, BY4741/BY4742 background, his3Δ1/his3Δ1 leu2Δ0/leu2Δ0 lys2Δ0/LYS2 MET15/met15Δ0 ura3Δ0/ura3Δ0 NCP1/ncp1Δ) (Transomic, Huntsville, US)
2. Plasmids: *pYeDP60-F5H1-2muCB5D-C4H, pYeDP60-F5H1-CB5D-C4H, pYeDP60-F5H1-CBR1-2muCB5D-C4H, pYeDP60-F5H1-CBR1-CB5D-C4H, pYeDP60-F5H1-ATR1-2muCB5D-C4H, pYeDP60-F5H1 -ATR1-CB5D-C4H, pYeDP60-F5H1-ATR2-2muCB5D-C4H* and *pYeDP60-F5H1-ATR2 -CB5D-C4H*
3. YPDA broth (Takara)
4. YPDA agar plate: YPDA broth supplemented with 2% agar
5. 200 mg/mL G418 stock solution
6. EZ-Yeast transformation kit (MP Biomedicals, Irvine, US)
7. SD-Ura agar plate containing 200 μg/mL G418
8. Eppendorf 5424R centrifuge with FA-45-24-11 rotor
9. New Brunswick Innova 42 incubator shaker

3.2 Procedure

1. Streak *cyb5*, *cbr1* and *ncp1* yeast strains on YPDA agar plates containing 200 μg/mL G418 sulfate (see note 1) and incubate at 28 °C for 2–3 days until single colonies can be observed (see note 2).
2. Inoculate a single colony into 4 mL YPDA medium containing 200 μg/mL G418 and incubate overnight at 28 °C with 250 rpm shaking
3. Pellet the yeast cells with centrifugation at 5000 × *g* (Eppendorf 5424R centrifuge) for 5 min
4. Resuspend yeast pellet with Ez-yeast transformation solution (MP Biomedicals) to prepare master mix competent cells (see note 3).
5. For each transformation, add 5 μL carrier DNA (MP Biomedicals), 200 μg plasmid DNA and 70 μL yeast master mix to a 1.5 mL centrifuge tube
6. Heat shock at 42 °C for 30 min; cool to 28 °C and keep for 30 min
7. Pellet yeast cells at 5000 × *g* (Eppendorf 5424R centrifuge) for 5 min and discard supernatant

8. Resuspend in 100 μL water and plate in auxotrophic yeast medium devoid of Ura
9. Incubate the plate at 28 °C for 2–3 days until single colonies can be observed

3.3 Notes

1. The adopted yeast mutant strains possess KanMX4 selection marker. The addition of G418 sulfate antibiotics can prevent potential contamination and maintain the purity of the incubated strains
2. The fresh yeast colonies from YPDA plate can be used for transformation directly. Scrape the fresh yeast colonies and resuspend in transformation solution to prepare master mix competent cells
3. Normally, 1 mL overnight incubated yeast cells can be used for 5 transformations. Too many yeast cells used in the transformation generates more but tiny colonies on plates. Therefore, it is important to use proper amount of yeast cells to obtain discrete and healthy yeast colonies

4. Yeast induction and in vivo feeding assays

4.1 Materials

1. Yeast minimal broth: SD-Ura-sugar, autoclaved at 121 °C for 15 min
2. 20% Glucose solution, autoclaved at 121 °C for 15 min or filtration sterilized
3. 20% Galactose solution, autoclaved at 121 °C for 15 min or filtration sterilized
4. TE buffer: 10 mM Tris-HCl (pH 8.0), 1 mM EDTA
5. Coniferyl alcohol (Sigma, #223735). 2 mM coniferyl alcohol solution was prepared by dissolving 3.6 mg powder in 2 mL methanol first, then adding water to final volume of 10 mL
6. *t*-Cinnamic acid (Sigma, #C80857). 2 mM solution is prepared in the same way as above
7. Water saturated ethyl acetate containing 0.1 mM chrysin (Sigma); chrysin is used as the extraction internal standard
8. LUNA™ Automated Cell Counter (Logosbio, Gyeonggi-do, South Korea)
9. Thermo Scientific Labquake Tube Rotator

10. GENESYS™ 10S UV–Visible Spectrophotometer (Thermo Scientific)
11. Centrivap benchtop vacuum concentrator (Labconco, Kansas, US)
12. Eppendorf 5424R centrifuge with FA-45-24–11 rotor
13. Beckman GS-6 centrifuge with G.H. 3.8 rotor
14. New Brunswick Innova 42 incubator shaker

4.2 Procedure

1. Inoculate about 10 colonies of fresh transformed yeast cells into 4 mL of SD-Ura + 2% glucose selective broth. Incubate yeast starting culture at 28 °C with 250 rpm shaking for overnight
2. Inoculate 0.5 mL starting culture into 15 mL SD-Ura + 2% glucose selective media in 50 mL flasks and incubate at 28 °C with 250 rpm shaking for about 6 h
3. Transfer the cultures to 50 mL centrifuge tubes and rinse the flasks with sterilized water once. Pellet yeast cells by centrifugation at $2000 \times g$ (Beckman GS-6 centrifuge) for 5 min. Resuspend the yeast cell pellets with 2 mL SD-Ura + 2% galactose inductive media
4. Measure the density of resuspended yeast cells with a LUNA™ Automated Cell Counter or with a UV–Visible spectrophotometer where 1 OD_{600} approximately equivale to 3×10^7 cells/mL
5. Inoculate appropriately resuspended yeast cells into 15 mL SD-Ura + 2% galactose inductive media in the original 50 mL flasks to a final cell density of about 0.6×10^7 cells/mL (see note 1) Alternatively, measure the cell density of the resuspended yeast cells first, then calculate the required amount/volume of the inoculation strain
6. Incubate at 28 °C with 250 rpm shaking for about 14 h, until the cell density reaches $6–9 \times 10^7$ cells/mL (see note 2).
7. Pellet the galactose-induced yeast cells by centrifugation at $2000 \times g$ (Beckman GS-6 centrifuge) for 5 min. Wash yeast cells once with TE buffer. Resuspend yeast cells with about 8 mL TE buffer. A cell density should be about 1.5×10^8 cells/mL
8. P450 activity whole cell assays. For C4H assay, aliquot 0.5 mL yeast cells to 2 mL tubes containing one 5 mm glass bead each, add 125 μL 2 mM *t*-cinnamic acid substrate, incubate the mixture at 28 °C with slow rotation (Thermo Scientific Labquake Tube Rotator) for 0.5 h (see note 3). For F5H1 assay, aliquot 1 mL yeast cells to 2 mL tubes containing one 5 mm glass bead each, add 125 μL 2 mM coniferyl alcohol substrate, incubate the reactions at 28 °C with slow rotation (Thermo Scientific Labquake Tube Rotator) for 3 h (see note 4).

9. Add 0.5 mL water saturated ethyl acetate containing 0.1 mM chrysin, vortex vigorously to stop the reactions. Partition reaction products through centrifugation at 20,000 g (Eppendorf 5424R centrifuge) for 5 min
10. 0.2 mL ethyl acetate solvent phase are taken, vacuum-dried, and stored in freezer

4.3 Notes

1. Although most of the protocols claim the starting yeast cell density at about 5×10^7 cells/mL or even higher (Nett, Lau, & Sattely, 2020; Urban et al., 1994), the best enzymatic activity monitored in our experiment was achieved with a lower starting cell density at about 0.6×10^7 cells/mL
2. The longer induction time increases cell density of yeast culture; but the monitored P450 enzymatic activity decreases
3. The catalytic efficiency of C4H is relatively high. Therefore, a relatively shorter incubation time was applied after the supplement of substrate to ensure adequate amount of substrate remaining during the reaction thus obtaining a more precise max velocity
4. The catalytic efficiency of F5H1 is about 10 times lower than that of C4H. To obtain detectable amount of reaction product of F5H1, increasing the yeast amount in the reaction mixture is a better choice than extending incubation time

5. Quantification of reaction products by UHPLC-MS

The P450 catalytic products were resolved and determined via an Ultra High Performance Liquid Chromatography (UHPLC) hyphenated to a Q-Exactive mass spectrometer. Retention times, ultraviolet-visible spectra and mass spectra were referred to confirm the F5H1-catalyzed product 5-hydroxyconiferyl alcohol and the C4H-catalyzed product *p*-coumaric acid. UV absorbance peak areas were acquired to quantify product abundance.

5.1 Materials

1. Thermofisher Scientific Dionex Ultimate 3000 UHPLC hyphenated to a Q-Exactive Plus mass spectrometer
2. C18 column (Luna, 150 × 2.1 mm, 1.6 μm, 00F-4742-AN, Phenomenex)
3. Guard cartridge (AJ0-9502, Phenomenex)

4. Mobile phase A: 0.1% acetic acid in water
5. Mobile phase B: 0.1% acetic acid in acetonitrile
6. A series of diluted coniferyl alcohol, coniferaldehyde and *p*-coumaric acid authentic standard samples
7. Eppendorf 5424R centrifuge with FA-45-24-11 rotor

5.2 Procedure

1. Redissolve the dried extracts in 100 μL 80% methanol and transfer samples to vial inserts sitting in 2 mL vials. Load experimental sample and standard sample vials in the autosampler tray maintaining at 4 °C
2. Set chromatographic program with a gradient of solvent B in solvent A as 5% B (0 min); 30% B (1 min); 80% B (19 min); 99% B (20 min); 99% B (22 min); 5% B (23 min) at flow rate of 0.1 mL/min. The mass spectrometer parameters are set as the follows: sheath gas flow rate 40 units (negative model), aux gas unit flow rate 10; capillary temperature 350 °C; aux gas heater temperature 250 °C; spray voltage 3.5 kV; S lens RF level 60. The UV–Vis diode array detector detection wavelengths are 265, 280, 310 and 330 nm. Set an acquisition sequence including an instrument method, a file name, a directory path, a vial position, and an injection volume of 1 μL. A blank sample (80% methanol) is loaded at the first position to equilibrate the column
3. Start the sequence run and data acquisition
4. Perform peak integration of authentic standard samples and define the individual peak areas for coniferyl alcohol and *p*-coumaric acid, respectively. Plot the peak areas vs injection amount of a series of diluted standard samples and obtain the linear regression standard curves (Fig. 3) (see note 1).
5. Perform peak integration of experimental samples and define absorptive peak areas for 5-hydroxyconiferyl alcohol (at 265 nm), *p*-coumaric acid (at 280 nm) and the internal standard chrysin (at 265 nm or 280 nm).
6. Calculate the catalytic efficiencies of F5H1 and C4H according to the equation below. Here, *A1* is the peak area of the product (5-hydroxyconiferyl alcohol or *p*-coumaric acid), *A2* is the peak area of the internal standard (chrysin). *c* is the y-intercept of peak area, and *m* is the slope of linear regression equition of standard curve. *V1* is the injection volume, *V2* is total sample volume, *t* is reaction time (min) and *n* is yeast cell number. *B1* is the total amount of ethyl acetate added to sample and *B2* is the amount of ethyl acetate taken from the partitioning

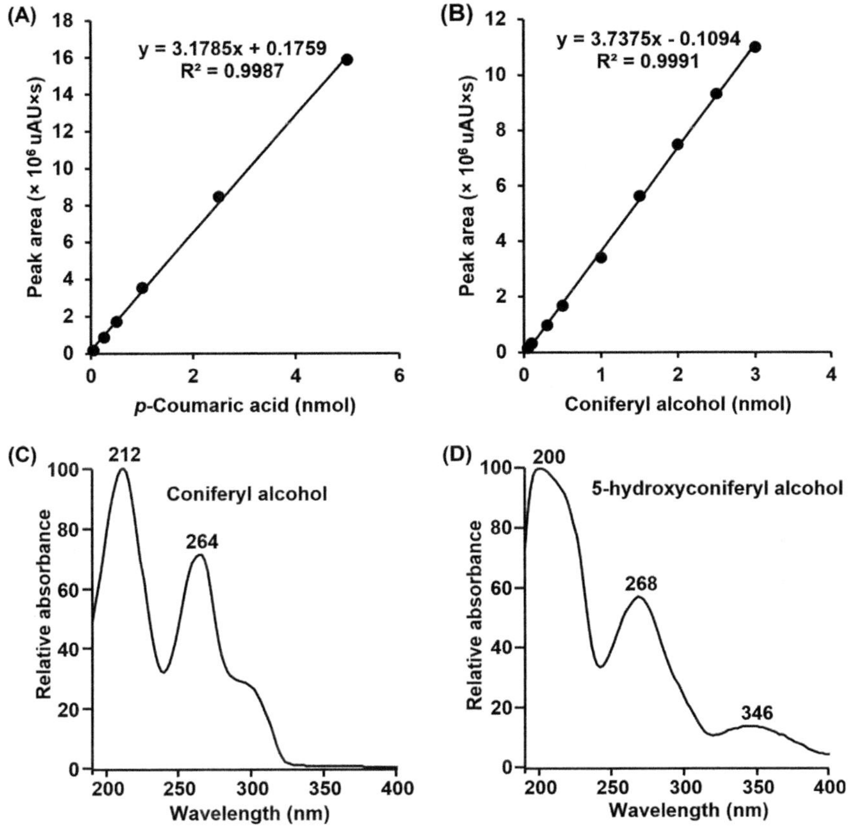

Fig. 3 Standard curves for quantifying P450 C4H and F5H reaction products. (A) *p*-Coumaric acid. (B) Coniferyl alcohol. The x-axis defines the amount of *p*-coumaric acid (A) or coniferyl alcohol (B), and the y-axis defines the peak area at 280 nm for *p*-coumaric acid (A) or 265 nm for coniferyl alcohol (B). Formula $y = mx + b$ was used to calculate the linear regression standard curve, where m and b represent the slope and intercept, respectively. (C) and (D) UV spectra of coniferyl alcohol (C) and 5-hydroxyconiferyl alcohol (D).

$$\text{Catalytic efficiency (pmol min}^{-1}\text{ cells}^{-1}) = \frac{\left(A1/\left(A2/\overline{A2}\right)-c\right)\times V2\times B1}{m\times V1\times t\times n\times B2}$$

5.3 Notes

1. Because of the commercial unavailability of 5-hydroxyconiferyl alcohol authentic compound, the standard curve of coniferyl alcohol was used instead to quantify F5H enzymatic product. This is to take advantage of both compounds sharing similar UV spectra at the wavelength of 264–268 nm (Fig. 3C and D).

6. Yeast induction and microsomal protein preparation

6.1 Materials

1. Yeast strains containing different expression cassettes: *pYeDP60-F5H1-2muCB5D-C4H, pYeDP60-F5H1-CB5D-C4H, pYeDP60-F5H1-2muCB5D-C4H-CBR1, pYeDP60-F5H1-CB5D-C4H-CBR1, pYeDP60-F5H1-2muCB5D-C4H-ATR1, pYeDP60-F5H1-CB5D-C4H-ATR1, pYeDP60-F5H1-2muCB5D-C4H-ATR2 and pYeDP60-F5H1-CB5D-C4H-ATR2*
2. Buffer A: 50 mM Tris/HCl (pH 7.4), 2 mM EDTA, 100 mM KCl
3. Buffer B: 50 mM Tris/HCl (pH 7.4), 2 mM EDTA, 0.6 M sorbitol
4. Buffer C: 50 mM Sodium Pipes (pH 7.0), 4 mM EDTA, 20% glycerol
5. 0.5 mm glass mill beads (BioSpec Products, # 11079105)
6. 100 μm cell strainers (Corning, # 07-201-432)
7. WHEATON® Tenbroeck Tissue Grinder, 2 mL
8. Eppendorf 5424R centrifuge with FA-45-24-11 rotor
9. GENESYS™ 10S UV–Visible Spectrophotometers (Thermo Scientific)
10. Beckman GS-6 centrifuge with G.H. 3.8 rotor
11. Sorvall RC 5C Plus centrifuge with SS-34 rotor
12. New Brunswick Innova 42 incubator shaker

6.2 Procedure

1. Streak the yeast strain stock on a SD-Ura + 200 μg/mL G418 agar plate. Incubate the plate at 28 °C for 2–3 days until single colonies are observed
2. Inoculate about 10 yeast colonies into 10 mL SD-Ura + 2% glucose selective medium. Incubate yeast starting culture at 28 °C with 250 rpm shaking for overnight
3. Inoculate the 10 mL yeast starting culture into 40 mL SD-Ura + 2% glucose selective medium. Incubate the secondary culture at 28 °C with 250 rpm shaking for about 8 h (see note 1).
4. Transfer the secondary culture to 50 mL centrifuge tubes and rinse the flasks with sterilized water once. Pellet yeast cells by centrifugation at 2000 × *g* (Beckman GS-6 centrifuge) for 5 min. Resuspend the yeast cell pellets with 2 mL SD-Ura + 2% galactose inductive medium

5. Inoculate appropriate resuspended yeast cells into 500 mL SD-Ura + 2% galactose inductive medium to a cell density about 0.6×10^7 cells/mL. Incubate at 28 °C with 250 rpm shaking for about 14 h
6. Pellet yeast cells at 2000 × *g* (Beckman GS-6 centrifuge) for 5 min and wash once with buffer A
7. Resuspend yeast pellet with appropriate cold buffer B (0.3–0.5 g wet cell/mL) and move yeast cells to 50 mL centrifuge tubes
8. Add same volume of glass beads
9. Disrupt yeast cell walls mechanically by handshaking (30 s of shaking and 30 s interval, 10 cycles) in cold room. Otherwise, if the procedure is done at room temperature, keep yeast cells on ice during interval time (see note 2).
10. Place a cell strainer on a 50 mL centrifuge tube and pass the crude extracts
11. Wash the glass beads for four times with 6 mL of buffer B each time and pass the strainer
12. Centrifuge at 7650 × *g* (Sorvall RC 5C Plus centrifuge) at 4 °C for 10 min and recover the supernatant
13. Repeat step 12 once and recover the supernatant (see note 3). Pool supernatant in step 12 and 13 together
14. Add polyethyleneglycol 4000 (at the final concentration of 0.1 g/mL) and NaCl (at final concentration of 8.8 mg/mL) to the supernatant
15. Incubate the extracts on ice with periodic mixing or rotating in cold room for 1 h
16. Centrifuge at 9682 × *g* (Sorvall RC 5C Plus centrifuge) for 20 min to collect the microsomal protein fractions. Discard the supernatant. Wash the pellet twice with 10 mL buffer B
17. Add 1.5 mL buffer C to the microsomal pellet and briefly resuspend the pellet. Transfer the microsomal pellet to a 2 mL Cole-Parmer Essentials PTFE tissue grinder. Add 0.5 mL buffer C to rinse the centrifuge tube to transfer all the pellet
18. Homogenize the pellet gently with about 10 strokes
19. Quantify the microsomal protein concentration using Bradford reagent. Adjust concentration to 10 mg/mL
20. 200 μL aliquots are snap-frozen in liquid nitrogen and stored at −80 °C until further use

6.3 Notes

1. The purpose of this protocol is to compare the catalytic efficiencies of P450s coupling with different electron transfer components. Therefore, it is critical to keep all the experimental strains under the same condition and treatment
2. Yeast cells are tough to lyse. The handshaking should be done vigorously
3. When transferring the supernatants to a new tube after centrifugation, do not try to pour out all the upper layer liquid. Otherwise, it will introduce yeast cell debris contamination

7. In vitro enzymatic assays with yeast microsomal preparation

NADH and NADPH are the reductant cofactors for CBR and CPR, respectively. Using the isolated microsomal proteins containing monolignol P450 enzymes and particular redox proteins, the preferences for NADH and NADPH in augmenting C4H- or F5H1-catalyzed reaction can be assessed.

7.1 Materials

1. Coniferaldehyde (Sigma, #382051), 2 mM solution is prepared in the same way as above
2. *t*-Cinnamic acid (Sigma, #C80857), 2 mM solution is prepared in the same way as above
3. Assay buffer: 50 mM Sodium Pipes (pH 7.0), 4 mM EDTA, 20% glycerol
4. 50 mM $NaHCO_3$ (pH 9)
5. NADH (Sigma, #10128023001), 100 mM solution was prepared in 50 mM $NaHCO_3$ (pH 9)
6. NADPH (Sigma, # 1010782400), 100 mM solution was prepared in 50 mM $NaHCO_3$ (pH 9)
7. Water saturated ethyl acetate containing 0.1 mM chrysin
8. Centrivap benchtop vacuum concentrator (Labconco)
9. Eppendorf 5424R centrifuge with FA-45-24-11 rotor
10. Beckman GS-6 centrifuge with G.H. 3.8 rotor
11. New Brunswick Innova 42 incubator shaker

7.2 Procedure

1. Label two rows of nine 1.5 mL centrifuge tubes with A1 to A9 for F5H assays and B1 to B9 for C4H assays, respectively
2. Add 1 μL of 50 mM $NaHCO_3$ (as control), 100 mM NADH, and 100 mM NADPH, respectively, to each of three tubes, until to all nine tubes, thus obtaining triplicates
3. For F5H enzymatic assays, mix 95 μL 2 mM coniferaldehyde and 190 μL microsomal protein (at the concentration of 10 mg/mL) in 6.65 mL assay buffer; then aliquot 100 μL of the mixture to each tube of the row A. The final concentration of coniferaldehyde is 0.2 mM and microsomal protein is 200 μg per reaction (see note 1).
4. For C4H enzymatic assays, mix 95 μL 2 mM *t*-cinnamic acid and 47.5 μL microsomal protein (10 mg/mL) in 8.07 mL assay buffer; then aliquot 100 μL of the mixture to each tube of the row B. The final concentration of *t*-cinnamic acid is 0.2 mM and microsomal protein is 50 μg per reaction (see note 2).
5. Incubate the reactions at 28 °C for 1 h
6. Add 0.2 mL water saturated ethyl acetate containing 0.1 mM chrysin to each tube and vigorously vortex to stop reactions. Partition the reaction products by centrifugation at 20,000 g (Eppendorf 5424R centrifuge) for 5 min. 150 μL extracts are vacuum-dried and stored in freezer
7. The reaction products are resuspended with 60 μL 80% methanol and quantified by UHPLC-MS (Fig. 4) following the same procedure as described for the whole cell biocatalytic assays (Section 6).
8. In LC-MS chromatographic profiles, a clear product peak, identified as *p*-coumaric acid, can be detected in the C4H-catalyzed reaction at the presence of 100 mM NADPH (Fig. 4A, C and D); while two product peaks are observed in the F5H1-catalyzed in vitro reaction with coniferaldehyde as substrate at the presence of either NADPH or NADH (Fig. 4B). They are the expected reaction product 5-hydroxconiferaldehyde and the non-specific product ferulic acid (Fig. 4E–H). Presumably, the endogenous aldehyde dehydrogenase activity in yeast microsome might cause the formation of ferulic acid. Note that 5-hydroxyconiferaldehyde and ferulic acid have the same molecular weight but different MS2 spectra, which enable to distinguish them from each other (Fig. 4E–H).

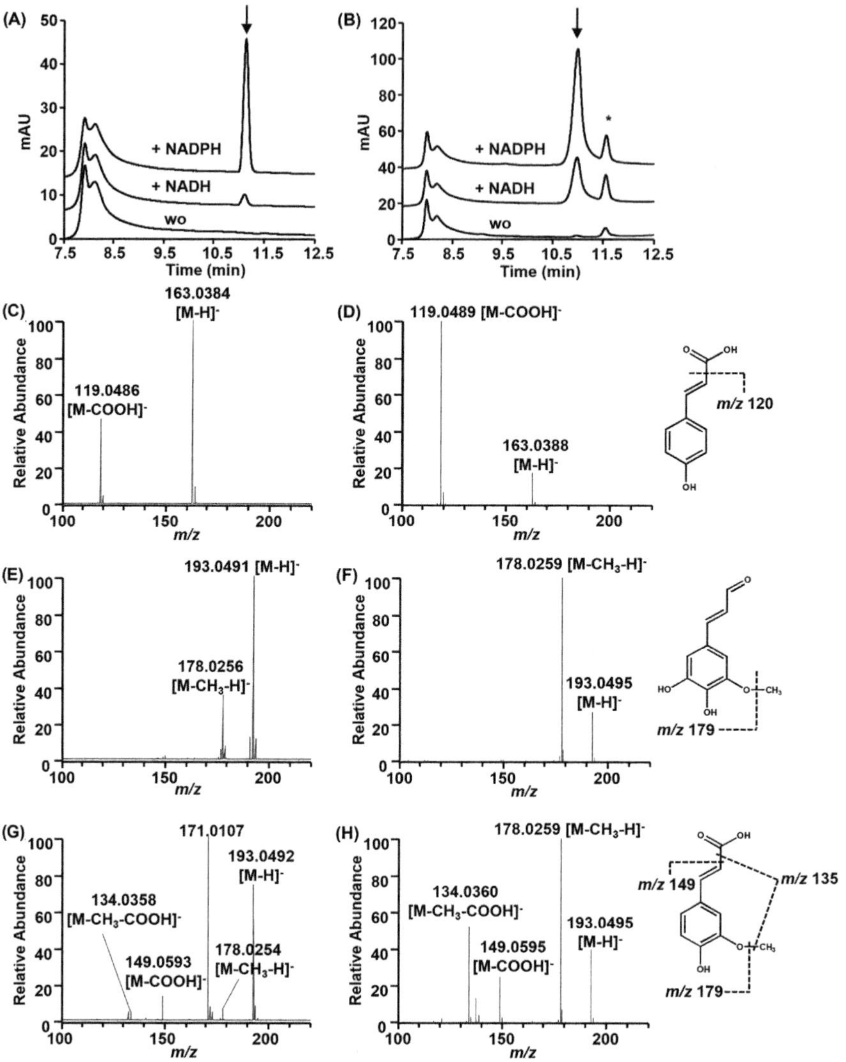

Fig. 4 Characterization of C4H- and F5H1-catalyzed reactions. (A) and (B) the portion of UHPLC-UV profiles of the C4H (A) and F5H1 (B)-catalyzed reactions. Microsomal proteins were extracted from the engineered *cyb5* yeast co-expressing C4H, F5H1, CBR1 and CB5D. The arrows in (A) and (B) indicate *p*-coumaric acid and 5-hydroxyconiferaldehyde products, respectively. The asterisk in (B) indicates the by-product ferulic acid. (C and D) Mass spectrum of *p*-coumaric acid (C) and MS2 spectrum of the molecular ion 163.0384 *m/z* (D). (E and F) Mass spectrum of 5-hydroxyconifer aldehyde product (E) and MS2 spectrum of the molecular ion 193.0491 *m/z* (F). (G and H) Mass spectrum of ferulic acid by-product (G) and MS2 spectrum of the molecular ion 193.0492 *m/z* (H). *Trans*-cinnamic acid and coniferaldehyde were used as the substrates for C4H (A) and F5H1 (B) respectively. NADH or NADPH were added to the reactions as reductants and a reaction without reductant (wo) was conducted as the control. After 1 h incubation at 28 °C, ethyl acetate extracts were monitored by UHPLC-MS.

7.3 Notes

1. Both coniferaldehyde and coniferyl alcohol can be used as substrates for the in vitro enzymatic assays of F5H1 activity, but coniferaldehyde appears to be a more preferred substrate for the enzyme
2. The catalytic efficiency of C4H is much higher than that of F5H1, which requires less microsomal proteins to obtain the detectable activity. Attention should be paid to ensure enough substrate in the reaction to obtain the precise max velocity

8. Summary

Plant P450 systems (monooxygenases and the related redox partners) are the ER-localized membrane-bound proteins, which makes their characterization challenging. The heterologous yeast expression system is conventionally the preferential choice to produce P450 proteins and to determine their activities. P450 monooxygenase is insufficient for its activity, and the redox components are necessarily required for delivering reducing power to support the P450-catalyzed oxidative reaction. The method described here employs polycistronic gene expression strategy to integrate monolignol biosynthetic P450 enzymes with different electron transfer components in yeast mutant cells. This strategy combines genes encoding P450 enzymes and redox partners into a same operon, thus ensuring the synchronized gene expression and protein production. With the developed whole cell biocatalytic assays and the in vitro enzymatic activity confirmation, the described method offers a relatively time- and labor-efficient procedure to examine and compare the functional associations of monolignol P450 enzymes with their preferred electron transfer systems. The described procedure can be applied to evaluate any given P450 monooxygenase-electron transport chain system.

Acknowledgment

This work was supported by the U.S. Department of Energy, Office of Science, Office of Basic Energy Sciences under contract number DE-SC0012704—specifically through the Physical Biosciences program of the Chemical Sciences, Geosciences and Biosciences Division (to C.-J.L.).

Author contributions

X. Z. and C.-J. L. designed the experiments. X.Z. conducted experiments. X.Z. and C.-J. L. analyzed the data and wrote the manuscript.

Competing interests

The authors declare that they have no competing interests.

Data and materials availability

All data and materials needed to evaluate the conclusions in the paper are present in the paper. Additional data and materials related to this paper may be requested from C.J.L.

References

Bak, S., Beisson, F., Bishop, G., Hamberger, B., Hofer, R., Paquette, S., et al. (2011). Cytochromes p450. *Arabidopsis Book*, *9*, e0144. https://doi.org/10.1199/tab.0144.

Bernhardt, R. (2006). Cytochromes P450 as versatile biocatalysts. *Journal of Biotechnology*, *124*(1), 128–145. https://doi.org/10.1016/j.jbiotec.2006.01.026.

Bhatt, M. R., Khatri, Y., Rodgers, R. J., & Martin, L. L. (2017). Role of cytochrome b5 in the modulation of the enzymatic activities of cytochrome P450 17alpha-hydroxylase/17,20-lyase (P450 17A1). *The Journal of Steroid Biochemistry and Molecular Biology*, *170*, 2–18. https://doi.org/10.1016/j.jsbmb.2016.02.033.

Boerjan, W., Ralph, J., & Baucher, M. (2003). Lignin biosynthesis. *Annual Review of Plant Biology*, *54*, 519–546. https://doi.org/10.1146/annurev.arplant.54.031902.134938.

Cowley, A. B., Altuve, A., Kuchment, O., Terzyan, S., Zhang, X., Rivera, M., et al. (2002). Toward engineering the stability and hemin-binding properties of microsomal cytochromes b5 into rat outer mitochondrial membrane cytochrome b5: Examining the influence of residues 25 and 71. *Biochemistry*, *41*(39), 11566–11581.

de Vetten, N., ter Horst, J., van Schaik, H. P., de Boer, A., Mol, J., & Koes, R. (1999). A cytochrome b5 is required for full activity of flavonoid 3′, 5′-hydroxylase, a cytochrome P450 involved in the formation of blue flower colors. *Proceedings of the National Academy of Sciences of the United States of America*, *96*(2), 778–783.

Degtyarenko, K. N., & Archakov, A. I. (1993). Molecular evolution of P450 superfamily and P450-containing monooxygenase systems. *FEBS Letters*, *332*(1–2), 1–8. https://doi.org/10.1016/0014-5793(93)80470-f.

Duggal, R., Liu, Y., Gregory, M. C., Denisov, I. G., Kincaid, J. R., & Sligar, S. G. (2016). Evidence that cytochrome b5 acts as a redox donor in CYP17A1 mediated androgen synthesis. *Biochemical and Biophysical Research Communications*, *477*(2), 202–208. https://doi.org/10.1016/j.bbrc.2016.06.043.

Enoch, H. G., & Strittmatter, P. (1979). Cytochrome b5 reduction by NADPH-cytochrome P-450 reductase. *The Journal of Biological Chemistry*, *254*(18), 8976–8981.

Estrada, D. F., Laurence, J. S., & Scott, E. E. (2016). Cytochrome P450 17A1 interactions with the FMN domain of its reductase as characterized by NMR. *The Journal of Biological Chemistry*, *291*(8), 3990–4003. https://doi.org/10.1074/jbc.M115.677294.

Fukuchi-Mizutani, M., Mizutani, M., Tanaka, Y., Kusumi, T., & Ohta, D. (1999). Microsomal electron transfer in higher plants: Cloning and heterologous expression of NADH-cytochrome b5 reductase from Arabidopsis. *Plant Physiology*, *119*(1), 353–362.

Gibson, D. G., Young, L., Chuang, R. Y., Venter, J. C., Hutchison, C. A., & Smith, H. O. (2009). Enzymatic assembly of DNA molecules up to several hundred kilobases. *Nature Methods*, *6*(5), 343–U341. https://doi.org/10.1038/Nmeth.1318.

Gou, M., Yang, X., Zhao, Y., Ran, X., Song, Y., & Liu, C. J. (2019). Cytochrome b5 is an obligate electron shuttle protein for syringyl lignin biosynthesis in Arabidopsis. *The Plant Cell*. https://doi.org/10.1105/tpc.18.00778.

Guo, M., Ran, X., Martin, D. W., & Liu, C.-J. (2018). The scaffold proteins of lignin biosynthetic cytochrome P450 enzymes. *Nature Plants*, *4*, 299–310. https://doi.org/10.1038/s41477-018-0142-9.

Hannemann, F., Bichet, A., Ewen, K. M., & Bernhardt, R. (2007). Cytochrome P450 systems—Biological variations of electron transport chains. *Biochimica et Biophysica Acta*, *1770*(3), 330–344. https://doi.org/10.1016/j.bbagen.2006.07.017.

Henderson, C. J., McLaughlin, L. A., & Wolf, C. R. (2013). Evidence that cytochrome b5 and cytochrome b5 reductase can act as sole electron donors to the hepatic cytochrome P450 system. *Molecular Pharmacology*, *83*(6), 1209–1217. https://doi.org/10.1124/mol.112.084616.

Hildebrandt, A., & Estabrook, R. W. (1971). Evidence for the participation of cytochrome b5 in hepatic microsomal mixed-function oxidation reactions. *Archives of Biochemistry and Biophysics*, *143*(1), 66–79. https://doi.org/10.1016/0003-9861(71)90186-X.

Hwang, Y. T., Pelitire, S. M., Henderson, M. P., Andrews, D. W., Dyer, J. M., & Mullen, R. T. (2004). Novel targeting signals mediate the sorting of different isoforms of the tail-anchored membrane protein cytochrome b5 to either endoplasmic reticulum or mitochondria. *Plant Cell*, *16*(11), 3002–3019. https://doi.org/10.1105/tpc.104.026039.

Ichinose, H., & Wariishi, H. (2012). Heterologous expression and mechanistic investigation of a fungal cytochrome P450 (CYP5150A2): Involvement of alternative redox partners. *Archives of Biochemistry and Biophysics*, *518*(1), 8–15. https://doi.org/10.1016/j.abb.2011.12.010.

Iyanagi, T. (2019). Molecular mechanism of metabolic NAD(P)H-dependent electron-transfer systems: The role of redox cofactors. *Biochimica et Biophysica Acta - Bioenergetics*, *1860*(3), 233–258. https://doi.org/10.1016/j.bbabio.2018.11.014.

Jensen, K., & Moller, B. L. (2010). Plant NADPH-cytochrome P450 oxidoreductases. *Phytochemistry*, *71*(2–3), 132–141. https://doi.org/10.1016/j.phytochem.2009.10.017.

Krest, C. M., Onderko, E. L., Yosca, T. H., Calixto, J. C., Karp, R. F., Livada, J., et al. (2013). Reactive intermediates in cytochrome p450 catalysis. *The Journal of Biological Chemistry*, *288*(24), 17074–17081. https://doi.org/10.1074/jbc.R113.473108.

Kumar, R., Tran, L. S. P., Neelakandan, A. K., & Nguyen, H. T. (2012). Higher plant cytochrome b5 polypeptides modulate fatty acid desaturation. *PLoS One*, 7(2). https://doi.org/10.1371/journal.pone.0031370. e31370.

Maggio, C., Barbante, A., Ferro, F., Frigerio, L., & Pedrazzini, E. (2007). Intracellular sorting of the tail-anchored protein cytochrome b5 in plants: A comparative study using different isoforms from rabbit and Arabidopsis. *Journal of Experimental Botany*, *58*(6), 1365–1379. https://doi.org/10.1093/jxb/erl303.

Nelson, D. R. (2009). The cytochrome p450 homepage. *Human Genomics*, *4*(1), 59–65. https://doi.org/10.1186/1479-7364-4-1-59.

Nett, R. S., Lau, W., & Sattely, E. S. (2020). Discovery and engineering of colchicine alkaloid biosynthesis. *Nature*, *584*(7819), 148–153. https://doi.org/10.1038/s41586-020-2546-8.

Niu, G., Zhao, S., Wang, L., Dong, W., Liu, L., & He, Y. (2017). Structure of the Arabidopsis thaliana NADPH-cytochrome P450 reductase 2 (ATR2) provides insight into its function. *The FEBS Journal*, *284*(5), 754–765. https://doi.org/10.1111/febs.14017.

Pompon, D., Louerat, B., Bronine, A., & Urban, P. (1996). Yeast expression of animal and plant P450s in optimized redox environments. In E. F. Johnson, & M. R. Waterman (Eds.), *Methods in enzymology* (Vol. 272, pp. 51–64). Academic Press.

Porter, T. D. (2002). The roles of cytochrome b5 in cytochrome P450 reactions. *Journal of Biochemical and Molecular Toxicology*, *16*(6), 311–316. https://doi.org/10.1002/jbt.10052.

Porter, T. D., & Coon, M. J. (1991). Cytochrome P-450. Multiplicity of isoforms, substrates, and catalytic and regulatory mechanisms. *The Journal of Biological Chemistry*, *266*(21), 13469–13472.

Schenkman, J. B., & Jansson, I. (2003). The many roles of cytochrome b5. *Pharmacology & Therapeutics*, *97*(2), 139–152.

Shockey, J. M., Dhanoa, P. K., Dupuy, T., Chapital, D. C., Mullen, R. T., & Dyer, J. M. (2005). Cloning, functional analysis, and subcellular localization of two isoforms of NADH:Cytochrome b5 reductase from developing seeds of tung (Vernicia fordii). *Plant Science*, *169*(2), 375–385. https://doi.org/10.1016/j.plantsci.2005.03.023.

Stiborová, M., Indra, R., Moserová, M., Frei, E., Schmeiser, H. H., Kopka, K., et al. (2016). NADH:Cytochrome b5 reductase and cytochrome b5 can act as sole electron donors to human cytochrome P450 1A1-mediated oxidation and DNA adduct formation by benzo[a]pyrene. *Chemical Research in Toxicology*, *29*(8), 1325–1334. https://doi.org/10.1021/acs.chemrestox.6b00143.

Sundin, L., Vanholme, R., Geerinck, J., Goeminne, G., Hofer, R., Kim, H., et al. (2014). Mutation of the inducible Arabidopsis thaliana cytochrome P450 reductase2 alters lignin composition and improves saccharification. *Plant Physiology*, *166*(4), 1956–1971. https://doi.org/10.1104/pp.114.245548.

Urban, P., Werck-Reichhart, D., Teutsch, H. G., Durst, F., Regnier, S., Kazmaier, M., et al. (1994). Characterization of recombinant plant cinnamate 4-hydroxylase produced in yeast. Kinetic and spectral properties of the major plant P450 of the phenylpropanoid pathway. *European Journal of Biochemistry*, *222*(3), 843–850. https://doi.org/10.1111/j.1432-1033.1994.tb18931.x.

Vergeres, G., & Waskell, L. (1995). Cytochrome b5, its functions, structure and membrane topology. *Biochimie*, *77*(7–8), 604–620.

Wayne, L. L., Wallis, J. G., Kumar, R., Markham, J. E., & Browse, J. (2013). Cytochrome b5 reductase encoded by CBR1 is essential for a functional male gametophyte in Arabidopsis. *Plant Cell*, *25*(8), 3052–3066. https://doi.org/10.1105/tpc.113.113324.

Yamazaki, H., Johnson, W. W., Ueng, Y. F., Shimada, T., & Guengerich, F. P. (1996). Lack of electron transfer from cytochrome b5 in stimulation of catalytic activities of cytochrome P450 3A4. Characterization of a reconstituted cytochrome P450 3A4/NADPH-cytochrome P450 reductase system and studies with apo-cytochrome b5. *The Journal of Biological Chemistry*, *271*(44), 27438–27444.

Yamazaki, H., Nakamura, M., Komatsu, T., Ohyama, K., Hatanaka, N., Asahi, S., et al. (2002). Roles of NADPH-P450 reductase and apo- and holo-cytochrome b5 on xenobiotic oxidations catalyzed by 12 recombinant human cytochrome P450s expressed in membranes of Escherichia coli. *Protein Expression and Purification*, *24*(3), 329–337. https://doi.org/10.1006/prep.2001.1578.

CHAPTER SIX

Programmable chemical actuator control of soluble and membrane-bound enzymatic catalysis

Parker Evans[a,*] and Katrina Cornish[b,c]

[a]Indiana Center for Regenerative Medicine & Engineering, Indiana University Health Comprehensive Wound Center, Department of Surgery, Indiana University School of Medicine, Indianapolis, IN, United States
[b]Department of Horticulture and Crop Science, College of Food Agricultural and Environmental Sciences, The Ohio State University, Wooster, OH, United States
[c]Department of Food, Agricultural and Biological Engineering, College of Food Agricultural and Environmental Sciences, The Ohio State University, Wooster, OH, United States
*Corresponding author: e-mail address: evanspa@iu.edu

Contents

Methods in Enzymology, Volume 676
ISSN 0076-6879
https://doi.org/10.1016/bs.mie.2022.07.022

Abstract

Enzyme catalyzed reactions can be started, stopped, and throttled by regulating ion concentration in solution. The flow of charge through solution is related to the flow of electrons through an electric circuit using electrochemistry. However, current electrochemical methods to vary ionic concentration are largely limited to the electrode surface, making it impractical to exploit ionic concentration as a control variable in traditional enzyme assays. This chapter presents an electrochemical cell, termed a programmable chemical actuator (PCA), that uses selective electrochemical REDOX reactions to regulate the concentration of ions that control enzyme catalyzed reaction kinetics. The PCA is a polypyrrole polymerized and equilibrated to impart Mg^{2+}, Mn^{2+}, or OH^- ion selectivity to working electrodes. Duplex pulse amperometry demonstrated reversible control over the pH-dependent kinetics of alkaline phosphatase, a soluble homodimeric enzyme, during a spectrophotometric assay. PCAs were also used to extend electrochemical control to the membrane-bound enzyme complex rubber transferase by modulating Mg^{2+} and Mn^{2+} concentrations during a radiometric enzyme assay. PCAs demonstrated tunable control over rubber transferase activity and rubber molecular weight without dislodging either the elongating rubber molecule or irreversibly inhibiting the rubber transferase complex. This chapter presents methods to fabricate and operate PCAs to facilitate insights into the roles of ions in biochemical pathways and environmental responses in plants.

1. Introduction

Biocatalysis is the use of enzymes to catalyze chemical synthesis. Biocatalysts are well documented in the synthesis of industrially important compounds for the chemical and pharmaceutical sectors (Delgove, Elford, Bernaerts, & Wildeman, 2018; Messiha et al., 2018; Woodley, 2020).

The use of enzymes provides various sustainability benefits that regular chemical treatments lack. These advantages frequently include greater product quality, less waste, and lower energy use. The field has used catalytic processes, kinetic characteristics, and enzyme structure-function connections to accomplish this.

Ionic composition of the enzyme-containing electrolyte is commonly used to control biocatalysis because the prominence of charged groups in proteins make electrostatic interactions highly influential over protein structure, stability, and activity (Gitlin, Carbeck, & Whitesides, 2006; Nelson, Lehninger, & Cox, 2008; Sandler, 2017). As depicted in Fig. 1, charged groups on proteins are concentrated on the protein surface but are also critical components of enzyme active sites and tertiary structures (Fu, Klibanov, & Langer, 2000; Güven, Prodanovic, & Schwaneberg, 2010; Hageman, 1988; Manning, Patel, & Borchardt, 1989; Nelson et al., 2008).

The most common method to alter ionic composition is by fluid exchange (Lane, 2010; Levin, 2012; Rodrigues & Curtis, 2010). This is accomplished via micro/nanofluidics, pipette, and circulation pumps, depending on scale of the process and system. Optical methods provide much higher spatial resolution than solution exchange methods but are

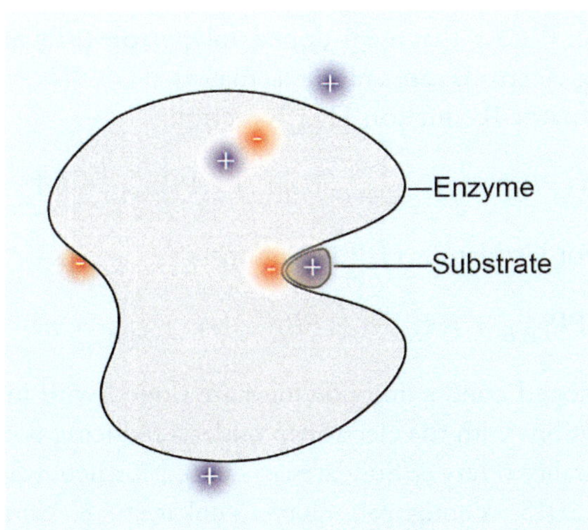

Fig. 1 Schematic diagram of an enzyme depicting charged groups at the surface, active site, and buried in the enzyme structure. *Adapted from Gitlin, I., Carbeck, J.D., & Whitesides, G.M., (2006). Why are proteins charged? Networks of charge–charge interactions in proteins measured by charge ladders and capillary electrophoresis.* Angewandte Chemie, International Edition *45, 3022–3060.*

currently limited to pH reactions with photoacids and photobases (Kagel, Bier, Frohme, & Glökler, 2019; Kohse, Neubauer, Pazidis, Lochbrunner, & Kragl, 2013). Bare metal electrodes can control additional ions, but often denature biology or affect limited volumes (Bard, Faulkner, Leddy, & Zoski, 1980; Ding, Hu, Paek, & Mitlin, 2018; Li & Anand, 2016). Coating electrodes with conducting polymers has been shown to improve the properties of electrochemical ion control and shows promise for enzyme control (Lee et al., 2020; Sundaresan, Northcutt, Venkatesh, & Evans, 2017).

Polypyrrole is a commonly used conducting polymer due to its biocompatibility and aqueous synthesis conditions (Shirakawa, 2001a, 2001b). Polypyrrole is electrically conducting in its charged state due to charge carriers along the backbone, referred to as dopants (Venugopal, 2016). Electronically reducing and oxidizing a polypyrrole membrane controls the doping level of the polymer (Iyoda, Ohtani, Shimidzu, & Honda, 1986; Sadki, Schottland, Brodie, & Sabouraud, 2000). Polymer conductivity varies with dopant level. The change in doping level is accompanied by ion exchange with an electrolyte to maintain charge neutrality (Northcutt, 2015; Northcutt & Sundaresan, 2015).

Conducting polymers doped with mobile anions are termed p-doped and exchange anions with the electrolyte under a reducing potential to maintain charge neutrality (Shinohara, Kojima, & Aizawa, 1989). Programmable chemical actuator (PCAs) that use p-doped polypyrrole (PPy in the equations) at their working electrodes are anionic actuators.

Anionic Actuator Reduction Half Reaction:

$$\left((PPy^{n+})(A^-)_n\right)_{WE} + ne^- \leftrightarrow PPy^{\circ}_{WE} + nA^-_{aq} \tag{1}$$

Anionic Actuator Oxidation Half Reaction:

$$PPy^{\circ}_{WE} + nA^-_{aq} \rightarrow \left((PPy^{n+})(A^-)_n\right)_{WE} + ne^- \tag{2}$$

Conversely, n-doped conducting polymers are doped with immobile anions and exchange cations with the electrolyte under a reducing potential to maintain charge neutrality (Hery & Sundaresan, 2016; Northcutt, 2015; Northcutt & Sundaresan, 2015; Venugopal, Hery, Venkatesh, & Sundaresan, 2017; Venugopal, Zhang, Northcutt, & Sundaresan, 2014). PCAs that use n-doped polypyrrole at their working electrodes are cationic actuators. Ion ingress is proportional to the electronic current measured by the potentiostatic circuit (Lee et al., 2020; Venugopal, 2016; Venugopal et al., 2017, 2014). Consequently, the Faradaic charge is assumed to be equal to the ion exchange for a monovalent

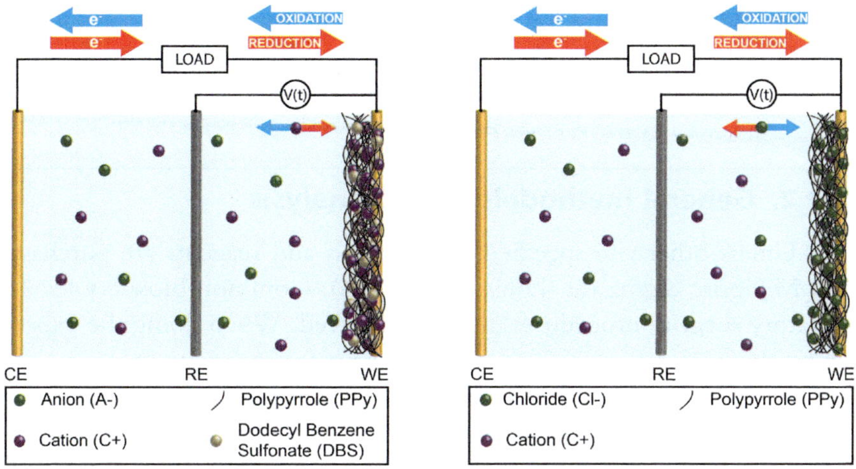

Fig. 2 (A) Schematic diagram of a reduced cationic programmable chemical actuator with a dodecylbenzenesulfonate-doped polypyrrole (PPy(DBS)) working electrode (WE), gold counter (CE) and silver-silver chloride (Ag/AgCl) reference electrodes (RE). Reducing the WE imparts an ion motive force (red arrows) that pulls cations into the polymer and oxidizing the polymer reverses the ion motive force (blue arrow) pushing cations out of the polymer. The chemical structure of a dopant site is shown below the working electrode. (B) Schematic diagram of an oxidized anionic programmable chemical actuator with a chloride-doped polypyrrole (PPy(Cl)) WE. The reduction of IMF pushes anions out of the polymer and the oxidative IMF pulls anions into the polymer. The chemical structure of a dopant site is shown below the working electrode. Please note, the RE is much smaller than the control volume and does not obstruct ion movement.

ion (Venugopal et al., 2017, 2014). Fig. 2A shows the operation of a cationic actuator and 2b depicts the operation of an anionic actuator.

Cationic Actuator Reduction Half Reaction:

$$\left(PPy^{n+}(DBS^{-})_{n}\right)_{WE} + nC^{+}_{aq} + ne^{-} \rightarrow \left((PPy)(DBS^{-}C^{+})_{n}\right)_{WE} \quad (3)$$

Cationic Actuator Oxidation Half Reaction:

$$\left((PPy)(DBS^{-}C^{+})_{n}\right)_{WE} \rightarrow \left(PPy^{n+}(DBS^{-})_{n}\right)_{WE} + nC^{+}_{aq} + ne^{-} \quad (4)$$

Polypyrrole presents itself as an interesting method to quantitatively control enzyme activity by electrochemically modulating ion concentration. In this chapter, we describe methods used for PCA fabrication, control of a soluble metalloenzyme, and control of a membrane bound metal activated enzyme. Specifically, alkaline phosphatase was controlled with hydroxyl ion actuators and rubber transferase was controlled with magnesium and manganese actuators. The procedure for enzymatic control using a PCA is as follows: (1) electrode fabrication, (2) PCA assembly, (3) electropolymerization, (4) program ion selection, and (5) enzyme activity control testing.

2. General methodology and analysis

Unless otherwise specified, all materials and reagents are purchased from Millipore Sigma (St. Louis, MO, USA). Common biosafety level-1 laboratory security procedures must be followed. When doing the experiments, the researcher must use personal protection equipment such as safety glasses, a lab coat, and chemically resistant gloves. MATLAB v2020b for MacOS isused to evaluate the results (Mathworks, Natick, MA, United States). The user can use alternative compatible applications based on their preferences. Kolmogorov-Smirnov tests, normal distribution fitting, means, standard deviations, F-tests, and analyses of variances are calculated in MATLAB using the statistics toolbox. Unless otherwise indicated, the statistical difference is assessed on triplicates with a significance level of $P=0.05$. The potentiostat/galvanostst used in this work is an Ivium PocketStat v1 (Ivium Technologies AJ Eindhoven, Netherlands).; however, the user can use alternative compatible hardware based on availability/preference.

2.1 Specific capacitance of dodecylbenzenesulfonate-doped polypyrrole (PPy(DBS))

The capacitance of a metal electrode can be increased up to 352 F/cm^3 by synthesizing a conducting polymer on the surface (Bard, 2001; Ding et al., 2018; Sahalianov, Singh, Tybrandt, Berggren, & Zozoulenko, 2019). The specific capacitance of a 100 mm^2 polypyrrole film deposited at 0.5C/cm^2, doped with dodecylbenzenesulfonate can be calculated using Eq. (5):

$$C = \frac{\delta \times F}{\Delta E \times Py_{MW}} = 593.23\ F/g \tag{5}$$

where δ is the doping level per monomer unit (0.33), F is the Faraday constant (96,485C/mol), ΔE is the operating voltage range (0.8 V), and Py_{MW} is the molecular weight of the pyrrole monomer (67.09 g/mol) (Ding et al., 2018; Feldberg, 1984; Sahalianov et al., 2019). The specific capacitance of a 1 cm^2 polypyrrole film deposited at 0.5C/cm^2 and doped with dodecylbenzenesulfonate is 593.23 F/g.

2.2 PPy(DBS) mass as a function of charge density

The relationship between polymer mass (m) and charge density accumulated during polymerization (Q_p) is calculated using Eq. (6):

$$m = \frac{Q_p \times (Py_{MW} + \gamma \times A^-{}_{MW} - 2H^+{}_{AW})}{F \times (2 + \gamma)} = 3.96 \times 10^{-4} g \quad (6)$$

where $A^-{}_{MW}$ *and* $H^+{}_{AW}$ are the molecular weight of the anion dopant (342.4 g/mol for DBS) and atomic weight of the proton (1 kg/mol). The mass of a 100 mm^2 polypyrrole film deposited at 0.5C/cm^2 is 396 μg (Dejeu et al., 2010).

2.3 Duty cycle

The frequency at which potential pulses are applied affects the kinetics of ion transport (Datta & Landolt, 1985; Green & Roy, 2017; Seitanidou, Tybrandt, Berggren, & Simon, 2019). The switching frequency can be defined in terms of a duty cycle where the duration at full reduction/oxidation is expressed as a percentage of the total oxidation cycle duration.

$$DutyCycle\% = \frac{t_{RED}}{t_{OX_1} + t_{OX_2}} \times 100 \quad (7)$$

3. Electrode fabrication

3.1 Alkaline phosphatase chip electrode fabrication

Standard lift-off photolithography methods are used to fabricate custom electrodes of precise dimensions. The basic process for chip electrode fabrication is illustrated in Fig. 3. All materials and reagents used for chip electrode fabrication can be provided by NanoTech West (Columbus, OH, USA). It is necessary to use similar clean room facilities to fabricate the chip electrodes. The working electrode is fabricated using reported methods from a 100 mm diameter by 1 mm thick silicon wafer with one side polished (Northcutt, 2015).

3.1.1 Equipment

- BlueM Resist Bake Oven (Blue M, New Columbia, PA, USA)
- Laurell Polymer Spin Coater (Laurell, North Wales, PA, USA)
- Vistec EBPG5000 Electron Beam Lithography System (Vistec Lithography, Watervliet, NY, USA)

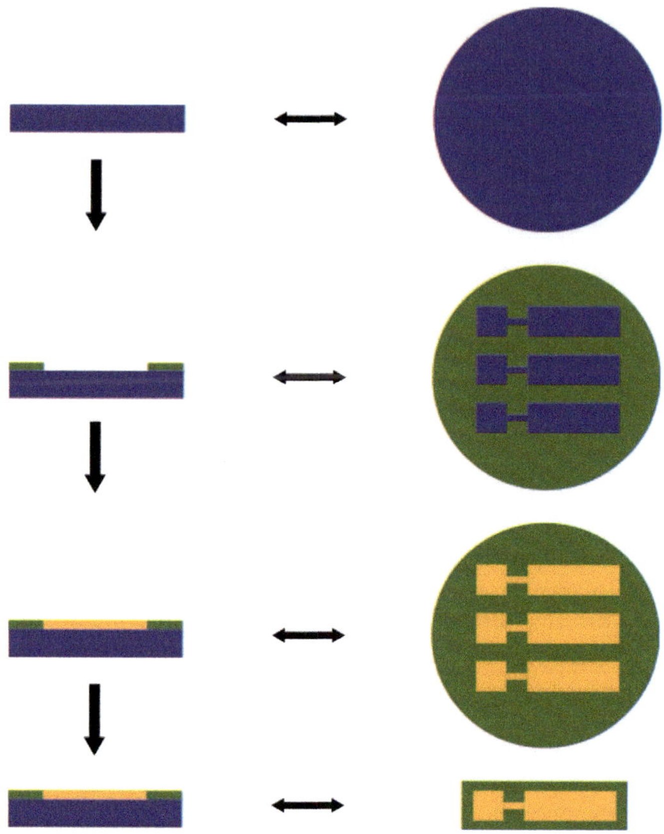

Fig. 3 Schematic of the lithography process used to fabricate the working electrodes used to control alkaline phosphatase activity. In the first steps, a clean wafer (blue) is patterned with photoresist (green), gold is evaporated onto the patterned surface, excess photoresist and gold are removed, and the electrodes are diced from the wafer for individual use.

- CHA Solution System *E*-Gun Evaporator (CHA Industries, South Elgin, IL, USA)
- Advanced Dicing Technologies 7122 Dicing Saw (Advanced Dicing Technologies, Horsham, PA, USA)

3.1.2 Consumables

- 1 mm thick silicon wafer single side polished (NanoTech West, Columbus, OH, USA)
- LOR2A photoresist (DuPont, Wilmington, DE, USA)
- SR1813 photoresist (DuPont, Wilmington, DE, USA)

- 1-Methyl-2-pyrrolidinone (NMP) (MilliporeSigma, St. Louis, MO, USA)
- Cr (NanoTech West, Columbus, OH, USA)
- Au (NanoTech West, Columbus, OH, USA)
- MF-319 developer (DuPont, DuPont, Wilmington, DE, USA)
- Sodium dodecylbenzenesulfonate (MilliporeSigma, St. Louis, MO, USA)
- Isopropyl alcohol (MilliporeSigma, St. Louis, MO, USA)
- N_2 gas (Ohio State Cryogenics Laboratory, Columbus, OH, USA)
- Copper tape (McMaster-Carr, Aurora, OH, USA)
- MarineWeld (McMaster-Carr, Aurora, OH, USA)

3.1.3 Procedure

1. The surface of the silicon-silicon nitride wafer is prepared for photoresist deposition by performing a dehydration bake at 190 °C for 300 s
2. Excess LOR2A photoresist film is deposited on the wafer and spun at a speed of 3000 RPM to form a 1 μm film
3. The wafer is baked at 190 °C for 300 s and SR1813 photoresist is deposited onto the wafer and spun at 3000 RPM to form a 1.4 μm film
4. The wafer with photoresist is baked at 115 °C for an additional 60 s
5. The wafer is exposed to UV light with an intensity of $15\,\mathrm{mW} \times \mathrm{cm}^{-2}$ for 2.4 s to pattern the photoresist then developed with MF-319 to remove the patterned photoresist
6. Metallization is performed using electron beam evaporation. First, an adhesive layer of 15 nm Cr is deposited. Subsequently a 60 nm Au layer is deposited at $1E^{-6}$ Torr in one small (5 × 5 mm) square and one large rectangle (15 × 5 mm) connected by a thin Au strip (500 μm × 5 mm).
7. Excess Cr and Au are removed using a lift-off technique with hot NMP
8. The individual chips are diced from the larger wafer with a diamond blade to 7 × 27 mm
9. Hydrophilization of the gold surfaces is accomplished by soaking the newly fabricated wafers in a 0.1 M solution of sodium dodecylbenzenesulfonate for at least 3 days
10. Following hydrophilization, the Au surface is cleaned using isopropyl alcohol and dried using N_2 gas
11. An electrical connection is made to the small Au pad using 5 cm of copper tape
12. The copper tape is insulated with MarineWeld
13. With a workable Au surface and an insulated electrical connection, the modified chips are referred to as chip electrodes

14. Chip electrodes should be stored covered, at room temperature, in the dark, until use

3.2 Alkaline phosphatase au foil electrode fabrication

Counter electrodes are fabricated by cutting gold foil to appropriate dimensions. High quality gold foil (99.99%) is difficult and expensive to obtain so we use the discarded sputter targets from our lab. This has the added benefit of plasma cleaning the electrode surface prior to beginning the procedures listed below. Alternatives are available, but care must be taken to clean the surface equally well.

3.2.1 Equipment

- Utility scissors (Fisher Scientific, Tewksbury, MA, USA)

3.2.2 Consumables

- Gold foil (lab-made or Fisher Scientific, Tewksbury, MA, USA)
- Isopropyl alcohol (MilliporeSigma, St. Louis, MO, USA)
- N_2 gas (Ohio State Cryogenics Laboratory, Columbus, OH, USA)
- Copper tape (McMaster-Carr, Aurora, OH, USA)
- MarineWeld (McMaster-Carr, Aurora, OH, USA)

3.2.3 Procedure

1. The Au surface is cleaned using isopropyl alcohol and dried using nitrogen gas
2. An electrical connection is made to the top using copper tape
3. As with the chip electrodes, the copper tape is insulated with MarineWeld
4. The Au surface is mechanically cleaned using isopropyl alcohol and dried under nitrogen
5. Gold foil electrodes is electrochemically cleaned, rinsed, and stored in the same manner as chip electrodes

3.3 Rubber transferase electrode fabrication

Smaller electrodes are used to accommodate the small volumes used in radiometric assays. Working and counter electrodes are fabricated from 500 μm diameter Au wire.

3.3.1 Equipment

- Utility scissors (Fisher Scientific, Tewksbury, MA, USA)

3.3.2 Consumables

- 500 μm diameter Au wire (Fisher Scientific, Tewksbury, MA, USA)
- Isopropyl alcohol (MilliporeSigma, St. Louis, MO, USA)
- N_2 gas (Ohio State Cryogenics Laboratory, Columbus, OH, USA)
- Sodium hydroxide (MilliporeSigma, St. Louis, MO, USA)
- Distilled Water (MilliQDirect16, MilliporeSigma, St. Louis, MO, USA)

3.3.3 Procedure

1. The Au wire is cut to 5 cm length
2. The Au surface is cleaned using isopropyl alcohol and dried
3. The Au surface is electrochemically cleaned by performing a potential sweep from −200 to +1200 mV in 50 mM NaOH, rinsed with DI water, rinsed with isopropyl alcohol, then dried under N_2
4. Following cleaning, the Au electrodes is stored covered, in air, at room temperature, in the dark, until use

3.4 Reference electrode fabrication

The alkaline phosphatase and rubber transferase actuators use the same kind of reference electrode. Reference electrodes are fabricated from 250 μm diameter Ag wire and chlorinated in sodium hypochlorite.

3.4.1 Equipment

- 10 mL beaker (McMaster-Carr, Aurora, OH, USA)
- Utility scissors (Fisher Scientific, Tewksbury, MA, USA)

3.4.2 Consumables

- 15% sodium hypochlorite (MilliporeSigma, St. Louis, MO, USA)
- 250 μm diameter Ag wire (Fisher Scientific, Tewksbury, MA, USA)
- Sodium chloride (MilliporeSigma, St. Louis, MO, USA)
- Distilled Water (MilliQDirect16, MilliporeSigma, St. Louis, MO, USA)

3.4.3 Procedure

1. 5 mL of sodium hypochlorite is poured into a 10 mL beaker
2. The wire is trimmed to 5 cm, straightened, and half of the wire is placed into solution and chlorinated for 12 h
3. Following chlorination, the silver-silver chloride wire is rinsed with distilled water and stored in a 0.1 M solution of sodium chloride at 4 °C in the dark until use

4. Programmable chemical actuator (PCA) assembly

4.1 Alkaline phosphatase PCA assembly

This procedure is simply the assembly of the electrodes in the cuvette. The electrodes must be kept out of the light path to preserve the integrity of the spectrophotometric assay.

4.1.1 Equipment

- N/A

4.1.2 Consumables

- Chip electrode (lab made)
- Au foil electrode (lab made)
- Ag/AgCl reference electrode (lab made)
- Semi microcuvette (Fisher Scientific, Tewksbury, MA, USA)
- MarineWeld (McMaster-Carr, Aurora, OH, USA)

4.1.3 Procedure

1. The counter and reference electrodes are adhered to the left side of the cuvette with MarineWeld. Care must be taken to ensure they are not touching and are out of the light path
2. The working electrode is adhered to the left side of the cuvette with MarineWeld. Care must be taken to ensure it is out of the light path

4.2 Rubber transferase PCA fabrication

PCAs are fabricated to fit the small reaction volume used for radiometric assays. Unlike spectrophotometric assays there is no light path to consider during cell assembly, but evaporation of the small reaction volume must be prevented.

4.2.1 Equipment

- N/A

4.2.2 Consumables

- 2 × Au wire electrodes (lab made)
- Ag/AgCl reference electrode (lab made)
- 1.5 mm O.D.; 0.86 mm I·D borosilicate capillary (Sutter, Atlanta, GA, USA)

- PTFE heat shrink (Component Supply Company, Sparta, TN, USA)
- MarineWeld (McMaster-Carr, Aurora, OH, USA)

4.2.3 Procedure

- A Ag/AgCl reference electrode is threaded through the middle of a borosilicate capillary and secured using MarineWeld
- The Au working and counter electrodes are secured to the outside of the capillary using PTFE heat shrink

5. Polypyrrole electropolymerization

Electro-oxidative polymerization of polypyrrole doped with an anion is most efficient in a three-electrode electrochemical cell in an aqueous electrolyte containing an anionic compound (Sadki et al., 2000). Electro-oxidative polymerization tightly adheres the conducting polymer to the electrode for thousands of REDOX cycles and creates a robust working electrode (Bengtsson, Nilsson, & Robinson, 2014; Erlandsson & Robinson, 2011; Lee et al., 2020). During polymerization, the spacing between the working and counter electrodes should be minimized while avoiding an electrical short. As shown in Table 1, cationic actuators and anionic actuators are fabricated by doping polypyrrole with dodecylbenzenesulfonate and chloride, respectively.

5.1.1 Equipment

- Ivium PocketStat (pocketSTAT1, Ivium Technologies, De Zaale, The Netherlands)
- 10 mL beaker (McMaster-Carr, Aurora, OH, USA)

5.1.2 Consumables

- Pyrrole (MilliporeSigma, St. Louis, MO, USA)
- Sodium dodecybenzenesulfonate (MilliporeSigma, St. Louis, MO, USA)
- Sodium chloride (MilliporeSigma, St. Louis, MO, USA)

Table 1 Composition of the polymerization solutions used to create PPy(DBS) and PPy(Cl) polymers. PPy(DBS) polymers are integral to cationic actuators and PPy(Cl) polymers are integral to anionic actuators.

Polymer	[Py Monomer]	[Dopant]
PPy(DBS)	200 mM	200 mM NaDBS
PPy(Cl)	200 mM	200 mM NaCl

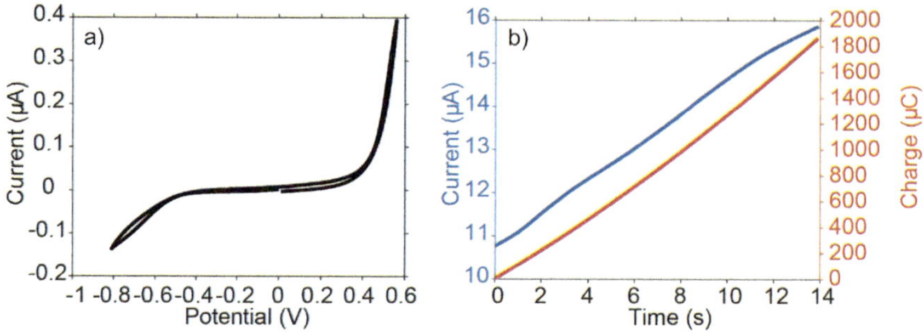

Fig. 4 (A) A typical pilot cyclic voltammogram used to identify the oxidation potential of the pyrrole monomer. (B) A typical chronoamperogram from the polymerization reaction, where the blue curve represents the current passed during polymerization and the red curve represents the charge passed during polymerization.

5.1.3 Procedure

1. Polypyrrole polymerization is performed in 0.2 M dopant solution with 0.2 M pyrrole monomer
2. The polymerization potential is determined by the pyrrole monomer's oxidation peak in the electrolyte
3. The monomeric oxidation peak is found potentiodynamically using cyclic voltammetry on an Ivium PocketStat by sweeping the potential from 0 V to −0.8 V to +0.8 V vs Ag/AgCl at a scan rate of 50 mV/s. Fig. 4A is a representative pilot cyclic voltammogram for pyrrole in 0.2 M NaDBS
4. Oxidation peaks for pyrrole in 0.2 M NaDBS are observed around +0.55 V vs. Ag/AgCl whereas oxidation peaks for pyrrole in 0.2 M NaCl are observed around +0.8 V vs. Ag/AgCl
5. Potentiostatic control during polymerization is carried out using chronoamperometry until the total charge passed during chronoamperometry reaches 0.5C/cm^2. Fig. 4B is a representative chronoamperogram for pyrrole in 0.2 M NaDBS

6. Programming ion selectivity

Ions do not often exist in isolation in biological systems. Table 2 summarizes the free concentration of common ions in bacterial, fungal, plant, mammalian cells, and blood plasma. The concentration of ions bound to biological macromolecules is given in parentheses when available. Concentrations can change significantly with physiological and environmental conditions. Therefore, programming selectivity is particularly important when

Table 2 Ionic concentrations (in mM) in bacterial, yeast, plant, and mammalian red blood cells and in the blood plasma. Concentrations include free and bound ions unless otherwise indicated. Adapted from (Milo & Phillips, 2015); Commelina cyanea data is from (Karley, Leigh, & Sanders, 2000; Karley & White, 2009; Maathuis, 2009).

	[K] mM	[Na] mM	[Mg] mM	[Ca] mM	[Cl] mM
Escherichia coli	30–300	10	0.1–1	$1E^{-4}$ (3)	10–200
Saccharomyces cerevisiae	300	30	50	(2)	10–200
Commelina cyanea	127	3	0.4–0.5	2	34
Red Blood Cell	100	10	0.5 (10)	$1E^{-5}$ - $1E^{-4}$	5–100
Blood Plasma	4	100–200	1	2	100

Table 3 Polypyrrole ion selectivity is based on equilibration solution and polymer dopant.

Equilibration Solution	PPy(DBS)	PPy
0.5 M HCl	H^+	Cl^-
0.5 M NaOH	Na^+	OH^-
0.5 M NaCl	Na^+	Cl^-
0.5 M $MgCl_2$	Mg^{2+}	Cl^-
0.5 M $MnSO_4$	Mn^{2+}	SO_4^{2-}

fabricating PCAs. The programming procedure is referred to as equilibration and Table 3 lists equilibration solutions for common ions.

6.1.1 Equipment

- Ivium PocketStat (Ivium Technologies, De Zaale, The Netherlands)
- 10 mL beaker (McMaster-Carr, Aurora, OH, USA)

6.1.2 Consumables

- N_2 gas (Ohio State Cryogenics Laboratory, Columbus, OH, USA)
- Distilled Water (MilliQDirect16, MilliporeSigma, St. Louis, MO, USA)
- Hydrochloric acid (MilliporeSigma, St. Louis, MO, USA)
- Sodium hydroxide (MilliporeSigma, St. Louis, MO, USA)
- Sodium chloride (MilliporeSigma, St. Louis, MO, USA)
- Magnesium chloride (MilliporeSigma, St. Louis, MO, USA)
- Manganese sulfate (MilliporeSigma, St. Louis, MO, USA)
- Equilibration solution (lab made)

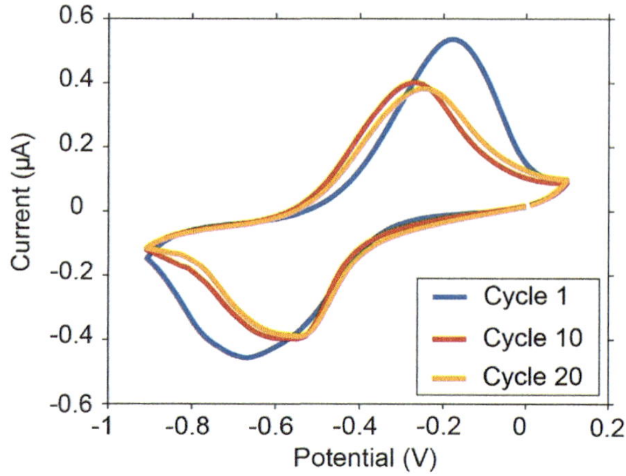

Fig. 5 The 1st, 10th, and 20th cycles of the hydroxyl equilibration process. The overlap between the later cycles indicates the polymer is reproducibly exchanging ions with solution.

6.1.3 Procedure

1. Make the equilibration solution
2. Rinse the electrodes using distilled water
3. Dry the electrodes under nitrogen gas
4. The electrolyte is exchanged with a 0.5 M equilibration solution
5. The potentiodynamic control strategy for equilibration is carried out via cyclic voltammetry; Fig. 5 shows that sweeping from 0 to −0.4 V to +0.4 V vs Ag/AgCl at a scan rate of 50 mV/s until the charge moved into the polymer is within 10% of the charge moving out of the polymer, usually around 20 cycles

7. Pulsed electric potential control strategy

Pulsing electric fields can overcome ion transport limitations in a variety of electrochemical systems. Alkaline phosphatase and rubber transferase have assay-specific waveforms described below. Both assays have four features in common: the initial phase of the control strategy is a no connection period to establish a baseline. During the reduction phase, the applied potential is pulsed between partially (−0.2 V) and fully (−0.4 V) reduced states. Similarly, the oxidation phase pulses the applied potential between partially (+0.2 V) and fully (+0.4 V) oxidized states. The final phase is a no connection period to allow diffusion to restore the initial solution composition. This is a form of chronoamperometry utilizing an uncommon waveform.

7.1 Equipment

- Ivium PocketStat (pocketSTAT1, Ivium Technologies, De Zaale, The Netherlands)
- Assembled PCA (lab made)

7.2 Consumables

- N_2 gas (Ohio State Cryogenics Laboratory, Columbus, OH, USA)
- Distilled Water (MilliQDirect16, MilliporeSigma, St. Louis, MO, USA)
- Equilibration solution (lab made)

7.3 Alkaline phosphatase procedure

1. Hold the electrochemical cell at open circuit potential (OCP) for 120 s
2. Apply duplex oxidation pulses, where the partial oxidation potential, +0.2 V vs Ag/AgCl, is held for 5 s and the full reduction potential, +0.4 V vs Ag/AgCl, is held for 5 s
3. This pulsing is repeated 24 times for a total oxidation duration of 240 s
4. The reduction potential pulses should mirror the oxidation pulses, where the partial reduction potential, −0.2 V vs Ag/AgCl, is held for 5 s and the full reduction potential, −0.4 V vs Ag/AgCl, is held for 5 s
5. This pulsing is repeated 24 times for a total reduction duration of 240 s
6. At the end of the protocol the cell should be returned to OCP for 120 s

7.4 Rubber transferase procedure

1. Hold the electrochemical cell at open circuit potential (OCP) for 60 s
2. Apply duplex reduction potential pulses where the partial reduction potential, −0.2 V vs Ag/AgCl, is held for 5 s and the full reduction potential, −0.4 V vs Ag/AgCl, is held for 5 s
3. This pulsing is repeated 720 times for a total reduction duration of 7200 s
4. Oxidation pulses should mirror the reduction pulses, where the partial oxidation potential, +0.2 V vs Ag/AgCl, is held for 5 s and the full reduction potential, +0.4 V vs Ag/AgCl, is held for 5 s
5. This pulsing is repeated 720 times for a total reduction duration of 7200 s
6. At the end of the protocol the cell should be returned to OCP before disconnecting

8. Control alkaline phosphatase enzyme activity

The cuvette PCA is used to control pH during alkaline phosphatase assays. Here, alkaline phosphatase is used to demonstrate the feasibility of

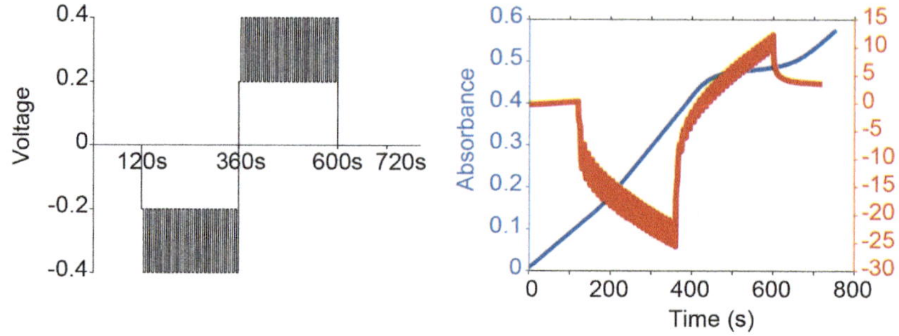

Fig. 6 (A) Applied potential waveform during the assay. (B) An overlayed plot of the absorbance (blue) and charge passed (red) during the assay.

bioelectronic control of a soluble enzyme. A common assay for detecting alkaline phosphatase activity is measuring the conversion of the artificial substrate p-nitrophenol phosphate to nitrophenol (Harris, 1990; Olsen, Øverbø, & Nilsen, 1991; Qi, Zheng, Qin, & Zhai, 2020). Nitrophenol absorbs 410 nm light, which facilitates spectrophotometric detection. Hydroxyl ion selectivity is imparted to the chloride-doped polypyrrole working electrode by equilibrating the polymer prior to operation. Fig. 6 shows that reducing the polymer increases the activity of alkaline phosphatase and subsequent oxidation of the polymer decreases enzyme activity. The functional interrelationships among the alkaline phosphatase superfamily suggest that methods successful in imparting bioelectronic control of alkaline phosphatase can be broadly applicable across enzymes.

8.1 Equipment

- Ivium PocketStat (pocketSTAT1, Ivium Technologies, De Zaale, The Netherlands)
- 10 mL beaker (McMaster-Carr, Aurora, OH, USA)
- PCA-modified semi-microcuvettes (lab made)
- Heated dry bath (Fisher Scientific, Tewksbury, MA, USA)
- Varian Cary 100 Bio UV–Visible Spectrophotometer (Varian, Palo Alto, CA, USA)
- Assembled PCA (lab made)
- Syringe (Hamilton, Reno, NV, USA)

8.2 Consumables

- N_2 gas (Ohio State Cryogenics Laboratory, Columbus, OH, USA)
- Distilled Water (MilliQDirect16, MilliporeSigma, St. Louis, MO, USA)

- Equilibration solution (lab made)
- Shrimp alkaline phosphatase (New England BioLabs, Ipswich, MA, USA)
- $MgCl_2$ (MilliporeSigma, St. Louis, MO, USA)
- TRIS-acetate buffer (MilliporeSigma, St. Louis, MO, USA)
- Citrate buffer (MilliporeSigma, St. Louis, MO, USA)
- TRIS-Cl buffer (MilliporeSigma, St. Louis, MO, USA)

8.3 Procedure

- Adjust buffers
  - **a.** TRIS-acetate buffer = pH 9
  - **b.** Citrate buffer = pH 7.0
  - **c.** TRIS-Cl buffer = pH 5.5
- Alkaline phosphatase assays are performed by adding 940 μL of 1 mM buffer and 50 μL of para-nitrophenyl phosphate (final concentration 25 mM) to a PCA-modified semi-microcuvette
- PCA-modified semi-microcuvettes are used for all alkaline phosphatase assays under the indicated potential control strategy
- For negative controls, the enzyme stock solution is thermally deactivated by incubating at 65 °C for 20 min to denature the enzyme (Nilsen, Øverbø, & Olsen, 2001; Olsen et al., 1991).
- The spectrophotometer is zeroed and 10 μL of enzyme stock is added using a syringe (final concentration 0.1 units).
- Immediately following enzyme injection, the electrolyte is stirred using the syringe for 3 s
- Absorbance monitoring at 410 nm and the electrochemical protocol are started at the same time
- Data can be exported as .csv files and plotted

9. Control rubber transferase activity

Natural rubber is synthesized at the surface of rudimentary uni-lamella membrane-bound organelles known as rubber particles (Hagel, Yeung, & Facchini, 2008). All rubber particles have a specialized *cis*-prenyl transferase, rubber transferase, that catalyzes the initiation and elongation of *cis*-1,4-polyisoprene (Cornish & Siler, 1996; Cornish & Xie, 2012). The rubber transferase complex spans the rubber particle membrane and adds isopentyl pyrophosphate (IPP) monomers from outside the particle to the rubber polymer growing inside the particle via magnesium-dependent sequential elongation reactions (Cornish, Siler, Grosjean, & Goodman, 1993). The most likely rubber elongation mechanism is given by Eq. (7) (da Costa, Keasling, McMahan, & Cornish, 2006):

$$\begin{aligned} &2C_{aq}^{2+} + IPP_{aq} \leftrightarrow \left(IPP \cdot 2C^{2+} \right)_{aq} + E_{rp} \\ &\leftrightarrow \left(IPP \cdot 2C^{2+} \cdot E \right)_{rp} \leftrightarrow \left(IPP \cdot E \right)_{rp} + 2C_{aq}^{2+} \end{aligned} \tag{7}$$

where, C_{aq}^{2+} is an activating divalent cation in aqueous solution, IPP_{aq} is an isopentyl pyrophosphate monomer in aqueous solution, $(IPP \cdot 2C^{2+})_{aq}$ is a cation monomer complexed to an IPP monomer, E_{rp} is rubber transferase on a rubber particle, $(IPP \cdot 2C^{2+} \cdot E)_{rp}$ represents the cation monomer complexed to an IPP monomer and the rubber transferase complex all associated with the rubber particle, and $(IPP \cdot E)_{rp}$ is the rubber transferase complexed to the IPP substrate following cation ejection during the condensation reaction. The active elongating end of the rubber molecule is in the form of an allylic pyrophosphate. Eq. (8) indicates that the divalent cations complexed with IPP are concomitantly released into solution with each condensation reaction (Chang, Ko, Chen, Wang, & Liang, 2004). PCAs exchange free divalent cations in solution according to Eq. (8) (Gupta et al., 2020; Venugopal et al., 2014, Venugopal et al., 2017; Venugopal & Sundaresan, 2016):

$$nC_{aq}^{2+} + \left[PPy^{n+}(DBS^{-})_{2n}\right]_{WE} \leftrightarrow \left[(PPy)\left(DBS_2^{-}C^{+}\right)_n\right]_{WE} - 2ne^{-} \tag{8}$$

where, C_{aq}^{2+} represents activating divalent cations in solution, n is a stoichiometric coefficient, $[PPy^{n+}(DBS^{-})_{2n}]_{WE}$ is polypyrrole doped with dodecyl benzene sulfonate in the oxidized state, e^{-} is an electron, and $[(PPy)(DBS_2^{-}C^{+})_n]_{WE}$ is PPy(DBS) in the reduced state.

9.1 *Ficus elastica* latex collection and washed rubber particle preparation

Washed rubber particles are isolated from latex collected from *F. elastica* using previously published methods (Cornish & Siler, 1996). Rubber particle isolation procedures are species specific. The literature should be consulted for isolation procedures for other species. Specifically, the Cornish lab also has developed procedures for *Taraxacum kok-saghyz* (Cornish et al., 1993), *Parthenium argentatum* (Cornish & Backhaus, 1990), *Helianthus annuus* (Pearson, Cornish, Rath, Brichta, & Van Fleet, 2010), two *Lactuca* species (Bushman et al., 2006), and *Hevea brasiliensis* (Cornish, 1993). This chapter will focus on *F. elastica* because the rubber particles are the most robust and easiest to isolate among these species.

9.1.1 Equipment

- Ivium PocketStat (pocketSTAT1, Ivium Technologies, De Zaale, The Netherlands)
- 10 mL beaker (McMaster-Carr, Aurora, OH, USA)
- *F. elastica* plants (purchased from a local nursery and grown in a greenhouse in Wooster Ohio for several years)

9.1.2 Consumables

- N_2 gas (Ohio State Cryogenics Laboratory, Columbus, OH, USA)
- Distilled Water (MilliQDirect16, MilliporeSigma, St. Louis, MO, USA)
- Equilibration solution (lab made)
- Tris-HCl (MilliporeSigma, St. Louis, MO, USA)
- Glycerol (MilliporeSigma, St. Louis, MO, USA)
- $MgSO_4$ (MilliporeSigma, St. Louis, MO, USA)

9.1.3 Procedure for Ficus elastica rubber particles

1. Prepare Collection Buffer by mixing 10 mM Tris-HCl at pH 7.5, 12% glycerol, and 2.5 mM $MgSO_4$ and chilling to 4 °C
2. Prepare Spin Buffer by mixing 10 mM Tris-HCl at pH 7.5 and 2.5 mM $MgSO_4$ and chilling to 4 °C
3. Remove leaves at the apical portions of the petioles at a 30-degree angle to the base of the petiole using a scalpel or razor blade leaving the cut surface almost horizontal. Collect latex in 10 mL of chilled collection buffer (10 mM Tris-HCl at pH 7.5, 12% glycerol, and 2.5 mM $MgSO_4$) on ice in 50 mL centrifuge tubes
4. Collection tubes are centrifuged at 2500 × *g* for 10 min at 4 °C. The buoyant fraction is transferred to SS-34 centrifuge tubes (50 mL capacity) containing 10 mL Spin Buffer and the sedimented fraction is discarded (inactive particles).
5. Tubes containing the buoyant fraction are centrifuged at 2500 × *g* for 10 min at 4 °C and the pelletized rubber particles are washed by discarding the supernatant, resuspending the pellet in Rinse Buffer (10 mM Tris-HCl at pH 7.5, and 2.5 mM $MgSO_4$), then re-spinning at 2500 × *g* for 10 min at 4 °C
6. Washing is repeated twice, and the final pellets are suspended in 1 mL Rinse Buffer
7. Glycerol is added to stock solutions to achieve a final concentration of 10% and the solution is gently mixed by inversion

8. Uniformly sized droplets of the solution are flash-frozen in liquid nitrogen and collected into precooled (−196 °C) cryovials (Cornish & Bartlett, 1997).

9.2 Siliconization of reaction tubes

Rubber particles may stick to the sides of stock tubes altering the quantification of rubber transferase activity later in the procedure. To preclude this, stock tubes are siliconized (also known as silanized) to provide a non-stick barrier between the electrolyte and the plastic tube.

9.2.1 Equipment

- Chemical fume hood

9.2.2 Consumables

- 1.5 mL microtubes (Fisher Scientific, Tewksbury, MA, USA)
- Sigmacote (MilliporeSigma, St. Louis, MO, USA)

9.2.3 Procedure

1. 1.5 mL microtubes (Fisher) are placed into a tube holder and filled with Sigmacote
2. Tubes are incubated for 25 h with caps off in a chemical fume hood. Excess Sigmacote is decanted into a clean reagent bottle and tubes are inverted and left to dry for an additional 25 h with caps off in a chemical fume hood
3. Prepared tubes are now referred to as "siliconized tubes" and stored in the dark, at room temperature, with the caps on

9.3 Rubber transferase reactions under static metal ion control

Rubber transferase activity is determined by tracking the incorporation of radiolabeled initiators and monomers into newly synthesized rubber polymers (Cornish & Siler, 1996).

9.3.1 Equipment

- Packard Tri-Carb 2100TR Liquid Scintillation Counter (Packard Bioscience, Meriden, CT, USA)
- Vacuum manifold
- Vacuum pump
- Positive displacement pipettes

9.3.2 Consumables

- Washed rubber particle preparation (lab made)
- 1.5 mL microcentrifuge tubes (Fisher Scientific, Tewksbury, MA, USA)
- Filter
- TRIS-Cl (MilliporeSigma, St. Louis, MO, USA)
- ^{3}H FPP (MilliporeSigma, St. Louis, MO, USA)
- FPP (MilliporeSigma, St. Louis, MO, USA)
- ^{14}C IPP (MilliporeSigma, St. Louis, MO, USA)
- IPP (MilliporeSigma, St. Louis, MO, USA)
- $MgSO_4$ (MilliporeSigma, St. Louis, MO, USA)
- $MnSO_4$ (MilliporeSigma, St. Louis, MO, USA)
- EDTA (MilliporeSigma, St. Louis, MO, USA)
- Ethanol (MilliporeSigma, St. Louis, MO, USA)
- EcoScint-A (MilliporeSigma, St. Louis, MO, USA)

9.3.3 Procedure

1. 25 μg (dry weight basis) of rubber particles are assayed in siliconized 1.5 mL microcentrifuge tubes with 100 μL of reaction mixture containing 10 mM TRIS-Cl (pH 7.5), 20 μM FPP, 200 μM IPP, and the concentrations of either Mg^{2+} or Mn^{2+} indicated in the figures
2. Reactions are started by adding rubber particles to premixed solutions and incubating for 4 h at 27 °C
3. Reactions are stopped by adding 50 μL of 500 mM EDTA to the reaction tubes. Rubber particles are harvested by filtration, washed with 1 M HCl to protonate and solubilize any unincorporated IPP, followed with three washes with 100% ethanol to remove any protonated, unincorporated substrate, and the radioactivity determined by liquid scintillation spectroscopy in EcoScint-A scintillation cocktail

9.4 PCA-controlled rubber transferase reactions

9.4.1 Equipment

- Packard Tri-Carb 2100TR Liquid Scintillation Counter (Packard Bioscience, Meriden, CT, USA)
- Ivium PocketStat (pocketSTAT1, Ivium Technologies, De Zaale, The Netherlands)
- Assembled PCA (lab made)

9.4.2 Consumables

- Washed rubber particle preparation (lab made)

- 1.5 mL microcentrifuge tubes (Fisher Scientific, Tewksbury, MA, USA)
- TRIS-Cl (MilliporeSigma, St. Louis, MO, USA)
- ^{3}H FPP (MilliporeSigma, St. Louis, MO, USA)
- FPP (MilliporeSigma, St. Louis, MO, USA)
- ^{14}C IPP (MilliporeSigma, St. Louis, MO, USA)
- IPP (MilliporeSigma, St. Louis, MO, USA)
- $MgSO_4$ (MilliporeSigma, St. Louis, MO, USA)
- $MnSO_4$ (MilliporeSigma, St. Louis, MO, USA)
- EDTA (MilliporeSigma, St. Louis, MO, USA)
- Ethanol (MilliporeSigma, St. Louis, MO, USA)
- EcoScint-A (MilliporeSigma, St. Louis, MO, USA)
- 2 mm ID O-ring (McMaster-Carr, Aurora, OH, USA)

9.4.3 Procedure

Fig. 7 shows how rubber transferase activity is affected by reducing and oxidizing a PCA in the reaction volume to control metal ion concentration.

1. The manufacturer's cap is replaced with a rubber O-ring to prevent evaporation during incubation while allowing the PCA to access the reaction volume
2. 25 μg (dry weight basis) of rubber particles are assayed in siliconized 1.5 mL microcentrifuge tubes with 100 μL of reaction mixture containing 10 mM

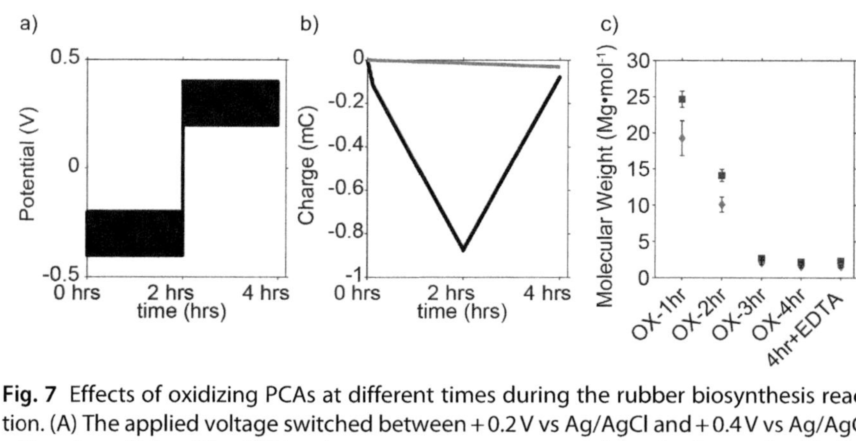

Fig. 7 Effects of oxidizing PCAs at different times during the rubber biosynthesis reaction. (A) The applied voltage switched between + 0.2 V vs Ag/AgCl and + 0.4 V vs Ag/AgCl with pulse widths of 5 s. (B) The charge responses to sequential reduction and oxidation duplex pulses. Black curves represent the response to the waveform in 100 mM $MgSO_4$, and the gray curves represent reduction pulsing in 100 mM $MgSO_4$ with 500 mM EDTA. (C) Mean molecular weight for each control regime. Squares represent Mg^{2+}-activated reactions and diamonds represent Mn^{2+}-activated reactions. The error bars represent the standard deviation of three replicates at each concentration.

TRIS-Cl (pH7.5), 20 μM FPP, 200 μM IPP, and either 100 mM Mg^{2+} or 10 mM Mn^{2+}

3. Reactions are started by adding rubber particles to premixed solutions and incubating for 4 h at 27 °C
4. Immediately following the addition of the rubber particles, the electrochemical protocol is started
5. Reactions are stopped by adding 50 μL of 500 mM EDTA to the reaction tubes to chelate the divalent cation activator
6. Rubber particles are harvested by vacuum filtration, washed with 1 M HCl to protonate and solubilize any unincorporated IPP, followed with three washes with 100% ethanol to remove any protonated, unincorporated substrate, and the radioactivity determined by liquid scintillation spectroscopy as described in the next section

9.5 Scintillation counting

Liquid scintillation counts are performed on a Packard Tri-Carb 2100TR Liquid Scintillation Counter, or similar counter.

9.5.1 *Equipment*

- Packard Tri-Carb 2100TR Liquid Scintillation Counter (Packard Bioscience, Meriden, CT, USA)
- Packard ^{3}H, ^{14}C, and ^{32}P standards (Packard Bioscience, Meriden, CT, USA)
- Counting racks (Packard Bioscience, Meriden, CT, USA)

9.5.2 *Consumables*

- EcoScint-A (MilliporeSigma, St. Louis, MO, USA)

9.5.3 *Procedure*

1. The machine is calibrated with the manufacturer's ^{3}H, ^{14}C, and ^{32}P standards
2. Samples are incubated with 10 mL EcoScintA until the filter dissolved
3. Samples are then loaded into counting racks and separate ^{14}C and ^{3}H counting protocols are run back-to-back
4. The molecular weight of natural rubber, MW_{NR}, is calculated according to:

$$MW_{NR} = \frac{\left(R_{^{14}C\cdot IPP} + \alpha \cdot \left(R_{^{3}H\cdot FPP}\right)\right) \cdot MW_{iso}}{R_{^{3}H\cdot FPP}} + MW_{DP} \qquad (9)$$

where $R_{^{14}C\cdot IPP}$ is the incorporation of ^{14}C-labeled isopentyl pyrophosphate in nmol of tracer · g dry weight of rubber^{-1} · 4^{-1} hours α is an initiator-specific constant (3 for FPP), MW_{iso} is the molecular weight of isoprene, $R_{^{3}H\cdot FPP}$ is the incorporation of ^{3}H-labeled farnesyl pyrophosphate in nmol of tracer · g dry weight of rubber^{-1} · 4^{-1} h, and MW_{DP} is the molecular weight of the pyrophosphate end group on the terminal IPP molecule which ranges from 174 to 176 Da depending on the ionization state.

10. Interpretation of the results

Enzymatic assays are assessed by spectrophotometry and liquid scintillation spectroscopy as described above. This section focuses on the interpretation of the electrochemical data.

10.1 Cyclic voltammetry

Cyclic voltammetry (CV) is a linear electric potential sweep technique used to find the potential at which electron transfer becomes favorable in an experimental setup (Elgrishi et al., 2018). The electric potential waveform applied to PPy(DBS) during a typical cyclic voltammogram is shown in Fig. 8A and the current-voltage response in 8b. The electric potential is applied at a starting potential, 0 V vs Ag/AgCl, and scanned at a linear rate

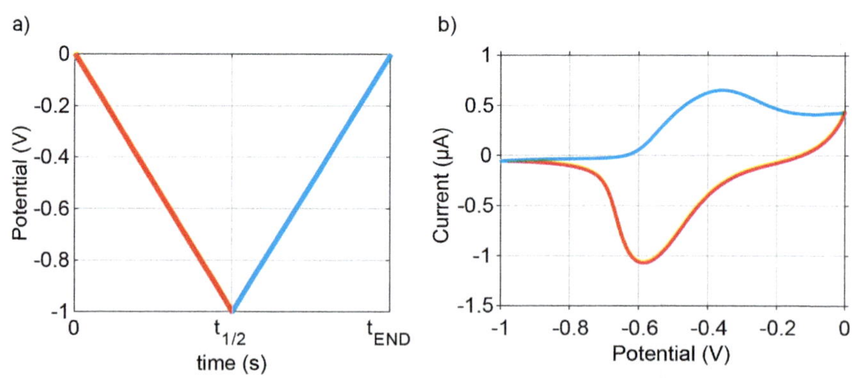

Fig. 8 (A) The potential waveform used for cyclic voltammetry. (B) A typical current-voltage response of PPy(DBS) in response to the waveform.

to the first vertex potential −1 V vs Ag/AgCl. Upon reaching −1 V vs Ag/AgCl, the scanning direction is reversed, and scanning continues until 0 V vs Ag/AgCl is reached.

When the potential is scanned during the CV experiment, the concentration of the species in solution near the electrode changes over time in accordance with the Nernst equation.

$$E = E^0 - \frac{RT}{F} ln\left(\frac{[Re\,d]}{[Ox]}\right) \quad (10)$$

where E^0 is the standard half-cell reduction potential, E is the reduction potential, R is the gas constant, T is the temperature (293 K), F is Faraday's constant (96,485C*mol^{-1}), [*Red*] is the reduced species' concentration and, and [*Ox*] is the oxidized species' concentration.

If the reduction process is chemically and electrochemically reversible, the difference between the anodic and cathodic peak potentials is 57 mV at 25 °C (Elgrishi et al., 2018). Electrochemical reversibility indicates there is a low barrier to electron transfer (Bard et al., 1980; Elgrishi et al., 2018; György, 2008). Electrochemically reversible processes are referred to as Nernstian, indicating that equilibrium is established immediately upon any change in applied potential (Bard et al., 1980; Elgrishi et al., 2018; Funt & Diaz, 1991; György, 2008). By contrast, electrochemical irreversibility indicates there is a high barrier to electron transfer giving rise to larger differences between the anodic and cathodic peaks (Elgrishi et al., 2018).

Although polypyrrole systems deviate from an ideal Nernstian behavior, as indicated by the offset of the anodic and cathodic peaks, the assumption that cathodic and anodic peak currents are approximately equal is considered valid (György, 2008; Northcutt, 2015; Smela, 1999; Venugopal, 2016; Venugopal et al., 2017, 2014; Venugopal & Sundaresan, 2016). The current response of PPy(DBS) to the applied potential is characterized by a broad reduction peak near −0.6 V vs Ag/AgCl and a mirrored oxidation peak around −0.4 V vs Ag/AgCl, and plateaus beyond the peaks (Feldberg, 1984; Gupta, 2020; Gupta, Ghatak, et al., 2020; Hery & Sundaresan, 2016). The plateaus are attributable to double layer charging and the peaks to faradaic charge transfer (Gupta, Ghatak, et al., 2020a; Venugopal, 2016). From −0.4 to −0.6 V vs Ag/AgCl and beyond, the rise in current is associated with undoping of polypyrrole and corresponding cation ingress to maintain charge neutrality (Northcutt & Sundaresan, 2015; Sundaresan et al., 2017b; Venugopal et al., 2014). At the cathodic peak potential, current reaches a maximum as the dopant ions in the electrode vicinity have ingressed into the polymer (Bard et al., 1980; Elgrishi et al., 2018; Gellett, Schumacher, Kesmez, Le, & Minteer, 2010; Rasmussen, Abdellaoui, & Minteer, 2016). Upon oxidation, doping and ion egress are observed (Bard et al., 1980; Elgrishi et al., 2018; Gellett et al., 2010; Rasmussen et al., 2016).

Experimental conditions are reflected in the CV response. Increasing scan rate in PPy(PSS^-) and PPy(ClO^{-4}) has been shown to shift the peak corresponding to ion ingress away from zero (Ren & Pickup, 1993). Decreasing electrolyte concentration shifts the peak corresponding to ion ingress toward zero and diminishes peak magnitude (Gupta, 2020; Gupta, Otero, Sundaresan, & Czeisler, 2020). The shape of the voltammogram reflects non-dopant ions in solution as changes in peak separation, width, and/or magnitude (Diaz, Kanazawa, & Gardini, 1979; John & Wallace, 1993).

10.2 Chronoamperometry

Chronoamperometry is a potential step technique (Bard et al., 1980; Kelly, Scully, Shoesmith, & Buchheit, 2002). The electric potential waveform applied to PPy(DBS) during a typical chronoamperogram is shown in Fig. 9A and the current response in 9b. Chronoamperometry often starts with an idle period where the cell is at open circuit potential, the experimental potential observed at zero current (Goodridge & Scott, 2013; György, 2008). After the idle period, the potential applied to the working electrode is stepped to the first potential ($t=0$ s to $t=5$ s). The potentiostat

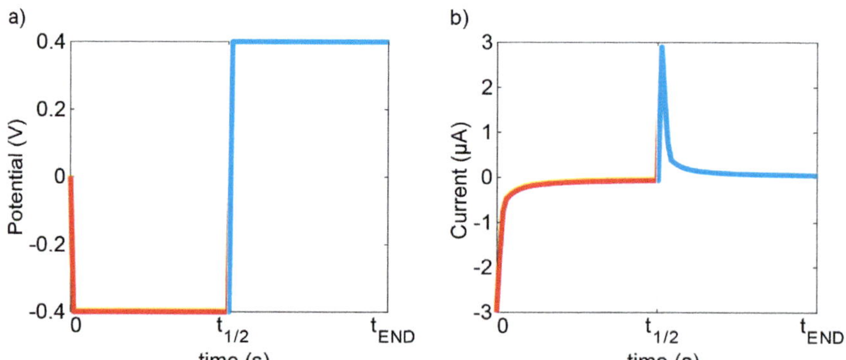

Fig. 9 (A) A potential waveform used for chronoamperometry. (B) A typical current-time response of PPy(DBS) to the waveform.

maintains the applied potential relative to the reference electrode while measuring the current at the working electrode (Goodridge & Scott, 2013; György, 2008).

Reduction and oxidation peaks identified in cyclic voltammetry are used to inform the magnitude of the potential steps in chronoamperometric studies of PPy(DBS) (Gupta, Ghatak, et al., 2020; Hery & Sundaresan, 2016; Venugopal et al., 2014; Venugopal & Sundaresan, 2016). In single potential step experiments the polymer is reduced first, as the polymer returns to the oxidized state in the absence of an electric field (Cosnier & Karyakin, 2011; Diaz et al., 1979; Northcutt, 2015; Sadki et al., 2000). More commonly, PPy(DBS) is studied in double potential step experiments, where the first, and second potentials are often −0.4 V vs Ag/AgCl and + 0.4 V vs Ag/AgCl, respectively. At potentials negative of the reduction potential, the net ion ingress is diminished. Increasing the potentials to well positive of +0.4 V vs Ag/AgCl degrades the polymer in a process termed over-oxidation (Krische & Zagorska, 1989).

The current response to a reduction pulse is characterized by an initial spike in current followed by an exponential decay. The spike is attributable to capacitive discharge and uncompensated solution resistance (Bard et al., 1980; Venugopal, 2016). The exponential region is a combination of double layer formation and Faradaic charge (Northcutt, 2015; Venugopal, 2016). The Faradaic charge is assumed to be equal to the ion ingress for a monovalent ion (Venugopal, 2016; Venugopal et al., 2017, 2014). When the applied potential is switched to +0.4 V vs Ag/AgCl, the polymer is oxidized and doped with DBS ions. To maintain charge neutrality, cations from the polymer egress into the solution. The current response to an oxidation pulse mirrors that of a reduction pulse where the Faradaic charge is equal to ion egress.

The decay in the current is attributed to the depletion of ions at the polymer-electrolyte interface. The current-time response for a diffusion-controlled current is described by the Cottrell equation (Cottrell, 1903; Kelly et al., 2002):

$$i = \frac{nFAc_j^0\sqrt{D_j}}{\sqrt{\pi t}} \tag{11}$$

where, n is the number of electrons transferred/molecule, F is Faraday's constant (96,500C*mol^{-1}), A is the electrode area (cm^2), D is the diffusion coefficient (cm^{2}*s^{-1}), and C is the concentration (mol*cm^{-3}).

10.3 Chronocoulometry

Chronoamperometry and chronocoulometry use the same potential waveform as shown in Figs. 9A and 10A. Chronocoulometry can be performed as a separate experiment, but it is most common to perform chronoamperometry and post-process the data to obtain the charge response as depicted in Fig. 10B (Bard et al., 1980; Kelly et al., 2002; Kissinger & Heineman, 2018).

The charge-time response is obtained by integrating the Cottrell equation with respect to time, to obtain the Anson equation (Bard et al., 1980; Venugopal, 2016):

$$Q = \frac{nFAC\sqrt{D}}{\sqrt{\pi t}} \tag{12}$$

where, Q is the charge in coulombs, n is the number of electrons transferred/molecule, F is Faraday's constant (96,485C*mol^{-1}), A is the electrode area (cm^2), C is the concentration (mol*cm^{-3}), D is the diffusion coefficient (cm^2 *s^{-1}), and t is time in s.

The charge response to the reduction potential (t=0 s – 5 s) increases with time, whereas the current response decreases in this same time. The accumulation of negative charge in the Anson plots is indicative of ion ingress into PPy(DBS) (Sadki et al., 2000; Sundaresan et al., 2017; Sundaresan & Leo, 2006). Oxidation ejects negative charges from the polymer as indicated by the reversal in charge accumulation from t=5 s to 10 s.

The dephosphorylation of para-nitro phenyl phosphate by alkaline phosphatase is started, stopped, and throttled by regulating pH with a PCA

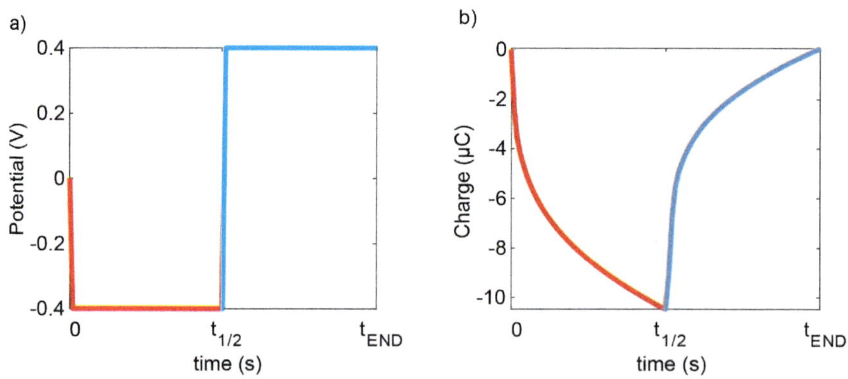

Fig. 10 (A) A potential waveform used in chronocoulometry (and chronoamperometry). (B) The typical charge-time response of PPy(DBS) to the waveform.

embedded in a semi-microcuvette. Reducing the polymer increases the activity of alkaline phosphatase and subsequent oxidation of the polymer decreases enzyme activity. Using a 50% duty cycle, an increase in dephosphorylation rate ≥ 1.5 times the initial rate occurs. Applying the 80% oxidation duty cycle to the PCA halts the dephosphorylation reaction. Multiple cycles are run sequentially demonstrating the biocompatibility of the chemical actuator when controlling alkaline phosphatase.

PCA control also is effective for membrane bound enzyme complexes. The electrochemical cycling of the PCA is correlated to the activity of rubber transferase from *Ficus elastica* in a radiometric enzyme assay. Reduction duty cycles remove Mg^{2+} and Mn^{2+} from solution and completely halt rubber transferase activity. Oxidation duty cycles injecting cations into solution maximize IPP incorporation. Cycling the PCA between reduced and oxidized states has the desired "off-on" behavior without irreversibly inhibiting the rubber transferase complex or dislodging the elongating rubber particle from the complex.

11. Conclusion

This chapter presents methods for the fabrication of programmable chemical actuators to selectively modulate electrochemical gradients in biological systems. pH-indicating dyes are used to measure the effects of potential application techniques on the temporal control over the gradient. When PCAs are operated under duplex pulse amperometry, the duty cycle can be studied as a control variable over alkaline phosphatase activity during spectrophotometric enzyme assays. PCAs can also be used to study the effects of modulating Mg^{2+} and Mn^{2+} concentrations on the membrane bound rubber transferase complex in radiometric enzyme assays.

PCAs have a variety of uses in studying the biochemical pathways that underly environmental responses in plants, especially salt tolerance. Specifically, PCAs can be developed as an automated control mechanism for enzymes with established chemistries, like alkaline phosphatase. They also have utility as a quantitative method of exploring enigmatic reaction chemistries, as in the case of rubber transferase. Finally, programmable chemical actuators hold promise in lowering the environmental impact of existing enzyme-based biocatalysis as they have been shown to significantly reduce the carbon footprint of enzyme-based processes.

The integration of chemical actuators with common laboratory techniques and reaction-ware facilitates the application of PCAs for control of a variety of enzymes. Consequently, this framework can be used to explore ionic interactions in biochemical pathways in plants. Applications in the fields of materials science and regenerative medicine are under exploration in the Evans lab.

Acknowledgments

This work is adapted from the methods used in the first author's dissertation: Evans, J. P. (2021). *Programmable Chemical Actuators Control Enzyme Activity* [Doctoral dissertation, Ohio State University]. This work received funding from the NSF CAREER Program #1055311. This work was supported in part by the United States Department of Agriculture National Institute of Food, Agriculture, Hatch project number OHO01417, accession number 1014257. PhD Fellowships from the Translational Plant Sciences Graduate Program in the Center for Applied Plant Sciences at The Ohio State University and from the College of Engineering, Department of Mechanical and Aerospace Engineering are acknowledged. The views and opinions expressed in this article are only those of the authors.

References

Bard, A. J. (2001). Introduction and principles. In *Scanning electrochemical microscopy* (pp. 1–10). CRC Press.

Bard, A. J., Faulkner, L. R., Leddy, J., & Zoski, C. G. (1980). *Electrochemical methods: Fundamentals and applications*. Wiley New York.

Bengtsson, K., Nilsson, S., & Robinson, N. D. (2014). Conducting polymer electrodes for gel electrophoresis. *PLOS ONE*, *9*(2), e89416.

Bushman, S. B., Scholte, A. A., Cornish, K., Scott, D. J., Brichta, J. L., Vederas, J. C., et al. (2006). Identification and comparison of natural rubber from two *Lactuca* species. *Phytochemistry*, *67*(23), 2590–2596.

Chang, S.-Y., Ko, T.-P., Chen, A. P.-C., Wang, A. H.-J., & Liang, P.-H. (2004). Substrate binding mode and reaction mechanism of undecaprenyl pyrophosphate synthase deduced from crystallographic studies. *Protein Science*, *13*, 971–978. https://doi.org/10.1110/ps.03519904.

Cornish, K. (1993). The separate roles of plant *cis* and *trans* prenyl transferases in *cis*-1,4-polyisoprene biosynthesis. *European Journal of Biochemistry*, *218*, 267–271.

Cornish, K., & Backhaus, R. A. (1990). Rubber transferase activity in rubber particles of guayule. *Phytochemistry*, *29*, 3809–3813.

Cornish, K., & Bartlett, D. L. (1997). Stabilisation of particle integrity and particle bound cis-prenyl transferase activity in stored, purified rubber particles. *Phytochemical Analysis: An International Journal of Plant Chemical and Biochemical Techniques*, *8*, 130–134.

Cornish, K., & Siler, D. (1996). Characterization of cis-prenyl transferase activity localised in a buoyant fraction of rubber particles from *Ficus elastica* latex. *Plant Physiology and Biochemistry-Paris*, *34*, 377–384.

Cornish, K., Siler, D. J., Grosjean, O. K., & Goodman, N. (1993). Fundamental similarities in rubber particle architecture and function in three evolutionarily divergent plant species. *Journal of Natural Rubber Research*, *8*, 275–285.

Cornish, K., & Xie, W. (2012). Natural rubber biosynthesis in plants: Rubber transferase. *Methods in Enzymology*, *515*, 63–82.

Cosnier, S., & Karyakin, A. (2011). *Electropolymerization: Concepts, materials and applications*. John Wiley & Sons.

Cottrell, F. (1903). Residual current in galvanic polarization, regarded as a diffusion problem. *Zeitschrift für Physikalische Chemie*, *42*, 385–431.

da Costa, B. M., Keasling, J. D., McMahan, C. M., & Cornish, K. (2006). Magnesium ion regulation of in vitro rubber biosynthesis by *Parthenium argentatum* Gray. *Phytochemistry*, *67*, 1621–1628.

Datta, M., & Landolt, D. (1985). Experimental investigation of mass transport in pulse plating. *Surface Technology*, *25*(2), 97–110.

Delgove, M. A., Elford, M. T., Bernaerts, K. V., & Wildeman, S. M. D. (2018). Toward upscaled biocatalytic preparation of lactone building blocks for polymer applications. *Organic Process Research & Development*, *22*(7), 803–812.

Diaz, A. F., Kanazawa, K. K., & Gardini, G. P. (1979). Electrochemical polymerization of pyrrole. *Journal of the Chemical Society, Chemical Communications*, 635–636.

Ding, J., Hu, W., Paek, E., & Mitlin, D. (2018). Review of hybrid ion capacitors: From aqueous to lithium to sodium. *Chemical Reviews*, *118*, 6457–6498.

Elgrishi, N., Rountree, K. J., McCarthy, B. D., Rountree, E. S., Eisenhart, T. T., & Dempsey, J. L. (2018). A practical beginner's guide to cyclic voltammetry. *Journal of Chemical Education*, *95*, 197–206.

Erlandsson, P. G., & Robinson, N. D. (2011). Electrolysis-reducing electrodes for electrokinetic devices. *Electrophoresis*, *32*(6–7), 784–790.

Feldberg, S. W. (1984). Reinterpretation of polypyrrole electrochemistry. Consideration of capacitive currents in redox switching of conducting polymers. *Journal of the American Chemical Society*, *106*, 4671–4674.

Fu, K., Klibanov, A., & Langer, R. (2000). Protein stability in controlled-release systems. *Nature Biotechnology*, *18*, 24–25.

Funt, B., & Diaz, A. (1991). *Organic electrochemistry: An introduction and a guide*. New York: Marcel Dekker.

Gellett, W., Schumacher, J., Kesmez, M., Le, D., & Minteer, S. D. (2010). High current density air-breathing laccase biocathode. *Journal of the Electrochemical Society*, *157*, B557.

Gitlin, I., Carbeck, J. D., & Whitesides, G. M. (2006). Why are proteins charged? Networks of charge–charge interactions in proteins measured by charge ladders and capillary electrophoresis. *Angewandte Chemie, International Edition*, *45*, 3022–3060.

Goodridge, F., & Scott, K. (2013). *Electrochemical process engineering: A guide to the design of electrolytic plant*. Springer Science & Business Media.

Green, T. A., & Roy, S. (2017). Application of a duplex diffusion layer model to pulse reverse plating. *Transactions of the IMF*, *95*(1), 46–51.

Gupta, S. (2020). *Development and utilization of a novel synaptic transistor to detect dynamic neuronal processes*. The Ohio State University.

Gupta, S., Ghatak, S., Hery, T., Khanna, S., El Masry, M., Sundaresan, V. B., et al. (2020). Ad hoc hybrid synaptic junctions to detect nerve stimulation and its application to detect onset of diabetic polyneuropathy. *Biosensors & Bioelectronics*, *169*, 112618.

Gupta, S., Otero, J. J., Sundaresan, V. B., & Czeisler, C. M. (2020). Near field non-invasive electrophysiology of retrotrapezoid nucleus using amperometric cation sensor. *Biosensors & Bioelectronics*, *151*, 111975.

Güven, G., Prodanovic, R., & Schwaneberg, U. (2010). Protein engineering–an option for enzymatic biofuel cell design. *Electroanalysis: An International Journal Devoted to Fundamental and Practical Aspects of Electroanalysis*, *22*, 765–775.

György, I. (2008). *Conducting polymers: A new era in electrochemistry*. Springer.

Hagel, J. M., Yeung, E. C., & Facchini, P. J. (2008). Got milk? The secret life of laticifers. *Trends in Plant Science*, *13*, 631–639.

Hageman, M. J. (1988). The role of moisture in protein stability. *Drug Development and Industrial Pharmacy*, *14*, 2047–2070.

Harris, H. (1990). The human alkaline phosphatases: What we know and what we don't know. *Clinica Chimica Acta*, *186*, 133–150.
Hery, T., & Sundaresan, V.-B. (2016). Ionic redox transistor from pore-spanning PPy (DBS) membranes. *Energy & Environmental Science*, *9*, 2555–2562.
Iyoda, T., Ohtani, A., Shimidzu, T., & Honda, K. (1986). Charge-controllable membrane. Polypyrrole-polyelectrolyte composite membrane through anodic doping process. *Chemistry Letters*, *15*, 687–690.
John, R., & Wallace, G. (1993). Doping-dedoping of polypyrrole: A study using current-measuring and resistance-measuring techniques. *Journal of Electroanalytical Chemistry*, *354*, 145–160.
Kagel, H., Bier, F. F., Frohme, M., & Glökler, J. F. (2019). A novel optical method to reversibly control enzymatic activity based on photoacids. *Scientific Reports*, *9*, 1–6.
Karley, A. J., Leigh, R. A., & Sanders, D. (2000). Where do all the ions go? The cellular basis of differential ion accumulation in leaf cells. *Trends in Plant Science*, *5*, 465–470.
Karley, A. J., & White, P. J. (2009). Moving cationic minerals to edible tissues: Potassium, magnesium, calcium. *Current Opinion in Plant Biology*, *12*, 291–298.
Kelly, R. G., Scully, J. R., Shoesmith, D., & Buchheit, R. G. (2002). *Electrochemical techniques in corrosion science and engineering*. CRC Press.
Kissinger, P., & Heineman, W. R. (2018). *Laboratory techniques in electroanalytical chemistry, revised and expanded*. CRC press.
Kohse, S., Neubauer, A., Pazidis, A., Lochbrunner, S., & Kragl, U. (2013). Photoswitching of enzyme activity by laser-induced pH-jump. *Journal of the American Chemical Society*, *135*, 9407–9411.
Krische, B., & Zagorska, M. (1989). Overoxidation in conducting polymers. *Synthetic Metals*, *28*(1–2), 257–262.
Lane, N. (2010). Why are cells powered by proton gradients. *Nature Education*, *3*, 2.
Lee, K., Evans, J. P., Satagopan, S., Sun, Y., Parquette, J. R., Sundaresan, V. B., et al. (2020). Carbon footprint of biomimetic carbon fixation by immobilizing nature's CO2-sequestering enzyme and regenerating its energy carrier. *ACS Sustainable Chemistry & Engineering*, *8*, 16833–16841.
Levin, M. (2012). Molecular bioelectricity in developmental biology: New tools and recent discoveries: Control of cell behavior and pattern formation by transmembrane potential gradients. *BioEssays*, *34*, 205–217.
Li, M., & Anand, R. K. (2016). Recent advancements in ion concentration polarization. *Analyst*, *141*, 3496–3510.
Maathuis, F. J. (2009). Physiological functions of mineral macronutrients. *Current Opinion in Plant Biology*, *12*, 250–258.
Manning, M. C., Patel, K., & Borchardt, R. T. (1989). Stability of protein pharmaceuticals. *Pharmaceutical Research*, *6*, 903–918.
Messiha, H. L., Ahmed, S. T., Karuppiah, V., Suardíaz, R., Ascue Avalos, G. A., & Scrutton, N. S. (2018). Biocatalytic routes to lactone monomers for polymer production. *Biochemistry*, *57*(13), 1997–2008.
Milo, R., & Phillips, R. (2015). What are the concentrations of different ions in cells. In Draft (Ed.), *Cell biology by the numbers* (pp. 127–130). Garland Science.
Nelson, D. L., Lehninger, A. L., & Cox, M. M. (2008). *Lehninger principles of biochemistry*. Macmillan.
Nilsen, I. W., Øverbø, K., & Olsen, R. L. (2001). Thermolabile alkaline phosphatase from Northern shrimp (Pandalus borealis): Protein and cDNA sequence analyses. *Comparative Biochemistry and Physiology Part B: Biochemistry and Molecular Biology*, *129*(4), 853–861.
Northcutt, R. G. (2015). *A mechanistic interpretation for charge storage in conducting polymers* (PhD Thesis). The Ohio State University.

Northcutt, R. G., & Sundaresan, V.-B. (2015). Mechanoelectrochemistry of PPy (DBS) from correlated characterization of electrochemical response and extensional strain. *Physical Chemistry Chemical Physics*, *17*, 32268–32275.

Olsen, R. L., Øverbø, K., & Nilsen, I. W. (1991). Alkaline phophatase from the hepatopancreas of shrimp (*Pandalus borealis*): A dimeric enzyme with catalytically active subunits. *Comparative Biochemistry and Physiology Part B: Comparative Biochemistry*, *99*, 755–761.

Pearson, C. H., Cornish, K., Rath, D. J., Brichta, J. L., & Van Fleet, J. E. (2010). Agronomic and natural rubber characteristics of sunflower as a rubber-producing crop. *Industrial Crops and Products*, *31*, 481–491.

Qi, S., Zheng, H., Qin, H., & Zhai, H. (2020). Development of a facile and sensitive method for detecting alkaline phosphatase activity in serum with fluorescent gold nanoclusters based on the inner filter effect. *Analyst*, *145*, 3871–3877. https://doi.org/10.1039/D0AN00052C.

Rasmussen, M., Abdellaoui, S., & Minteer, S. D. (2016). Enzymatic biofuel cells: 30 years of critical advancements. *Biosensors & Bioelectronics*, *76*, 91–102.

Ren, X., & Pickup, P. G. (1993). Ion transport in polypyrrole and a polypyrrole/polyanion composite. *The Journal of Physical Chemistry*, *97*, 5356–5362.

Rodrigues, G., & Curtis, R. (2010). Instrument performance verification: Micropipettes. In *Practical approaches to method validation and essential instrument qualification* (pp. 327–346). New York, New York: John Wiley sons.

Sadki, S., Schottland, P., Brodie, N., & Sabouraud, G. (2000). The mechanisms of pyrrole electropolymerization. *Chemical Society Reviews*, *29*, 283–293. https://doi.org/10.1039/A807124A.

Sahalianov, I., Singh, S. K., Tybrandt, K., Berggren, M., & Zozoulenko, I. (2019). The intrinsic volumetric capacitance of conducting polymers: Pseudo-capacitors or double-layer supercapacitors? *RSC Advances*, *9*(72), 42498–42508.

Sandler, S. I. (2017). *Chemical, biochemical, and engineering thermodynamics*. John Wiley & Sons.

Seitanidou, M., Tybrandt, K., Berggren, M., & Simon, D. T. (2019). Overcoming transport limitations in miniaturized electrophoretic delivery devices. *Lab on a Chip*, *19*(8), 1427–1435.

Shinohara, H., Kojima, J., & Aizawa, M. (1989). Electrically controlled ion transfer and pH change near a conducting polymer-coated electrode. *Journal of Electroanalytical Chemistry and Interfacial Electrochemistry*, *266*, 297–308.

Shirakawa, H. (2001a). Nobel lecture: The discovery of polyacetylene film—The dawning of an era of conducting polymers. *Reviews of Modern Physics*, *73*, 713.

Shirakawa, H. (2001b). The discovery of polyacetylene film: The dawning of an era of conducting polymers (Nobel lecture). *Angewandte Chemie, International Edition*, *40*, 2574–2580.

Smela, E. (1999). Microfabrication of PPy microactuators and other conjugated polymer devices. *Journal of Micromechanics and Microengineering*, *9*, 1.

Sundaresan, V. B., & Leo, D. J. (2006). Chemo-mechanical model for actuation based on biological membranes. *Journal of Intelligent Material Systems and Structures*, *17*, 863–870.

Sundaresan, V. B., Northcutt, R., Venkatesh, V., & Evans, J. P. (2017). Mechanoelectrochemistry of soft electroactive materials via surface-tracked scanning electrochemical microscopy. In *Behavior and mechanics of multifunctional materials and composites 2017* (p. 1016506).

Venugopal, V. (2016). *Kinetics of ion transport in conducting polymers* (PhD Thesis). The Ohio State University.

Venugopal, V., Hery, T., Venkatesh, V., & Sundaresan, V. B. (2017). Mass and charge density effects on the saturation kinetics of polypyrrole doped with dodecylbenzene sulfonate. *Journal of Intelligent Material Systems and Structures*, *28*, 760–771.

Venugopal, V., & Sundaresan, V. B. (2016). Polypyrrole-based amperometric cation sensor with tunable sensitivity. *Journal of Intelligent Material Systems and Structures*, *27*, 1702–1709.
Venugopal, V., Zhang, H., Northcutt, R., & Sundaresan, V. B. (2014). A thermodynamic chemomechanical constitutive model for conducting polymers. *Sensors and Actuators B: Chemical*, *201*, 293–299.
Woodley, J. M. (2020). Advances in biological conversion technologies: New opportunities for reaction engineering. *Reaction Chemistry & Engineering*, *5*(4), 632–640.

Further reading

Cornish, K., Xie, W., Kostyal, D., Shintani, D. K., & Hamilton, R. A. (2015). Immunological analysis of the alternate natural rubber crop *Taraxacum kok-saghyz* indicates multiple proteins cross-reactive with *Hevea brasiliensis* latex allergens. *Journal of Biotechnology and Biomaterials*, *5*, 201–207.

SECTION II

Metabolite

CHAPTER SEVEN

Measurement of flux through sulfate assimilation using [^{35}S]sulfate

Süleyman Günal and Stanislav Kopriva*
Institute for Plant Sciences, Cluster of Excellence on Plant Sciences (CEPLAS), University of Cologne, Cologne, Germany
*Corresponding author: e-mail address: skopriva@uni-koeln.de

Contents

Abstract

Sulfur metabolism provides a number of compounds that are essential for plant survival and fitness and that affect the yield and quality of crops. Sulfur metabolism is a dynamic process, responding to a number of external cues. Because of this dynamics and rapid turnover, steady-state levels of sulfur-containing compounds do not always fully reflect plant responses to such cues. Therefore, measurements of the flux through sulfate assimilation may give a more precise estimate of the effects of environmental stimuli or metabolic disturbances on sulfur metabolism. The flux can be determined after feeding plants with sulfate labelled with an isotopic tracer. Here we describe a protocol for using [^{35}S]sulfate to measure flux through sulfate assimilation in Arabidopsis. The protocol can be adapted for any plant species and growth conditions, and does not require any special equipment beyond a standard high performance liquid chromatograph. We hope that the protocol will support our colleagues in a more frequent use of flux measurements to answer new biological questions in plant sulfur metabolism.

1. Introduction

Sulfate assimilation is an essential pathway of plant primary metabolism, providing great number of vital compounds, including the amino acids

Methods in Enzymology, Volume 676
ISSN 0076-6879
https://doi.org/10.1016/bs.mie.2022.07.021

cysteine and methionine, as well as co-enzymes, such as iron sulfur centers, lipoic acid, thiamine, S-adenosylmethionine, coenzyme A, molybdenum coenzyme, and many others (Takahashi, Kopriva, Giordano, Saito, & Hell, 2011). Sulfur is also incorporated into many specialized products, the best known examples being the glucosinolates in the Brassicaceae or alliin and other organosulfur compounds in onion and garlic. The source of sulfur for plants is mainly inorganic sulfate, which needs to be assimilated into the bioorganic compounds.

Sulfate is taken up to the cells via sulfate transporters localized in the plasma membrane and is distributed to vacuole for storage or plastids and cytosol for further metabolism (Gigolashvili & Kopriva, 2014). For assimilation, the inert sulfate needs to be activated by adenylation, at the expense of ATP, in a reaction catalyzed by ATP sulfurylase (Murillo & Leustek, 1995). The product, adenosine 5′-phosphosulfate (APS) is a branching point to reductive and oxidative assimilation, often called primary and secondary (Kopriva et al., 2012). In the reductive branch, APS is reduced by APS reductase to sulfite, which is subsequently reduced by sulfite reductase to cysteine, the first product of primary sulfate assimilation (Fig. 1). Cysteine serves as a donor of reduced sulfur for other S-containing metabolites,

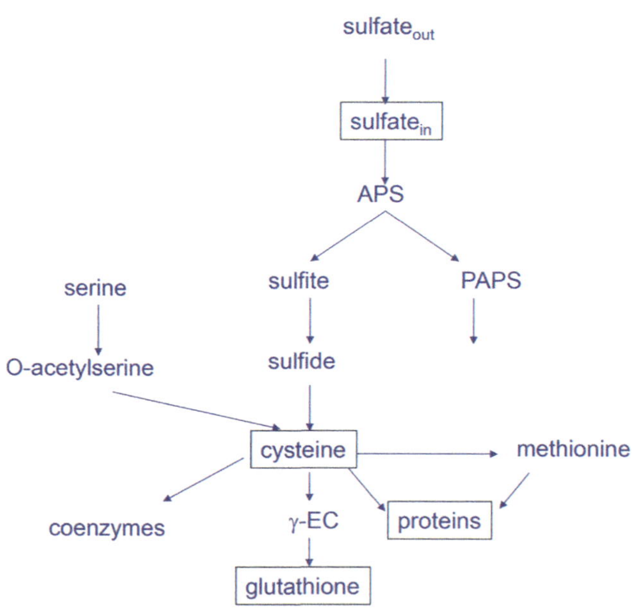

Fig. 1 Scheme of sulfate assimilation. The pools measured in the flux protocol are boxed.

including methionine and the tripeptide glutathione (GSH) (Kopriva, Malagoli, & Takahashi, 2019). In the oxidative branch of sulfate assimilation APS is phosphorylated to 3′-phosphoadenosine 5′-phosphosulfate (PAPS), which is a donor of activated sulfate for synthesis of various secondary metabolites, or sulfated peptides (Mugford et al., 2009).

Sulfate assimilation is tightly regulated by plant demand for reduced sulfur and by sulfate supply (Koprivova & Kopriva, 2014; Lappartient & Touraine, 1996; Takahashi et al., 1997). Thus, numerous reports described changes in transcriptome and metabolome of plants under sulfur deficiency (Dietzen et al., 2020; Hirai et al., 2003, 2005, 2007; Maruyama-Nakashita, Inoue, Watanabe-Takahashi, Yamaya, & Takahashi, 2003; Nikiforova et al., 2003, 2005) or under conditions changing the demand for sulfur molecules, such as oxidative or biotic stress (Ball et al., 2004; Kruse et al., 2007; Van Hoewyk et al., 2008; Vauclare et al., 2002). The molecular mechanisms responsible for these regulatory processes are, however, still largely unknown (Ristova & Kopriva, 2022). In particular, it is not always clear which alterations in steady-state levels of transcripts and metabolites actually affect the sulfur fluxes in the plants (Calderwood, Morris, & Kopriva, 2014). This knowledge is even more imperative since many S-containing compounds are beneficial for the plant as well as for humans, therefore, manipulation of their content is an important biotechnological aim.

Indeed, determining fluxes through metabolic pathways and determining their control are the prerequisites for understanding metabolic control (Calderwood et al., 2014). For sulfur metabolism this is possible through feeding plants sulfate labelled either with radioactive ^{35}S or stable ^{34}S. Labelling with the stable isotope ^{34}S is usually used for long term experiments to determine fluxes of sulfur throughout the plant, remobilization, and seed filling (D'Hooghe, Bataille, Trouverie, & Avice, 2013; Dubousset et al., 2009). On the other hand, labelling with ^{35}S has been used since the early days of plant sulfur research to quantify in vivo sulfate assimilation and identify regulatory mechanisms (Brunold & Suter, 1984; Huseby et al., 2013; Mugford, Lee, Koprivova, Matthewman, & Kopriva, 2011; Neuenschwander, Suter, & Brunold, 1991; Rennenberg, 1983; Scheerer et al., 2010; Vauclare et al., 2002). The flux measurements explained, for example, the accumulation of sulfate in Arabidopsis mutants in genes for sulfate assimilation enzymes *apr2* and *atps1* (Chao et al., 2014; Koprivova et al., 2013) or accumulation of glutathione in mutants in APS kinase (Mugford et al., 2011). They were also used for metabolic control analyses to reveal an importance of APS reductase and sulfate uptake in the control of the pathway (Vauclare et al., 2002) and to reveal dynamics in this control and its dependence on environmental conditions (Feldman-Salit, Veith, Wirtz, Hell, & Kummer, 2019). Since ^{35}S flux measurements are the best way to determine disturbances in plant sulfate assimilation, in this chapter we provide an optimized protocol for determination of the flux into thiols and proteins in *A. thaliana*, that is easily adaptable to other plant (and non-plant) species.

2. Measurement of flux through sulfate assimilation

2.1 Plant growth

2.1.1 Nutrient solution

2.1.1.1 Materials

0.6 M $Ca(NO_3)_2 \times 4H_2O$
0.4 M KNO_3
0.3 M KH_2PO_4
0.3 M $MgSO_4 \times 7H_2O$
40 mM Na Ethylenediaminetetraacetic acid Fe(III)
10 mM $MnCl_2 \times 4H_2O$
50 mM H_3BO_3
1.75 mM $ZnCl_2$
0.5 mM $CuCl_2$
0.8 mM Na_2MoO_4
0.1 mM $CoCl_2 \times 6H_2O$
1 mM KI
4-Morpholineethanesulfonic acid hydrate (MES)
sucrose

2.1.1.2 Procedure

Add 2.5 mL of macronutrient solutions ($Ca(NO_3)_2$, KNO_3, KH_2PO_4, $MgSO_4$, Fe-EDTA) and 1 mL of micronutrient solutions ($MnCl_2$, H_3BO_3, $ZnCl_2$, $CuCl_2$, Na_2MoO_4, $CoCl_2$, KI) to 900 mL deionized water. Add 0.8 g MES and 5 g sucrose. Mix and adjust pH to 5.7 with 1 M KOH.

2.1.2 Preparation of plates

2.1.2.1 Materials

Arabidopsis seeds
Desiccator
5% (w/v) sodium hypochlorite (NaOCl)
Concentrated HCl

12-well plates (Sarstedt)
Polypropylene mesh 14 mm × 14 mm (Franz Eckert) autoclaved and dried
Nutrient solution from 2.1.1
0.1% (w/v) Agarose, sterilized by autoclaving

2.1.2.2 Procedure

1. For seed sterilization by chlorine gas, put ca. 20 μL Arabidopsis seeds in 0.5 mL Eppendorf tube and place inside a desiccator kept in a laminar flow hood. Add 125 mL of 5% NaOCl into a 250 mL beaker and place inside the desiccator. Add 2.5 mL concentrated HCl to the beaker and quickly close the desiccator lid. Keep in the laminar flow hood for 4 h, carefully open the desiccator lid and close the tubes with sterile seeds.
2. For a biological replicate plant material from two wells of the 12-well plate is combined. In a typical experiment 6 genotypes will be compared, or 3 genotypes at two conditions, with 4 biological replicates, i.e., 24 samples, which will need 48 wells. In a laminar flow hood fill each well of four 12-well plates with 1 mL of nutrient solution, either autoclaved or sterile filtered. Place one piece of sterile polypropylene mesh cut to 14 × 14 mm squares onto the nutrient solution. Add 500 μL of sterile 0.1% agarose to the seeds, re-suspend by pipetting, and with a pipette place 30–50 seeds onto the membranes in each well.
3. Close the lids of the 12-well plates, wrap them in aluminum foil and place to 4 °C for 2–3 days for stratification.
4. Place the wrapped plates to plant growth cabinets at 22 °C for 72 h. The germination in dark leads to prolonged hypocotyls which prevents contact of the leaves with nutrient solution.
5. Unwrap the plates and keep them in plant growth cabinets at 22 °C, at long day conditions 16 h light 8 h night for 17 days, change the nutrient solution after 7 and 14 days. The setup is displayed in Fig. 2.

Note: The plant growth conditions—temperature, day-length—can be adapted depending on available growth cabinets. Similarly, sterilization in solution by ethanol and hypochlorite can be used depending on local standard operation procedures.

2.1.3 Feeding [^{35}S]sulfate

2.1.3.1 Materials

[^{35}S]sulfate (Hartmann analytic, S-RA-1)
Nutrient solution as in Section 2.1.1 with $MgSO_4$ concentration adjusted to 0.2 mM and without sucrose

Fig. 2 Example of the Arabidopsis growth system in 12-well plates, the plates were photographed 12 days after unwrapping. Bar = 1 cm.

Scintillation cocktail, e.g., ROTISZINT eco plus (Roth)
Scintillation counter
Balances
Liquid nitrogen

2.1.3.2 Procedure

1. Prepare the labelling solution, containing the nutrient solution adjusted to sulfate concentration of 0.2 mM. To a 50 mL falcon tube place at least 1 mL of the nutrient solution per well, for the typical experiment 50 mL.
2. From now on, further protocol steps have to be performed in a room certified for work with radioactive substances. Add 0.5 µL of the [^{35}S]sulfate stock per ml of nutrient solution to the labelling solution in Falcons, close and mix. Add 2 mL scintillation cocktail into a scintillation vial, add 10 µL of the labelling solution, mix, and determine the radioactivity by scintillation counting. The activity should reach at least 100,000 cpm per 10 µL of the labelling solution; if it is too low, add additional volume of the [^{35}S]sulfate stock, mix and re-measure. This measurement gives the specific activity of sulfate in the experiment (cpm_{sulf}) as the measured cpm/2 per nmol sulfate. (10 µL of 0.2 mM solution correspond to 2 nmol sulfate).
3. Carefully remove the nutrient solution from the 8 wells with plants of the first genotype using a 1 mL pipette and discard. Carefully add 1 mL of the labeling solution so that the plants on the mesh float on the surface. After 5 min repeat with second genotype, etc., always leaving 5 min between the genotypes to allow for handling during sample collection.

4. Keep the plates under a light source for 4 h (ideally in a growth cabinet if available, otherwise on bench under light is sufficient).
5. Prepare and pre-weigh 1.5 mL tubes, for roots and shoots of each sample (in the typical experiment 24 for shoots, 24 for roots). Prepare a beaker with at least 1.5 L water to wash the plants.
6. After a 4 h incubation, remove the mesh with the plants by forceps, wash it in the beaker with water, and place onto tissue paper. Take the roots by another forceps and collect them from the bottom of the mesh with a gentle ripping. Collect shoots and through the empty mesh into radioactive waste. Combine roots and shoots from two wells for one biological replicate of either organ, dry the plant material with the tissue, place in the pre-weighed tubes, weigh, record the fresh weight (FW), and freeze the tubes in liquid nitrogen.

Note: The experiment should yield ca. 50 mg FW of the shoots (FWS) and 20 mg FW of the roots (FWR).

2.2 Measurement of sulfate uptake and flux

2.2.1 Homogenisation and determination of sulfate uptake

2.2.1.1 Materials

0.1 M HCl
Polypropylene micro pestles 1.5 mL (Roth, CXH9.1)
Homogenizer
Scintillation cocktail

2.2.1.2 Procedure

1. Homogenize the plants in 10-fold (*V/w*) volume of 0.1 M HCl using micro pestles. For volumes above 300 μL homogenize first in a smaller volume (300 μL), add the rest volume, and shortly homogenize with the full volume. After homogenization keep tubes on ice till all extracts are finished.
2. Centrifuge 5 min maximum speed in a microfuge.
3. Add 2 mL scintillation cocktail into a scintillation vial, add 10 μL of the supernatant, mix, and determine the radioactivity by scintillation counting. This measurement gives the radioactivity of [^{35}S]sulfur in 1 mg of shoot (cpm_S) or root (cpm_R) (10 μL after 1:10 extraction correspond to 1 mg FW).
4. For determination of [^{35}S] in proteins pipette 100 μL of supernatant into a new tube. For determination of [^{35}S] in thiols pipette 50 μL of supernatant into a new tube (Fig. 3).

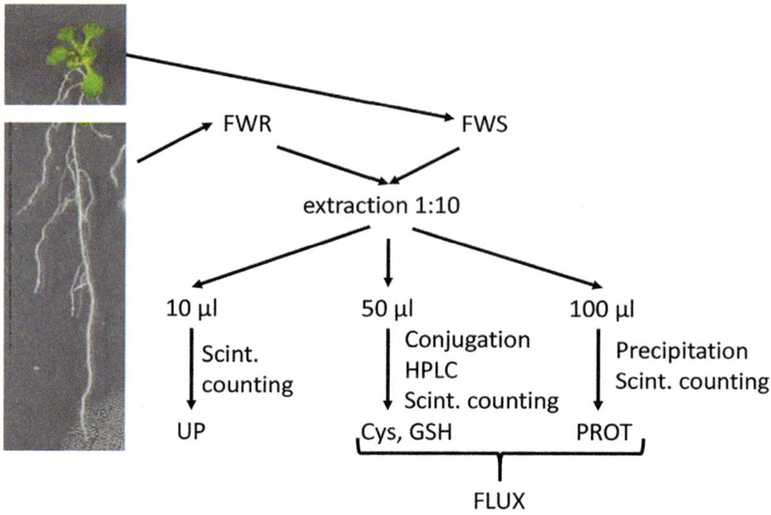

Fig. 3 Simplified workflow of the flux measurements.

5. To calculate sulfate uptake in the plants, determine first the uptake into roots (UP-R) and shoots (UP-S) in nmol/h/g FW according to following formula: $1000 \times cpm_R/cpm_{sulf}/4$ or $1000 \times cpm_S/cpm_{sulf}/4$, respectively. From these values calculate the total amount of sulfate taken up into the plant per hour (TOT): (UP-R x FWR) + (UP-S x FWS). This value divided by FWR gives the sulfate uptake in nmol/h/g root.
6. The UP-S value represents the rate of sulfate root-to-shoot translocation.
7. The allocation of sulfate in the shoot can be calculated as percentage of sulfate taken up found in the shoot, i.e., (UP-S x FWS)/TOT.

2.2.2 Determination of incorporation into proteins

2.2.2.1 Materials

100% (w/v) Trichloroacetic acid (TCA)
1% (w/v) TCA
96% (v/v) ethanol
0.1 M NaOH
Scintillation cocktail

2.2.2.2 Procedure

1. Add 25 µL 100% TCA solution to the 100 µL extract from 2.2.1 and keep on ice for 15 min.

2. Centrifuge for 10 min at maximum speed in a microfuge, discard the supernatant.
3. Wash the pellet in 100 µL 1% TCA, centrifuge for 5 min, discard supernatant.
4. Wash the pellet in 200 µL 96% EtOH, centrifuge for 5 min, discard supernatant, and leave air dry for 10 min.
5. Dissolve the pellet in 100 µL 0.1 M NaOH.
6. Add 2 mL scintillation cocktail into a scintillation vial, add the whole 100 µL protein solution, mix, and determine the radioactivity by scintillation counting. This measurement, after division by cpm_{sulf}, division by 4 (for 4 h), division by 10 (for 10 mg tissue), and multiplication by 1000 (to adjust for g FW) gives the incorporation of [^{35}S]sulfur per hour and g of shoot ($PROT_S$) or root ($PROT_R$). Relative incorporation into proteins can be calculated as $PROT_S$/UP-S or $PROT_R$/UP-R.

2.2.3 Determination of incorporation into thiols

2.2.3.1 Materials

0.1 M NaOH
0.1 M 1,4-Dithiothreitol (DTT), freshly made
1 M Tris HCl, pH 8.0
100 mM monobromobimane
50% (v/v) acetic acid
10% (v/v) methanol, 0.25% (v/v) acetic acid, pH 3.9
90% (v/v) methanol, 0.25% (v/v) acetic acid, pH 3.9

2.2.3.2 Equipment

HPLC column Spherisorb™ ODS2, 250 × 4.6 mm, 5 µm, Waters.
HPLC system including a fluorescent detector.
HPLC Radio detector LB513 FlowStar, Berthold Technologies.

2.2.3.3 Procedure

1. To the 50 µL extract from 2.2.1. add 50 µL 0.1 M NaOH to neutralize and 1 µL 0.1 M DTT to reduce the thiols and incubate in dark at 37 °C for 15 min.
2. Add 11.5 µL 1 M Tris-HCl pH 8.0 and 5 µL 100 mM monobromobimane to label the thiols, mix and incubate in dark at 37 °C for 15 min.
3. Add 11.5 µL 50% acetic acid, mix, centrifuge for 10 min at maximal speed and transfer the supernatant into HPLC vials.

4. Inject 100 µL into high-performance liquid chromatography (HPLC) system, where the thiol conjugates are separated and quantified. Any HPLC system can be used with a reverse phase column (Spherisorb™ ODS2, 250 × 4.6 mm, 5 µm), fluorimetric detector (excitation: 390 nm, emission:480 nm) and radiodetector (e.g., FlowStar). Two solvents are used in a linear gradient from 5 to 18% B (90% methanol, 0.25% acetic acid, pH 3.9) in A (10% methanol, 0.25% acetic acid, pH 3.9) over 20 min with a constant flow rate of 1 mL/min. The first radioactive peak coming with the front represents sulfate, later peaks correspond to cysteine and glutathione and another peak is detected after the column is washed with 100% solvent B. The output of the Chromatography Data System Software should be set to peak relative area.
5. To determine the incorporation into the individual sulfur pools (in nmol S/h/g FW), use the peak percentage area (PPA), using this guide: $GSH_{R\ or\ S} = 1000 * PPA_{GSH} \times cpm_{R\ or\ S}/cpm_{sulf}/4$ and analogically, $Cys_{R\ or\ S} = 1000 * PPA_{Cys} \times cpm_{R\ or\ S}/cpm_{sulf}/4$.
6. The flux through primary assimilation is calculated by addition of the incorporations into proteins and thiols—$PROT_{R\ or\ S} + Cys_{R\ or\ S} + GSH_{R\ or\ S}$. The flux is dependent both on sulfate uptake and assimilation rate.
7. The relative flux is the percentage of the primary flux in the total ^{35}S uptake. The relative flux is dependent only on the assimilation rate.

Note 1: If no radio detector is available, it is possible to connect the HPLC to a fraction collector and collect fractions of 1 mL from the HPLC runs, add 4 mL scintillation cocktail and determine the radioactivity by scintillation counting. This determines directly the incorporation into the different pools, but limits the number of samples to be processed automatically.

Note 2: This protocol does not include flux measurement into the de novo synthesis of methionine. Previous work showed that under standard conditions, in WT Arabidopsis accession Col-0 the flux to methionine corresponds to ca. 7% of that to thiols (Mugford et al., 2011), so unless the research question is focused on Met, this flux might be ignored.

3. Conclusions

Sulfur-containing metabolites underlie a number of important traits in crop plants (Koprivova & Kopriva, 2016). Nevertheless, metabolic flux analysis to support genetic engineering of such traits has been very rarely used, if at all. The flux measurements can also help to better understand, for example, natural variation in sulfur homeostasis, as initiated in a few Arabidopsis accessions (Koprivova et al., 2013; Loudet et al., 2007). The methodology provided here can easily be adapted to different plant species, the limitation being the volume of [^{35}S]sulfate-containing solution in which the plants are incubated, as to achieve the necessary specific activity, great amount of radioactive tracer is necessary. Therefore, alternative plant material might be used, such as cell cultures or leaf disks. The core of the method can also be used with the stable ^{34}S isotope, the difference being in the detection of the labelled sulfur pool, LC-MS instead of scintillation counting, and thus a need for more expensive and sophisticated instruments. Thus, we believe that because of its simple, cheap, and quick nature, the method for flux measurement will continue to be used in the future.

Acknowledgments

Research in SK's lab is funded by the Deutsche Forschungsgemeinschaft (DFG) under Germany's Excellence Strategy—EXC 2048/1—project 390686111. SG was supported by DFG individual grant Nr. 372353760.

References

Ball, L., Accotto, G. P., Bechtold, U., Creissen, G., Funck, D., Jimenez, A., et al. (2004). Evidence for a direct link between glutathione biosynthesis and stress defense gene expression in Arabidopsis. *Plant Cell*, *16*(9), 2448–2462. https://doi.org/10.1105/tpc.104.022608.

Brunold, C., & Suter, M. (1984). Regulation of sulfate assimilation by nitrogen nutrition in the duckweed *Lemna minor* L. *Plant Physiology*, *76*(3), 579–583.

Calderwood, A., Morris, R. J., & Kopriva, S. (2014). Predictive sulfur metabolism - a field in flux. *Frontiers in Plant Science*, *5*, 646. https://doi.org/10.3389/fpls.2014.00646.

Chao, D. Y., Baraniecka, P., Danku, J., Koprivova, A., Lahner, B., Luo, H., et al. (2014). Variation in sulfur and selenium accumulation is controlled by naturally occurring isoforms of the key sulfur assimilation enzyme ADENOSINE 5'-PHOSPHOSULFATE REDUCTASE2 across the Arabidopsis species range. *Plant Physiology*, *166*(3), 1593–1608. https://doi.org/10.1104/pp.114.247825.

D'Hooghe, P., Bataille, M. P., Trouverie, J., & Avice, J. C. (2013). A specific method of 34S labelling provides evidence that sulphate assimilation occurs in developing seeds and pod walls of Brassica napus L. subjected to ample or limited S nutrition. *Rapid Communications in Mass Spectrometry*, *27*(24), 2737–2744. https://doi.org/10.1002/rcm.6738.

Dietzen, C., Koprivova, A., Whitcomb, S. J., Langen, G., Jobe, T. O., Hoefgen, R., et al. (2020). The transcription factor EIL1 participates in the regulation of sulfur-deficiency response. *Plant Physiology*, *184*(4), 2120–2136. https://doi.org/10.1104/pp.20.01192.

Dubousset, L., Abdallah, M., Desfeux, A. S., Etienne, P., Meuriot, F., Hawkesford, M. J., et al. (2009). Remobilization of leaf S compounds and senescence in response to restricted sulphate supply during the vegetative stage of oilseed rape are affected by mineral N availability. *Journal of Experimental Botany*, *60*(11), 3239–3253. https://doi.org/10.1093/jxb/erp172.

Feldman-Salit, A., Veith, N., Wirtz, M., Hell, R., & Kummer, U. (2019). Distribution of control in the sulfur assimilation in Arabidopsis thaliana depends on environmental conditions. *New Phytologist*. https://doi.org/10.1111/nph.15704.

Gigolashvili, T., & Kopriva, S. (2014). Transporters in plant sulfur metabolism. *Frontiers in Plant Science*, *5*, 442. https://doi.org/10.3389/fpls.2014.00442.

Hirai, M. Y., Fujiwara, T., Awazuhara, M., Kimura, T., Noji, M., & Saito, K. (2003). Global expression profiling of sulfur-starved Arabidopsis by DNA macroarray reveals the role of O-acetyl-l-serine as a general regulator of gene expression in response to sulfur nutrition. *Plant Journal*, *33*(4), 651–663.

Hirai, M. Y., Klein, M., Fujikawa, Y., Yano, M., Goodenowe, D. B., Yamazaki, Y., et al. (2005). Elucidation of gene-to-gene and metabolite-to-gene networks in arabidopsis by integration of metabolomics and transcriptomics. *Journal of Biological Chemistry*, *280*(27), 25590–25595. https://doi.org/10.1074/jbc.M502332200.

Hirai, M. Y., Sugiyama, K., Sawada, Y., Tohge, T., Obayashi, T., Suzuki, A., et al. (2007). Omics-based identification of Arabidopsis Myb transcription factors regulating aliphatic glucosinolate biosynthesis. *Proceedings of the National Academy of Sciences of the United States of America*, *104*(15), 6478–6483. https://doi.org/10.1073/pnas.0611629104.

Huseby, S., Koprivova, A., Lee, B. R., Saha, S., Mithen, R., Wold, A. B., et al. (2013). Diurnal and light regulation of sulphur assimilation and glucosinolate biosynthesis in Arabidopsis. *Journal of Experimental Botany*, *64*(4), 1039–1048. https://doi.org/10.1093/jxb/ers378.

Kopriva, S., Malagoli, M., & Takahashi, H. (2019). Sulfur nutrition: Impacts on plant development, metabolism, and stress responses. *Journal of Experimental Botany*, *70*(16), 4069–4073. https://doi.org/10.1093/jxb/erz319.

Kopriva, S., Mugford, S. G., Baraniecka, P., Lee, B. R., Matthewman, C. A., & Koprivova, A. (2012). Control of sulfur partitioning between primary and secondary metabolism in Arabidopsis. *Frontiers in Plant Science*, *3*, 163. https://doi.org/10.3389/fpls.2012.00163.

Koprivova, A., Giovannetti, M., Baraniecka, P., Lee, B. R., Grondin, C., Loudet, O., et al. (2013). Natural variation in the ATPS1 isoform of ATP Sulfurylase contributes to the control of sulfate levels in Arabidopsis. *Plant Physiology*, *163*(3), 1133–1141. https://doi.org/10.1104/pp.113.225748.

Koprivova, A., & Kopriva, S. (2014). Molecular mechanisms of regulation of sulfate assimilation: First steps on a long road. *Frontiers in Plant Science*, *5*, 589. https://doi.org/10.3389/fpls.2014.00589.

Koprivova, A., & Kopriva, S. (2016). Sulfur metabolism and its manipulation in crops. *Journal of Genetics and Genomics*, *43*(11), 623–629. https://doi.org/10.1016/j.jgg.2016.07.001.

Kruse, C., Jost, R., Lipschis, M., Kopp, B., Hartmann, M., & Hell, R. (2007). Sulfur-enhanced defence: Effects of sulfur metabolism, nitrogen supply, and pathogen lifestyle. *Plant Biology*, *9*(5), 608–619. https://doi.org/10.1055/s-2007-965432.

Lappartient, A. G., & Touraine, B. (1996). Demand-driven control of root ATP Sulfurylase activity and SO42- uptake in intact canola (the role of phloem-translocated glutathione). *Plant Physiology*, *111*(1), 147–157.

Loudet, O., Saliba-Colombani, V., Camilleri, C., Calenge, F., Gaudon, V., Koprivova, A., et al. (2007). Natural variation for sulfate content in *Arabidopsis thaliana* is highly controlled by APR2. *Nature Genetics*, *39*(7), 896–900. https://doi.org/10.1038/Ng2050.

Maruyama-Nakashita, A., Inoue, E., Watanabe-Takahashi, A., Yamaya, T., & Takahashi, H. (2003). Transcriptome profiling of sulfur-responsive genes in Arabidopsis reveals global effects of sulfur nutrition on multiple metabolic pathways. *Plant Physiology*, *132*(2), 597–605. https://doi.org/10.1104/pp.102.019802.

Mugford, S. G., Lee, B. R., Koprivova, A., Matthewman, C., & Kopriva, S. (2011). Control of sulfur partitioning between primary and secondary metabolism. *Plant Journal*, *65*(1), 96–105. https://doi.org/10.1111/j.1365-313X.2010.04410.x.

Mugford, S. G., Yoshimoto, N., Reichelt, M., Wirtz, M., Hill, L., Mugford, S. T., et al. (2009). Disruption of Adenosine-5 '-Phosphosulfate kinase in Arabidopsis reduces levels of sulfated secondary metabolites. *Plant Cell*, *21*(3), 910–927. https://doi.org/10.1105/tpc.109.065581.
Murillo, M., & Leustek, T. (1995). Adenosine-5′-triphosphate-sulfurylase from Arabidopsis thaliana and Escherichia coli are functionally equivalent but structurally and kinetically divergent: Nucleotide sequence of two adenosine-5′-triphosphate-sulfurylase cDNAs from Arabidopsis thaliana and analysis of a recombinant enzyme. *Archives of Biochemistry and Biophysics*, *323*(1), 195–204. https://doi.org/10.1006/abbi.1995.0026.
Neuenschwander, U., Suter, M., & Brunold, C. (1991). Regulation of sulfate assimilation by light and O-acetyl-l-serine in *Lemna minor* L. *Plant Physiology*, *97*(1), 253–258.
Nikiforova, V., Freitag, J., Kempa, S., Adamik, M., Hesse, H., & Hoefgen, R. (2003). Transcriptome analysis of sulfur depletion in *Arabidopsis thaliana*: Interlacing of biosynthetic pathways provides response specificity. *Plant Journal*, *33*(4), 633–650.
Nikiforova, V. J., Kopka, J., Tolstikov, V., Fiehn, O., Hopkins, L., Hawkesford, M. J., et al. (2005). Systems rebalancing of metabolism in response to sulfur deprivation, as revealed by metabolome analysis of Arabidopsis plants. *Plant Physiology*, *138*(1), 304–318. https://doi.org/10.1104/pp.104.053793.
Rennenberg, H. (1983). Role of o-acetylserine in hydrogen sulfide emission from pumpkin leaves in response to sulfate. *Plant Physiology*, *73*(3), 560–565.
Ristova, D., & Kopriva, S. (2022). Sulfur signaling and starvation response in Arabidopsis. *iScience*, *25*(5), 104242. https://doi.org/10.1016/j.isci.2022.104242.
Scheerer, U., Haensch, R., Mendel, R. R., Kopriva, S., Rennenberg, H., & Herschbach, C. (2010). Sulphur flux through the sulphate assimilation pathway is differently controlled by adenosine 5′-phosphosulphate reductase under stress and in transgenic poplar plants overexpressing gamma-ECS, SO, or APR. *Journal of Experimental Botany*, *61*(2), 609–622. https://doi.org/10.1093/Jxb/Erp327.
Takahashi, H., Kopriva, S., Giordano, M., Saito, K., & Hell, R. (2011). Sulfur assimilation in photosynthetic organisms: Molecular functions and regulations of transporters and assimilatory enzymes. *Annual Review of Plant Biology*, *62*(62), 157–184. https://doi.org/10.1146/annurev-arplant-042110-103921.
Takahashi, H., Yamazaki, M., Sasakura, N., Watanabe, A., Leustek, T., Engler, J. A., et al. (1997). Regulation of sulfur assimilation in higher plants: A sulfate transporter induced in sulfate-starved roots plays a central role in *Arabidopsis thaliana*. *Proceedings of the National Academy of Sciences of the United States of America*, *94*(20), 11102–11107.
Van Hoewyk, D., Takahashi, H., Inoue, E., Hess, A., Tamaoki, M., & Pilon-Smits, E. A. (2008). Transcriptome analyses give insights into selenium-stress responses and selenium tolerance mechanisms in Arabidopsis. *Physiologia Plantarum*, *132*(2), 236–253. https://doi.org/10.1111/j.1399-3054.2007.01002.x.
Vauclare, P., Kopriva, S., Fell, D., Suter, M., Sticher, L., von Ballmoos, P., et al. (2002). Flux control of sulphate assimilation in *Arabidopsis thaliana*: Adenosine 5 '-phosphosulphate reductase is more susceptible than ATP sulphurylase to negative control by thiols. *Plant Journal*, *31*(6), 729–740.

CHAPTER EIGHT

Isoprene measurements to assess plant hydrocarbon emissions and the methylerythritol pathway

Sarathi M. Weraduwage[a,b], Bahtijor Rasulov[c], Abira Sahu[a], Ülo Niinemets[c], and Thomas D. Sharkey[a,*]

[a]MSU-DOE Plant Research Laboratory, Plant Resilience Institute, and Department of Biochemistry and Molecular Biology, Michigan State University, East Lansing, MI, United States
[b]Department of Biology/Biochemistry, Bishop's University, Sherbrooke, QC, Canada
[c]Crop Science and Plant Biology, Estonian University of Life Sciences, Tartu, Estonia
*Corresponding author: e-mail address: tsharkey@msu.edu

Contents

Abstract

Isoprene is the most abundant non-methane hydrocarbon emitted to the atmosphere and a target of biotechnology as a source of biofuels or chemical feedstock. Measurements of the amount of isoprene or the rate of production of isoprene are important for atmospheric chemistry, evaluating biotechnology processes, and can provide information on the capacity and regulation of the methyl erythritol 4-phosphate pathway found in plants and bacteria. In this chapter we discuss techniques, and their

Methods in Enzymology, Volume 676
ISSN 0076-6879
https://doi.org/10.1016/bs.mie.2022.07.020

strengths and weaknesses, of methods in common use for measuring isoprene. There are many sources of isoprene for measurements including emissions from leaves and head space analysis of reactions involving recombinant enzymes or bacterial/fungal cultures. Similarly, there are a variety of detection methods including several mass spectrometer methods that are useful for examining rates of labeling of isoprene when carbon isotopes are used.

1. Introduction

Isoprene is the root member of the isoprenoid family of compounds. Isoprenoids (also known as terpenoids) have one to many isoprene units, a five-carbon branched chain. Isoprene in the strict sense is 2-methyl-1,3-butadiene but is not the precursor to higher order isoprenoids. The primary precursor of isoprenoids is dimethylallyl diphosphate (DMADP) with one- to many- isopentenyl diphosphate (IDP) molecules added. Molecules with two isoprenoid units are called monoterpenes, three units = sesquiterpenes, and four = diterpenes. Therefore, isoprene is a hemiterpene, i.e., half of a monoterpene. Other hemiterpenes are isoamylene (2-methyl-2-butene) and 2-methyl-3-buten-2-ol, the latter being emitted from several evergreen conifers (Lehnert, Perreca, Gershenzon, Pohnert, & Trumbore, 2020; Schade, Goldstein, Gray, & Lerdau, 2000).

Isoprene is made from DMADP by isoprene synthase (IspS) (Miller, Oschinski, & Zimmer, 2001; Silver & Fall, 1991). The enzyme found in angiosperm (flowering) plants is related to monoterpene synthases in the TpsB family, especially β-ocimene synthase (Li et al., 2017; Sharkey et al., 2005). Plants other than angiosperms do not have TpsB genes so the isoprene synthases responsible for significant isoprene emissions from ferns and mosses are unknown. Some bacteria make isoprene, especially *Bacillus* species. In at least one case, isoprene is made by hydroxymethylbutenyl diphosphate (HMBDP) reductase (also known as IspH and LytB). Normally this enzyme converts HMBDP to a mix of DMADP and IDP but there is a report that the enzyme can also convert HMBDP directly to isoprene (and convert DMADP to isoamylene) (Ge, Xue, & Ma, 2016). Humans and other animals exhale isoprene in their breath ($\sim$40 mg day^{-1} for an average human) but the mechanism is unknown (Karl et al., 2001; Miekisch, Schubert, Vagts, & Geiger, 2001; Mochalski et al., 2011; Sharkey, 1996; Trovarelli et al., 2001).

In plants, DMADP for isoprene synthesis is made by the methyl erythritol 4-phosphate (MEP) pathway. There is significant interest in engineering the MEP pathway in part because it is more efficient than the mevalonic acid pathway used by animals to make DMADP. Protocols for measuring metabolites of the MEP pathway have recently been published (González-Cabanelas, Hammerbacher, Raguschke, Gershenzon, & Wright, 2016). Isoprene measurements can supplement LC-MS/MS methods for measuring DMADP and another key pathway intermediate methylerythritol cyclodiphosphate (MEcDP) using post-illumination isoprene emission characteristics. Recombinant isoprene synthase can convert all DMADP in a sample to isoprene, which can then be measured by one of several methods described below. Adding isopentenyl diphosphate isomerase will ensure conversion of all DMADP and IDP to isoprene and allow separate determination of IDP and DMADP (Zhou, Li, Wiberley-Bradford, Weise, & Sharkey, 2013). Here we focus on how measuring isoprene can be used to gain insight into the MEP pathway. Other methods for measuring these compounds will not be presented here.

2. Methods for detecting and measuring isoprene

There are many methods for measuring isoprene (Cao & Hewitt, 1995) but we will cover three methods in common usage in biology laboratories. Isoprene is highly volatile and so is handled in the gas phase for most measurements. It is generally stable and does not partition into water (and buffers, etc.) to a great degree (Niinemets et al., 2010). It also does not stick to walls of containers as much as many other isoprenoids.

2.1 Gas chromatography

Isoprene-containing gas samples can be introduced into a gas chromatograph (GC). Typically, some form of cryofocusing is used in which air is passed through a cold trap, which is then heated to quickly release all the isoprene into the GC. This can be done external to the GC making use of a six-way valve (Hills, Fall, & Monson, 1992; Loreto & Sharkey, 1990). This allows very large air samples to be processed increasing the system sensitivity. On-column cryofocusing is often used when samples are introduced using solid phase microextraction systems (SPME). Cartridges are often used with GC (and other detection methods). Some care is required when using cartridges to ensure that the packing material will retain the isoprene at ambient temperature and release it upon moderate heating (Niinemets et al., 2011). GC relies on the chromatography to separate isoprene from other gases. Detection can be by flame ionization (FID), photoionization detection (PID), or mass spectrometry (MS). PID is typically more sensitive than FID and modern mass spectrometers are even more sensitive. Both FID and PID are very quantitative and discreet measurements can be made rapidly (less than 3 min per sample). MS detectors provide the mass spectrum of individual compounds and as such provide a benchmark method for confirming the identity of the given compound. FID and PID cannot separate among interfering compounds that have similar retention time as isoprene in widely used GC columns for biogenic volatile separation.

2.2 Fast Isoprene Sensor

An instrument designed to measure isoprene using chemiluminescence is sold commercially by Hills Scientific and is called the Fast Isoprene Sensor (FIS) (Hills et al., 1992; Hills & Zimmerman, 1990) (http://hills-scientific.com/). This instrument combines a flow of oxygen with very high ozone content with air being pulled through the instrument with a small air pump. Isoprene in the air reacts with the ozone to make a chemiluminescent product in front of a photomultiplier tube. The selectivity for isoprene depends on the wavelength of emitted light and the timing of the reaction. This instrument was developed for measurements related to atmospheric chemistry but is easily adapted to laboratory measurements. It has some sensitivity to water vapor that is problematic when air humidity is varying during the measurements as is often the case with plant measurements. The problem with water vapor sensitivity can be handled by passing the air through an ice trap to maintain a constant low humidity in the gas stream going into FIS or humidifying the air to a constant humidity (Rasulov, Copolovici, Laisk, & Niinemets, 2009). The FIS also has some cross-sensitivity to other hydrocarbons, for example propene (Hills & Zimmerman, 1990), but little cross reactivity to hydrocarbons likely to be present in isoprene air samples. It has an especially strong cross-sensitivity to some sulfur gases. This limits how much dithiothreitol can be used in enzyme assays when head space analysis using an FIS is planned. Thus, regularly checking for the magnitude of other hydrocarbons by gas-chromatography is recommended. The FIS has a very wide dynamic range and is very sensitive to isoprene (Cao & Hewitt, 1995; Toda & Dasgupta, 2008). One of the key advantages of using the FIS to measure isoprene is the fact that it provides high time resolution (as fast as 0.1 s sampling time).

2.3 Proton-transfer-reaction mass-spectrometry

Proton transfer reaction mass spectrometry (PTR-MS) uses chemical ionization of volatile molecules with protonated water vapor (H_3O^+) followed by subsequent detection of protonated molecules with a mass spectrometer (Hansel et al., 1995; Jordan, Haidacher, Hanel, Hartungen, Herbig, et al., 2009a; Lindinger, Hansel, & Jordan, 1998). Currently, two principal spectrometric detectors are in use, quadrupole mass spectrometer (QMS, PTR-QMS) and time-of-flight mass spectrometer (TOF-MS, PTR-TOF-MS). PTR-MS measurements are not confined to isoprene, but all volatiles with a proton affinity greater than that of water vapor (691 kJ mol^{-1}, (Hunter & Lias, 1998)) can be measured. Several PTR-MS models also include selective reagent ion (SRI) option that allows use of additional reagent ions, NO^+, O_2^+, NH_4^+, thereby extending the range of volatile compounds that can be measured (Jordan, Haidacher, Hanel, Hartungen, Märk, et al., 2009b; Lehnert et al., 2020). The time resolution of PTR-MS systems is on the order of 0.1 s. For the highest sensitivity, the sampling rate may need to be decreased. In practice, in laboratory measurements with a certain ambient air background isoprene concentration, a time resolution of 0.5–1 s is typically used to measure plant isoprene emissions (Rasulov, Talts, & Niinemets, 2019).

In the case of PTR-QMS, the protonated masses measured are defined a priori before the measurements, and they are measured by the QMS in sequence. Thus, the time-resolution of the instrument depends on the number of compounds measured. In contrast, all protonated masses present can be measured simultaneously with PTR-TOF-MS. The detection limit of PTR-QMS systems is on the order of 0.1–0.5 ppb (Warneke et al., 2015). Modern PTR-TOF-MS instruments are characterized by superior sensitivity, typically a few ppt (Jordan, Haidacher, Hanel, Hartungen, Märk, et al., 2009b), and the sensitivity of the newest instruments even extends to less than 0.1 ppt (FUSION PTR-TOF, www.ionicon.com). Furthermore, PTR-TOF-MS has a superior mass resolution, better than 0.1 amu that is important for distinguishing among compounds with similar molecular mass.

PTR-MS systems only measure protonated ion concentrations with isoprene detected as a protonated parent mass (m/z) of 69^+ (69.1^+ for PTR-TOF-MS). Although there are no typical plant volatiles that could provide the same parent ion, several plant species are significant constitutive emitters of another 5-carbon (C5) DMADP pathway compound 2-methyl-3-buten-2-ol (MBO Gray, Breneman, Topper, & Sharkey, 2011; Gray, Goldstein, & Lerdau, 2006). The protonated parent ion of MBO has a *m/z* of 87^+; however, even upon soft ionization, it fragments to yield a main fragment ion with *m/z* of 69^+ identical to isoprene (Karl et al., 2013). In addition, stressed plants might emit the C5 green leaf volatile pentenol (e.g., 1-penten-3-ol) (Fisher, Grimes, & Fall, 2003; Rasulov et al., 2019) and some plants species also emit significant amounts of another C5 volatile pentanone (Jardine et al., 2010). Both pentenol and pentanone partly fragment to the ion with *m/z* 69^+. Thus, for compound identification, regular checks with GC–MS are advisable, especially when starting experiments with new species or conducting experiments with stressed plants. Furthermore, in addition to the ion *m/z* 69^+, it is important to simultaneously monitor the ions of relevant parent ions (e.g., *m/z* 87^+ for MBO) that could yield fragments ions *m/z* 69^+, and also confirm the identity of detected compounds by GC–MS. Provided there is only one interfering compound in the plant volatile mixture, the share of the ion *m/z* 69^+ between isoprene and the interfering compound can be estimated based on the degree of fragmentation of the interfering volatile. For example, in the case of MBO, the share of the ion *m/z* 69^+ is ca. 75% of total (Karl et al., 2013), while it is on the order of 1% for 3-pentanone (Malásková et al., 2019). Simultaneous emission of both isoprene and MBO by a C5 isoprenoid-emitting species is very rare (Lehnert et al., 2020). Thus, unless heavily stressed plants are measured, interference due to other volatiles is typically not a major issue in isoprene emission measurements by PTR-MS.

Compared with other methods of isoprene detection, a major advantage of PTR-MS measurements is that it allows conduction of real-time quantitative isoprene ^{13}C-labeling experiments (Karl et al., 2002). Upon $^{13}CO_2$-feeding, isoprene molecules become progressively enriched with ^{13}C, starting from *m/z* 70^+ (one ^{13}C atom and four ^{12}C atoms) to *m/z* 74^+ (fully ^{13}C-labeled). Such measurements provide detailed insight into relationships among photosynthesis, use of alternative carbon sources and MEP pathway activity (Sharkey, Preiser, Weraduwage, & Gog, 2020).

3. Strategies for isoprene measurements

Isoprene is measured for several reasons and specific measurement systems are better for some strategies than others. On the other hand, there is a great deal of flexibility in matching sample generation with detection methods.

3.1 Continuous measurements

A common reason to measure isoprene is to determine how fast plant leaves are emitting isoprene. This is typically done with continuous measurements. Both FIS and PTR-MS continuously sample air to provide real-time, continuous data for isoprene emission. This is particularly helpful for observing transients in isoprene emission, especially from leaves. Continuous sampling is also used by the atmospheric community together with sensitive wind measurements to measure isoprene emission from forests by eddy covariance (Guenther & Hills, 1998; Sarkar et al., 2020). Other methods used by the atmospheric community include canister collection (Wiedinmyer et al., 2001) and even measurements from space (Wells et al., 2022) but these methods will not be covered here.

3.2 Discrete measurements

In many instances it is best to make discreet measurements of isoprene, for example when sampling head space above an enzyme assay or bacterial production assay. To assess candidate genes for improving the capacity for the MEP pathway or isoprene synthase, the amount of isoprene accumulating in the head space above a closed culture can be measured at a specific time point. However, this may miss important information about how quickly gene expression is induced. A better measure is to leave the bacterial culture open (or sparged with desired gas composition, for example with low oxygen or no oxygen). Then at specific time points, a small amount of the culture is removed and put into a sealed vial. This is incubated at a specific temperature for a specified time and then the head space is sampled for isoprene. This provides a measure of the rate of isoprene production by the culture.

4. Examples

Some examples will be described here but detection methods can be varied. Most flexible is detection by PTR-MS because it can make continuous or discreet measurements, can distinguish among isotopologues, and can measure other molecules in the gas sample. Least flexible is GC-FID. Example protocols for using these methods applied to specific questions are given here.

4.1 Analyzing isoprene production by bacterial cultures using GC–MS

When a mass-selective detector is used, both the chromatography and mass selection are used to identify the isoprene signal. In addition, various isotopologues (differences in ^{13}C amount) can be distinguished (Sharkey et al., 2020). Below is a protocol for measuring isoprene production from a purified protein, such as putative isoprene synthases (Li et al., 2017) and insertion engineered isoprene synthase (Gonzalez-Esquer et al., 2021), by GC–MS using a SPME fiber to introduce the sample into the GC. This is a good method when the goal is to determine the degree of labeling. For quantitation, an alternative making use of an FIS for detection, is presented.

4.1.1 Materials required

- Purified protein
- Dimethylallyl diphosphate (DMADP) (Echelon Biosciences)
- Borosilicate glass vials (we use 2 mL volume vials)
- Aluminum crimp top with red rubber septa
- 20 mm Kebby Standard Crimper (20001-00-C01A) for aluminum seals
- Water bath
- Solid-phase microextraction (SPME) fused-silica fiber coated with Carboxen/Polydimethlysiloxane (Cat # 57318, Supelco, PA)
- SPME fiber holder (Cat # 577330-U, Supelco, PA)
- Ring stand
- Agilent 7010B Triple Quadrupole GC/MS (Agilent, CA)
- EZ guard column (VF5 CP9013, Agilent, 30 m length, 10 m guard length, EZ Guard, 7 in cage, 0.25 mm inner diameter)

4.1.2 Procedure

1. Before sampling, the SPME fiber needs to be conditioned. After the GC oven reaches 230 °C, place the SPME fiber holder in the GC injection port and carefully push the fiber into the port. Make sure to lock the fiber in place.
2. Let the fiber condition for 20 min. Then retract the fiber and remove it from the GC injection port.
3. Prepare the assay buffer that contains 50 mM HEPES buffer (pH 8.0), 10 mM $MgCl_2$, 20 mM KCl, 2 mM DTT, and 1 mM EDTA.

4. Put buffer into the glass vials, add DMADP to make the desired concentration, then start the assay by adding the desired amount of enzyme solution. Adjust the amount of buffer added to make the total volume 300 μL.
5. Crimp seal the vial with an aluminum crimp top immediately and then vortex the vial to ensure complete mixing.
6. Put the vial in 40 °C water bath for 10 min.
7. Take out the vial from the water bath after 10 min
8. Insert the needle at the end of the SPME fiber holder through the rubber septum of the crimp top
9. Insert the SPME fiber into the headspace making sure it doesn't touch the liquid and lock it in place. Allow the fiber to absorb the analytes in the headspace for 10 min
10. Clip the upper part of the SPME fiber holder to a ring stand to keep it steady
11. At the end of 10 min, retract the fiber and immediately insert it into the GC injection port
12. Allow it to desorb for 2 min at 230 °C. During this process, isoprene is collected in a cryotrap in the GC cooled to −10 °C using CO_2
13. At the end of 2 min, start the Agilent QQQ/MassHunter program (which also warms the cryotrap) and run it for 6.75 min. Isoprene should elute around 1.4 min
14. At the end of the program, retract the SPME fiber and remove it from the GC injection port
15. Repeat the procedure for the rest of the gas samples

4.1.3 Alternative

Instead of analyzing with GC–MS, the amount of isoprene can be analyzed with an FIS. Starting at step 8, the head space gas is displaced into a 1 mL syringe by simultaneously withdrawing the plunger of an empty syringe and injecting an equal amount of water to keep the pressure inside the vial constant (Fig. 1) (Weise et al., 2013). This 1 mL sample is then injected into the air stream of an FIS instrument. The peak of isoprene is integrated by summing the counts from the FIS for 15 s before and 15 s after the peak (background) and subtracting that from the 30 s period that encompasses the signal from the injection. This is compared to a standard curve to determine the amount of isoprene in the head space. The standard curve can be established by mixing different concentrations of isoprene in air or nitrogen

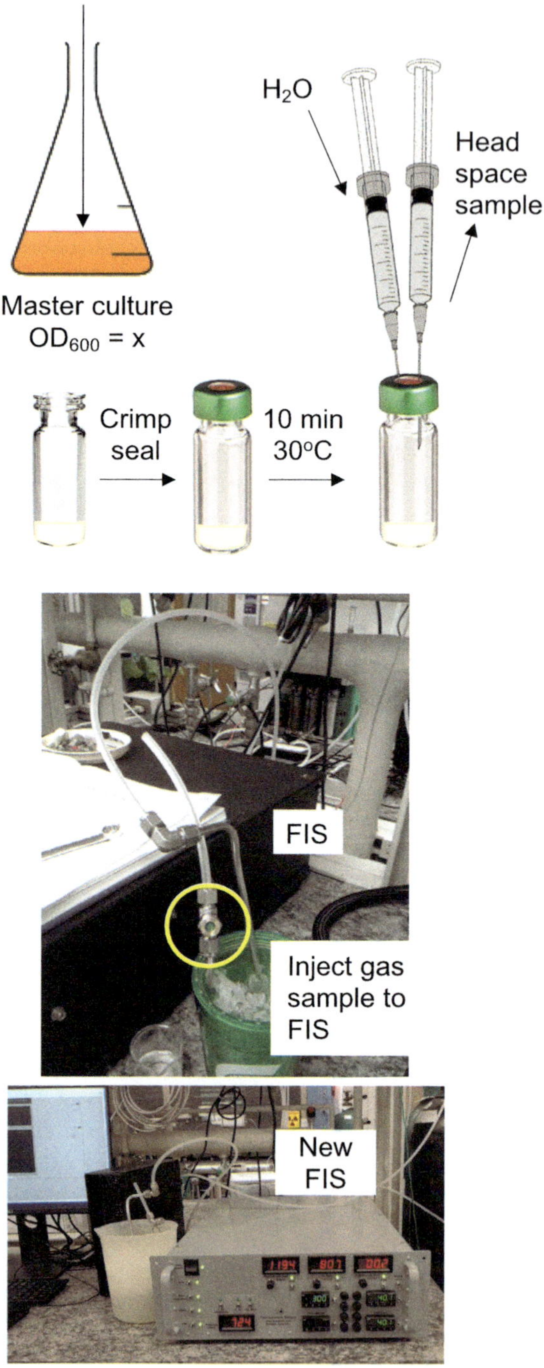

Fig. 1 See figure legend on opposite page.

and injecting 1 mL of these standards into the FIS air stream. We have used glass flasks with a septum to mix standards (e.g., https://chemglass.com/gas-sampling-tubes-ptfe-stopcocks-with-sampling-portand) or Tedlar bags (e.g., ESS GD0707–7000 Sampling Bags With Combination Valve, 1 L, from Cole Parmer).

Isoprene dissolves in water better than some monoterpenes (Copolovici & Niinemets, 2005) and this can be accounted for using the Henry's constant ($7780\,\text{Pa}\,\text{m}^3\,\text{L}^{-1}$at 25 °C) (Copolovici & Niinemets, 2005; Weise et al., 2013). An example of this calculation is shown below.

Example—In the gas phase:

$$I = \chi_I \cdot \frac{v_H}{v_M}$$

where I is the number of moles of isoprene, χ_I is the mole fraction of isoprene in the headspace air (mole fraction is interchangeably nmol isoprene/mol air or nPa isoprene/Pa air), v_H is the volume of the head space and v_M is the molar volume (22.4 L/mol corrected for temperature and pressure). For example, in a 2 mL vial with 200 μL liquid medium containing reactants, 1.8 mL gas phase, and assuming the FIS returns a reading of 100 ppb, and the temperature is 25 °C

$$100 \bullet 10^{-9}\frac{\text{mol}}{\text{mol}} \bullet 1.8 \bullet 10^{-3}\text{L} \bullet \frac{\text{mol}}{22.8\ \text{L}} = 7.89 \bullet 10^{-12}\text{mol isoprene.}$$

In the liquid phase:

$$I = \chi_I \bullet \frac{v_l \bullet P_H}{H}$$

where v_l is the volume of the liquid phase, P_H is the pressure in the head space, assumed to be 1 atm or 101 kPa, and H is the Henry's constant in units of Pa m^3 mol^{-1}. In the 2 mL vial with 200 μL liquid phase

Fig. 1 Measuring isoprene production by bacterial cultures. A bacterial culture is grown until a desired concentration is achieved. If required, the culture is induced to express enzymes of interest. At any time point 200 μL of culture are removed and put into a 2 mL crimp seal vial. The vial is incubated for a set time (we often use 10 min) and then the headspace is removed with a 1 mL syringe while water is injected to prevent formation of a vacuum. The sample is then injected into the airstream of a Fast Isoprene Sensor (FIS) instrument (a PTR-MS either quadrupole or time-of-flight would also work well). We often use an ice bath to bring the water vapor in the gas sample to 0 °C dew point. Because these are discreet measurements gas chromatographs would also work.

$$100 \bullet 10^{-9} \frac{\text{Pa}}{\text{Pa}} \bullet \frac{200 \bullet 10^{-6} \text{L} \bullet 101 \bullet 10^3 \text{Pa mol m}^3}{7780 \text{ Pa m}^3 \text{ } 10^3 \text{ L}} = 0.26 \bullet 10^{-12} \text{mol isoprene.}$$

Thus 3.3% (0.26/7.89) of the isoprene is in the liquid phase and the total isoprene in the vial is 7.89 + 0.26 = 8.15 pmol. This can then be expressed relative to the incubation time and amount of material (e.g., amount of protein) in the reaction.

4.2 Analyzing isoprene emission from leaves using an FIS

The rate of isoprene emission can be used as a proxy to measure the activity of the MEP pathway and its regulation. An FIS coupled with the LI-6800 Portable Photosynthesis System (LI-COR Biosciences, Lincoln, NE) can be used to measure isoprene and photosynthesis rates simultaneously in leaves detached from plants (**Fig. 2**) or intact leaves (**Fig. 3**).

4.2.1 Materials and equipment

- A plant
- Bev-a-line tubing (BEV-IV 1/8" ID X1/4" OD, EW-06490-12 and 1/2" ID × 5/8" OD EW-06490-19). Alternatives include Teflon tubing. Teflon is more inert but stiffer and a little more difficult to handle. The differences are minor for isoprene but can be significant if larger terpenoids are also being measured. The high ozone tube from the ozonizer to the FIS must be Teflon
- Swagelock connectors
- LI-6800 Portable Photosynthesis System (LI-COR Biosciences, Lincoln, NE). We most often use a Multiphase FlashTM Fluorometer (6800- 01A) chamber but other chambers work fine
- A Fast Isoprene Sensor (FIS) with ozone generator
- Ultra-high purity oxygen

4.2.2 Procedure

4.2.2.1 Setting up the FIS

First, turn on the oxygen supply to the FIS, and power up the FIS and the ozonizer following the manufacturer's instructions. After the ozone has been flowing for several minutes, turn on the PMT and open the LabVIEW software program. We normally select the 5 s averaging period, but a higher sampling rate is used when following transients. Start the program. A file name is suggested that includes the date and time and can be modified so that you can organize your data. The FIS has a flow controller that allows mixing in known amounts of an isoprene standard gas (we use a 3 PPM

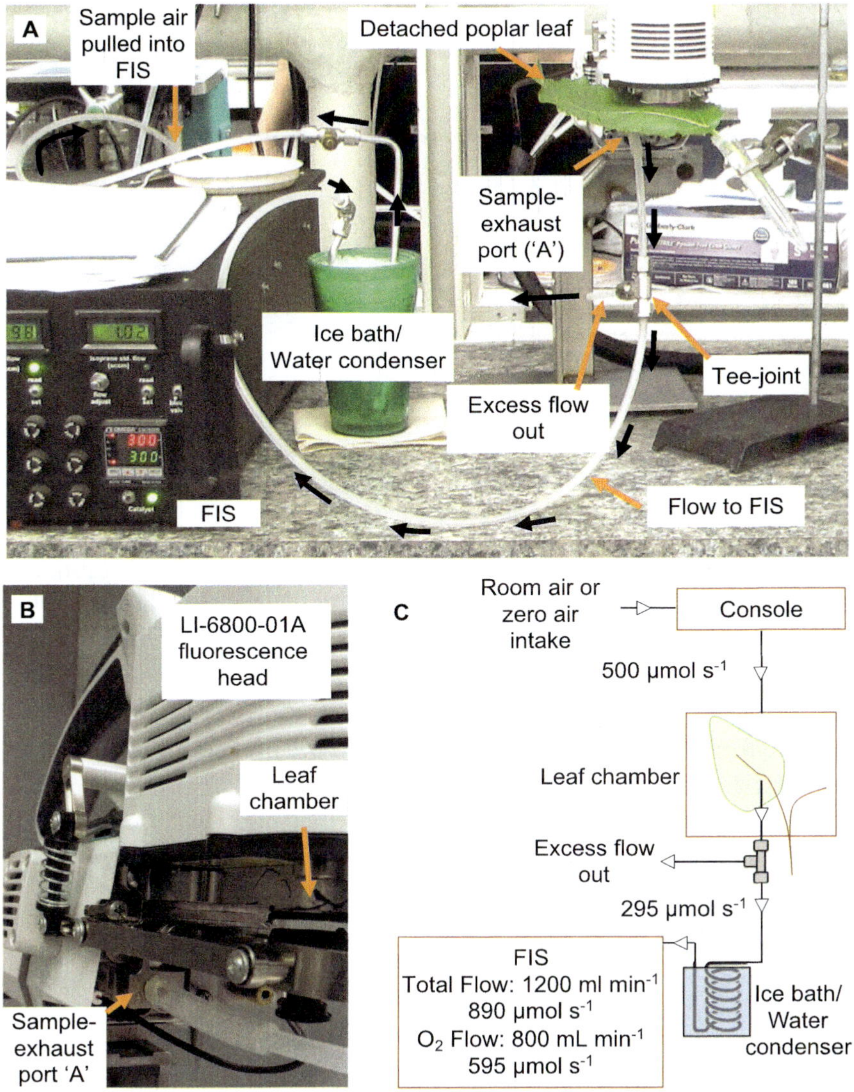

Fig. 2 System for measuring isoprene emission from a leaf. In Panel (A), a leaf detached from the plant is shown. This allows feeding poisons such as fosmidomycin or potential substrates through the transpiration stream. For example, feeding labeled substrates such as ^{13}C-pyruvate can provide information on carbon sources for the MEP pathway. Panel (B) shows a closeup of the leaf chamber and the sample exhaust that is used for measurements as shown in panel (C). PTR-MS instruments would also work well but gas chromatographs would be less convenient.

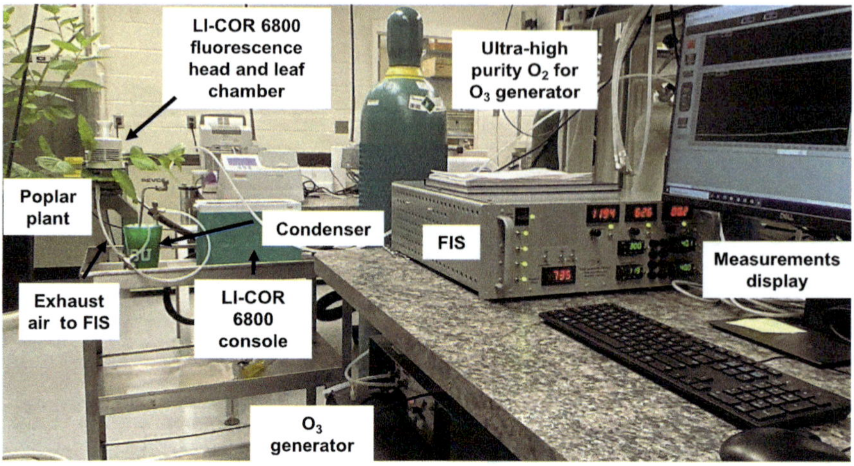

Fig. 3 Measuring isoprene emitted from a leaf attached to the plant. This allows for repeated measurements in the days following a treatment. Many plants show little effect of detaching leaves but using attached leaves circumvents any possible problem associated with cutting the leaf off the plant. The flow path is the same as shown in Fig. 2C including the LI-COR 6800, which is clamped on an attached poplar leaf, the exhaust air from the 6800 that flows to the Fast Isoprene Sensor (FIS) through a condensoer to remove water vapor. The FIS requires a high level of ozone. An FIS with its display is also shown.

standard from Airgas). Plot the photon counts per ppb of isoprene. You will need this slope to analyze data from leaf measurements. It is important to use the same flow rate during the calibration that you will use during the measurements.

4.2.2.2 Setting up the LI-6800

Switch on the LI-6800 following the manufacturer's instructions. Go to the 'Environment' window and set environmental conditions. It is very helpful to synchronize the clocks on the FIS and LI-COR so that gas exchange data and isoprene emission data can be correlated.

Standard environmental conditions for isoprene measurements are: leaf chamber flow rate[a]—500 μmol s^{-1}; temperature—30 °C; sample CO_2 concentration - 420 μmol mol^{-1}; light intensity—1000 μmol m^{-2} s^{-1} is the preferred default.

[a] Flow rate in the LI-COR is expressed in μmol s^{-1} while flow meters in the FIS report standard cubic centimeters per minute (SCCM), also expressed as mL min^{-1}. There are 22.4 standard liters per mole (do not adjust for temperature and pressure because the flow meter has already made this adjustment). One SCCM = 0.744 μmol s^{-1}.

The LI-COR 6800 instrument allows two methods for sampling the air that has passed over the leaf. At the back of the head there is an outlet that allows the user to split the air leaving the leaf chamber between the LI-COR analyzer and an external analyzer such as an FIS or PTR-MS. At the front, users can take all the air that has flowed over the leaf and through the analyzer. When sampling from the front it is important to impose the least possible resistance to flow since this can change the pressure in the analyzer and introduce errors. To measure in the exhaust air, assemble a sample-exhaust line using Bev-A-Line IV or Teflon tubing, and a Swagelok union tee-joint (Fig. 2A).

Tee-joint-1 (Fig. 2A) connects the LI-6800 sample-exhaust port A (Fig. 2A, B) with the "sample in" port at the back of the FIS. We typically use a total flow rate of the FIS of 1200 $mL\,min^{-1}$ and the O_2 flow rate of 800 $mL\,min^{-1}$, and so the sample flow rate into the FIS is 400 $mL\,min^{-1}$ (Fig. 2C). The LI-6800 sample flow rate when set at 500 $\mu mol\,s^{-1}$ is 672 $mL\,min^{-1}$). The excess flow from the leaf chamber will vent from the 'tube for excess flow out' at the tee-joint (Fig. 2A, C). A length of tubing on this exhaust will prevent back diffusion which would dilute the air entering the FIS, but it should not be too long to avoid back pressure on the LI-COR analyzers. The FIS pulls in 400 $mL\,min^{-1}$ or $\sim$295 $\mu mol\,s^{-1}$ and so consumes 295/500 or approximately 60% of the air flow out of the LI-COR (but see below for the effect of transpiration on total flow rate exiting the chamber). The excess flow protects against the FIS pulling in room air. Increasing the flow to the FIS increases the signal but runs the risk of pulling in room air, which would spoil the measurement and would require a recalibration at the new flow rates.

4.2.3 Measuring isoprene

Clamp a leaf in the LI-6800 leaf chamber and close the chamber (Fig. 2A and 3). Zero readings can be made by switching the LI-COR to "Match" mode. There is little drift in sensitivity of the FIS but significant zero drift. The background reading taken while in Match mode will account for the baseline signal of the FIS, any isoprene in the supply air, and any contaminating/cross reacting components of the air supplied to the LI-COR. It is important to make the zero reading frequently, especially early in the day.

To measure isoprene and photosynthesis simultaneously, log photosynthesis on the 6800 manually or use the auto-program feature in the LI-6800. Record time units in the FIS and the LI-6800 so that the FIS reading can be matched with the corresponding photosynthesis measurements. Once isoprene and photosynthesis reach steady state, record start and end time in the FIS for a one-minute period during which isoprene measurement is stable. For each sample, record the start and stop times of the sampling period.

If using a large leaf that covers the entire area of the LI-6800-01A leaf chamber, the area is 6 cm^2. If the leaf is smaller, photographing the section of the leaf that was inserted to the leaf chamber, followed by analysis using imaging software will have to be carried out to accurately estimate leaf area.

4.2.4 Data analysis

The slope (counts/isoprene ppb) from the calibration data of the FIS will be needed for data analysis. Subtract the average number of photon counts of the background from the sample. Using the slope (counts/isoprene ppb) of the calibration curve, divide the number of counts by the calibration factor (slope) to give the mole fraction of isoprene in ppb (nanomoles of isoprene per mole of air). Multiply the mole fraction of isoprene in the air exiting the chamber by the flow out of the leaf chamber. This will be different from the flow entering the leaf chamber reported by the LI-COR because transpiration is adding gas to the total flow. This is the same effect that affects gas exchange measurements of photosynthesis (von Caemmerer & Farquhar, 1981). Niinemets et al. (2011) provides the following equation that should be used:

$$F_{out} = F_{in} + \lambda \bullet A$$

where F is flow out or flow in, λ is transpiration rate and A is leaf area. With large leaf areas, high transpiration rates, or slow flow rates, the flow rate out can be 20% more than that reported by the LI-COR but the effect is more typically 2% to 5% (Niinemets et al., 2011). In East Lansing we sometimes use an ice trap before the FIS because of minor water sensitivity and this will also affect the conversion of ingoing chamber flow rate to outgoing flow rate.

$$F_{out} = F_{in} \cdot (1 + W_{out} - W_{in}).$$

where W is the mole fraction of water vapor in the air entering the chamber (in) or leaving the chamber (out). With an ice trap in place, W_{out} will be 0.006 Pa/Pa and if the dew point of the ingoing air is 15 °C then W_{in} is 0.016 and F_{out} (flow into the FIS) will be 1% less than the flow reported by the LI-COR.

Use the concentration of isoprene in mol isoprene mol^{-1} air and the LI-COR flow (corrected for water vapor) (mol air s^{-1}) to calculate the rate of isoprene emission in mol isoprene s^{-1}. Then divide by leaf area. Typical values are 1–60 nmol isoprene $m^{-2}\ s^{-1}$. Niinemets et al. (2011) provides much additional information on measuring isoprene (and other terpenes).

4.3 Labeling isoprene by feeding $^{13}CO_2$

4.3.1 Method

To feed $^{13}CO_2$ to observe the rate and degree of labeling of isoprene by photosynthesis, the LI-COR console is set to provide CO_2 free air to the LI-COR head. Different isotopes are then fed through mass flow controllers and switched using a four-way valve (Fig. 4A–C). Most CO_2 analyzers have much reduced sensitivity to $^{13}CO_2$ compared to $^{12}CO_2$ so it is not possible to measure the rate of photosynthesis during feeding. Also, it is necessary to calculate air flows to set the rate of flow of $^{13}CO_2$ during the feeding. This should be set so that the concentration of CO_2 does not change regardless of which isotope is being fed. The flow out of the chamber can be collected in a Tedlar bag (for example "ESS GD0707-7000 Sampling Bags With Combination Valve" from Cole-Parmer). Be careful not to impose back pressure on the LI-COR.

The degree of label in isoprene can be determined by GC–MS using a SPME fiber or PTR-MS (Fig. 5D and E). A PTR-MS provides real time readings and simplifies measuring time courses.

Many publications have assumed that labeling to less than 100% indicates an alternative carbon source for isoprene emission and hence the MEP pathway, but this assumes that the Calvin Benson cycle labels to 100%. It does not (Sharkey et al., 2020). Isoprene, and so the MEP pathway, label to the same degree as the Calvin-Benson cycle intermediates and so labeling of isoprene can provide a window on photosynthetic carbon metabolism.

4.4 Measuring DMADP and MEcDP using post-illumination isoprene measurements

In vivo estimation of DMADP and MEcDP pools in plant leaves of isoprene-emitting species rests on the observation that when light is switched off isoprene emission continues for about 10 min. at the expense of MEP pathway metabolites synthesized during the previous light period (Li, Ratliff, & Sharkey, 2011; Rasulov et al., 2009; Rasulov, Hüve, Laisk, & Niinemets, 2011; Weise et al., 2013). The post-illumination isoprene emission is biphasic. The first phase continues for about 300 s after switching off the light, followed by a second rise of isoprene emission between about 300–1000 s. The first phase is used to estimate DMADP pool size and the

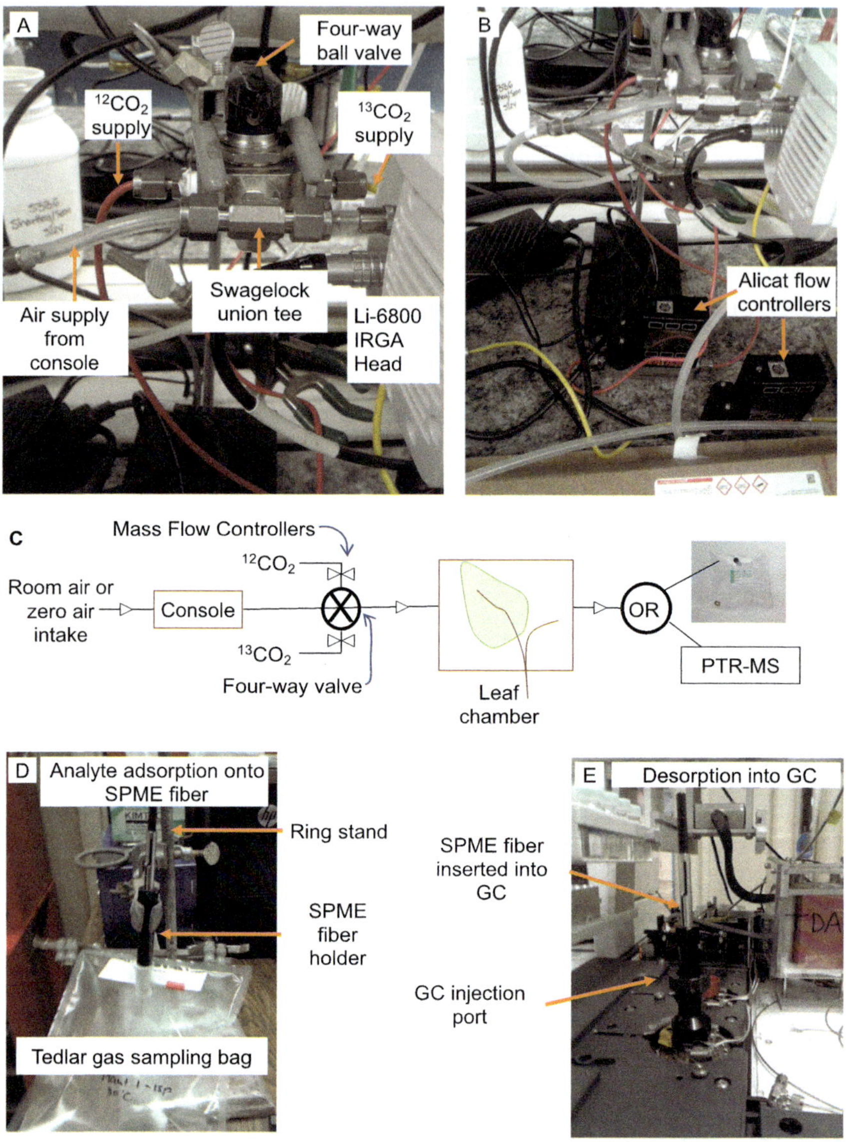

Fig. 4 Measuring labeling of isoprene when feeding $^{13}CO_2$. A system for rapid switching between $^{12}CO_2$ and $^{13}CO_2$ at the back of the LI-COR head is shown in panel (A). Panel (B) shows the mass flow controllers that are used to set the concentration of the two isotopes so that there is no change in CO_2 concentration when switching from one isotope to the other. Panel (C) shows exposing a Solid Phase Micro Extraction (SPME) element to a sample in a Tedlar bag and Panel D shows desorbing the sample into a GC–MS. This measurement is easier with a PTR-MS and the high resolution of a PTR-TOF-MS can help ensure that isoprene is being measured.

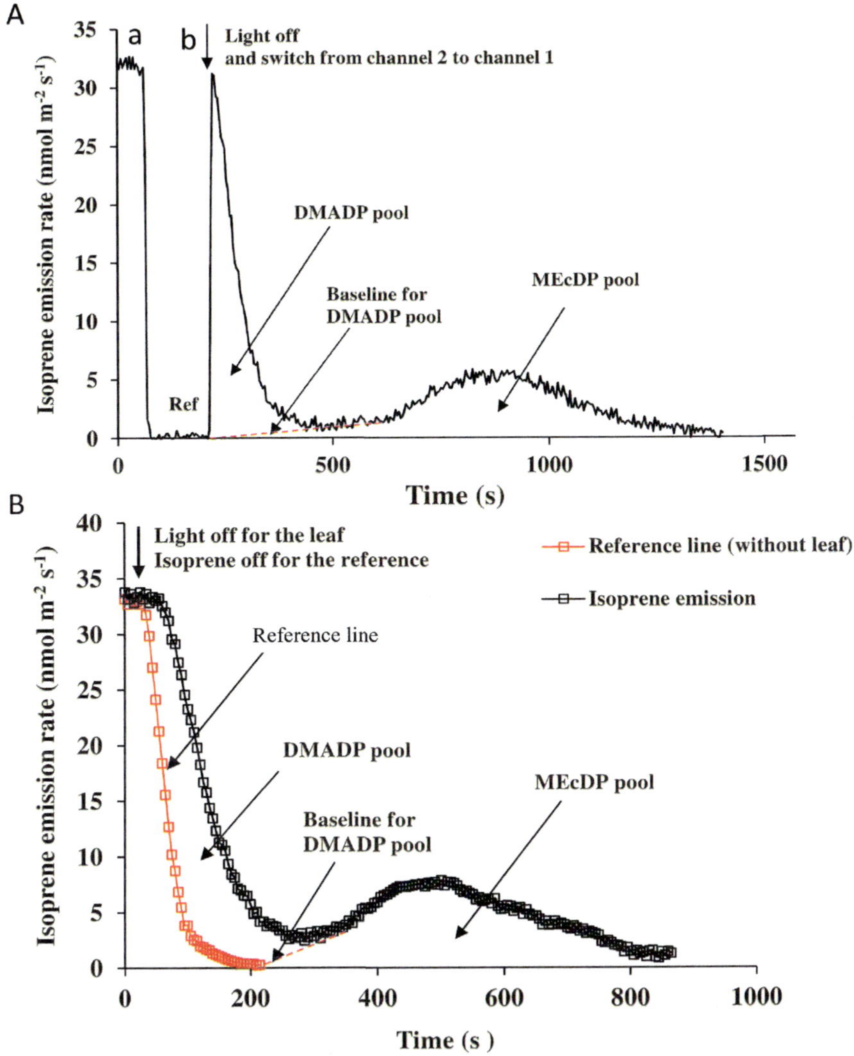

Fig. 5 Measurements of the pool sizes of isoprene precursors DMADP and MEcDP using a two-channel ultra-fast gas exchange system (Rasulov, Talts, & Niinemets, 2016) (chamber volume 2.4 mL, flow rate 0.67 L min^{-1}, system half-time 0.15 s) that does not require consideration of system delay effects (A) and a slower system where the system delay (amount of isoprene present within the chamber and system gas lines) needs to be taken into account (Rasulov et al., 2009) (chamber volume 1 L, flow rate 1.5 L min^{-1}, system half-time 28 s). In panel (A), the leaf was first stabilized until steady-state conditions were reached, at time a, the reference line was measured (leaf switched from line 1 to line 2), and at time b, the leaf was switched from line 2 to line 1 and light was simultaneously switched off and the dark decay kinetics of isoprene release was followed until the emission reached to the background level. In panel (B), the measurement

second phase to estimate MEcDP pool size. The intermediate pool sizes estimated by the in vivo method are in very good agreement with separate destructive chemical measurements of the pool sizes (Rasulov et al., 2009; Weise et al., 2013).

For in vivo estimation of isoprene precursor pool size, a real-time isoprene sensor is needed (either FIS or PTR-MS), whereas the protocol of measurements depends on the system response time, which depends on the chamber volume and air flow rate. For fast systems, the measurements do not need to consider the chamber response time (as, e.g., in Rasulov et al., 2016), whereas for slow systems, the system response time should be separately estimated (as, e.g., in Rasulov et al., 2009). For example, the Rasulov et al. (2016) ultra-fast gas-exchange system has a chamber volume of 2.4 mL and flow rate of 0.67 L min^{-1}, yielding a system half-time (τ) of only 0.15 s and time to reach a steady state (4τ) of 0.6 s, whereas the volume of LI-6400 standard 2 cm × 3 cm (6 cm^2 window area) chamber is 80 mL, and for the same flow rate, 4τ is 20 s, and for the conifer chamber of LI-COR (volume of 155 mL) 4τ is 39 s (Niinemets, 2012).

The gas exchange systems can have one (measurements switched between reference measurement and sample measurement) or two measurement lines (reference and sample measured continuously). For one-line systems, the time for switching and stabilization of gas flows between the reference and sample measurement can further add to the whole system response time.

4.4.1 Use of a fast gas-exchange system for estimation of intermediate pool sizes

Here we provide a sample protocol with the ultrafast gas-exchange system of Rasulov et al. (2016) that includes two identical parallel gas lines that allow independent measurement of the background and chamber isoprene

---

protocol was similar. After measurement of the dark decay kinetics of isoprene release, the plant was removed from the chamber, and a stable isoprene flow was established through a calibrated capillary. After reaching a stable value, the supply of isoprene was interrupted and the artificial "isoprene release" was recorded again. This line was superimposed on the post-illumination emission kinetics. The baselines for DMADP pool sizes were determined as shown in the figure. In (a), the DMADP pool size was estimated as the integral of the first peak and the MEcDP pool size as the integral of the second peak. In (b), the DMADP pool size was estimated as the integral of the first peak of the plant emissions minus the integral of artificial isoprene release, and the MEcDP pool size as the integral of the second peak with plant emissions.

concentrations. Isoprene concentration can be measured by both FIS (see above for settings), or PTR-MS as explained here.

4.4.1.1 Materials

- A plant
- A PTR-TOF-MS
- An ultra-fast gas-exchange system with a 2.4 mL circular leaf chamber (3 mm height, 32 mm diameter) (Rasulov et al., 2016)
- Isoprene calibration standard (3.43 ppm isoprene in N_2)

4.4.1.2 Procedure

4.4.1.2.1 Setting up PTR-TOF-MS Use the following PTR-TOF-MS setup through the measurements: the inlet and drift chamber temperature: 60 °C, inlet flow 100: $mL\,min^{-1}$, flow of water vapor: 5.0 $mL\,min^{-1}$, ion current: 4 mA, drift chamber pressure: 2.1 mbar, drift tube field density ratio: 140 Td, pressure of the TOF-MS module: $2.4 \cdot 10^{-7}$ mbar (Rasulov et al., 2019). Calibrate the PTR-MS-TOF instrument using the flow of the isoprene standard into the leaf measurement chamber.

4.4.1.2.2 Setting up the gas exchange system In vivo MEP pool sizes can be estimated for any combination of environmental drivers. First, enclose the leaf in the chamber and establish the desired environmental conditions in the chamber (e.g., leaf temperature of 30 °C, air humidity of 60%, CO_2 concentration of 410 $\mu mol\,mol^{-1}$ and light intensity of 1000 $\mu mol\,m^{-2}\,s^{-1}$, air flow rate of 0.67 $L\,min^{-1}$). Establish identical ingoing gas concentrations in both lines of the gas exchange system.

4.4.1.2.3 Measurement procedure Measure isoprene background concentration in the reference line with the PTR-TOF-MS at 400 ps sample interval and record averages at 1 Hz interval. Switch the channels and measure isoprene concentration in the gas exchange chamber exhaust air. Wait until leaf isoprene emission reaches a steady state, typically for 10–20 min after leaf enclosure. Switch back to reference line and measure again isoprene background concentration. Simultaneously switch to the sample line and switch off the light. Measure isoprene emission through both phases of post-illumination kinetics (Fig. 5A) for about 10–15 min after switching off the light. Switch back to the reference line and measure again the background isoprene concentration. Calculate isoprene emission rates through the post-illumination kinetics as explained above. Establish the baseline through the post-illumination measurements using the reference line measurements at different time points. Estimate the baseline for the first phase of the post-illumination decay curve by extrapolating from the start of the emission rise (at ca. 570 s in Fig. 5A) to the end of the reference line measurement (at ca. 200s in Fig. 5A). Integrate the first phase of the dark decay kinetics above the baseline (between ca. 200 to 410s in Fig. 5A) and the second phase using the trapezoidal rule. The first integral is the DMADP pool size and the second MEcDP pool size supporting the isoprene emission rate prior to leaf darkening.

4.4.2 Use of a slower gas-exchange system for estimation of intermediate pool sizes

In the case of the slower system, all measurements are carried out identically to fast system measurement. The only difference is the need to consider system delay effects. At the end of the measurement, the leaf or plant is removed from the chamber and a flow of isoprene is fed into the carrier air flow or directly into the leaf chamber **(**Fig. 5B**)** (Li et al., 2011). Once a stable isoprene flow from the chamber is achieved, isoprene flow into the chamber is stopped and isoprene flow from the chamber is measured until it reaches the background level. The kinetics of isoprene release from the empty chamber is scaled such that the value of isoprene release prior to stopping isoprene flow is scaled to the level corresponding to steady-state isoprene emission rate. The chamber isoprene kinetics is superimposed on the plant measurements and the baseline for the first phase of the post-illumination kinetics is established as shown in Fig. 5B. For plant measurements, the two peaks of isoprene emission are integrated as explained in Section 4.4.1 and the isoprene release without plant is also integrated. The DMADP pool size is the difference between the integral of the first peak of plant measurements minus the integral of isoprene release without the plant (Fig. 5B).

5. Safety considerations and standards

Many of these techniques require handling pressurized gases and care must be exercised in handling gas tanks. At low concentrations in air, isoprene is not toxic (humans generally exhale air with 25 to 100ppb isoprene) but higher concentrations in air should be avoided. If liquid isoprene is stored, it should be in an explosion-proof container and refrigerated because the boiling point is 32 °C. Care should be taken to avoid contact with liquid isoprene. Long-term storage of liquid isoprene can lead to formation of impurities due to condensation reactions, including formation of monoterpenes.

6. Summary

Measuring isoprene in air contributes to studies in atmospheric chemistry, plant physiology, biotechnology, and biochemistry. There are many methods used to measure isoprene and different methods of detection. Isoprene measurements can be made with bacterial cultures, isolated enzymes, and attached or detached leaves. Isotopic labeling can add to the utility of isoprene measurements. The Proton Transfer Reaction mass spectrometers are the most versatile instruments for detection but also the most expensive (Table 1). Isoprene is relatively easy to handle, it is relatively stable and does not stick to tubing and chamber walls as much as other terpenes. It also is emitted from leaves as soon as it is made allowing insight into the working of the methyl erythritol 4-phosphate pathway, the pathway in plants responsible for isoprene emission.

Table 1 Pros and cons of the methods discussed here.

	Pros	**Cons**
Continuous	Can follow transients	Requires PTR-MS to follow isotopologues
Discrete	Flexibility, simplicity	Difficult to follow transients
Detection		
GC–PID	High sensitivity, good quantitation Easily adapted to large samples Other gases can be measured in single run	Less selective Discreet measurements only
GC–MS (SPME)	Good selectivity Other gases can be measured in single run	Discreet measurements only Poor quantitation
FIS	Large dynamic range Good quantitation High temporal resolution	Specialized instrument for isoprene measurements
PTR-MS (both quadrupole and time-of-flight)	High sensitivity and time resolution Many other gases can be measured simultaneously Continuous monitoring Isotopologue-specific	Expensive to acquire Some compound fragments might interfere with isoprene detection

Acknowledgments

Isoprene research at MSU is supported by the US National Science Foundation under Grant No. 2022495. Partial salary support for TDS comes from Michigan AgBioResearch.

References

Cao, X.-L., & Hewitt, C. N. (1995). Detection methods for the analysis of biogenic non-methane hydrocarbons in air. *Journal of Chromatography A*, *710*(1), 39–50. https://doi.org/10.1016/0021-9673(95)00427-O.

Copolovici, L. O., & Niinemets, Ü. (2005). Temperature dependencies of Henry's law constants and octanol/water partition coefficients for key plant volatile monoterpenoids. *Chemosphere*, *61*(10), 1390–1400. https://doi.org/10.1016/j.chemosphere.2005.05.003.

Fisher, A. J., Grimes, H. D., & Fall, R. (2003). The biochemical origin of pentenol emissions from wounded leaves. *Phytochemistry*, *62*(2), 159–163. https://doi.org/10.1016/s0031-9422(02)00521-6.

Ge, D., Xue, Y., & Ma, Y. (2016). Two unexpected promiscuous activities of the iron-sulfur protein IspH in production of isoprene and isoamylene. *Microbial Cell Factories*, *15*, 79. https://doi.org/10.1186/s12934-016-0476-9.

González-Cabanelas, D., Hammerbacher, A., Raguschke, B., Gershenzon, J., & Wright, L. P. (2016). Chapter ten—Quantifying the metabolites of the methylerythritol 4-phosphate (MEP) pathway in plants and bacteria by liquid chromatography–triple quadrupole mass spectrometry. In S. E. O'Connor (Ed.), *Vol. 576. Methods in enzymology* (pp. 225–249). Academic Press. https://doi.org/10.1016/bs.mie.2016.02.025.

Gonzalez-Esquer, C. R., Ferlez, B., Weraduwage, S. M., Kirst, H., Lantz, A. T., Turmo, A., et al. (2021). Validation of an insertion-engineered isoprene synthase as a strategy to functionalize terpene synthases. *RSC Advances*, *11*(48), 29997–30005. https://doi.org/10.1039/D1RA05710C.

Gray, D. W., Breneman, S. R., Topper, L. A., & Sharkey, T. D. (2011). Biochemical characterization and homology modeling of methylbutenol synthase and implications for understanding hemiterpene synthase evolution in plants. *The Journal of Biological Chemistry*, *286*, 20582–20590.

Gray, D. W., Goldstein, A. H., & Lerdau, M. (2006). Thermal history regulates methylbutenol basal emission rate in *Pinus ponderosa*. *Plant, Cell and Environment*, *29*, 1298–1308.

Guenther, A. B., & Hills, A. J. (1998). Eddy covariance measurement of isoprene fluxes. *Journal of Geophysical Research*, *103*, 13145–13152. https://doi.org/10.1029/97jd03283.

Hansel, A., Jordan, A., Holzinger, R., Prazeller, P., Vogel, W., & Lindinger, W. (1995). Proton transfer reaction mass spectrometry: On-line trace gas analysis at the ppb level. *International Journal of Mass Spectrometry and Ion Processes*, *149-150*, 609–619.

Hills, A. J., Fall, R., & Monson, R. K. (1992). Methods for the analysis of isoprene from leaves. In H. F. Linskins, & J. F. Jackson (Eds.), *Vol. 13. Plant toxin analysis. Modern methods of plant analysis* (pp. 297–315). Springer-Verlag. https://doi.org/10.1007/978-3-662-02783-7_12.

Hills, A. J., & Zimmerman, P. R. (1990). Isoprene measurement by ozone-induced chemiluminescence. *Analytical Chemistry*, *62*, 1055–1060. https://doi.org/10.1021/ac00209a017.

Hunter, E. P., & Lias, S. G. (1998). Evaluated gas phase basicities and proton affinities of molecules: An update. *Journal of Physical and Chemical Reference Data*, *27*(3), 413–656. https://doi.org/10.1063/1.556018.

Jardine, K., Abrell, L., Kurc, S. A., Huxman, T., Ortega, J., & Guenther, A. (2010). Volatile organic compound emissions from *Larrea tridentata* (creosotebush). *Atmospheric Chemistry and Physics*, *10*(24), 12191–12206. https://doi.org/10.5194/acp-10-12191-2010.

Jordan, A., Haidacher, S., Hanel, G., Hartungen, E., Herbig, J., Märk, L., et al. (2009a). An online ultra high sensitivity proton-transfer-reaction mass-spectrometer combined with switchable reagent ion capability (PTR+SRI−MS). *International Journal of Mass Spectrometry*, *286*(1), 32–38. https://doi.org/10.1016/j.ijms.2009.06.006.

Jordan, A., Haidacher, S., Hanel, G., Hartungen, E., Märk, L., Seehauser, H., et al. (2009b). A high resolution and high sensitivity proton-transfer-reaction time-offlight mass spectrometer (PTR-TOF-MS). *International Journal of Mass Spectrometry*, *286*, 122–128.

Karl, T., Fall, R., Rosenstiel, T. N., Prazeller, P., Larsen, B., Seufert, G., et al. (2002). On-line analysis of the $^{13}CO_2$ labeling of leaf isoprene suggests multiple subcellular origins of isoprene precursors. *Planta*, *215*, 894–905.

Karl, T., Hansel, A., Cappellin, L., Kaser, L., Herdlinger-Blatt, I., & Jud, W. (2013). BVOC measurements based on NO^+ ionization. In A. Hansel, & J. Dunkl (Eds.), *6th International conference on proton transfer reaction mass spectrometry and its applications, Obergurgl, Austria February 3rd–February 8th, 2013* (pp. 84–88). Innsbruck University Press.

Karl, T., Prazeller, P., Mayr, D., Jordan, A., Rieder, J., Fall, R., et al. (2001). Human breath isoprene and its relation to blood cholesterol levels: New measurements and modeling. *Journal of Applied Physiology*, *91*, 762–770.

Lehnert, A.-S., Perreca, E., Gershenzon, J., Pohnert, G., & Trumbore, S. E. (2020). Simultaneous real-time measurement of isoprene and 2-methyl-3-buten-2-ol emissions from trees using SIFT-MS [methods]. *Frontiers in Plant Science*, *11*, 578204. https://doi.org/10.3389/fpls.2020.578204.

Li, Z., Ratliff, E. A., & Sharkey, T. D. (2011). Effect of temperature on postillumination isoprene emission in oak and poplar. *Plant Physiology*, *155*(2), 1037–1046. https://doi.org/10.1104/pp.110.167551.

Li, M., Xu, J., Algarra Alarcon, A., Carlin, S., Barbaro, E., Cappellin, L., et al. (2017). In planta recapitulation of isoprene synthase evolution from ocimene synthases. *Molecular Biology and Evolution*, *34*(10), 2583–2599. https://doi.org/10.1093/molbev/msx178.

Lindinger, W., Hansel, A., & Jordan, A. (1998). Proton-transfer-reaction mass spectrometry (PTR-MS): On-line monitoring of volatile organic compounds at pptv levels. *Chemical Society Reviews*, *27*, 347–354.

Loreto, F., & Sharkey, T. D. (1990). A gas-exchange study of photosynthesis and isoprene emission in *Quercus rubra* L. *Planta*, *182*, 523–531.

Malásková, M., Olivenza-León, D., Piel, F., Mochalski, P., Sulzer, P., Jürschik, S., et al. (2019). Compendium of the reactions of H_3O^+ with selected ketones of relevance to breath analysis using proton transfer reaction mass spectrometry. *Frontiers in Chemistry*, *7*, 401. https://doi.org/10.3389/fchem.2019.00401.

Miekisch, W., Schubert, J. K., Vagts, D. A., & Geiger, K. (2001). Analysis of volatile disease markers in blood. *Clinical Chemistry*, *47*, 1053–1060. http://www.clinchem.org/content/47/6/1053.full.pdf.

Miller, B., Oschinski, C., & Zimmer, W. (2001). First isolation of an isoprene synthase gene from poplar and successful expression of the gene in *Escherichia coli*. *Planta*, *213*(3), 483–487. https://doi.org/10.1007/s004250100557.

Mochalski, P., King, J., Kupferthaler, A., Unterkofler, K., Hinterhuber, H., & Amann, A. (2011). Measurement of isoprene solubility in water, human blood and plasma by multiple headspace extraction gas chromatography coupled with solid phase microextraction. *Journal of Breath Research*, *5*(4), 046010. http://stacks.iop.org/1752-7163/5/i=4/a=046010. http://iopscience.iop.org/1752-7163/5/4/046010/pdf/1752-7163_5_4_046010.pdf.

Niinemets, Ü. (2012). Whole plant photosynthesis. In J. Flexas, F. Loreto, & H. Medrano (Eds.), *Terrestrial photosynthesis in a changing environment. A molecular, physiological and ecological approach* (pp. 399–423). Cambridge University Press.

Niinemets, Ü., Kuhn, U., Harley, P. C., Staudt, M., Arneth, A., Cescatti, A., et al. (2011). Estimations of isoprenoid emission capacity from enclosure studies: Measurements, data processing, quality and standardized measurement protocols. *Biogeosciences*, *8*(8), 2209–2246. https://doi.org/10.5194/bg-8-2209-2011.

Niinemets, Ü., Monson, R. K., Arneth, A., Ciccioli, P., Kesselmeier, J., Kuhn, U., et al. (2010). The leaf-level emission factor of volatile isoprenoids: Caveats, model algorithms, response shapes and scaling. *Biogeosciences*, 7(6), 1809–1832. https://doi.org/10.5194/bg-7-1809-2010.

Rasulov, B., Copolovici, L., Laisk, A., & Niinemets, Ü. (2009). Postillumination isoprene emission: In vivo measurements of dimethylallyldiphosphate pool size and isoprene synthase kinetics in aspen leaves. *Plant Physiology*, *149*, 1609–1618.

Rasulov, B., Hüve, K., Laisk, A., & Niinemets, Ü. (2011). Induction of a longer-term component of isoprene release in darkened aspen leaves: Origin and regulation under different environmental conditions. *Plant Physiology*, *156*, 816–831.

Rasulov, B., Talts, E., & Niinemets, Ü. (2016). Spectacular oscillations in plant isoprene emission under transient conditions explain the enigmatic CO_2 response. *Plant Physiology*, *172*, 2275–2285.

Rasulov, B., Talts, E., & Niinemets, Ü. (2019). A novel approach for real-time monitoring of leaf wounding responses demonstrates unprecedently fast and high emissions of volatiles from cut leaves. *Plant Science*, *283*, 256–265. https://doi.org/10.1016/j.plantsci.2019.03.006.

Sarkar, C., Guenther, A. B., Park, J. H., Seco, R., Alves, E., Batalha, S., et al. (2020). PTR-TOF-MS eddy covariance measurements of isoprene and monoterpene fluxes from an eastern Amazonian rainforest. *Atmospheric Chemistry and Physics*, *20*(12), 7179–7191. https://doi.org/10.5194/acp-20-7179-2020.

Schade, G. W., Goldstein, A. H., Gray, D. W., & Lerdau, M. T. (2000). Canopy and leaf level 2-methyl-3-buten-2-ol fluxes from a ponderosa pine plantation. *Atmospheric Environment*, *34*(21), 3535–3544. http://www.sciencedirect.com/science/article/B6VH3-40D0KTV-6/2/1b127c2c8fa624efa6ffb348e490ec73.

Sharkey, T. D. (1996). Isoprene synthesis by plants and animals. *Endeavor*, *20*, 74–78.

Sharkey, T. D., Preiser, A. L., Weraduwage, S. M., & Gog, L. (2020). Source of ^{12}C in Calvin-Benson cycle intermediates and isoprene emitted from plant leaves fed with $^{13}CO_2$. *Biochemical Journal*, *477*(17), 3237–3252. https://doi.org/10.1042/BCJ20200480.

Sharkey, T. D., Yeh, S., Wiberley, A. E., Falbel, T. G., Gong, D., & Fernandez, D. E. (2005). Evolution of the isoprene biosynthetic pathway in kudzu. *Plant Physiology*, *137*, 700–712. https://www.ncbi.nlm.nih.gov/pmc/articles/PMC1065370/pdf/pp1370700.pdf.

Silver, G. M., & Fall, R. (1991). Enzymatic synthesis of isoprene from dimethylallyl diphosphate in aspen leaf extracts. *Plant Physiology*, *97*, 1588–1591.

Toda, K., & Dasgupta, P. K. (2008). Chapter 22—Environmental applications: Atmospheric trace gas analyses. In S. D. Kolev, & I. D. McKelvie (Eds.), *Vol. 54. Comprehensive analytical chemistry* (pp. 639–683). Elsevier. https://doi.org/10.1016/S0166-526X(08)00622-3.

Trovarelli, G., Brunori, F., De Medio, G. E., Timio, M., Lippi, G., Pelli, M. A., et al. (2001). Onset, time course, and persistence of increased haemodialysis-induced breath isoprene emission. *Nephron*, *88*, 44–47.

von Caemmerer, S., & Farquhar, G. D. (1981). Some relationships between the biochemistry of photosynthesis and the gas exchange of leaves. *Planta*, *153*(4), 376–387. https://doi.org/10.1007/BF00384257.

Warneke, C., Veres, P., Murphy, S. M., Soltis, J., Field, R. A., Graus, M. G., et al. (2015). PTR-QMS versus PTR-TOF comparison in a region with oil and natural gas extraction industry in the Uintah Basin in 2013. *Atmospheric Measurement Techniques*, *8*(1), 411–420. https://doi.org/10.5194/amt-8-411-2015.

Weise, S. E., Li, Z., Sutter, A. E., Corrion, A., Banerjee, A., & Sharkey, T. D. (2013). Measuring dimethylallyl diphosphate available for isoprene synthesis. *Analytical Biochemistry*, *435*, 27–34. https://doi.org/10.1016/j.ab.2012.11.031.

Wells, K. C., Millet, D. B., Payne, V. H., Vigouroux, C., Aquino, C. A. B., De Mazière, M., et al. (2022). Next-generation isoprene measurements from space: Detecting daily variability at high resolution. *Journal of Geophysical Research-Atmospheres*, *127*(5), e2021JD036181. https://doi.org/10.1029/2021JD036181.

Wiedinmyer, C., Friedfeld, S., Baugh, W., Greenberg, J., Guenther, A., Fraser, M., et al. (2001). Measurement and analysis of atmospheric concentrations of isoprene and its reaction products in central Texas. *Atmospheric Environment*, *35*, 1001–1013.

Zhou, C., Li, Z., Wiberley-Bradford, A. E., Weise, S. E., & Sharkey, T. D. (2013). Isopentenyl diphosphate and dimethylallyl diphosphate/isopentenyl diphosphate ratio measured with recombinant isopentenyl diphosphate isomerase and isoprene synthase. *Analytical Biochemistry*, *440*(2), 130–136. https://doi.org/10.1016/j.ab.2013.05.028.

CHAPTER NINE

Using targeted metabolomics to elucidate the indole auxin network in plants

Jerry D. Cohen[a,*], Qian Tang[a], and Adrian D. Hegeman[a,b]
[a]Department of Horticultural Science and Microbial and Plant Genomics Institute, University of Minnesota, Saint Paul, MN, United States
[b]Department of Plant and Microbial Biology, University of Minnesota, Saint Paul, MN, United States
*Corresponding author: e-mail address: cohen047@umn.edu

Contents

Methods in Enzymology, Volume 676
ISSN 0076-6879
https://doi.org/10.1016/bs.mie.2022.07.038

Abstract

The plant hormone auxin plays important roles throughout the entire life span of a plant and facilitates its adaptation to a changing environment. Multiple metabolic pathways intersect to control the levels and flux through indole-3-acetic acid (IAA), the primary auxin in most plant species. Measurement of changes in these pathways represents an important objective to understanding core aspects of auxin signal regulation. Such studies have become approachable through the technologies encompassed by targeted metabolomics. By monitoring incorporation of stable isotopes from labeled precursors into proposed intermediates, it is possible to trace pathway utilization and characterize new biosynthetic routes to auxin. Chemical inhibitors that target specific steps or entire pathways related to auxin synthesis aid these techniques. Here we describe methods for obtaining stable isotope labeled pathway intermediates necessary for pathway analysis and quantification of compounds. We describe how to use isotope dilution with methods employing either gas chromatography or high performance liquid chromatography mass spectrometry techniques for sensitive analysis of IAA. Complete biosynthetic pathway analysis in seedlings using multiple stable isotope-labeled precursors and chemical inhibitors coupled with highly sensitive liquid chromatography-mass spectrometry methods are described that allow rapid measurement of isotopic flux into biochemical pools. These methods should prove to be useful to researchers studying aspects of the auxin metabolic network in vivo in a variety of plant tissues and during various environmental conditions.

1. Introduction

Research into plant hormones dates back over 150 years (Tivendale & Cohen, 2015) with the classic work of Ciesielski (1872), in which he removed plant root tips and observed the loss of gravitropism and its restoration by the replacement of the severed tip. This signaling history includes many examples of attempts to find or classify hormonal signals, especially related to auxin, that are the master regulators of specific plant processes (Bohn-Courseau, 2010; Hayashi et al., 2012; Rosquete, Barbez, & Kleine-Vehn, 2012; Saini, Sharma, Kaur, & Pati, 2013). Plant growth control is regulated by both changes in the levels of active hormones and the sensitivity of plant tissues to these concentration changes (Bou-Torrent et al., 2014; Li et al., 2013), although this concept was debated some decades ago (Weyers & Paterson, 2002).

Auxins are a central player in the regulation of embryo development, the establishment of the polar axis, and early aspects of seedling growth, as well as during later stages of plant development. Auxin signaling is now best characterized by interlocking networks that regulate auxin levels (Tillmann, Tang, Gardner, & Cohen, 2022), a dynamic protein network responsible for auxin response regulation (Das, Weijers, & Borst, 2021; Farcot, Lavedrine, & Vernoux, 2015), and a nuclear auxin signaling network that controls auxin regulated gene expression (Paque & Weijers, 2016; Radoeva et al., 2019; Truskina et al., 2021). The traditional approach to visualizing auxin level control in plants was to examine several different metabolic events. This then led to the consideration of the combination of metabolic regulatory processes as leading to "auxin homeostasis" to delineate the expected high degree of control that plants exercise to maintain the appropriate rate of growth control (Cohen & Gray, 2006; Cooke, Poli, Sztein, & Cohen, 2002; Ljung et al., 2002; Tivendale & Millar, 2022).

A common view of auxin homeostatic regulation involves an interactive network of simple pathways for biosynthesis (TAA1/YUC), conjugation (GH3 and UGTs), conjugate hydrolysis (ILR1) and oxidation (DAO1/2). Therefore, this view suggests free (active) auxin homeostasis can be explained through understanding the inputs/outputs (Fig. 1) contributed by the processes for synthesis, auxin conjugation, hydrolysis of auxin conjugates, and auxin degradation (Iglesias, Sellaro, Zurbriggen, & Casal, 2018; Kasahara, 2016). This concept, that was insightful and novel when proposed over 40 years ago (Bandurski, 1980), provided a context for understanding environmental effects on plant growth regulation.

Recent advances in metabolomics have enabled the global scale analyses of metabolism, often untargeted methods can detect more than a thousand low molecular weight molecules in each experiment (Xu, Freund, Hegeman, & Cohen, 2022). These advances have great promise for a global understanding of metabolism, which has been called "the chemical engine that drives the living process" (Schilling, Schuster, Palsson, & Heinrich, 1999) and is the most direct predictor of plant phenotype. Metabolomics approaches, in common with genomics and proteomics, have the advantage of being driven by both the need to answer critical hypotheses as well as having the ability to provide unexpected insight that makes it a hypothesis generating approach as well. Defined pathways, however, represent prior knowledge and thus the interpretation of results is largely dependent on existing knowledge of metabolic pathways. Such a detailed understanding of metabolism remains far from comprehensive for plants. Thus, the ab initio identification of plant pathways remains an important and often cited challenge for modern metabolomics studies (Menikarachchi, Hamdalla, Hill, & Grant, 2013). The collective statistical power of a properly designed targeted metabolomic study can be used to identify patterns that are missed at individual metabolite level and can be applied to perform pathway discovery analysis. Stable isotope

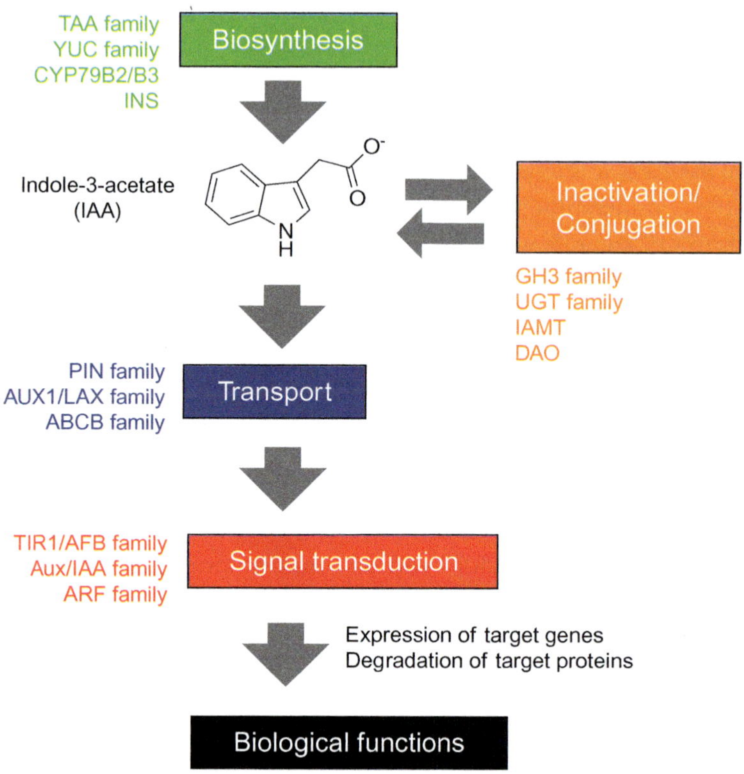

Fig. 1 A schematic model of auxin biosynthesis, inactivation/conjugation, transport and signal transduction in plants. Biosynthesis: The biosynthesis of IAA precursors, such as indole-3-glycerol phosphate, indole, and tryptophan, are obtained via the shikimic acid pathway. Inactivation/conjugation: formation of sugar/cyclitol or amino acid conjugates provide a potentially reversible method for sequestration, whereas ring oxidation of IAA itself or an amino acyl conjugate to form an derivative of 2-oxindole-3-acetic acid are the first steps in irreversible catabolism. Transport: Spatial patterns of auxin are important determinants of plant development and the process of polar auxin transport IAA and the movement of IBA and various conjugates are important in establishment of auxin levels in target tissues. Signal transduction: Auxin binding to the TIR1/AFB and Aux/IAA co-receptor complex promotes ubiquitin-mediated degradation by the 26S proteasome of Aux/IAA repressors, allowing ARF to facilitate transcriptional activation of auxin-regulated genes. Biological functions: Growth related to apical dominance, tropisms, cell proliferation, cellular differentiation, cellular elongation and a plethora of other auxin-mediated responses are related to specific auxin genes being activated or gene suppression. *Modified from Kasahara, H. (2016). Current aspects of auxin biosynthesis in plants.* Bioscience, Biotechnology, and Biochemistry, 80*(1):34–42. https://doi.org/10.1080/09168451.2015.*

(non-radioactive) labels as metabolic tracers are a well-established approach for the determination of the fate of metabolites, but in its most traditional application requires targeted analysis of known metabolites (Previs & Downes, 2020). One common instrumental approach that underpins both stable isotope tracer approaches and plant metabolomics, mass spectrometry, provides unique opportunities to marry these two approaches at high sensitivity, and as instrumentation has advanced, now permitting an ever more complex network-wide investigation of metabolic pathways.

Although there are notable exceptions, the vast bulk of plant hormone metabolism studies rely on measurements of plant hormone levels and/or levels of a metabolite to infer a metabolic origin, fate or regulation (see Mashiguchi et al., 2011, for example). These studies, while informative, are limited static "windows" that are often extrapolated to predict time-dependent metabolic activities. The full limitations of such approaches are difficult to predict, but it is clear that measurements of entire pathways over time, in discrete tissues and using isotopic tracer techniques offers significantly more detailed information about a complex interacting network of metabolic reactions. In the methods protocol we will describe current advances that facilitate a more global target metabolomics analysis of the network of reactions that result in formation of indole-3-acetic acid (IAA). This type of study may be used to answer different questions about auxin metabolism, and we will thus divide this protocol into sections that alone or in various combinations can be used to approach many different and varied analytical needs.

2. Instrumentation

Although there has been some debate over time as to the "best" way to measure auxin levels in plants, from chemical methods, bioassay to immunoassay, the application of mass spectrometry (MS) in the early 1970s brought about an impressive increase in the ability to accurately quantify auxin from plant tissues (Tivendale & Cohen, 2015). While MS dates for the turn of the 20th century, there were at least four important advances in the last half century that made gas chromatography-mass spectrometry (GC-MS) and liquid chromatography-mass spectrometry (LC-MS) the center of 21st century auxin analytical studies. These include the development of capillary GC-MS, the development of quadrupole tandem mass spectrometry (MS/MS) hardware, the application of liquid chromatography-electrospray ionization mass spectrometry (LC-ESI-MS), and the rapid evolution of high-resolution high sensitivity mass spectrometry using either orbitrap or time-of-flight architectures.

Currently, MS has been used in countless studies on plant biochemistry where its high sensitivity and information-rich spectra are invaluable. MS has proven critical to the studies of plant hormones, including auxin determinations because of its sensitivity and specificity. To obtain meaningful MS data starting with a plant extract, the various plant constituents usually must be separated prior to entry into the MS analyzer. Coupling of either GC or LC to the MS (called "hyphenated techniques," GC-MS or LC-MS) has greatly advanced the speed and reliability of modern plant hormone analysis but has also resulted in the common debate about the decision between GC-MS and LC-MS. There are many factors in this consideration, but the main things to consider appear to be (1) that for many compounds that lack volatility, to be analyzed by GC some form of derivatization is required (and the type of derivative, the completeness of the process, and the recovery); (2) the issue of thermal breakdown of some sensitive compounds during GC analysis; (3) efficiency of ionization and ion transfer efficiency in LC-MS. The appeal of GC-MS/MS is the high efficiency of transfer of the target compound directly into the vacuum system of the MS before ionization. Mass spectrometry is constantly evolving, but at this writing GC-MS/MS appears to be ideal for ultra-high sensitivity analysis of simple acids like IAA and indole-3-butyric acid (IBA), while LC-MS/MS has advantages for targeted metabolomics where many compounds with different functional groups are to be analyzed.

3. Stable isotope standards

Stable isotopes have several critical roles in auxin analysis, however their primary need is as internal standards that are required for absolute quantification (Cohen, Baldi, & Slovin, 1986) and as metabolic tracers, or for both uses in the same experiment (Tillmann et al., 2022). Because of the unique properties of the indole ring, IAA and its precursors and conjugates, the ease for loss of indoles during extraction and purification has a long history. The loss is typically in variable amounts (Chen, Miller, Patterson, & Cohen, 1988; Little, Heald, & Browning, 1978), likely through oxidation, adherence to surfaces and molecular rearrangements, making simple correction impossible. Thus from its first use more than 60 years ago (Turian & Hamilton, 1960) isotope dilution analysis (IDA) has become the preferred method for auxin quantification (Barkawi et al., 2008). An important transition occurred in the 1980s with the movement from radioisotope tracers to the use of stable isotopes and mass spectrometry (Tivendale & Cohen, 2015).

For the IDA quantification of IAA from plant extracts different stable isotope labeled forms have been used with success. The ring labeled [$^2H_5$]IAA and [$^2H_4$]IAA standards (Magnus, Bandurski, & Schulze, 1980) are still sometimes used, either for economy or for specialized needs. However following the introduction of [$^{13}C_6$]IAA in 1986 (Cohen et al., 1986) and, a short time later its commercial availability, it has become more commonly used for IDA. The incorporation of six ^{13}C atoms that label the six-membered ring of indole results in having covalent heavy atom labels within the molecule. The standard is added at the beginning of the extraction procedure and then serves to quantify the endogenous unlabeled content following MS analysis. Because the isotopic-labeled IAA is structurally or near-structurally identical to IAA, it performs identically throughout the extraction procedure and, dependent on the isotopic label used, identically or nearly identically during chromatographic separation (deuterium-labeled standards have slightly different run times than the unlabeled compound on chromatographic columns (Cohen et al., 1986)).

4. Common sources of concern

4.1 Extraction

Likely, the most critical step in any analysis is the extraction step. Although early studies used apolar solvents like diethyl ether, these were shown unsuitable when auxin biology entered its chemical era mid-20th century (Tivendale & Cohen, 2015). Also, while there were early advocates of water or buffer extraction who thought it removed only "active" auxin (Gorter, 1932), it was shown to have the disadvantage that enzymatic activity both degraded as well as produced "new" auxin from conjugates and precursors even at reduced temperatures (Tivendale & Cohen, 2015). There remain more recent aqueous extraction advocates for certain plant materials, especially those that lack significant quantities of the more labile ester conjugates (Pilet & Saugy, 1987; Sundberg, 1990) but there are also recent reports of significant problems (Gélinas-Marion, Nichols, & Ross, 2020). Acetone is an excellent solvent for many classes of organic compounds and was early shown to denature many enzymes such that solvent systems with aqueous acetone were used for auxin extraction. However, its UV cutoff at 330–340 nm created problems for liquid scintillation counting, UV quantitation and HPLC and the use of various alcohols were adopted. Alcohols used for extraction have two problems worthy of consideration, first is the formation of esters with the alcohol during extraction especially if the extract itself is acidic, or acids are used in the process, or acid SPE resins are used. A second, related issue is transesterification from IAA ester or acyl anhydrides like IAA-glucose or IAA-inositol. These can be problematic to use if the goal of extraction is to analyze either the endogenous esters or IAA methyl or ethyl esters. These issues are greatly reduced by employing secondary alcohols, for example using a solvent-soluble anti-oxidant buffer like imidazole with aqueous 2-propanol for extraction (Liu, Hegeman, Gardner, & Cohen, 2012).

4.2 Isotope exchange

The use of IAA with deuterium in the side chain was an early issue in isotope dilution analysis. The use of stable isotope-labeled IAA as an internal standard began with the use of side-chain labeled [2,2-$^2H_2$]IAA (Caruso, Smith, Smith, Cheng, & Daves, 1978; Iino & Carr, 1982; McDougall & Hillman, 1978). However, with the availability of better internal standards, such a ring pentadeutero IAA, it was shown that an order of magnitude error could result from the easy exchange of such labels (Caruso & Zeisler, 1983). Unfortunately, [2,2-$^2H_2$]IAA is relatively cheap, easy to prepare (Kai, Nakamura, Wakasa, & Miyagawa, 2007) and thus still finds its way into the modern literature, either as a label for IAA or as a label to form labeled conjugates. Notably, however, that deuterium exchange in conjugates has not been studied well and side-chain labels in tryptophan do not seem to exchange are freely as IAA (Jensen & Bandurski, 1996). It should also be noted that storage of deuterated compounds in acid or basic solutions should generally be avoided (Davison, Milan, & Dutton, 2013). While there may be specific uses where [2,2-$^2H_2$]IAA would prove useful, many reports have documented its easy deuterium exchange (Cohen et al., 1986; Magnus et al., 1980) and it is likely best to avoid using it for serious quantitative studies.

4.3 Hydrolysis methods

A plethora of analytical studies have demonstrated that the majority of cellular auxin is conjugated to simple sugars, cyclitols, glycans, amino acids, and proteins. A long held analytical practice had been to use "selective base hydrolysis" to measure these sets of compounds. Typically, 1 N sodium hydroxide treatment of a plant extract for 1 h at room temperature was found to hydrolyze IAA-ester and ester-like conjugates and release free IAA without affecting IAA-amide conjugates (Baldi, Maher, & Cohen, 1989; Bandurski & Schulze, 1977). In contrast, 7 N sodium hydroxide treatment at 100 °C for 3 h completely hydrolyzed all conjugated forms of IAA (Bialek & Cohen, 1989) and this was termed "total IAA." By simple subtraction, the amount of free IAA, ester conjugates, and amide conjugates were then calculated. This method, however, has come under increased scrutiny as analytical methods have advanced and potential errors documented (Ilić, Normanly, & Cohen, 1996). More recent critical evaluation has demonstrated quite clearly that these previous methods of estimating the amount of IAA conjugates, base hydrolysis, can be significantly complicated by chemical artifacts rendering the techniques unadvisable (Yu, Lor, Ludwig-Müller, Hegeman, & Cohen, 2015).

5. Isotope dilution analysis (IDA)

IDA (Hevesy & Paneth, 1913; Rittenberg & Foster, 1940) has become an important tool for quantification of plant hormones since its early use (Turian & Hamilton, 1960). The use of isotopomers as internal standards and the calculations with the IDA equation, which assumes chemical identity between the labeled and unlabeled form, obviate the need for continual use of calibration plots. However, some investigators (Pagliano, Mester, & Meija, 2015) prefer such plots or find them as a useful "second check" and calibration plots can elucidate rare instances where mass spectrometer malfunctions compromise the linearity of the response, or, if using deuterated standards, when there is unanticipated proton-deuterium exchange. Internal standard $[^{13}C_6]$IAA and endogenous IAA have the same chromatographic retention times but can be distinguished from each other in the mass spectral analysis because of the additional six mass units. This mass difference is important in that it is a large enough spread to avoid the natural isotope distribution of ^{13}C in endogenous IAA and yields an initial isotopic enrichment in which no unlabeled IAA is found in the internal standard. This simplifies IDA of IAA when using the isotopomer $[^{13}C_6]$IAA as an internal standard. $[^{13}C_6]$IAA is added, in known amounts, early in the processing of unlabeled plant extracts. After purification and MS analysis, the amount of $[^{13}C_6]$IAA added (x) and its isotopic enrichment (Co), as well as the isotopic enrichment of the isolated mix of plant IAA and $[^{13}C_6]$IAA (Cf), the amount of IAA in the original plant extract (y) can be determined.

$$y = \left(\frac{Co}{Cf} - 1\right)x$$

Co, the specific enrichment of the internal standard, is usually 1 when using [$^{13}C_6$]IAA because its mass fragmentation does not overlap with that of unlabeled IAA for the ions typically monitored. *Cf* is the new specific enrichment of IAA after isolation (a mixture of the internal standard and endogenous IAA), the abundance of the m/z for the internal standard [$^{13}C_6$]IAA (*Amzs*) divided by the sum of *Amzs* plus the abundance of m/z of the endogenous [^{12}C]IAA (*Amze*):

$$Cf = \frac{Amzs}{Amzs + Amze}$$

Although the direct application of the isotope dilution equation gives a quite adequate measurement of IAA levels in plants by measuring the ratio of the monoisotopic ion peak for the endogenous and standard, highly accurate determinations of the total IAA pool requires a slight correction that is frequently used. This correction (*R*) is included to account for minor isotopologues not analyzed at the selected monoisotopic ion monitored by the MS (Cohen et al., 1986). However, if the amount of standard was determined by reverse isotope dilution analysis without correction for the isotopic envelope (Liu et al., 2012), the calculations are self-correcting. R must be determined empirically for each lot of [$^{13}C_6$]IAA purchased or synthesized, although it has remained relatively constant for the last 30 years. It is determined by running standards of IAA and [$^{13}C_6$]IAA separately using the scan mode of the MS instrument or using multiple ion monitoring for nine ions. For both compounds, the abundance of the ion used for quantification (typically either the base peak of each compound or its molecular ion) is divided by the sum of the ions that represent its isotopic envelope (i.e., m/z, $m/z+1$, $m/z+2$, …). This isotopic envelope is formed primarily in the endogenous IAA by the relative abundance for carbon-12 at 98.93% and for carbon-13 1.07% in each of the 10 carbons in IAA and in [$^{13}C_6$]IAA by the lower than perfect enrichment. The limits to the range of the spread can be set empirically and should represent ions, the abundance (*A*) of which is significantly higher than the noise. For plant-derived IAA and the [$^{13}C_6$]IAA standard, we (Cohen et al., 1986) determined the values to be m/z 130–134 (where $Ae = (A_{m/z130}\ /\Sigma\ A_{m/z130} + A_{m/z131} + A_{m/z132} + A_{m/z133} + A_{m/z134})$ and m/z 134–139 (where $As = (A_{m/z136}\ /\Sigma\ A_{m/z134} + A_{m/z135} + A_{m/z136} + A_{m/z137} + A_{m/z138} + A_{m/z139})$, respectively. In this case, the calculation for *R* is:

$$R = \frac{Ae}{As}$$

The resulting values were *Ae* 0.89 and *As* 0.79 giving an *R* of 1.13. The full isotope dilution equation is then:

$$y = \left(\frac{Co}{Cf} - 1\right)\left(\frac{x}{R}\right)$$

Thus, with very simple arithmetic equations, it is possible to measure minute amounts of auxin in plant extracts with extreme precision and certitude.

6. Preparation of stable isotope internal standards

Many of the labeled indole and precursor compounds required as internal standards and for labeling studies are readily available from commercial suppliers (Table 1). However, many other compounds can be of utility, and we will describe several routes (Fig. 2) we have found useful for our work. This list is not exhaustive, but the advantages of those we describe are that they are one or two steps, usually high yields and one or more precursor is readily available with the required label.

6.1 Labeled indole-3-acetic acid (IAA) and substituted indole-3-acetic acids

One of the most generally applicable methods for the preparation of labeled indole-3-acetic acid and substituted indole-3-acetic acids is from phenylhydrazine or halogen substituted phenylhydrazines (Baldi, Slovin, & Cohen, 1985) using [$^{13}C_5$]2-oxoglutaric acid ([$^{13}C_5$]α-ketoglutarate). This method can label the five-member ring, the side chain, or both.

a. Dissolve 0.8 μmol of [$^{13}C_5$]2-oxoglutaric acid in 35 μL pyridine in an ignition tube or other container that can be well sealed

b. Add 20 μmol of the appropriate phenylhydrazine by first making a 0.13 M suspension in cold HCl (conc):85% phosphoric acid (2:1). Add in 50 μL aliquots with cooling and mixing between additions

c. Flush with N_2 gas, seal under vacuum, heat at 85 °C for 17–20 h in a dry block heater

d. Chill on ice, open, acidify with 1 mL 0.1 M HCl

e. Extract 3 × with 1 mL diethyl ether, dry over anhydrous sodium sulfate for 1 h to remove residual acidic water

f. Filter, add 0.5 mL water, evaporate the diethyl ether leaving the water phase

Table 1 A list of current "catalog-listed" stable isotope labeled compounds recommended for studies of indole-3-acetic acid quantitative analysis and for metabolic studies.

Compound	Labeling	Listing supplier
IAA and related		
Indole-3-acetic acid	[$^{13}C_6$] ring label	CIL, OLC, TRC, BOC, MII
	[$^2H_4$] ring label	TRC, BOC
	[$^2H_5$] ring label	CIL, CDN, ISO, OLC, TRC, BOC
IAA methyl ester	[$^2H_5$] ring label	OLC
Indole-3-acetamide	[$^2H_5$, ^{15}N] ring label	OLC
	[$^2H_5$] ring label	CDN, CRC
Indole-3-acetaldoxime	[$^2H_5$] ring label	TRC
Indole-3-acetonitrile	[$^2H_4$] ring label	OLC
	[$^2H_5$] ring label	TRC
	[$^{13}C_2$,^{15}N] side chain	TRC, CRC
Indole-3-butyric acid	[$^{13}C_6$]	MII
	[^{15}N]	MII
	[$^{13}C_8$,^{15}N]	MII
DL-Indole-3-lactic acid	[$^2H_5$] ring label	CRC
Indole-3-pyruvic acid	[$^2H_5$] ring label	TRC, BOC
Indole-3-pyruvic acid methyl ester	[$^2H_5$] ring label	TRC
IAA-L-alanine	[$^2H_5$] ring label, [^{15}N]ala	OLC
IAA-L-aspartate	[$^2H_5$] ring label, [^{15}N]Asp	OLC
	[$^{13}C_6$] ring label, [^{15}N]asp	OLC
IAA-L-glutamate	[$^2H_5$] ring label, [^{15}N]glu	OLC
	[$^{13}C_6$] ring label, [^{15}N]glu	OLC
IAA-L-glycine	[$^{13}C_2$,^{15}N]	MII
IAA-L-isoleucine	[$^2H_5$] ring label, [^{15}N]Ile	OLC
IAA-L-leucine	[$^2H_5$] ring label, [^{15}N]Ieu	OLC

Table 1 A list of current "catalog-listed" stable isotope labeled compounds recommended for studies of indole-3-acetic acid quantitative analysis and for metabolic studies.—cont'd

Compound	Labeling	Listing supplier
IAA-L-phenylalanine	[$^2H_5$] ring label, [^{15}N]phe	OLC
IAA-L-tryptophan	[$^2H_5$] ring label, [^{15}N]trp	OLC
IAA-L-valine	[$^2H_5$] ring label, [^{15}N]val	OLC
Indolc-3-ethanol	[$^2H_5$] ring label	OLC
Oxindole-3-acetic acid	[$^2H_4$] ring label	TRC, MII
	[$^{13}C_6$] ring label	TRC, CRC
Precursors		
Anthranilic acid	[^{15}N]	CIL, BOC
	[$^{13}C_6$] ring label	CIL, ISO, BOC, ENT
Indole	[$^{13}C_8$,^{15}N]	CIL, BOC, MII
	[$^{13}C_8$]	CIL
	[^{15}N]	CIL, ISO, BOC
	[2-^{13}C]	CIL, BOC, MII
	[4-^{13}C]	CIL, BOC, MII
	[$^2H_4$]	TRC, CRC
	[$^2H_5$]	TRC
	[$^2H_6$]	TRC, BOC, ENT, CRC
	[$^2H_7$]	CDN, ISO, TRC, BOC
L-Serine	[$^{13}C_3$,^{15}N]	CIL, COR, ISO
	[$^{13}C_3$]	CIL, ISO
	[2-^{13}C,^{15}N]	ISO,COR
	[$^{13}C_3$, β2-^2H,^{15}N]	COR
	[$^{13}C_3$,^{15}N,2,3,3-$^2H_3$]	CIL, COR
	[2,3-$^{13}C_2$]	ISO, COR
	[3-^{13}C]	CIL, ISO, COR
	[2-^{13}C]	CIL, ISO, COR
	[2-^{13}C, 2,3,3-$^2H_3$]	CIL

Continued

Table 1 A list of current "catalog-listed" stable isotope labeled compounds recommended for studies of indole-3-acetic acid quantitative analysis and for metabolic studies.—cont'd

Compound	Labeling	Listing supplier
L-Tryptophan	[$^{13}C_{11}$,$^{2}H_{8}$,$^{15}N_{2}$]	CIL, ISO, BOC, MII
	[$^{13}C_{11}$,$^{15}N_{2}$]	CIL, ISO, BOC, MII
	[$^{13}C_{6}$] ring label	BOC
	[$^{13}C_{11}$]	BOC, MII
	[$^{15}N_{2}$]	CIL, ISO, BOC
	[indole-^{15}N]	CIL, BOC
	[indole-2-^{13}C]	CIL, ISO, COR, BOC, MII
	[indole-3-^{13}C]	CIL, BOC, MII
	[indole-4-^{13}C]	CIL, BOC, MII
	[$^{2}H_{8}$]	CIL, BOC
	[$^{2}H_{8}$,$^{15}N_{2}$]	CIL, BOC
	[indole-$^{2}H_{5}$]	CIL, ISO, CDN, BOC
D-Tryptophan	[indole-$^{2}H_{5}$]	CIL, CDN, BOC
DL-Tryptophan	Side chain [2-^{13}C]	ISO
4-F-DL-Tryptophan	[2,4,6,7-$^{2}H_{4}$]	ISO, BOC
N-Acetyl-L-tryptophan	[indole-$^{2}H_{5}$]	ISO, BOC
2-oxoglutaric acid	[$^{13}C_{5}$]	CIL, ISO, BOC
	[$^{13}C_{4}$]	MII

Some of the listed compounds may not actually be available from the listed suppliers since a listing does not assure availability. Nevertheless, this serves to guide interest and experimental design. In addition, note that some of these compounds are only available in extremely small quantities useful as internal standards and other similar uses at the μg or lower level and some vary in price significantly among suppliers. BOC: BOC Sciences (https://isotope.bocsci.com/); CDN: CDN Isotopes (https://cdnisotopes.com/); CIL: Cambridge Isotope Laboratories (https://www.isotope.com/); CRC: Coompo Research Chemicals (http://www.coompo.com/); COR: CortecNet (https://cortecnet.com/); ENT: Entegris IsoSciences (https://isosciences.com/), ISO: ISOTEC (Sigma-Aldrich) (https://www.sigmaaldrich.com); MII: Medical Isotopes, Inc. (https://www.medicalisotopes.com); OLC: Olchemim.cz (https://www.olchemim.cz); TRC: Toronto Research Chemicals (https://www.trc-canada.com/).

g. Purification can be by Sephadex LH-20 or by semi-preparative C18-HPLC using a gradient of acetonitrile-water appropriate for IAA or the substituted form

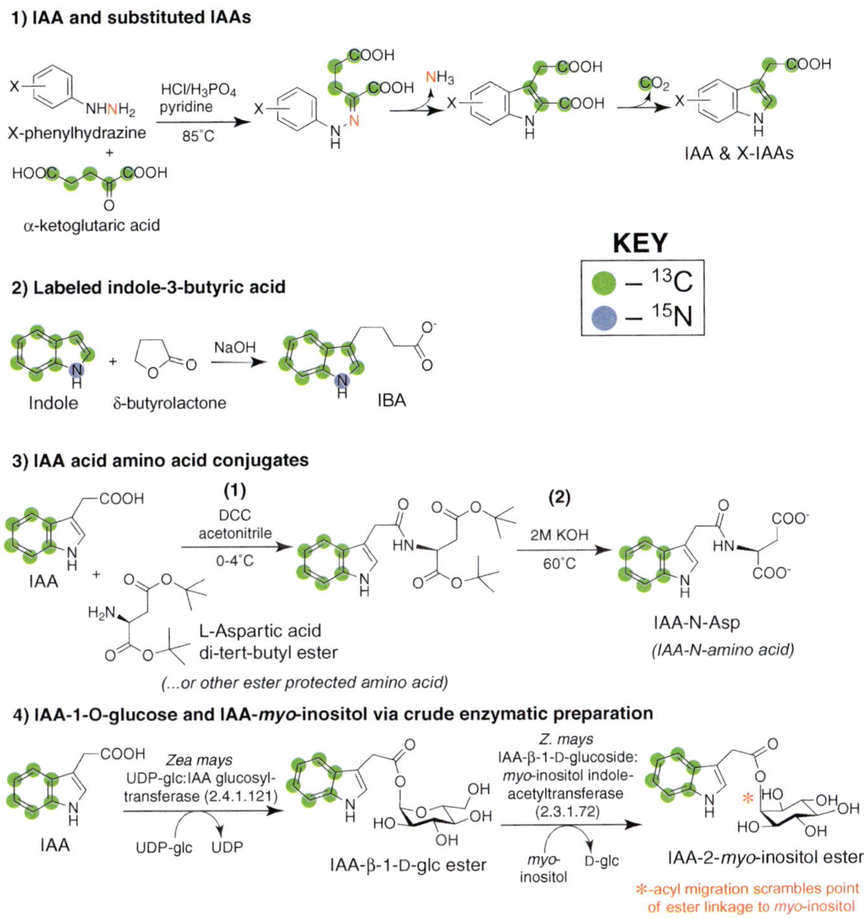

Fig. 2 Schemes for the synthesis of labeled indoles for the quantitative analysis by isotope dilution and for intermediate labeling studies relate to IAA biosynthesis, IAA-IBA metabolic studies and transport, and IAA sugar/cyclitol and amino acid conjugates. (1) IAA and 4-Cl-IAA can be synthesized using fully [^{13}C]-labeled α-ketoglutarate, or alternately, specifically labeled α-ketoglutarate. (2) IBA can be made using a variety of labeled indoles, the most effective is fully labeled [$^{13}C_8$,^{15}N]indole, as shown (see Liu et al., 2012; Sutter & Cohen, 1992). (3) IAA-amino acid conjugates are readily synthesized in high yield by a simple dicyclohexylcarbodiimide (DCC) coupling to an esterified amino acid followed by base hydrolysis. Many different stable-labeled IAA isotopomers can be used and labeled amino acids are also suitable after esterification. (4) The chemical synthesis of IAA-β-1-O-glucose and IAA-myo-inositol are typically difficult to accomplish on a small scale, but both compounds can be formed quite efficiently by an enzyme preparation obtained from a fresh ear of sweet corn using UDP-glucose as the sugar donor. *Step (1) Modified from Baldi, B.G., Slovin, J.P. & Cohen, J.D. (1985). Synthesis of ^{14}C-labeled halogen substituted indole-3-acetic acids.* Journal of Labelled Compounds and Radiopharmaceuticals, 22, *279–285. 10.1002/jlcr.2580220309. Step (3) Modified from Cohen, J.D. (1981). Synthesis of ^{14}C-labeled indole-3-acetylaspartate.* Journal of Labelled Compounds and Radiopharmaceuticals, 18*(9):1393–1396. doi:10.1002/jlcr.2580180916. Step (4) Modified from Iyer, M., Slovin, J.P., Epstein, E. & Cohen, J.D. (2005). Transgenic tomato plants with a modified ability to synthesize indole-3-acetyl-β-1-O-D-glucose.* Journal of Plant Growth Regulation, 24*:142–152. https://doi.org/10.1007/s00344-004-0007-5.*

6.2 Labeled indole-3-butyric acid (IBA)

A slight modification of the method of Cohen and Schulze (1981), described for the synthesis of [^{14}C]-labeled IBA, works well for the production of [^{13}C]IBA (Sutter & Cohen, 1992) now with some modernization of the procedures (Liu et al., 2012).

a. The reaction can be scaled as required, but most commonly it is done at the 12 mg (0.1 mmol) scale. A variety of [^{13}C] labels or [^{15}N]indole can be used, with [$^{13}C_8$,^{15}N]- or [$^{13}C_8$]-indole being particularly useful, but costly

b. Labeled indole, 50 mg (0.1 mmol), is added to a 25 mL Teflon lined hydrothermal autoclave reactor. Freshly broken NaOH pellets (3.2 g) and 6.09 mL of δ-butyrolactone (Sigma) are added to the reactor (note: not all the solids will dissolve)

c. The autoclave reactor is sealed and placed in an appropriately sized beaker-style heating mantle, then slowly brought to a temperature of 220 °C at the rate of 2 °C/min, and then held at that temperature for a total of a 24-h period. It is critical that the start of the reaction proceed with slow gradual heating

d. The reaction is stopped by allowing the reaction container to cool. The product was then dissolved slowly by adding 2- to 5-mL aliquots of distilled water and shaking the mixture on a Vortex mixer each time, typically requiring 50 mL to fully dissolve

e. After the reaction mixture is fully dissolved, it is partitioned twice against 120 mL chloroform. The aqueous phase was then adjusted to pH 2.5 with 6 N HCl and partitioned three times against ethyl acetate. The ethyl acetate extracts are combined

f. The ethyl acetate fractions are dried over anhydrous sodium sulfate for 1 h, filtered, and evaporated to dryness and dissolved in a minimum volume of 50% 2-propanol. This concentrated [$^{13}C_8$,^{15}N]IBA solution can be stored at −20 °C for a few years

g. If a more refined product is desired, then the solution is purified by normal phase on a 4 g column with a Reveleris automated flash chromatography instrument or alternatively by C18 preparative HPLC

h. Final concentration was determined by reverse isotope dilution analysis using a mixture of a known volume of the product solution and a known amount of unlabeled IBA

i. A recent report for the preparation of [$^2H_5$]IAA and [$^2H_5$]IBA via acid-catalyzed hydrogen-deuterium exchange may be a useful method for making these compound for use as internal standards (Yamada, Arai, Kikuchi, & Okamoto, 2021). This facile method appears far simpler than prior routes, but deuterated compounds are problematic for metabolic studies

6.3 Labeled indole-3-acetic acid (IAA) amino acid conjugates

The synthesis of unlabeled lAA amino acids has been accomplished utilizing a variety of methods, but most were found to be unsuccessful for small-scale synthesis of IAA amino acids. We first described the synthesis of radioactive IAA-L-aspartate by dicyclohexylcarbodiimide (DCC) coupling of IAA to an ester of aspartate followed by base hydrolysis (Cohen, 1981). This is also the simplest route to a wide variety of stable isotope labeled IAA amino acids. This procedure can be used to make products with stable labeled IAA, stable labeled amino acids or labels in both. Related procedures, for example using water-soluble carbodiimides, have also appeared (Revelou & Constantinou-Kokotou, 2019) and should be suitable for synthesis of labeled amino acid conjugates.

a. To a 2 mL vial equipped with a teflon-lined screw cap and containing 1.5 mg (6.0 μmol) L-aspartic acid-bis-t-butyl ester HCl (or other amino acid ester, 6.0 μmol) is added 1 mL of an acetonitrile solution of [$^{13}C_6$] IAA (1 mg, 5.5 μmol).

b. After chilling on ice to 0 °C, 2.0 mg (10.0 μmol) of DCC is added

[While usually not necessary, the progress of the reaction can by monitored by thin layer chromatography on Silica Gel 60 (E. Merck) using chloroform: methanol:water (85:14:1) as solvent and Ehmann's reagent for indole detection (Ehmann, 1977)].

c. The reaction was complete after 8 h at 0–4 °C. The acetonitrile is the removed under a stream of N_2 in a 45 °C sand bath

d. The dried residue is washed several times with a minimum volume of methanol (3 × 50 μL) and transferred to a 5 mL Teflon vial and 2 mL of 2 N K0H in 50% methanol/water is added and heated in a sealed heavy-walled vial to 60 °C

e. After 2 h at 60 °C, ester hydrolysis was complete and the mixture was chilled to 4 °C

f. The solution can be acidified to approximately pH 2.0 with phosphoric acid and chromatographed on a 2 × 14 cm column of Sephadex LH-20 using 2-propanol:water (1:1) as solvent

g. Alternatively (Tang, Yu, Tillmann, Cohen, & Slovin, 2019), it is neutralized, reduced in volume to 200 μL and purified by C18 preparative HPLC with distilled water and methanol used as a solvent system with a gradient starting with methanol at 10% and ending at 90% over 20 min interval. [Removal of the t-butyl blocking groups can be monitored by silica gel 60 thin layer chromatography with development in methyl ethyl ketone:ethyl acetate: ethanol:water (3:5:1:1)].

h. When using labeled amino acids most of the available labeled amino acids need to be converted to their ester forms before the above steps can be performed. We typically use diazomethane or diazoethane to accomplish this cleanly (Cohen, 1984) but there are other methods (Anand & Vimal, 1998; Li & Sha, 2008) that should also work well

6.4 Labeled IAA-1-O-glucose and IAA-myo-inositol via crude enzymatic preparation

The reaction catalyzed by IAA-glucose synthase, IAA + UDP-Glucose <====≥IAA-Glucose + UDP, yields a product in which the acyl alkyl acetal bond between IAA and the aldehydic oxygen of glucose is approximately 1.4 kcal above that of the bond between glucose and UDP. This energetically unfavorable reaction is "pulled" in the direction of conjugate formation by the presence in the cell of high levels of UDP-glucose and by the second step of the process, the energetically favored transacylation of the IAA moiety to *myo*-inositol. We can take advantage of these two enzymes to prepare only IAA-glucose or only IAA-*myo*-inositol (Iyer, Slovin, Epstein, & Cohen, 2005), two compounds that are otherwise quite difficult to make on a small scale suitable of isotopic labeling.

$$\text{IAA} - \text{glucose} + \mathit{myo} - \text{inositol} \rightarrow \text{IAA} - \mathit{myo} - \text{inositol} + \text{glucose}$$

6.4.1 Maize endosperm preparation

1. Peel off the corn husk and wash the corn cob under running tap water.
2. Place two 1-L beakers on ice.
3. Use a razor-blade to lacerate the rows of corn kernels.
4. Scrape the corn kernels along the edge of the beaker, and let the liquid endosperm from the lanced kernels run into the beaker.
5. Place a piece of four-layer cheese cloth on the other beaker and pour the pieces of broken kernels as well as the liquid endosperm onto the cheese cloth.

6. Envelop the kernels by the cheese cloth, squeeze the pack of kernels, and allow the liquid to flow into the beaker.
7. The liquid collected is the crude maize endosperm. It can be used immediately for enzyme preparation or stored at −80 °C for further use.

6.4.2 Enzyme preparation

1. Take 5.0 mL of maize endosperm and add (on ice) 5.0 mL of cold 50 mM Tris buffer (pH 7.6).
2. Gradually add $(NH_4)_2SO_4$ to 85% saturation (3.05 g/5 mL). Add while stirring and watching pH. The pH will slowly go down to ~7.0, but a sharp pH change should be avoided.
3. Centrifuge the liquid at 10,000 *g*, 4 °C for 10 min (in five Eppendorf tubes).
4. Pour off the supernatant and re-suspend the precipitated proteins in cold 2 mL 0.01 M Tris buffer (pH 7.1). The re-suspended solution contains the enzyme for synthesis of IAA-glucose and IAA-inositol.

6.4.3 Enzyme synthesis

1. Add 0.3 mL of a solution in 50 mM Tris buffer (pH 7.6) containing 8.3 mM $MgCl_2$ + 8.3 mM glutathione into one vial (1) containing 10 mg UDP-glucose, and/or (2) a vial containing 10 mg UDP-glucose +15 mg *myo*-inositol, respectively.
2. Add 0.5 mL enzyme preparation into each vial.
3. Add 10 μL [$^{13}C_6$]IAA (10 mg/mL, in 50% 2-propanol) into each vial.
4. Gently shake the vials and incubate them at 35 °C for 1.5 h.
5. Stop the reaction by adding 2 mL 2-propanol.
6. Spin at 10,000 *g* for 5 min at 4 °C and decant the solution containing the conjugate from the small protein pellet.

7. Identification and quantitation of unknown IAA-conjugates and other 3-substituted indoles

Both in order to identify IAA-amino acid conjugates and discover new indolic compounds, procedures have been needed that decrease bias against unexpected chemical forms. Each step in a procedure inevitably results in some degree of sample loss, as well as potentially selects against specific classes of compounds. Thus, it was necessary to develop a protocol that minimized such sample preparation steps to decrease compound losses. Key to enable indole specific compound discovery was to use a method that monitored the quinolinium signature ion (*m/z* 130.0651) with the LC-MS because that unique mass is quite specific for two and three substituted indoles (Marx & Djerassi, 1968). To obtain the required specificity necessitated using high mass accuracy such as is possible on a hybrid quadrupole Orbitrap Q Exactive mass spectrometer. This protocol has detected quite unusual chemical forms, such as modified tryptophan (Yu, Hegeman, & Cohen, 2014), that were shown to be present and when novel plant species are investigated (Tang et al., 2019) discovery of other new indoles are likely.

a. Thus, the basic procedure uses only 50 mg or less of plant tissue, and involves only sample extraction, liquid clarification and dilution prior to LC–MS analysis. Dry seed powder or fresh plant material are extracted in 50% 2-propanol (accounting for the water in the plant material; kept at −20 °C) in a 1.5 mL microcentrifuge tube. Dry powered material is constantly vortexed for 30 min while fresh materials are homogenized for 4 min using a bead mill with three stainless steel balls in a tissue homogenizer (Retsch MM300).

b. The sample was then centrifuged at 25,000 *g*, at 25 °C for 20 min, then 100 μL of the supernatant was diluted 10 times with water. The diluted sample was sufficient for 50–100 analytical determinations

c. The extract was analyzed by UPLC on a 2.1 × 100 mm Kinetex XB-C18, 1.7 μm, column (Phenomenex, Torrance, CA, USA) coupled to a high-resolution, accurate-mass (HR/AM) hybrid quadrupole Orbitrap mass spectrometer (Q Exactive, Thermo Scientific, San Jose, CA, USA). Each injection of 20 μL extract was eluted with a solvent gradient formed from solvent A: 0.1% formic acid in water, B: 0.1% formic acid in acetonitrile. Flow rate was 0.2 mL min^{-1}. Gradient was: 0–3.99 min: 3% B, 3.99–4.00 min: 3–16% B, 4–15 min: 16–46.3% B, 15–15.5 min: 46.3–80% B, 15.5–18 min: 80% B. A heated electrospray ionization probe was used to interface the UPLC and MS. Ion source conditions were: spray voltage 3.6 kV, capillary temperature 350 °C, probe heater temperature 300 °C; sheath gas 35 arbitrary units, aux gas eight arbitrary units, S-lens RF level 47. MS was set to acquire one full MS scan followed by eight targeted-MS2 in positive ionization mode. Full MS conditions were: 1 microscan at 70,000 resolution with an automatic gain control target of 1 × 106. Maximum accumulation time was 50 ms. Scan range was from *m/z* 216–370. MS2 conditions were: 1 microscan at 17,500 resolution with automatic gain control target of 2 × 105. Isolation width was 10 *m/z* with multiplexing count of 2. A normalized collision energy of 40% was used with fixed first mass at 60 *m/z*. All data were saved in profile mode

d. Using a Q Exactive mass spectrometer, the quinolinium signature ion ($m/z = 130.0651$) produced from indolic compounds was monitored with high mass accuracy (Yu et al., 2014). The mass spectrometer was set to acquire a limited full scan (m/z 216–370) followed by a series of data independent MS/MS scans. When discovery is focused on IAA amino acid conjugates only a mass range of m/z 216–370 is targeted. Eight data independent scans with a 20 m/z isolation window (segment) each are collected

e. Mass spectra were then extracted for m/z 130.0651 signatures. Segments containing the signatures were then divided into smaller segments and analyzed for the signature ion in subsequent reinjections. This process was continued until unambiguous precursor ions were identified. The exact mass and fragmentation patterns of these precursors were obtained for compound identification

f. To improve further the sensitivity and speed of the analysis of suspected compounds, a targeted MS/MS approach was used. We examined m/z values corresponding to those predicted for each of the IAA amino acid conjugates to determine whether they were present in situ. The compound identifications were based on HPLC retention time and MS spectra as compared with their respective standards. Once the identity of the compounds had been established, similar procedures were employed except that [$^{13}C_6$]-labeled quantitative standards were added to allow isotope dilution quantitative calculations (Barkawi and Cohen, 2010).

8. GC-MS/MS analysis of IAA, indole, and IBA by IDA

A commonly made comparison is between the advantages of LC-MS over GC-MS or in some cases the other way around (He & Aga, 2019; Krone et al., 2010; Perez et al., 2016). We will avoid entering into that comparison because instrumentation improvements continue to make such comparisons quickly outdated. While there are excellent methods for analysis of IAA precursors and related compounds (Liu et al., 2012), recent progress has expanded the range of compounds that can be analyzed by using LC-MS/MS (Tillmann, Tang, & Cohen, 2021). The use of diazomethane to form the methyl esters of IAA and IBA, a method that is essentially quantitative and where gaseous diazomethane leaves behind little residue (Barkawi & Cohen, 2010), and the innate volatility of indole makes GC-MS/MS an extremely sensitive, sub-attomole (below 10^{-18} mol), analytical procedure for these compounds (Fig. 3) when employing the most modern and sensitive equipment. Such procedures have important applications for understanding

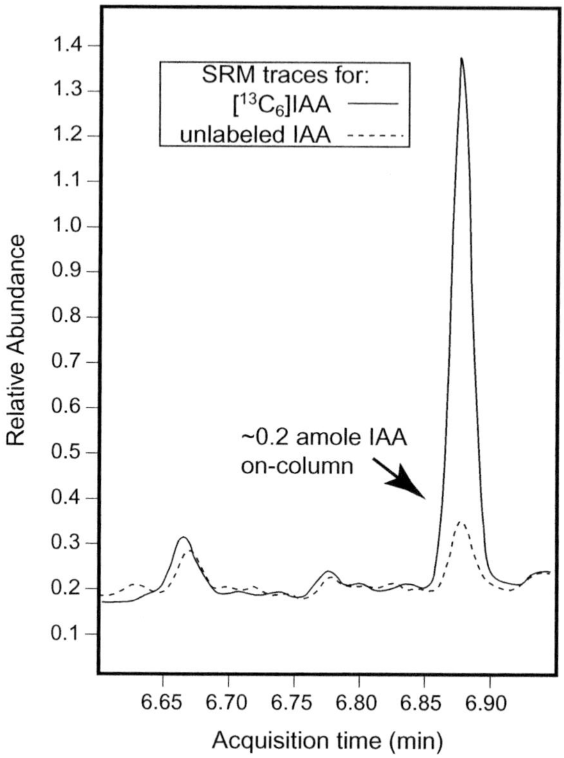

Fig. 3 The analytical sensitivity possible by using selective reaction monitoring (SRM) on a modern GC-triple quadrupole mass spectrometer is shown. The sample from an Arabidopsis seedling was purified as described in the text and analyzed on an Agilent 7010B tandem GC-MS/MS. The sample, 2×10^{-19} mol (0.35 fg) IAA, allows analysis of plant materials with high spatial resolution. The data was supplied by Dr. Molly Tillmann.

the spatial relationships of auxin distribution in small plant samples, especially where developmental events progress such as the development of root or shoot primordia.

a. Collect plant material (typically 2–10 mg/sample) in 0.5-mL microcentrifuge tubes, determine the exact fresh weight (FW), and quickly freeze the tubes in liquid N_2
b. Keep tubes on dry ice or, for longer periods, store them in a −80 °C freezer
c. Add 20 μL/10 mg tissue of homogenization buffer (65% isopropanol and 35% 0.2 M imidazole, pH 7.0) and a 2.4 mm tungsten carbide bead to each tube

d. Add known amounts of internal standard(s) (for example, [$^{13}C_6$]IAA; [$^{13}C_8$,^{15}N]IBA; [$^{13}C_8$,^{15}N]indole) together with the homogenization buffer. To make sure that the detector's response is well within the linear range for both endogenous and internal standard, the amount of internal standards should not exceed 10 × of the endogenous level, or below 1/10 of the endogenous amount. As a guide, usually for 20 mg of plant tissue, the addition of 100–500 pg of each internal standard is appropriate

e. Samples need to be kept on dry ice before being homogenized for 4 min using a bead mill at 20–25 Hz (Retsch MM300). For difficult tissues, the number of tungsten carbide beads can be increased and a higher shaking frequency used

f. After homogenization, leave the plant homogenate on ice for at least 1 h for isotopic equilibration

g. Spin samples a short time to pellet plant debris to the bottom of the tubes and decant

h. Dilute the sample 10 × by adding 180 μL distilled water and mix

i. For analysis of indole: Add 80 μL hexane and vortex

j. For analysis of indole: Spin samples a short time to separate the organic and aqueous layers

k. For analysis of indole: Transfer ~60 μL of the upper organic layer to a glass insert. Place the glass insert into a crimp vial, and cap the vial. This sample contains indole and can be analyzed by GC-MS/MS without derivatization

l. Steps i, j, k and l are designed to extract indole. These steps can be omitted if indole is not a targeted analyte. The samples of indole can be stored in well-sealed vials at 4 °C in the dark for 1 week. Because indole is volatile, it is critical to not let the sample dry

m. To prepare for analysis of the acid auxins, load 20 μL of a 25% aqueous suspension of NH_2 resin (Agilent Bondesil NH_2 bulk sorbent, 40 μm) into empty TopTips (PolyLC, Columbia, MD) that are inserted in adaptors that are placed on top of 2-mL screw capped micro tubes. Pre-wash the TopTips containing NH_2 resin sequentially with 60 μL of hexane, acetonitrile, ethyl acetate, 0.2 M imidazole, followed by three times with 100 μL of distilled water

n. Transfer the TopTips to new 2-mL tubes to collect the flow-through from NH_2 TopTips

o. Spin the diluted samples at 12,000 *g* for 10 min to remove any residual cellular debris

p. Load the supernatant of each sample onto individual TopTips and spin at 2000 *g*. Samples will pass through slowly and within a few minutes completely
q. The flow-through can be discarded
r. Wash the TopTips with 60 μL each of hexane and methanol
s. Transfer the TopTips to new 2-mL tubes, each of which contains 25 μL of 0.1 M succinic acid (SA), pH 6.0
t. Elute IAA and IBA from the TopTips sequentially using 50, 100, and 50 μL 0.25% phosphoric acid (PA)
u. Discard the NH_2 TopTips. Keep the 2-mL tubes containing the acidic eluate
v. Prepare new TopTips by loading 80 μL of a 25% aqueous suspension of polymethylmethacrylate epoxide resin (PMME, Macro-Prep; Bio-Rad, cat. no. 156–0000) into empty TopTips that are inserted in adaptors that are placed on top of 2-mL screw capped micro tubes
w. Prewash the new TopTips containing PMME resin two times with 100 μL methanol, followed by two times with 100 μL PA:SA (5:1, v:v), pH 3.0
x. Load the acidic eluate onto the PMME TopTips
y. Spin TopTip at 2000 *g* to allow liquid to slowly pass through
z. Wash the TopTip three times with 60 μL PA:SA
aa. Wash the TopTips with 10 μL methanol to remove water
bb. Spin the TopTips at 10,000 *g* for 1–2 min
cc. Remove the plastic spring from the bottom of the 250-μL GC glass inserts and put the glass inserts into a 2-mL screw capped micro tubes
dd. Transfer the PMME TopTips to the micro tube with glass inserts
ee. Elute IAA and IBA two times with 50 μL methanol. The eluate contains both IAA and IBA and is collected in the glass inserts in the micro tubes
ff. Discard the PMME TopTips
gg. Use forceps to take the glass inserts out of the micro tubes. Re-install the plastic spring at the bottom of the glass inserts, and put the glass inserts into autosampler crimp vial
hh. Derivatization of IAA and IBA: Add ethereal diazomethane into each glass insert until the insert is full, and incubate for ~5 min. The safe preparation of small quantities of ethereal diazomethane and appropriate safety warnings have been published (Barkawi & Cohen, 2010; Cohen, 1984)

ii. Evaporate the solvents (methanol and diethyl ether) until completely dry using a gentle N_2 gas stream in a sand bath heated to 55 °C in a solvent hood

jj. Add 15 µL of ethyl acetate into each glass insert to re-suspend the methylated IAA and IBA. Cap the vials

9. GC-selected ion monitoring (SIM)-MS/MS analysis of indole; GC-selected reaction monitoring (SRM)-MS/MS analysis of the methyl esters of IAA and IBA

a. The GC was equipped with a split/splitless capillary inlet, and the standard septum was replaced a Merlin Microseal high pressure seal. A 10-µL syringe with a Merlin appropriate needle was installed in the autosampler. An inert quartz liner was used in the injector and the MS was equipped with an electron impact (EI) source with an electron emission of 70 eV. The GC was fitted with an HP-5 ms, 30 m, 0.25 mm diameter, 0.25 µm film thickness (Agilent Technologies, cat. no. 19091S-433UI) column

b. The autosampler was configured for a sample injection volume of 1.0 µL. The GC initial oven temperature was 70 °C with a post-injection hold time of 2 min followed by a 20 °C/min ramp until the temperature reached 280 °C, and then held at 280 °C. The MS total acquire time was 17.50 min

c. SIM is used for indole analysis and only the molecular ions are monitored because few contaminating compounds are present and fragment ions are weak and not diagnostic. For the unlabeled indole, ion *m/z* 117 is monitored; for the [$^{13}C_8$,^{15}N]indole, ion *m/z* 126 is monitored; for other types of labeled indole, the *m/z* of the monitored ion equals the nominal molecular mass

d. GC-SRM-MS/MS analysis of the methyl esters of IAA and IBA is used as the acquisition mode and the transition from the molecular ions to the resulting quinolinium ions are monitored. For unlabeled methyl-IAA, the molecular ion is *m/z* 189 and the product mass is *m/z* 130; for the methyl-[$^{13}C_6$]IAA internal standard, the molecular ion is *m/z* 195 and the product mass is *m/z* 136. For unlabeled methyl-IBA, the molecular ion is *m/z* 217 and the product mass is *m/z* 130, while for methyl-[$^{13}C_8$,^{15}N]IBA internal standard, the molecular ion is *m/z* 226 and the product mass is *m/z* 139

10. Targeted metabolomic analysis

The development of methods utilizing new metabolic inhibitors coupled with a modified approach of isotope dilution/tracing has opened up new possibilities for imaging metabolic events related to auxin metabolism (Tillmann et al., 2021, 2022). Using liquid chromatography-high resolution-mass spectrometry (LC-HR-MS) for qualitative and quantitative analysis of a comprehensive set of IAA precursors and IAA itself enables very informative characterization of auxin biosynthesis in small seedlings such as Arabidopsis. A distinct advantage of this method approach (Fig. 4) is the ability to resolve precursors and related compounds by chromatographic retention, absolute mass and by elemental composition, enabling complex mixtures derived from varied precursors labeled with different stable isotopes (for example, multiple labeled compounds with ^{13}C and ^{15}N can be resolved) to be used in the experimental procedures.

The procedures we developed for studies using Arabidopsis, but should also be applicable to other small seeded plants and, with modification, to larger seeded species. We have used wild-type Columbia-0 ecotype *Arabidopsis thaliana* seeds or specific Arabidopsis metabolic mutant lines. Seeds need to be surface sterilized with 1.2% sodium hypochlorite (soaking in 20% commercial bleach with 20 μL/100 mL Tween 80 for 5 min and then rinsing four times with sterile water). In order to facilitate manipulations needed for chemical treatments, seeds are sown in a single row onto 20 μm nylon mesh covering the agar growth medium (see unlabeled and [^{15}N]-media discussed below). Basically, 20 μm nylon mesh (Sefar, 03–20/14) is cut into 9 cm × 9 cm squares, sterilized in the autoclave, and moisten with sterile water. Sterile forceps are used to place squares flat on top of solid media in a 10 cm square Petri plate. The nylon mesh allows all the seedlings to be moved as a unit from one agar medium to another as required. Once "planted" seeds are then imbibed for 5–10 days at 4 °C to promote uniform germination. Plates are then moved from the cold and placed vertically in a growth chamber with the desired conditions (light, temperature, photoperiod).

For metabolomics studies, two basic methods are employed, depending on the quantitative data required. The first approach is used for "tracer" studies where one or more labeled intermediates are provided to the plants and the movement of label into the pools of specific compounds are determined over time. Quantification is limited to compounds that are available

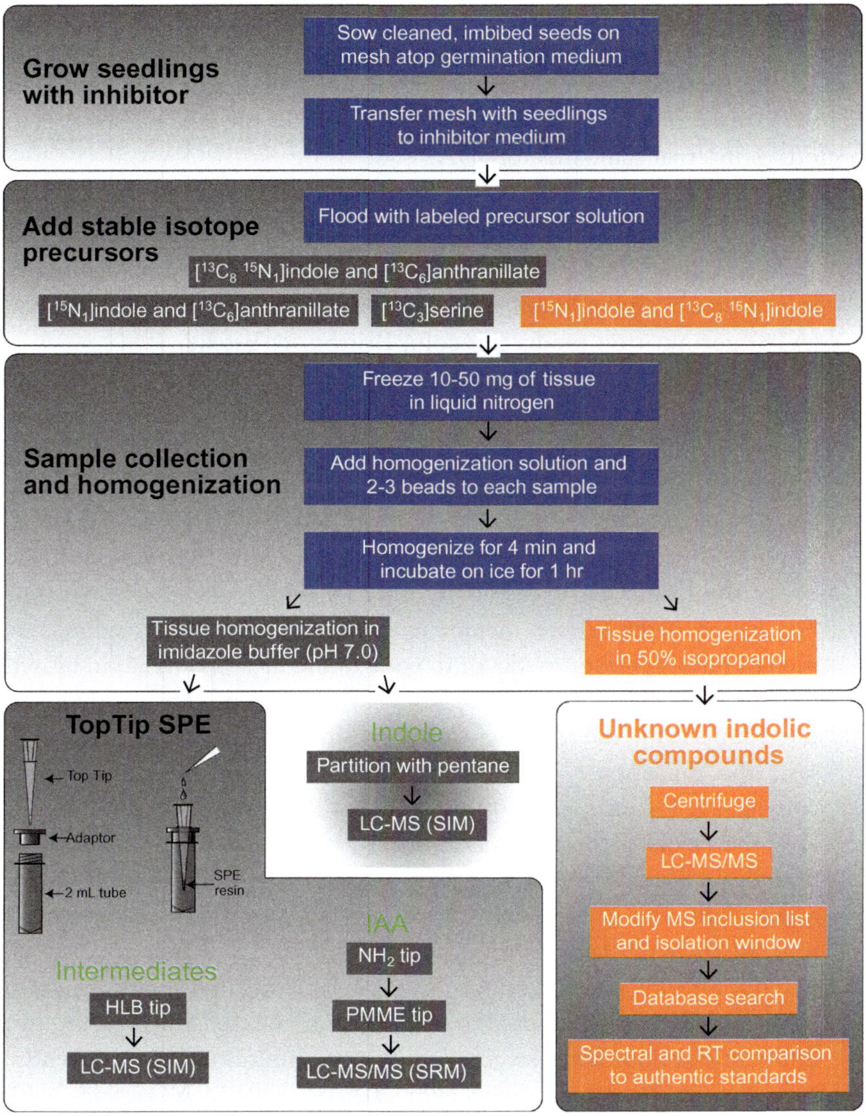

Fig. 4 Summary of the workflow for stable isotope labeling and subsequent analysis of compounds of interest for tracing the auxin metabolic network. As discussed in the text, different labeling and analysis techniques can be employed to investigate different aspects of auxin metabolism. For absolute quantitation using isotope dilution techniques, the internal standards are added to the samples prior to homogenization. *Modified from Tillmann, M.A., Tang, Q. & Cohen, J.D. (2021). Protocol: Analytical methods for visualizing the indolic precursor network leading to auxin biosynthesis.* Plant Methods, 17, 63. *https://doi.org/10.1186/s13007-021-00763-0.*

as labeled standards. A second approach is where each component is measured quantitatively as well as potentially the use of specific labeled precursors. In this method the seedlings are grown on [^{15}N]-media, thus labeling all indolic compounds in the plant and allowing the use of unlabeled compounds as internal standards for IDA.

10.1 Plant growth, labeling and inhibitors

a. Cold and disinfected pretreated seeds are planted, as described above on top of sterile nylon mesh on unlabeled or on [^{15}N]-*A. thaliana* salts (ATS) germination media (Lincoln, Britton, & Estelle, 1990).

b. Unlabeled or labeled media are made with 1% agar and 10% sucrose

c. For the labeled media, the [^{15}N]-salts are substituted to make [^{15}N]ATS: $K^{15}NO_3$ and $Ca(^{15}NO_3)_2$. Arabidopsis seedlings should be grown for 8 days or longer to assure nearly complete [^{15}N]-labeling. If using other plant species, this should be determined by growing plants on [^{15}N]-media. Measure by mass spectrometry the IAA isotopologues over time until IAA is essentially fully labeled

d. For experiments where metabolic inhibitors are used, seedlings are moved onto inhibitor media or to control media to begin auxin biosynthesis studies. In a laminar flow hood, forceps are used to gently lift the mesh supporting the seedlings from germination plates and then laid flat onto the appropriate [^{15}N]-ATS labeled or the unlabeled ATS media containing, for example: 100 μM 5-[2,6-difluorophenyl]-2,4-dihydro-[1,2,4]-triazole-3-thione (YDF; CAS# 1094690-87-5, dissolved in dimethyl sulfoxide (DMSO) and filter sterilized) or solvent control treatment (Tillmann et al., 2022). Plates are covered and placed vertically into the growth chamber for typically 20–30 h

e. Timing for inhibitor treatments, isotopic labeling, and subsequent sample collection can be adapted to study their metabolic transformations at different developmental stages. Significant inhibition of IAA biosynthesis occurs in 12-day seedlings after 20 h on 100 μM YDF and 30 μM PVM2153 media with as short as a 30 min labeling treatment

f. The stable isotopic precursor labeling treatments are begun by flooding the plates with 3 mL of 500 μM [$^{13}C_6$]anthranilate, 500 μM [$^{13}C_8$,^{15}N] indole, 3 mM [$^{13}C_3$]serine, or mixtures of two or more precursors. The plates are gently rocked back and forth 5–10 times to ensure labeling solution covers the entire mesh surface area. The plates are covered and placed flat in the growth chamber under experimental conditions

10.2 Homogenization and extraction

a. Plant tissue samples, typically 20–50 mg, are homogenized for 4 min using a bead mill with three stainless steel balls and 40–100 µL of the extraction buffer/solvent: 65% isopropanol, 35% 0.2 M imidazole (pH 7.0), 100 mM freshly prepared methoxyamine hydrochloride ($CH_3ONH_2{\cdot}HCl$) for targeted analysis of IAA and biosynthesis intermediates.

b. For absolute quantitation from unlabeled medium, stable isotope internal standard(s) are added into the homogenization buffer. The amount of internal standard added to each sample should be similar to the amount of endogenous compound expected in the plant tissue. For reverse isotope dilution quantitation in plants germinated on [^{15}N]ATS media, use unlabeled internal standards as endogenous compounds are [^{15}N]-labeled. Add 50 nM anthranilate, 500 nM indole, 5 µM tryptophan, 1 nM indole-3-acetamide (IAM), 2.5 µM indole-3-acetonitrile (IAN), 100 nM indole-3-pyruvate (IPyA), 10 nM indole-3-acetaldehyde (IAAld), 10 nM IAA, 1 nM tryptamine (TAM), 10 nM indole-3-acetaldoxime (IAOx) (synthesized as described by Ahmad, Eelnurme, & Spenser, 1960) into the homogenization buffer containing 100 mM methoxylamine hydrochloride (see Tillmann et al., 2021).

c. Although for IAA quantification, as in the previous step, unlabeled IAA is added. We also have, instead, in place of the unlabeled compound, added [$^2H_4$]IAA to allow for analysis of endogenous unlabeled IAA resulting from stored conjugate hydrolysis (Tillmann et al., 2022) or IBA conversion to IAA (Strader et al., 2011) remaining from seed reserves. This method also gives the highest sensitivity for IAA measurements

d. Homogenized samples are incubated on ice for ~1 h to allow isotopic or unlabeled standards equilibration with the endogenous compounds in the residual solids. Samples are then diluted tenfold with water, such that ion exchange will be effective, then centrifuged to remove residual solids

e. When using the [$^2H_4$]IAA standard for analysis of only IAA, the sample is loaded onto two consecutive SPE micro spin columns (TopTips).

f. First ion exchange on an amino phase and then on an epoxide resin. Bondesil-NH_2 resin is suspended in water, 1:4 w:v IAA extraction solvents and solutions are as described by Liu et al. (2012): hexane, acetonitrile, ethyl acetate, methanol, 0.2 M imidazole (pH 7.0), distilled water, 0.25% phosphoric acid (PA), 0.1 M succinic acid (SA, pH 6.0), 5:1 PA:SA solution

g. The next step uses a macro-prep epoxide support resin (PMME, Bio-Rad, 156–0000), suspended in 0.1 M sodium bicarbonate (pH 7.0), 1:4 w:v. Steps 14–31 in the GC-MS protocol purifies IAA for LC-MS analysis (this is also detailed in Tillmann et al., 2021)

h. For the extraction of proposed IAA biosynthesis intermediates, anthranilate, serine, IAAld, IPyA, IAOx, IAN, IAM, indole, as well as IAA the homogenized samples are incubated on ice for 50 min to allow isotopic or unlabeled standard equilibration with the endogenous compounds

  1. TopTips are prepared with RENSA hydrophilic-lipophilic balanced (HLB) resin (MIP Technologies) suspended in methanol, 1:5 w:v then prewashed with acetonitrile diluted with glass-distilled water: 100%, 80%, 20%, and 5%.

  2. The sample is diluted 10-fold with glass-distilled water to allow proper interaction with the solid phase, then centrifuged to remove solids

  3. The sample is loaded onto the SPE micro spin column (TopTips) containing the RENSA HLB resin that been conditioned immediately prior to use with acetonitrile followed by 20% acetonitrile in water

  4. After loading, the spin columns are washed with 5% acetonitrile and compounds are eluted with 80% acetonitrile

i. Indole extraction: Indole is a very lipophilic and a somewhat volatile compound that cannot be purified using the techniques used for the other compounds. Thus, its purification involves a simple solvent partitioning of the extract. We found pentane to be well-suited as its boiling point is 36.0 °C, well below the indole melting point of 52.5 °C, and the more polar indolic intermediates do not partition into pentane

j. Transfer 200 μL of the extract supernatant into a new microtube

k. Add 100 μL pentane and vortex mix (avoid warming) for 5 min

l. Spin samples for a few seconds to separate the organic and aqueous phases

m. Transfer the upper organic layer into a new tube. Save the aqueous phase for extraction of IAA or auxin biosynthesis intermediates

n. Evaporate pentane with a vacuum concentrator and suspend the sample in 20 μL of acetonitrile

10.3 LC-MS analysis

UPLC utilizes a column with an end-capped octadecylsilane fully porous 1.8 μm silica resin with high carbon loading (20%) in order to obtain highest sensitivity for indolic compounds. This is likely quite important because in our testing of different columns those with lesser loading give much reduced analytical sensitivity. Although a variety of instrumentation systems may be suitable, and the types and capabilities increases with time and developments, this is our current main configuration in our research laboratories:

a. We typically use amber autosampler vials with 50 μL glass inserts for our small volume samples

b. Liquid chromatograph/mass spectrometer system consisting of a Dionex UltiMate 3000 UHPLC, Q Exactive mass spectrometer running Xcalibur software (Thermo Scientific). The system is fitted with a C_{18} UHPLC column, 50 × 2.1 mm (Force, Restek) with a 0.2 μm precolumn filter (UltraShield, Restek).

c. Mobile phase: A, 0.1% formic acid in water; B, 0.1% formic acid in acetonitrile. Different LC-MS methods are used for specific target compounds:

1. IAA analysis: Mobile phase gradient of 5% B (−1 to 0 min), 5–20% B (0–3 min), 20–80% B (3–6 min), 80% B (6–6.5 min) at a flow rate of 0.4 mL min^{-1}
2. Mass spectra are collected in positive ion mode in a parallel reaction monitoring (PRM) scan and the inclusion list contains ions of *m/z* 176, 177, 178, and 182
3. For analysis of all the listed intermediates except indole, inject 5–10 μL plant extract into the LC system. Mobile phase: 5% B (−2 to 1 min), 5–15% B (1–3 min), 15–30% B (3–3.5 min), 30% B (3.5–5 min), 30–39% B (5–7.5 min), 39–80% B (7.5–8 min), 80% B (8–8.5 min) at a flow rate of 0.4 mL min^{-1}
4. Mass spectra are collected in selected ion monitoring (SIM) mode. MS is set to acquire several segments of mass lists each targeting 1–3 compounds. The segments are: *m/z* 200–217 (0–2.1 min), *m/z* 157–173 (2.1–3 min), *m/z* 133–150 (3–3.74 min), *m/z* 170–188 (3.74–5.4 min), *m/z* 152–170 (5.4–6 min), *m/z* 227–245 (6–6.7 min), *m/z* 184–201 (6.7–8.5 min).
5. Experiments with two differently labeled forms of indole provides a way to identify indole-derived compounds, as downstream intermediates will incorporate both labels. These samples can be analyzed by

repeated injections of the same sample. They are initially scanned broadly for formation of labeled quinolinium ions, and then narrowed in on specific ions in repeated injections until a molecular ion is identified and fragmentation analysis provides required structural information. This same basic strategy can be applied after treatment with multiple differently labeled precursors, again using a series of methods with different MS parameters. An example of this approach is detailed in Tillmann et al. (2021) for experiments simultaneously using [^{15}N]- and [$^{13}C_8$,^{15}N]indole

6. For indole analysis, 5–10 μL plant extract is injected with the following LC mobile phase gradient: 5% B (−1 to 1 min), 5–30% B (1–3 min), 30–39% B (3–5.5 min), 39–80% B (5.5–6.5 min), 80% B (6.5–7 min) at a flow rate of 0.4 mL·min^{-1}. Mass spectra are collected in SIM mode using a mass list of *m/z* 110–132

10.4 Data analysis

a. Unlabeled molecular ion and spectral data should be searched against mass spectral databases to identify potential compound identities and we find the most useful are MassBank, METLIN, NIST Tandem Mass Spectral Library, *m/z* Cloud, MS-DIAL. We also recommend examination of library spectra in more than one library. It is important to check such known spectra because there are published spectra where both the "authentic standard" and isolated compound gave the same incorrect spectral scan

b. Authentic standards should be used to confirm compound identities by comparing retention time and mass spectral data against isolated compounds

c. IAA analysis: Ion chromatograms of labeled and unlabeled quinolinium ions generated by fragmentation of labeled internal standard and endogenous IAA are generated. Narrow mass ranges are used to filter out background noise. Two or more mass ranges are used for IAA analysis: *m/z* 130.0641-130.0661 (corresponding to unlabeled quinolinium ion), and *m/z* 136.0843-136.0863 ([$^{13}C_6$]quinolinium ion produced from the [$^{13}C_6$]IAA internal standard). In some cases other ions, such as *m/z* 131.0612–131.0632 (corresponding to [$^{15}N_1$]quinolinium ion from IAA produced in plants grown on [^{15}N]-salts), are also monitored

d. Depending on the experiment, several other ions from different isotopologues can be acquired, as required. For example, for analysis of a labeling experiment the following set of ions may be used: *m/z* 131.0612–131.0632 (corresponding to [^{15}N]quinolinium ion formed on [^{15}N]-media), *m/z* 134.0892–134.1912(corresponding to [$^{2}H_4$] quinolinium ion from [$^{2}H_4$]IAA internal standard) and *m/z* 136.0843–136.0863 ([$^{13}C_6$]quinolinium ion produced from after stable isotope label incorporation from a [$^{13}C_6$]-labeled precursor).

e. Peaks at the correct *m/z* and retention times are selected and the software is used to calculate peak areas. IAA levels for all isotopomers are calculated using the isotope dilution equation based on the internal standard

f. Targeted indolic precursor analysis: Peak areas from the ion chromatograms from multiple compounds are determined. Mass ranges near the exact masses of ions produced from the compounds of interest, as well as their labeled forms synthesized from the supplied labeled precursors, are kept within a narrow window to exclude background noise

g. The raw data files from the mass spectrometer are converted to mzXML. Quantitative data for each indolic compound is extracted using the Metabolite-Turnover script (https://github.com/HegemanLab/Metabolite-Turnover, Evans, Freund, Sondervan, Cohen, & Hegeman, 2018). In this script, the ProteinTurnover (Fan et al., 2016) and the XCMS package (Smith, Want, O'Maille, Abagyan, & Siuzdak, 2006) are employed to extract the ion chromatograms for each isotopomer of IAA and indolic intermediates. This quantification approach using linear regression (Huttlin, Hegeman, Harms, & Sussman, 2007) is preferred over that using peak area when the MS data has high background noise due to low analyte abundance

h. Exact masses for isotopomers of interest are calculated using the University of Wisconsin—Madison Biological Magnetic Resonance Data Bank exact mass calculator (http://www.bmrb.wisc.edu/metabolomics/mol_mass.php). Isotopomers of proposed IAA biosynthetic intermediates derived from several isotopic labeling strategies are listed in Tillmann et al. (2021)

i. In the data output csv files, the slope of each linear regression line represents the ratio of the respective isotopic trace to its monoisotopomer. This ratio is used to calculate the relative abundance of labeled compounds, allowing the tracking of label incorporation from upstream precursors into IAA intermediates through multiple pathways

11. Exploration and the future

In this review we have outlined methods facilitating the movement toward a comprehensive targeted metabolomics understanding of IAA biosynthesis and protocols for the discovery of new compounds needed to understand IAA metabolism. We have also outlined important techniques that enable in-depth analysis by detailing methods that provide facile approaches for the synthesis for rare, labeled compounds and the use of these and other isotopomers for elucidation of metabolic events.

Because these procedures allow sampling at rapid time intervals, they also enable the measurement of accurate metabolic rates of conversion. Such metabolic flux analysis is a measure of the net flow of matter through a plant's network of metabolic pathways, compartments and tissues and is thus a key metric for understanding biological function at a molecular level. Metabolic flux information describes how signaling messengers are made, how different metabolic pathways may be activated, and how rates of biosynthesis, inactivation and catabolism are used to generate and accumulate products required for growth, development, and differentiation, integrated with responses to environmental conditions. Correspondingly, metabolic flux analysis of the auxin network can provide a direct route for understanding spatial/organism scale flux that is the molecular footing for any number of complex plant behaviors.

While we have described a series of protocols that provides the underpinnings of a substantial array of measurements of compounds that form the auxin metabolic network, there is still progress that will be important to allow an ever-fuller understanding. This has recently been highlighted by reports providing new information linking auxin conjugation with ring oxidation (Guo et al., 2022; Hayashi et al., 2021; Tuominen, Östin, Sandberg, & Sundberg, 1994) such that methods that allow the simultaneous and rapid measurement of labeling, precursor-product relationships, and metabolic flux in these more polar members of the metabolic network become desirable targets of metabolomic analysis by amplifications on the methods presented.

Acknowledgments

The work in the authors' laboratories was supported by NSF MCB 2225057, IOS-PGRP 2025297 and IOS-PGRP 1238812, by grants from USDA National Institute of Food and Agriculture, No. 2018-67011-28056, 2018-67013-27503 and 2019-51181-30025,

Minnesota Department of Agriculture grant CON000000092336_2021 SCBG and by funds from the Minnesota Agricultural Experiment Station and the Gordon and Margaret Bailey Endowment for Environmental Horticulture. We thank Dr. Molly Tillmann, whose work contributed in important ways to the methods described. This paper is dedicated to the memory of Dr. Axel Ehmann whose early vision for quantitative mass spectrometry and his mentoring of the first author will be long remembered.

References

Ahmad, A., Eelnurme, I., & Spenser, I. D. (1960). Indolyl-3-acetaldoxime. *Canadian Journal of Chemistry*, *38*(12), 2523. https://doi.org/10.1139/v60-342.

Anand, R. C., & Vimal. (1998). A mild and convenient procedure for the esterification of amino acids. *Synthetic Communications*, *28*(11), 1963–1965. https://doi.org/10.1080/00397919808007170.

Baldi, B. G., Maher, B. R., & Cohen, J. D. (1989). Hydrolysis of indole 3 acetic acid esters exposed to mild alkaline conditions. *Plant Physiology*, *91*, 9–12. https://doi.org/10.1104/pp.91.1.9.

Baldi, B. G., Slovin, J. P., & Cohen, J. D. (1985). Synthesis of ^{14}C-labeled halogen substituted indole-3-acetic acids. *Journal of Labelled Compounds and Radiopharmaceuticals*, *22*, 279–285. https://doi.org/10.1002/jlcr.2580220309.

Bandurski, R. S. (1980). Homeostatic control of concentrations of indole-3-acetic acid. In F. Skoog (Ed.), *Plant growth substances 1979* (pp. 37–49). Berlin: Springer-Verlag. https://doi.org/10.1007/978-3-642-67720-5_3.

Bandurski, R. S., & Schulze, A. (1977). Concentration of indole-3-acetic acid and its derivatives in plants. *Plant Physiology*, *60*, 211–213. https://doi.org/10.1104/pp.60.2.211.

Barkawi, L. S., & Cohen, J. D. (2010). A method for concurrent diazomethane synthesis and substrate methylation in a 96-sample format. *Nature Protocols*, *5*, 1619–1626. https://doi.org/10.1038/nprot.2010.119.

Barkawi, L. S., Tam, Y. Y., Tillman, J. A., Pederson, B., Calio, J., Al-Amier, H., et al. (2008). A high-throughput method for the quantitative analysis of indole-3-acetic acid and other auxins from plant tissue. *Analytical Biochemistry*, *372*, 177–188. https://doi.org/10.1016/j.ab.2007.08.009.

Bialek, K., & Cohen, J. D. (1989). Quantitation of indoleacetic acid conjugates in bean seeds by direct tissue hydrolysis. *Plant Physiology*, *90*, 398–400. https://doi.org/10.1104/pp.90.2.398.

Bohn-Courseau, I. (2010). Auxin: A major regulator of organogenesis. *Comptes Rendus Biologies*, *333*(4), 290–296. https://doi.org/10.1016/j.crvi.2010.01.004.

Bou-Torrent, J., Galstyan, A., Gallemí, M., Cifuentes-Esquivel, N., Molina-Contreras, M. J., Salla-Martret, M., et al. (2014). Plant proximity perception dynamically modulates hormone levels and sensitivity in Arabidopsis. *Journal of Experimental Botany*, *65*(11), 2937–2947. https://doi.org/10.1093/jxb/eru083.

Caruso, J. L., Smith, R. G., Smith, L. M., Cheng, T. Y., & Daves, G. D. (1978). Determination of indole-3-acetic acid in Douglas fir using a deuterated analog and selected ion monitoring: Comparison of microquantities in seedling and adult tree. *Plant Physiology*, *62*(6), 841–845. https://doi.org/10.1104/pp.62.6.841.

Caruso, J. L., & Zeisler, C. S. (1983). Indole-3-acetic acid in Douglas fir seedlings: A reappraisal. *Phytochemistry*, *22*(2), 589–590. https://doi.org/10.1016/0031-9422(83)83053-2.

Chen, K.-H., Miller, A. N., Patterson, G. W., & Cohen, J. D. (1988). A rapid and simple procedure for purification of indole-3-acetic acid prior to GC-SIM-MS analysis. *Plant Physiology*, *86*(3), 822–825. https://doi.org/10.1104/pp.86.3.822.

Ciesielski, T. (1872). Untersuchungen über die Abwärtskrümmung der Wurzel. *Beitr Biol Pflanzen*, 1–30. *1*(1870-5) Heft 2 https://www.zobodat.at/pdf/Beitr-Biologie-Pflanzen_1_2_0002-0030.pdf.

Cohen, J. D. (1981). Synthesis of ^{14}C-labeled indole-3-acetylaspartate. *Journal of Labelled Compounds and Radiopharmaceuticals*, *18*(9), 1393–1396. https://doi.org/10.1002/jlcr.2580180916.

Cohen, J. D. (1984). Convenient apparatus for the generation of small amounts of diazomethane. *Journal of Chromatography*, *303*(193), 196. https://doi.org/10.1016/S0021-9673(01)96061-3.

Cohen, J. D., Baldi, B. G., & Slovin, J. P. (1986). $^{13}C_6$ [Benzene ring] indole-3-acetic acid: A new internal standard for quantitative mass spectral analysis of indole-3-acetic acid in plants. *Plant Physiology*, *80*, 14–19. https://doi.org/10.1104/pp.80.1.14.

Cohen, J. D., & Gray, W. M. (2006). Auxin metabolism and signaling. In P. Hedden, & S. Thomas (Eds.), *Annual plant reviews: Vol. 24. Plant hormone signaling* (pp. 37–66). Oxford: Blackwell Publishing. https://doi.org/10.1002/9780470988800.ch2.

Cohen, J. D., & Schulze, A. (1981). Double standard isotope dilution assay I. Quantitative assay of indole 3 acetic acid. *Analytical Biochemistry*, *112*, 249–257. https://doi.org/10.1016/0003-2697(81)90290-6.

Cooke, T. J., Poli, D. B., Sztein, A. E., & Cohen, J. D. (2002). Evolutionary patterns in auxin action. *Plant Molecular Biology*, *49*, 319–338. https://doi.org/10.1023/A:1015242627321.

Das, S., Weijers, D., & Borst, J. W. (2021). Auxin response by the numbers. *Trends in Plant Science*, *26*(5), 442–451. https://doi.org/10.1016/j.tplants.2020.12.017.

Davison, A. S., Milan, A. M., & Dutton, J. J. (2013). Potential problems with using deuterated internal standards for liquid chromatography-tandem mass spectrometry. *Annals of Clinical Biochemistry*, *50*(3), 274. https://doi.org/10.1177/0004563213478938.

Ehmann, A. (1977). The Van Urk-Salkowski reagent—a sensitive and specific chromogenic reagent for silica gel thin-layer chromatographic detection and identification of indole derivatives. *Journal of Chromatography A*, *132*(2), 267–276. https://doi.org/10.1016/S0021-9673(00)89300-0.

Evans, E. M., Freund, D. M., Sondervan, V. M., Cohen, J. D., & Hegeman, A. D. (2018). Metabolic patterns in *Spirodela polyrhiza* revealed by ^{15}N stable isotope labeling of amino acids in photoautotrophic, heterotrophic, and mixotrophic growth conditions. *Frontiers in Chemistry*, *6*, 191. https://doi.org/10.3389/fchem.2018.00191.

Fan, K.-T., Rendahl, A., Chen, W.-P., Freund, D., Gray, W., Cohen, J. D., et al. (2016). Proteome scale-protein turnover analysis using high resolution mass spectrometric data from stable-isotope labeled plants. *Journal of Proteome Research*, *15*, 851–867. https://doi.org/10.1021/acs.jproteome.5b00772.

Farcot, E., Lavedrine, C., & Vernoux, T. (2015). A modular analysis of the auxin signalling network. *PLoS One*, *10*(3), e0122231. https://doi.org/10.1371/journal.pone.0122231.

Gélinas-Marion, A., Nichols, D. S., & Ross, J. J. (2020). Conversion of unstable compounds can contribute to the auxin pool during sample preparation. *Plant Physiology*, *183*(4), 1432–1434. https://doi.org/10.1104/pp.20.00251.

Gorter, C. J. (1932). *Groeistofproblemen bij Wortels*. 123 pp. Diss. Utrecht: Kemink. https://objects.library.uu.nl/reader/ftxt.php?obj=1874-308730&format=1&pagenum=1.

Guo, R., Hu, Y., Aoi, Y., Hira, H., Ge, C., Dai, X., et al. (2022). Local conjugation of auxin by the GH3 amido synthetases is required for normal development of roots and flowers in Arabidopsis. *Biochemical and Biophysical Research Communications*, *589*, 16–22. https://doi.org/10.1016/j.bbrc.2021.11.109.

Hayashi, K., Arai, K., Aoi, Y., Tanaka, Y., Hira, H., Guo, R., et al. (2021). The main oxidative inactivation pathway of the plant hormone auxin. *Nature Communications*, *12*, 6752. https://doi.org/10.1038/s41467-021-27020-1.

Hayashi, K., Neve, J., Hirose, M., Kuboki, A., Shimada, Y., Kepinski, S., et al. (2012). Rational design of an auxin antagonist of the SCF(TIR1) auxin receptor complex. *ACS Chemical Biology*, 7(3), 590–598. https://doi.org/10.1021/cb200404c.

He, P., & Aga, D. S. (2019). Comparison of GC-MS/MS and LC-MS/MS for the analysis of hormones and pesticides in surface waters: Advantages and pitfalls. *Analytical Methods*, *11*, 1436–1448. https://doi.org/10.1039/C8AY02774A.

Hevesy, G. V., & Paneth, F. (1913). Die Löslichkeit des Bleisulfids und Bleichromats. *Zeitschrift für Anorganische und Allgemeine Chemie*, *82*(1), 323–328. https://doi.org/10.1002/zaac.19130820125.

Huttlin, E. L., Hegeman, A. D., Harms, A. C., & Sussman, M. R. (2007). Comparison of full versus partial metabolic labeling for quantitative proteomics analysis in Arabidopsis thaliana. *Molecular & Cellular Proteomics*, *6*(5), 860–881. https://doi.org/10.1074/mcp.m600347-mcp200.

Iglesias, M. J., Sellaro, R., Zurbriggen, M. D., & Casal, J. J. (2018). Multiple links between shade avoidance and auxin networks. *Journal of Experimental Botany*, *69*(2), 213–228. https://doi.org/10.1093/jxb/erx295.

Iino, M., & Carr, D. J. (1982). Sources of free IAA in the mesocotyl of etiolated maize seedlings. *Plant Physiology*, *69*(5), 1109–1112. https://doi.org/10.1104/pp.69.5.1109.

Ilić, N., Normanly, J., & Cohen, J. D. (1996). Quantification of free plus conjugated indole-3-acetic acid in Arabidopsis requires correction for the non-enzymatic conversion of indolic nitriles. *Plant Physiology*, *111*, 781–788. https://doi.org/10.1104/pp.111.3.781.

Iyer, M., Slovin, J. P., Epstein, E., & Cohen, J. D. (2005). Transgenic tomato plants with a modified ability to synthesize indole-3-acetyl-β-1-O-D-glucose. *Journal of Plant Growth Regulation*, *24*, 142–152. https://doi.org/10.1007/s00344-004-0007-5.

Jensen, P. J., & Bandurski, R. S. (1996). Incorporation of deuterium into indole-3-acetic acid and tryptophan in *Zea mays* seedlings grown on 30% deuterium oxide. *Journal of Plant Physiology*, *147*(6), 697–702. https://doi.org/10.1016/S0176-1617(11)81480-X.

Kai, K., Nakamura, S., Wakasa, K., & Miyagawa, H. (2007). Facile preparation of deuterium-labeled standards of indole-3-acetic acid (IAA) and its metabolites to quantitatively analyze the disposition of exogenous IAA in *Arabidopsis thaliana*. *Bioscience, Biotechnology, and Biochemistry*, *71*(8), 1946–1954. https://doi.org/10.1271/bbb.70151.

Kasahara, H. (2016). Current aspects of auxin biosynthesis in plants. *Bioscience, Biotechnology, and Biochemistry*, *80*(1), 34–42. https://doi.org/10.1080/09168451.2015.

Krone, N., Hughes, B. A., Lavery, G. G., Stewart, P. M., Arlt, W., & Shackleton, C. H. L. (2010). Gas chromatography/mass spectrometry (GC/MS) remains a pre-eminent discovery tool in clinical steroid investigations even in the era of fast liquid chromatography tandem mass spectrometry (LC/MS/MS). *The Journal of Steroid Biochemistry and Molecular Biology*, *121*, 496–504. https://doi.org/10.1016/j.jsbmb.2010.04.010.

Li, J., & Sha, Y. (2008). A convenient synthesis of amino acid methyl esters. *Molecules*, *13*(5), 1111–1119. https://doi.org/10.3390/molecules13051111.

Li, W., Zhou, Y., Liu, X., Yu, P., Cohen, J. D., & Meyerowitz, E. M. (2013). LEAFY controls auxin response pathways in floral primordia formation. *Science Signaling*, *6*, ra23. https://doi.org/10.1126/scisignal.2003937.

Lincoln, C., Britton, J. H., & Estelle, M. (1990). Growth and development of the axr1 mutants of Arabidopsis. *Plant Cell*, *2*(11), 1071–1080. https://doi.org/10.1105/tpc.2.11.1071.

Little, C. H. A., Heald, J. K., & Browning, G. (1978). Identification and measurement of indoleacetic and abscisic acids in the cambial region of *Picea sitchensis* (Bong.) Carr. by combined gas chromatography-mass spectrometry. *Planta*, *139*, 133–138. https://doi.org/10.1007/BF00387138.

Liu, X., Hegeman, A. D., Gardner, G., & Cohen, J. D. (2012). Protocol: High-throughput and quantitative assays of auxin and auxin precursors from minute tissue samples. *Plant Methods*, *8*, 31. https://doi.org/10.1186/1746-4811-8-31.

Ljung, K., Hull, A. K., Kowalczyk, M., Marchant, A., Celenza, J., Cohen, J. D., et al. (2002). Biosynthesis, conjugation, catabolism and homeostasis of indole-3-acetic acid in *Arabidopsis thaliana*. *Plant Molecular Biology*, *50*, 309–332. https://doi.org/10.1023/a:1016024017872.

Magnus, V., Bandurski, R. S., & Schulze, A. (1980). Synthesis of 4, 5, 6, 7 and 2, 4, 5, 6, 7 deuterium-labeled indole-3-acetic acid for use in mass spectrometric assays. *Plant Physiology*, *66*(4), 775–781. https://doi.org/10.1104/pp.66.4.775.

Marx, M., & Djerassi, C. (1968). Mass spectrometry in structural and stereochemical problems. CXLIX. question of ring expansion in the fragmentation of carbon-13 nitrogen heterocycles. *Journal of the American Chemical Society*, *90*, 678–681. https://doi.org/10.1021/ja01005a022.

Mashiguchi, K., Tanaka, K., Sakai, T., Sugawara, S., Kawaide, H., Natsume, M., et al. (2011). The main auxin biosynthesis pathway in Arabidopsis. *Proceedings of the National Academy of Sciences of the United States of America*, *108*(45), 18512–18517. https://doi.org/10.1073/pnas.1108434108.

McDougall, J., & Hillman, J. R. (1978). Analysis of indole-3-acetic acid using GC-MS techniques. In J. R. Hillman (Ed.), *Society for experimental biology seminar series: Vol. 4. Isolation of plant growth substances* (pp. 1–25). London: Cambridge University Press.

Menikarachchi, L. C., Hamdalla, M. A., Hill, D. W., & Grant, D. F. (2013). Chemical structure identification in metabolomics: Computational modeling of experimental features. *Computational and Structural Biotechnology Journal*, *5*(6), e201302005. https://doi.org/10.5936/csbj.201302005.

Pagliano, E., Mester, Z., & Meija, J. (2015). Calibration graphs in isotope dilution mass spectrometry. *Analytica Chimica Acta*, *896*, 63–67. https://doi.org/10.1016/j.aca.2015.09.020.

Paque, S., & Weijers, D. (2016). Q&A: Auxin: The plant molecule that influences almost anything. *BMC Biology*, *14*(1), 67. https://doi.org/10.1186/s12915-016-0291-0.

Perez, E. R., Knapp, J. A., Horn, C. K., Stillman, S. L., Evans, J. E., & Arfsten, D. P. (2016). Comparison of LC-MS-MS and GC-MS analysis of benzodiazepine compounds included in the drug demand reduction urinalysis program. *Journal of Analytical Toxicology*, *40*(3), 201–207. https://doi.org/10.1093/jat/bkv140.

Pilet, P. E., & Saugy, M. (1987). Effect on root growth of endogenous and applied IAA and ABA: A critical reexamination. *Plant Physiology*, *83*(1), 33–38. https://doi.org/10.1104/pp.83.1.33.

Previs, S. F., & Downes, D. P. (2020). Key concepts surrounding studies of stable isotope-resolved metabolomics. *Methods in Molecular Biology*, *2104*, 99–120. https://doi.org/10.1007/978-1-0716-0239-3_6.

Radoeva, T., Lokerse, A. S., Llavata-Peris, C. I., Wendrich, J. R., Xiang, D., Liao, C.-Y., et al. (2019). A robust auxin response network controls embryo and suspensor development through a basic helix loop helix transcriptional module. *Plant Cell*, *31*(1), 52–67. https://doi.org/10.1105/tpc.18.00518.

Revelou, P.-K., & Constantinou-Kokotou, V. (2019). Preparation of synthetic auxin-amino acid conjugates. *Synthetic Communications*, *49*(13), 1708–1712. https://doi.org/10.1080/00397911.2019.1605446.

Rittenberg, D., & Foster, G. L. (1940). A new procedure for quantitative analysis by isotope dilution, with application to the determination of amino acids and fatty acids. *The Journal of Biological Chemistry*, *133*, 737–744. https://doi.org/10.1016/S0021-9258(18)73304-8.

Rosquete, M. R., Barbez, E., & Kleine-Vehn, J. (2012). Cellular auxin homeostasis: Gatekeeping is housekeeping. *Molecular Plant*, *5*(4), 772–786. https://doi.org/10.1093/mp/ssr109.

Saini, S., Sharma, I., Kaur, N., & Pati, P. K. (2013). Auxin: A master regulator in plant root development. *Plant Cell Reports*, *32*(6), 741–757. https://doi.org/10.1007/s00299-013-1430-5.

Schilling, C. H., Schuster, S., Palsson, B. O., & Heinrich, R. (1999). Metabolic pathway analysis: Basic concepts and scientific applications in the post-genomic era. *Biotechnology Progress*, *15*(3), 296–303. https://doi.org/10.1021/bp990048k.

Smith, C. A., Want, E. J., O'Maille, G., Abagyan, R., & Siuzdak, G. (2006). XCMS: Processing mass spectrometry data for metabolite profiling using nonlinear peak alignment, matching, and identification. *Analytical Chemistry*, *78*(3), 779–787. https://doi.org/10.1021/ac051437y.

Strader, L. C., Wheeler, D. L., Christensen, S. E., Berens, J. C., Cohen, J. D., Rampey, R. A., et al. (2011). Multiple facets of Arabidopsis seedling development require indole-3-butyric acid-derived auxin. *Plant Cell*, *23*, 984–999. https://doi.org/10.1105/tpc.111.083071.

Sundberg, B. (1990). Influence of extraction solvent (buffer, methanol, acetone) and time on the quantification of indole-3-acetic acid in plants. *Physiologia Plantarum*, *78*(2), 293–297. https://doi.org/10.1111/j.1399-3054.1990.tb02094.x.

Sutter, E. G., & Cohen, J. D. (1992). Measurement of indolebutyric acid in plant tissues by isotope dilution gas chromatography mass spectrometry analysis. *Plant Physiology*, *99*(1719), 1722. https://doi.org/10.1104/pp.99.4.1719.

Tang, Q., Yu, P., Tillmann, M., Cohen, J. D., & Slovin, J. P. (2019). Indole-3-acetylaspartate and indole-3-acetylglutamate, the IAA-amide conjugates in the diploid strawberry achene, are hydrolyzed in growing seedlings. *Planta*, *249*(4), 1073–1085. https://doi.org/10.1007/s00425-018-3061-0.

Tillmann, M. A., Tang, Q., & Cohen, J. D. (2021). Protocol: Analytical methods for visualizing the indolic precursor network leading to auxin biosynthesis. *Plant Methods*, *17*, 63. https://doi.org/10.1186/s13007-021-00763-0.

Tillmann, M. A., Tang, Q., Gardner, G., & Cohen, J. D. (2022). Complexity of the auxin biosynthetic network in Arabidopsis hypocotyls is revealed by multiple stable-labeled precursors. *Phytochemistry*, *200*, 113219. https://doi.org/10.1016/j.phytochem.2022.113219.

Tivendale, N. D., & Cohen, J. D. (2015). Analytical history of auxin. *Journal of Plant Growth Regulation*, *34*, 708–722. https://doi.org/10.1007/s00344-015-9519-4.

Tivendale, N. D., & Millar, A. H. (2022). How is auxin linked with cellular energy pathways to promote growth? *New Phytologist*, *233*(6), 2397–2404. https://doi.org/10.1111/nph.17946.

Truskina, J., Han, J., Chrysanthou, E., Galvan-Ampudia, C. S., Lainé, S., Brunoud, G., et al. (2021). A network of transcriptional repressors modulates auxin responses. *Nature*, *589*(7840), 116–119. https://doi.org/10.1038/s41586-020-2940-2.

Tuominen, H., Östin, A., Sandberg, G., & Sundberg, B. (1994). A novel metabolic pathway for indole-3-acetic acid in apical shoots of *Populus tremula* (L.) × *Populus tremuloides* (Michx.). *Plant Physiology*, *106*, 1511–1520. https://doi.org/10.1104/pp.106.4.1511.

Turian, G., & Hamilton, R. H. (1960). Chemical detection of 3-indolylacetic acid in *Ustilago zeae* tumors. *Biochimica et Biophysica Acta*, *41*, 148–150. https://doi.org/10.1016/0006-3002(60)90381-4.

Weyers, J. D. B., & Paterson, N. W. (2002). Plant hormones and the control of physiological processes. *New Phytologist*, *152*, 375–407. https://doi.org/10.1046/j.0028-646X.2001.00281.x.

Xu, Y., Freund, D., Hegeman, A., & Cohen, J. D. (2022). Metabolic signatures of Arabidopsis thaliana abiotic stress responses elucidate patterns in stress priming, acclimation and recovery. *Stress Biology*, *2*, 11. https://doi.org/10.1007/s44154-022-00034-5.

Yamada, T., Arai, K., Kikuchi, R., & Okamoto, S. (2021). Deuteration of indole compounds: Synthesis of deuterated auxins, indole-3-acetic acid-d5 and indole-3-butyric acid-d5. *ACS Omega*, *6*(30), 19956–19963. https://doi.org/10.1021/acsomega.1c02940.

Yu, P., Hegeman, A. D., & Cohen, J. D. (2014). A facile means for the identification of indolic compounds from plant tissues. *Plant Journal*, *79*, 1065–1075. https://doi.org/10.1111/tpj.12607.

Yu, P., Lor, P., Ludwig-Müller, J., Hegeman, A. D., & Cohen, J. D. (2015). Quantitative evaluation of IAA conjugate pools in *Arabidopsis thaliana*. *Planta*, *241*, 539–548. https://doi.org/10.1007/s00425-014-2206-z.

CHAPTER TEN

Tag you're it: Application of stable isotope labeling and LC-MS to identify the precursors of specialized metabolites in plants

Jeffrey Simpson[a,b] and **Clint Chapple**[a,b,*]
[a]Department of Biochemistry, Purdue University, West Lafayette, IN, United States
[b]Purdue Center for Plant Biology, Purdue University, West Lafayette, IN, United States
*Corresponding author: e-mail address: chapple@purdue.edu

Contents

Abstract

Untargeted liquid chromatography/mass spectrometry (LC-MS) can contribute a comprehensive and unbiased picture of the metabolic space of plants. These data can be used to quantify natural metabolite variation for genome wide association studies, to compare global metabolic responses from environmental or genetic perturbations, and to identify previously undescribed metabolites in Nature. A major limitation with untargeted metabolomics is the classification and identification of the thousands of metabolite features that can be detected in a single analytical run. Isotopic labeling improves the informational value of these datasets by categorizing metabolites as being

Methods in Enzymology, Volume 676
ISSN 0076-6879
https://doi.org/10.1016/bs.mie.2022.07.039

derived from specific upstream precursors and/or to known metabolic pathways. When a ^{13}C-labeled precursor is fed to either a plant or tissue, the downstream metabolites produced from it have a higher *m/z* value than the molecules in the pre-existing pool, generating an *m/z* peak pair that can be specifically identified within the MS data. This paper outlines methods and principles to consider when supplementing untargeted MS data with isotopic labeling, including how to choose the appropriate isotopic label, grow and feed plant tissues to maximize label uptake and incorporation into derivatives, optimize LC-MS methods, and interpret the resulting labeling data. Although the focus here is on annotation of amino acid-derived metabolites using LC-MS, we anticipate that the methods are generally adaptable to other precursors, plant species, and chromatographic approaches.

1. Introduction

Plants synthesize a diverse array of specialized metabolites that are critical to their survival and adaptation. At the same time, many natural products are of value to humans as pharmaceuticals, as sources of flavors and aromas, for their nutritional and caloric value, and as industrial products. The detection of new natural products and identification of the pathways of their biosynthesis and degradation is necessary to guide the improvement of output traits in crops, drug discovery, and to fuel innovation in synthetic biology. To make advances in compound and gene identification, it is necessary to capture and annotate the metabolic space of plants. Although current technologies can achieve the former with relative ease, the subsequent annotation of metabolites remains a bottleneck to swift progress.

Untargeted metabolomics can document the presence of most metabolites in plants without any prior knowledge of their existence or identity (Alseekh & Fernie, 2018; De Vos et al., 2007). In the most common applications, plant soluble metabolite extracts are separated by either liquid or gas chromatography (LC or GC, respectively), and then a mass spectrometer (MS) is used to determine the mass to charge (*m/z*) values and abundance of the metabolites and their fragments. Untargeted metabolomics protocols can be adapted to capture metabolites with specific properties by targeting specific tissues, by choice of extraction solvent polarity, through pre-chromatographic purification (e.g., solid phase extraction), and by selection of LC or GC column chemistry and separation method, as well as MS polarity (Perez de Souza, Alseekh, Naake, & Fernie, 2019). The data collected by untargeted MS can document metabolomic differences between genotypes, species, and responses to experimental treatments or environmental changes (Allwood et al., 2021; Boutet et al., 2022; Mannochio-Russo et al., 2022; Oh, Park, Kim, Choi, & Kim, 2021; Tohge et al., 2016; Zhou et al., 2019).

Unfortunately, *m/z*, retention time and ion abundance information alone do not result in unambiguous structural annotations of metabolites because many different metabolites have similar or even identical masses. The output also contains fragments of metabolites, generally termed metabolite features, due to soft in-source fragmentation and adduct formation that are inherent to MS-analysis, which can pose additional challenges in interpretation. Although we can speculate on compound identities based on prior information, familiarity with the metabolites typically found in a specific organism, or the analysis of verified standards, most features from an untargeted MS analysis are essentially "unknowns." The lack of annotation prevents us from understanding possible relationships between metabolites and assigning putative structures.

The interpretation of an untargeted metabolomic dataset can be enhanced by information on the biochemical pathway or precursor of origin of the MS features it contains. UV absorption spectra collected during LC can in some cases tentatively assign metabolites to a biochemical pathway, or even branches of a pathway. For example, using such spectra, phenylalanine- and tryptophan-derived molecules can often be annotated as such, even if we do not know their exact chemical identity (Hemm, Ruegger, & Chapple, 2003; Zhou et al., 2019). In contrast, many plant metabolites do not have distinctive UV absorption spectra, and interpretation of the UV-spectrum of a single metabolite can be confounded by co-eluting metabolites. A related, but more sensitive, targeted, and universal approach is to identify the precursor of origin for metabolites by labeling. In this method, a plant tissue is fed a metabolite that contains a heavy isotope on one or more of its atoms. This approach was pioneered using the radioactive isotopes ^{14}C and ^{3}H (Benson et al., 1950; Brown & Neish, 1955, 1956). The newly produced metabolites derived from the precursor will be radioactive and are detected and quantified by chromatography coupled with scintillation counting or autoradiography. Although radioactive labeling is very sensitive, it provides no further information on metabolite structures other than precursor of origin. In contrast, when stable isotope (e.g., ^{13}C) labeling is combined with MS-detection, labeled metabolites can be identified because they have a mass difference corresponding to the number of heavy atoms that was incorporated from the label (Fig. 1) (Feldberg, Venger, Malitsky, Rogachev, & Aharoni, 2009; Huang et al., 2014; Weng, Li, Mo, & Chapple, 2012). With that information, the precursor of the metabolite is identified, and the *m/z* value of the co-chromatographing unlabeled metabolite can be used to predict the identity of the metabolite with greater confidence.

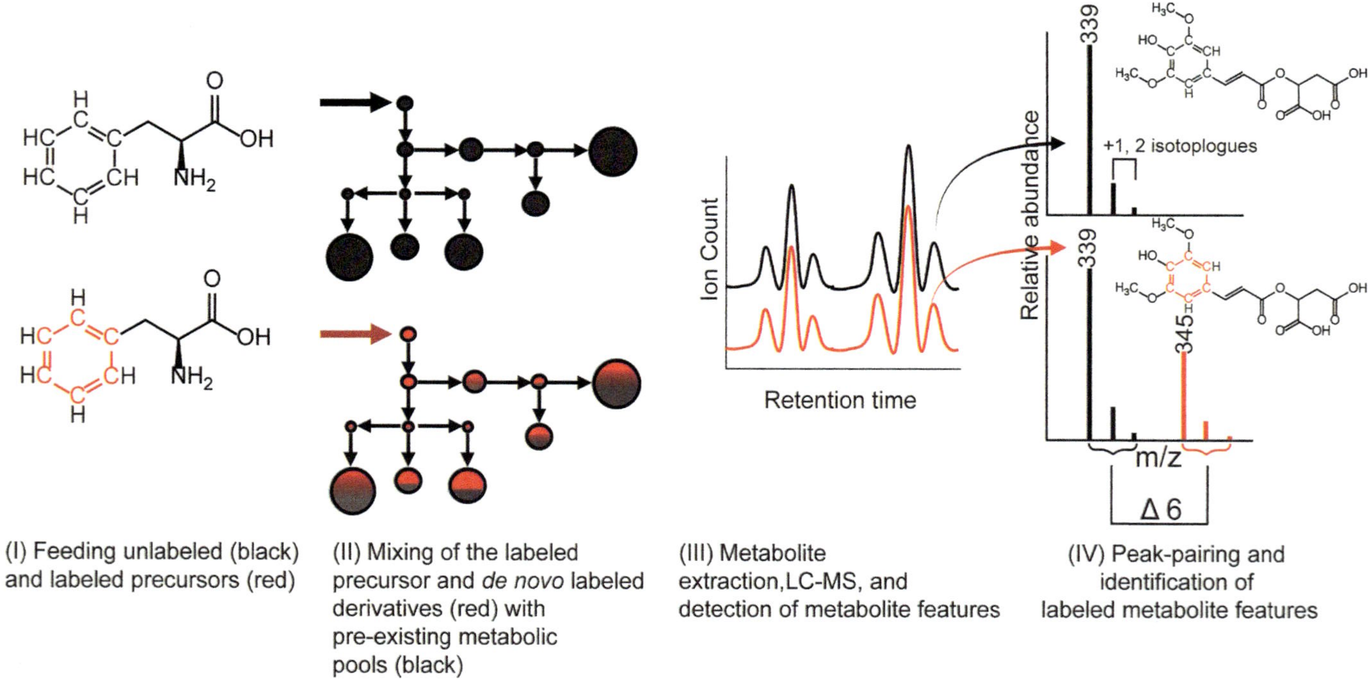

Fig. 1 Overview of the labeling procedure. The first step involves choosing a suitable labeled precursor, depicted here as ^{13}C ring-labeled phenylalanine to identify phenylpropanoids. Throughout the figure, ^{12}C features are colored black and labeled features are colored red. The second step depicts the incorporation of the labeled precursor into derivatives of a metabolic pathway. The circles represent metabolite pools and their relative sizes, and their color gradient in the ^{13}C feeding reflects differential incorporation of the label into downstream metabolites. The third step illustrates the extraction and analysis of metabolites by LC-MS, and the identification of co-eluting compounds between the labeled and unlabeled samples. The fourth step explains the identification of a labeled compound by PODIUM. In this example, sinapoylmalate was identified as phenylalanine-derived by co-elution of a feature of *m/z* 339 in the ^{12}C-fed tissue and *m/z* 345 in the ^{13}C-fed sample.

Stable isotope labeling to characterize plant metabolism is widely applied in plant research, often in the context of metabolic flux analysis (Allen, 2016; Allen, Bates, & Tjellstrom, 2015). In typical isotope feeding experiments, only a few metabolites of known identity are tracked, although dozens to hundreds of other metabolites may have also incorporated the label. Recently, we reported on an approach that uses this broad label incorporation and untargeted MS to annotate all metabolites and features that incorporated a ^{13}C-labeled precursor that was fed to a plant (Simpson, Wunderlich, et al., 2021). We used this approach to identify phenylalanine-derived compounds in Arabidopsis stems and also phenylalanine- and tyrosine-derived compounds in sorghum tissues (Simpson, Olson, Dilkes, & Chapple, 2021; Simpson, Wunderlich, et al., 2021). More than 500 Arabidopsis metabolites were annotated as being derived from phenylalanine, which included known Arabidopsis soluble phenylpropanoids and many phenylalanine-derived compounds that had not been described previously. We also explored the plasticity of the Arabidopsis phenylalanine-derived metabolome in mutants that were blocked at specific steps in the pathway (Fraser & Chapple, 2011). Those metabolite annotations were also applied to retrospectively annotate metabolite features in a genome wide association dataset, enabling the identification of natural genetic variation associated with phenylalanine-derived metabolite variation in Arabidopsis.

Here we describe our experimental and analytical approaches for using heavy isotope labeled substrates to annotate the precursor of origin for metabolites collected through untargeted LC-MS. The development and execution of the computational portions of this protocol was described extensively in Simpson, Wunderlich, et al. (2021). This paper expands on that work and provides protocols and considerations related to designing appropriate isotopic labeling experiments and how to interpret the resulting LC-MS datasets.

2. Methods

2.1 Objectives

The objective of this method is to acquire information about precursors of origin for plant metabolites based on their incorporation of a heavy isotope precursor that had been administered to the plant. The following methods sections describes how to design isotope feeding experiments, including plant growth and feeding procedures, an LC-MS protocol to detect amino acid-derived metabolites, and how to apply the isotope detection program,

PODIUM (Precursor Of Origin Determination In Untargeted Metabolomics) (Simpson, Wunderlich, et al., 2021) to detect labeled metabolites. Although this paper focuses on PODIUM, the principles outlined here should be applicable to similar programs that identify isotopically labeled compounds in untargeted MS data (Bueschl et al., 2012, 2017; Capellades et al., 2016; Chokkathukalam et al., 2013; Dong, Feldberg, & Aharoni, 2019; Feldberg, Dong, Heinig, Rogachev, & Aharoni, 2018; Feldberg et al., 2009; Huang et al., 2014).

2.2 Equipment

- Appropriate plant growth facilities (e.g., greenhouse, growth chamber, light shelves) to grow desired plant tissue and to incubate material during the labeling step
- Vacuum pump and flask (optional)
- pH meter
- Hot water bath
- Tabletop centrifuge
- Speedvac (optional)
- Lyophilizer (optional)
- Ultra-high pressure liquid chromatograph (UPLC) and MS capable of accurate mass determination
- Appropriately sized container for incubation of plant tissue during feeding (e.g., 2.0, 15, or 50 mL tubes; Petri dishes; 6–12 well multi-well plates; 100 mL specimen cups)
- Double edge stainless steel razor blades (e.g., Personna brand from Electron Microscopy Sciences)

2.3 Reagents

- Supplies for growing targeted plant species to a specific stage of development (e.g., agar plates; soil, or soilless potting mixture (e.g., Turface); appropriately sized pots)
- Appropriate ^{12}C and ^{13}C labeled amino acid to label metabolites from targeted pathway. Most isotopes can be purchased from Cambridge Isotope Laboratories (Tewksbury, MA; www.isotope.com), Sigma-Aldrich (St Louis, MO; www.sigma-aldrich.com) and PerkinElmer (Waltham, MA; www.perkinelmer.com). Sigma-Aldrich will also have corresponding ^{12}C compounds

- Sterile 0.5 × Murashige and Skoog (Murashige & Skoog, 1962) media for preparation of labeling solution (0.25 mM MES pH 5.8, 0.5 × Murashige and Skoog salts, 1% sucrose)
- Appropriately sized container to harvest plant material following labeling treatments (e.g., 2 mL Eppendorf tube for <0.1 g fresh weight; 15 mL Falcon tube for 0.1–1 g fresh weight; 50 mL Falcon tube for 1–5 g fresh weight)
- Liquid nitrogen or dry ice to harvest plant material
- Metabolite extraction solution (for most polar and semi polar metabolites 50% aqueous methanol (v/v) is appropriate)
- Supplies for LC-MS (e.g., sample vials, column)

2.4 Procedure steps

1. Determine the plant species to examine.
2. Determine a class of metabolites to annotate, identify an upstream precursor that will label that class of metabolite with specificity, and then determine whether that stable isotope is commercially available and affordable to use at the necessary scale. See Note 1 for a discussion of suitable isotopic labels to use.
3. Identify a tissue to label that is highly engaged in metabolite production. The depth of the resulting library is dependent on the rate of synthesis of the labeled metabolites. This can be affected by multiple factors, including the age and size of the plant, the tissue under investigation, and the sizes of any pre-existing metabolic pools that the label will mix with. Moreover, the tissue in which a metabolite accumulates may not always be the same as the tissue where the metabolite is produced, for example, the alkaloids berberine and nicotine are produced in roots but accumulate in leaves (Hashimoto & Yamada, 2003). Thus, when beginning to examine a new species, tissue, or metabolic precursor, it is good practice to initially verify that label was incorporated into known derivatives of the precursor (if any are known), and to compare the number of labeled compounds and ^{13}C incorporation between different types of tissues. Depending on the experimental objectives, the tissue that incorporated the most total ^{13}C, had the most labeled compounds, or incorporated the most ^{13}C in compounds of interest, would be used for subsequent experimentation.
4. Determine the optimal feeding method for tissue of interest. Our feeding methods involve growing plants to a desired developmental stage in

growth medium and then transferring them, in whole or in part, to liquid medium containing the labeled precursor. Important considerations are identifying a suitable growth medium that allows for a clean and efficient transfer of the plant to the labeling media, growing enough plants for at least three biological replicates per isotope treatment (i.e., labeled and unlabeled precursor), and pooling 3–5 individuals per replicate to quench biological variability. Fig. 2 illustrates the growth and labeling media we used with Arabidopsis leaves and stems and maize seedlings. See Note 2 for additional considerations related to growing plants with respect to different labeling protocols.

5. On the day of labeling, either whole plants are removed from their growth medium, or a specific organ that will be labeled is excised from the plant. The plant material is then cleaned of media if necessary and placed in a suitable container with 0.5 × Murashige and Skoog media and the ^{12}C or ^{13}C-labeled precursors (Fig. 2). The required control for the heavy-isotope feeding is a parallel feeding with the corresponding $[^{12}C]$ (or light isotope), rather than a tissue that was not fed.

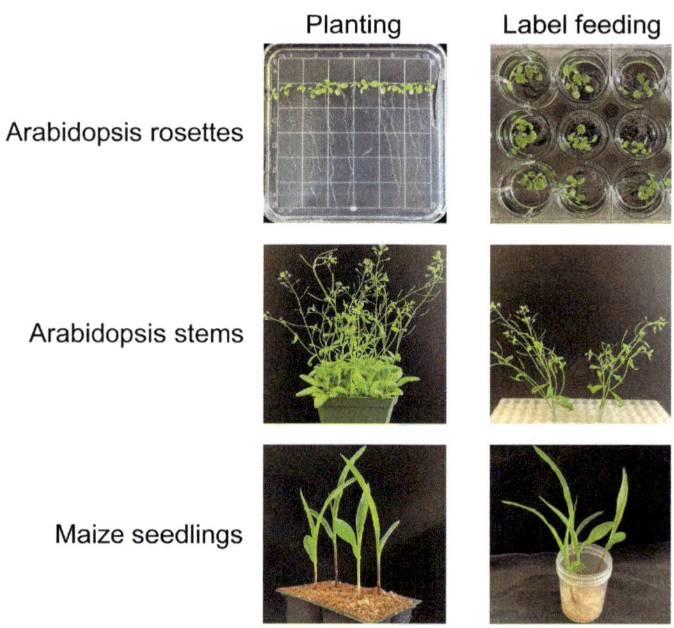

Fig. 2 Plant growth and isotope feeding procedures for Arabidopsis rosettes, stems and monocot seedlings.

6. Depending on experimental objectives, the label is administered either by mechanically forcing it through the surface pores and wounding sites via vacuum or syringe infiltration or by relying on the plant's transpiration stream to take up the label from solution. Note 2 provides more information on these labeling techniques, including information on how to avoid cavitation of the transpiration stream during tissue preparation if using approach (ii).
7. The tissue is incubated in the labeling media for a fixed time, usually 24 h. This is usually done under standard light and temperature regimes, unless a specific environmental treatment, such as high light, is to be applied to induce a certain set of metabolites. Note 3 describes additional considerations for optimal feeding.
8. Following the labeling experiment, the tissue is removed from the labeling media, rinsed to remove excess media, and then immediately placed into an appropriately sized harvesting tube and flash frozen in liquid nitrogen or dry ice.
9. Metabolites from the frozen plant tissue are extracted by a method suitable to the metabolic class of interest. Prior to metabolite extraction, frozen tissue may be lyophilized, although this not essential if the equipment is not available. Metabolite extracts are then frozen until they are to be analyzed. Note 4 describes a general method for extraction of semi-polar metabolites, but other methods can be used to enrich for different classes of compounds.
10. Extracted metabolites are separated by high resolution liquid chromatography (e.g., UPLC), and metabolites are detected and quantified using an MS with a high scan rate and accurate mass determination. If compounds are poorly separated, then there is a high likelihood of detecting false positive peak pairs among many co-eluting compounds. Similarly, if the mass accuracy of the MS is poor, then the power to identify a labeled-unlabeled peak-pair based on a specific m/z difference is diminished. A robust LC-MS protocol for examining label incorporation into semi-polar metabolites (such as those derived from amino acids) is described in Table 1, along with considerations of how the method can be adjusted to detect a broader diversity of labeled compounds.
11. Labeled peak pairs are identified and quantified using one of several methods to detect labeled metabolites. All of the programs capable of this have different procedures for pre-processing the raw MS data, initial peak picking, and identifying labeled-unlabeled peak-pairs

Table 1 Sample LC-MS method to detect semi-polar plant natural products.

Parameter	Methodology		
C18 reverse phase column	Shimadzu Shim-pack XR-ODS (2.2 μ particle size; 3.0 × 75 mm) with UltraShield pre-column filter or equivalent		
LC-MS	UPLC: Agilent 1290 infinity binary pump or equivalent MS: Agilent 6545A quadrupole time-of-flight (QTOF) or equivalent		
Mobile phases	A = 0.1% formic acid in water B = 0.1% formic acid in acetonitrile		
Gradient	**Time (min)**	**% B**	**Suitability of each phase in the chromatography gradient for the detection of labeled metabolites**
	0 0–1	2 2	Highly polar compounds that do not bind strongly to the column elute before 5 min. Chromatography is not optimized for their separation, thus there is higher likelihood for detecting false positive peak pairs
	1–5	6	
	5–15	46	During this time period, there is a linear increase in the concentration of acetonitrile in the mobile phase. This results in good separation of most semi-polar compounds. Labeled compounds are best identified through this time range
	15–21.5	95	At this time, there is a rapid increase in the polarity of the mobile phase to release compounds strongly bound to the column. Chromatography is not optimized for their separation, thus there is a higher likelihood of detecting false positives labeled compounds here
	21.5–23.5	95	The high acetonitrile content elutes remaining non-polar compounds from the column. Labeled compounds are not identified here
	23.5–24.5	2	Re-equilibration of the column for the next run
	24.5–28.5	2	
Flow rate	0.6 mL min^{-1}		
Column temperature	40 °C		
Injection volume	5 μL		

Table 1 Sample LC-MS method to detect semi-polar plant natural products.—cont'd

Parameter	Methodology
Fragmentor voltage	130
Mass range (m/z)	80–1100
Scan rate	5 spectra s^{-1}

(Bueschl et al., 2012, 2017; Capellades et al., 2016; Chokkathukalam et al., 2013; Dong et al., 2019; Feldberg et al., 2018, 2009; Huang et al., 2014). We focus here on how to execute and interpret our PODIUM method (Simpson, Wunderlich, et al., 2021).

For PODIUM, mass features collected by untargeted MS are first identified by a modified XCMS script (Smith, Want, O'Maille, Abagyan, & Siuzdak, 2006), and then that data is used by a second procedure that finds the unlabeled-labeled peak pairs. To initiate PODIUM, the raw MS data files are first converted into .mzXML, or .mzData formats so that can be read by XCMS (Smith et al., 2006). This can be done through Mass Hunter Qualitative Analysis software (for files generated from Agilent equipment) or msConvert within ProteoWizard (Adusumilli & Mallick, 2017). To reduce file size, profile spectra are converted into centroid spectra and the raw data are filtered to retain a specific number of peaks (4000–8000 usually). We do not deisotope or deconvolute the raw data to avoid inadvertently removing peaks that arise through labeling. The datafiles are then organized into folders that are then organized by experimental groups. Within each experimental group folder, files representing samples labeled with ^{12}C or ^{13}C are then placed into folders named "12C" and "13C." Next, the PODIUM program is executed by an R script, which is available for download through GitHub (github.com/chapple-lab/podium). Running XCMS and the isotope detection procedure requires a significant amount of computer memory, and thus for large data sets (i.e., greater than three experimental groups analyzed), we recommend using a high performance computing cluster, which should be available at most research institutions. Settings for the execution of PODIUM, and suggestions for its optimization are summarized in Table 2. Additional details on the development of PODIUM and specifics on its implementation, are contained in the Supplemental Data File in Simpson, Wunderlich, et al. (2021).

Table 2 Sample XCMS and PODIUM parameters to detect labeled compounds.

Processing step	Purpose	Function	Suggested Parameters	Comment
Peak picking	Processes MS data to find and quantify peaks	xcmsset ()	method = centwave; ppm = 30; peakwidth = c (3,30); snthresh = 10; prefilter = c (2,10); integrate = 1	Based upon chromatography, MS systems and analyte composition, the user will optimize these peak picking parameters
Retention time correction	Aligns retention times across multiple samples	retcor.obiwarp	response = 100	Metabolite peaks should align from samples that were analyzed under similar LC conditions. If chromatograms are not well aligned, then this retention time correction step will fail, but this should not occur if one set of samples are analyzed together
Grouping	Peaks are grouped across samples	group.density	minfrac = 0; minsamp = 3; mzwid = 0.01; bw = 5; max = 50	Minfrac/minsamp set the minimum number of samples a peak must be present to be considered reproducible
Peak filling	If a peak was not initially detected, this method fills in missing peak data for peak groups	fillpeaks. chrom	Expand.mz = 1; Expand.rt. = 1; Min. width.mz = 0.005; Min.width.rt. – 1	FillPeaks step is essential here so that each peak group has a complete set of peak intensity data for the statistical analysis by PODIUM
Labeled-unlabeled peak pairing	PODIUM specific program to identify labeled mass features	labeledPeaks_ GroupPairing_ tTest	mzppm = 10; mzabs = 0; rterror = 2; alpha = 0.05; nLabel = 6; mLabel = 1.0033548	PODIUM pairs peaks based on the number of labeled carbons in the heavy isotope-derived metabolite (nLabel), mass difference between the light and heavy isotope (mLabel), and alpha for one tailed *t*-test (alpha). A peak pair is returned as valid as long as the labeled peak group has $P \leq \alpha$ in at least one experimental group

2.5 Procedure notes

Note 1. Selection of a class of compounds to annotate and identification of a labeled precursor. PODIUM and related methods are most informative when using a labeled substrate that is a direct precursor to compounds of interest. In contrast, data interpretation and analysis will be complicated if the substrate is a common metabolic intermediate (e.g., glucose) or that can be readily routed into central metabolic pathways (e.g., glycine, alanine, aspartate, serine, and glutamate following transamination). Table 3 lists amino acids that might be expected to generate optimal results in these experiments and examples of the plant natural products produced from each, but other compounds, such as sugars that are not part of core metabolism, could also be employed where appropriate.

The labeled precursor should also contain heavy atoms at the positions anticipated to be incorporated into downstream derivatives. For example, to identify phenylpropanoids, it is convenient to use C_6(ring)-^{13}C-phenylalanine because the phenyl ring is generally incorporated intact into products of the pathway. In contrast, in other metabolic pathways, one or more labeled carbons are often lost from the precursor during metabolism. For example, the carboxyl group is lost from tryptophan, lysine, phenylalanine and methionine during glucosinolate biosynthesis; thus, if the labeled substrates contain ^{13}C at all positions, their glucosinolate derivatives will contain one less ^{13}C atom than was in the administered precursor, and solely carboxy-labeled versions of these amino acids would not label glucosinolates at all. Also, if possible, the labeled precursor should contain more than two heavy atoms. Labeled precursors of this type will ensure that the mass difference between the newly produced labeled compounds and the pre-existing unlabeled compounds will be large enough as to not be complicated by natural isotopologues.

These methods can also detect label incorporation from heavy hydrogen, nitrogen, sulfur, or oxygen (^{2}H, ^{15}N, ^{34}S, ^{18}O). Compared to ^{13}C, these heavier atoms may be less effective in labeling all products of a pathway because they might be more subjected to loss/exchange and, with the exception of ^{2}H, those atoms are less abundant in plant natural products. Instead of using non-^{13}C isotopic labels in isolation, ^{13}C labeling can be combined with other heavy elements to determine metabolite structures (Feldberg et al., 2009).

Note 2. Plant growth conditions for labeling. We have successfully fed Arabidopsis rosettes and stems, as well as sorghum and maize seedlings, with different methods. Label can be administered through cut sites in a

Table 3 Possible plant natural products derived ^{13}C-containing amino acid substrates.

Amino acid	Location of ^{13}C on molecule	Common Δ mass to detect via PODIUM	Known/putative products
Phe	Ring	6, 5, 12, 24	Hydroxycinnamic acids, lignans, neolignans, flavonoids, tannins, aromatic glucosinolates
Trp	Whole molecule	11, 10, 9, 8	Indole phytoalexins, indole glucosinolates
Tyr	Ring	6, 5, 12, 24	Aromatic glucosinolates, tocopherols, betalains, alkaloids, cyanogenic glycosides, and in monocots all ring hydroxylated Phe-derived metabolites
Leu	Aliphatic Chain	5 and 6	Aliphatic glucosinolates
Ile	Aliphatic Chain	6 and 5	Aliphatic glucosinolates
Val	Aliphatic Chain	5 and 4	Aliphatic glucosinolates
Met	Aliphatic Chain	5 and 4	Aliphatic glucosinolates
Pro	Whole molecule	5 and 4	Unclear
Lys	Aliphatic Chain	6 and 5	Diamines, polyamines
Arg	Aliphatic Chain	6 and 5	Agmatine conjugates
His	Whole molecule	6, 5, 4	Unclear
Asn	Aliphatic Chain	4	Unclear
Cys	Aliphatic Chain	3	Glutathione conjugates
Thr	Aliphatic Chain	4	Unclear

tissue, through the roots, or via vacuum infiltration. Implementation of each method has its own considerations that are tied to the plant tissue under investigation and the initial growth of the plant.

(i) Label is fed to excised tissue. This approach works well for stems because it has a minimal impact on their physiology and results in uniform distribution of the labeled substrate (Wang et al., 2018). In this procedure, whole rosettes are pulled from the soil and the base of the stem is submerged under water. The stems are then cut from the rosette using a Personna razor blade and immediately transferred into the labeling medium, usually in a 1.5 mL Eppendorf or 15 mL Falcon tube. Excising stems under water is important to prevent cavitation of the xylem and loss of transpiration stream integrity. In our experiments, 20 cm tall stems readily take up several mL of labeling media over 24 h of feeding. A similar process can be used for leaves by cutting them at the petiole underwater (Chapple, Vogt, Ellis, & Somerville, 1992); however, the integrity of the transpiration stream is more prone to failure, resulting in wilting. Thus, if leaf labeling needs to be done, it is advisable to transfer an abundance of excised leaves first into Eppendorf tubes containing water to identify those that will wilt. The remaining viable samples can then be transferred to labeling medium for continued incubation.

(ii) Label is applied through the roots. The root systems of intact plants can be submerged in a buffer containing the label, which results in efficient labeling of leaves and roots. This method of feeding is most amenable to relatively small plants grown in a medium that allows them to be transferred into the feeding solution with minimal tissue damage. For example, small plant species (e.g., Arabidopsis) can be initially grown on MS plates with 0.8% agar and under sterile conditions. At the time of labeling, tweezers can be used to liberate the plant with roots intact from the agar. The whole plant is then immediately placed in the labeling solution. Larger plants (e.g., corn, sorghum) can be initially cultivated in a soilless media (e.g., Turface or GreensGrade or perlite) (Fig. 2). Here, seedlings grown in plastic cell packs are liberated by first submerging the cell packs in a bucket of water, then cutting the plastic to release the media and the plant into the water. Excess growth media is then massaged off the roots, and the plant's roots are then submerged in the labeling medium.

(iii) Label is administered via vacuum infiltration. For vacuum infiltration, tissue is cut into pieces in the range of one to four square cm and submerged in a vacuum flask containing an aqueous solution of the labelled precursor. Vacuum is then applied until the flask liquid bubbles, and the vacuum is then released. After these steps, the tissue should appear darker, indicating uptake of the label solution. If uptake was not optimal during the first infiltration, the process can be repeated. The tissue is then incubated in the solution with label for the desired amount of time. Although it is likely that most of the label that will be used by the plant enters the tissue during the infiltration step, the tissue is kept fully submerged in the labeling solution through the treatment to prevent the tissue from drying out and to allow for additional uptake by diffusion. One advantage of vacuum infiltration is that it ensures complete saturation of the tissue with the labeled substrate, but, because the tissue is cut and submerged, its normal metabolism may be disrupted.

The choice of container to perform the labeling in depends upon the method of feeding, size of plant species and tissue, and cost/availability of the labeled substate. For short stems (<20 cm), a 1.5 or 15 mL Eppendorf tube containing 1–5 mL of the labeling solution is sufficient. For small plants (e.g., Arabidopsis rosettes) a 6 or 12 well plate containing 2–3 mL of the labeling solution is suitable. For larger plants (e.g., grasses), a 100 mL specimen cup containing 30 mL of labeling solution can be used. Although ^{12}C substrates are relatively inexpensive, uniformly labeled ^{13}C substrates can cost hundreds of dollars for less than 25 mg. Thus, when choosing the container for labeling, it is advisable to determine the minimum volume of medium needed to submerge/feed the tissue.

Note 3. Labeling conditions to maximize depth of metabolic libraries. The objective of PODIUM, and related methods, is to establish a comprehensive library of metabolites derived from a labeled precursor. A heavy isotope is fed for long enough and at a sufficiently high concentration to allow for its concentration to exceed that of its pre-existing unlabeled counterpart. This is in contrast to metabolic flux analysis where incubation time and the concentration of the labeled substrate are often varied (Allen, 2016). Long term (e.g., 24 h) labeling at a high concentration (e.g., 1 mM in labeling medium) maximizes the chances of detecting a heavy isotope signal in downstream products. Unfortunately, some substrates applied at high concentrations and for long periods of time may be toxic. For example, Arabidopsis seeds germinated on tryptophan, tyrosine or phenylalanine at greater than 1 mM concentration exhibit reduced growth and die, likely due to feedback inhibition of shikimate pathway enzymes (de Oliveira et al., 2019; Voll, Allaire, Fiene, & Weber, 2004), although, excised

Arabidopsis stems showed no significant growth defects when fed with 3 mM phenylalanine for over 24 h (Wang et al., 2018). Nevertheless, even if toxicity does not cause a visual phenotype, it may negatively impact labeling. Thus, optimization of labeling conditions may be necessary for new tissues and species and substrates. For example, prior to the isotopic label treatment, the unlabeled substrate could be fed at 1 mM for 24 h, and if negative impacts are observed relative to untreated samples, then the label concentration and/or incubation time can be reduced.

Note 4. Extraction of metabolites. The metabolite extraction protocol to be used depends on the metabolites and tissues under investigation. Here, we focus on extraction of semi-polar water-soluble compounds, which encompass many amino acids-derived specialized metabolites in plants. These compounds are readily extracted from ground tissue in 50–80% aqueous methanol at 65 °C for 1 h. To normalize metabolite concentration across the samples, metabolites are extracted at a consistent tissue fresh weight (FW) or dry weight (DW) to extraction solvent ratio, (i.e., 0.1 g FW or 0.01 g DW to 1 mL solvent). Following extraction, insoluble molecules (e.g., lipids, nucleic acids, proteins) and insoluble cell contents (e.g., starch, cell wall polysaccharides) are pelleted in a high speed microcentrifuge (max speed; 5 min) and the supernatant is transferred into a new tube and stored at −20 or −80 °C until analysis by LC-MS. For a more concentrated sample, solvent may be removed in a speedvac and redissolved in a reduced volume of extraction solvent; however, excessively concentrated samples can lead to saturation of the MS detector, so the final sample concentration should be optimized in preliminary experiments. Sometimes it can be desirable to enrich/eliminate certain metabolites by phase partitioning (e.g., extraction of an aqueous extract with ethyl acetate) or solid phase extraction.

3. Interpretation of PODIUM data and applications

3.1 PODIUM output

In PODIUM, and related programs, a mass feature is identified as being labeled if it meets three criteria. First, the mass feature must be present and co-elute across replicated heavy and light isotope-treated samples. Second, it exhibits an exact m/z difference that corresponds to the combined mass of the heavy atoms that were incorporated into the labeled feature. Third, there is a statistically significant higher ion signal for the feature in the heavy isotope-fed sample compared to signal at the same retention time and m/z in the light isotope-fed samples. The result is a table of unlabeled-labeled peak pairs detected across all experimental groups, along with the standard XCMS output for each feature. Fig. 3 shows labeling data for four selected phenylalanine-derived metabolites previously analyzed in Simpson, Wunderlich, et al. (2021). In that study, ^{13}C-ring labeled phenylalanine was fed to the stems of wild-type Arabidopsis and the *fah1-2* mutant that is defective in FERULATE 5-HYDROXYLASE, an enzyme that catalyzes a step in the phenylpropanoid pathway. Mutants were included in that study to demonstrate the ability of PODIUM to identify metabolic changes that arise from defects in the pathway. In this illustrative dataset, the four phenylalanine-derived mass features are designated with specific "Cluster Numbers" (Column 1). Each cluster contains one ^{12}C-containing mass feature and one (or more in the relatively rare case that the mass feature is a dimer or trimer of the administered precursor) ^{13}C-labeled mass feature. The *m/z* shift that defines the peak pair is reported following the underscore in the "Isotope" column (i.e., 13C_6 means that the compound incorporated six ^{13}C atoms). The columns to the right are the standard XCMS output, including a name for the feature in the format MNNNTNNN based on the median *m/z* and retention time of the mass feature, followed by accurate values for the *m/z* and retention time for the mass feature. The ion intensity values for each feature in each of the three biological replicates are provided in the next set of columns.

An initial survey of the data can indicate the overall effectiveness of the labeling. First, in each cluster, there is a higher ion count for the ^{13}C-containing feature in samples fed with the ^{13}C isotope, as compared to the same compounds in samples fed with ^{12}C. Although there are ion abundances reported for the ^{13}C-containing features in ^{12}C-fed samples, those values can be considered background and were likely generated by the "fillpeaks" function of XCMS. A value (as opposed to a zero) for "labeled features" in the unlabeled samples is necessary for the statistical analysis in PODIUM. Second, in samples fed with the ^{12}C-containing precursor, there is often a modest increase in the abundance of ^{12}C features, compared to samples fed with ^{13}C, which are reflective of metabolite abundance prior to the administration of the label. Presumably these increases occur because precursor availability often limits flux and the abundance of precursor provided during the experiment enhances end-product accumulation. Thus, this increase in abundances of ^{12}C features can be used as additional data supporting the identification of the precursor of origin and a preliminary indication that the feeding was effective.

MS Feature Information					Ion counts in samples fed with ^{12}C Phe or $^{13}C_6$ Phe											
					Wild type						*fah1-2*					
Cluster Number	Isotope	Feature name	m/z median	rt (s) median	WT_12C-fed_R1	WT_12C-fed_R2	WT_12C-fed_R3	WT_13C-fed_R1	WT_13C-fed_R2	WT_13C-fed_R3	fah1-2_12C-fed_R1	fah1-2_12C-fed_R2	fah1-2_12C-fed_R3	fah1-2_13C-fed_R1	fah1-2_13C-fed_R2	fah1-2_13C-fed_R3
1661	12C	M163T714	163.0395	714.5	35979	8050	24908	9063	9514	4722	76918	9710	42434	6807	7781	4728
1661	13C_6	M169T714	169.0599	713.8	347	376	284	38166	27119	7859	269	289	346	15536	19409	6563
1811	12C	M577T729	577.1554	729.5	4058527	4241266	4575512	3555225	3395000	3273135	3702773	3786828	4044180	3461991	3464724	2926113
1811	13C_6	M583T729	583.1774	729.5	580	564	517	1924	717	919	472	543	537	3164	6496	7923
1851	12C	M339T736	339.0719	735.9	3917576	5848367	3115620	3246593	3082891	4210827	5823	6585	6907	8069	5994	5967
1851	13C_6	M345T735	345.0918	735.2	1654	4321	2978	127404	287608	196882	1558	7672	2857	4724	4117	2579
1974	12C	M505T750	505.1349	749.5	125068	474169	348027	308485	264117	212328	149593	161510	150826	255280	113170	83863
1974	13C_6	M511T750	511.1561	750.2	1793	1345	2369	5250	4716	3638	1652	1443	2176	11971	7563	7728
1974	13C_12	M517T748	517.178	748.4	1404	1473	1652	6218	3035	4481	1130	1266	1448	7013	7122	3926

Post hoc metabolite identifications: 1661- *p*-coumaric acid; 1811- kaempferol-3-*O*-rhamnoside-7-*O*-rhamnoside; 1851-sinapoylmalate; 1974- guaiacyl (8-*O*-4) feruloyl malate

Fig. 3 Representative output from PODIUM. Arabidopsis wild-type and *fah1-2* mutant stems were fed with ^{12}C or $^{13}C_6$ Phe for 24h. The "Cluster Number" (first column) is the ^{12}C–$^{13}C_6$ peak-pair identifier. The "Isotope" column tells whether the detected MS feature is ^{12}C-derived (12C) or derived from the $^{13}C_6$-fed amino acid, and the *m/z* shift that was detected (13C_6). The next three columns provide the MS feature name (derived from *m/z* and retention time), and the exact *m/z* and exact retention time of the feature. The final 12 columns report the ion counts for a feature in each replicate (R#). Data was previously presented in Simpson, Wunderlich, et al. (2021).

The interpretation of the data in Fig. 3 was previously described in Simpson, Wunderlich, et al. (2021); but is presented here in more detail to exemplify the application of this method. The peak-pair in Cluster Number 1661 represents the pathway intermediate *p*-coumaric acid. Generally, pools of intermediates are small and turn over rapidly, which is why the ion count for the labeled *p*-coumaric acid in the ^{13}C-fed samples is comparable to the unlabeled *p*-coumaric acid in the ^{12}C-fed samples. Cluster Number 1811 represents a flavonoid glucoside that has a high relative preexisting ion count. Although a ^{13}C-derived compound was detected in the label-fed samples, it increased only two- to fourfold above background. This trend is in sharp contrast to that of *p*-coumaric acid and is consistent with the flavonoid glucoside being a much more downstream product of the pathway and suggests that it is derived from larger intermediate pools that incorporate label more slowly. Cluster Number 1851 represents sinapoylmalate, which also has a high relative ion count; however, unlike the flavonoid glycoside, there is a $>$40-fold difference between the ^{13}C-labeled feature and the background, suggesting that sinapoylmalate is made rapidly and represents much of the flux into soluble metabolites in wild type. *fah1-2* does not produce sinapoylmalate (Chapple et al., 1992) and accordingly, the ion count for the preexisting ^{12}C compound (M339T736) and the *de novo* ^{13}C-derived compound (M345T735) are at background levels in *fah1-2* plants. Cluster Number 1974 is an example of a metabolite that contains two label-derived components, guaiacyl (8-O-4) feruloyl malate. The 13C_6 derived feature was synthesized from one labeled and one unlabeled phenylalanine-derived precursor, whereas the 13C_12 feature represents the incorporation of two labeled phenylalanine molecules into the final structure.

3.2 Quantification and normalization

In MS, ionization efficiency varies between compounds, thus absolute quantification of individual features requires the generation of a standard curve using authentic standards. However, for most plant metabolites, standards are not available, and it is only possible to make comparisons of relative ion abundances of a given compound. In untargeted MS, internal standards can be used to normalize the total ion count between samples to account for variation in sample preparation. This source of error can arise in metabolite extraction procedures that have multiple purification steps, such as liquid-liquid partition or solid phase extraction, but are less of an issue in single-step extraction procedures like those described here where metabolites are extracted at a fixed concentration of extraction solvent to DW or FW. Thus, peak intensities can be reported as "normalized to DW/FW." Alternatively, they can be reported as a percentage of total ion count or total labeled ion count.

3.3 Identification and interpretation of natural isotopologues

For the most part, how the labeled precursor will be modified by metabolism (such as by decarboxylation) will be unclear to the experimenter *a priori*. For this reason, PODIUM should be run to detect multiple mass shifts. For example, if the labeled precursor contains six ^{13}C atoms, the user should query their data for compounds that have incorporated +6, +5, +4 and +3 heavy atoms. Unfortunately, interpretation of the resulting data can be complicated by natural isotopologues. For example, many Arabidopsis tryptophan-derived glucosinolates, such as glucobrassicin (indol-3-ylmethylglucosinolate; $C_{16}H_{19}N_2O_9S_2$), contain only 10 of the 11 carbons from tryptophan ($C_{11}H_{12}N_2O_2$) due to a decarboxylation step during biosynthesis. If the ^{13}C tryptophan used in the labeling experiment is uniformly labeled (i.e., has eleven ^{13}C atoms), then glucobrassicin will be represented by an M-H peak pair of 447/457 *m/z* when PODIUM is set to search for +10 mass shifts. However, if PODIUM is set to search for +11 mass shifts to identify compounds that incorporated all carbons from tryptophan, pairing will also be observed between the ^{12}C glucobrassicin (*m/z* 447) and glucobrassicin where 10 of its ^{13}Cs are derived from ^{13}C tryptophan and one natural abundance ^{13}C is present within the thioglucose moiety. Similarly, if PODIUM is set to search for +9 mass shifts, a peak pair will be observed between the +1 natural isotopologue of ^{12}C glucobrassicin (*m/z* 448) and ^{13}C tryptophan derived glucobrassicin (*m/z* 457). Thus, even though 10 carbons from tryptophan are incorporated into glucobrassicin, the presence of natural isotopologues may complicate interpretation of +9 and +11 peak pairs. That said, these isotopologue pairs should be readily identified in the dataset because the isotopologues will be less abundant.

3.4 Metabolite identification

The libraries that result from these analyses will identify metabolite precursors and can facilitate metabolite identification. Exact masses can be compared to existing mass spectral databases, and where exact matches to known metabolites are found, literature searches can reveal whether the putative metabolite is, in fact, derived from the labeled precursor that had been administered. Nevertheless, even high confidence exact mass matches cannot provide unambiguous identities because MS alone cannot distinguish between structural isomers, or determine whether a feature represents either an intact metabolite or a fragment generated in the MS. For example, a feature corresponding to sinapic acid is typically generated from sinapoylmalate, and it would be an error to conclude that the feature represented sinapic acid itself. To help mitigate this problem, MS/MS can be applied retrospectively to use molecular fragment mass shifts to aid in metabolite identification. For MS/MS, compounds of interest are targeted based on the m/z values of the ^{12}C and isotope-derived ^{13}C forms. Targeting ^{12}C and ^{13}C-containing compounds with separate analyses may be necessary because the higher abundance of the ^{12}C-containing compound may reduce the number of scans obtained from the ^{13}C-contaning compound. The resulting fragment peaks of the targeted ^{12}C and ^{13}C metabolites can then be compared and fragments of the metabolite that were derived from the labeled precursor can be detected by comparing m/z differences between fragments. In some cases, this comparison can lead to a structural identification, or, at minimum, new knowledge about what portions of a metabolite are and are not derived from the precursor.

4. Summary

This protocol outlines a general approach and factors to consider when designing and executing isotopic labeling to annotate the precursor of origin for plant metabolites. Although this protocol is based on our previous work annotating phenylalanine- and tyrosine-derived metabolites identified by PODIUM, we anticipate that the principles described here are adaptable other types of metabolites and other computational pipelines that have been reported. Several items bear repeating if optimal results are to be obtained. First, using precursors that are immediately upstream to a specific class of metabolites and that are uniformly labeled with ^{13}C is ideal, but this consideration will be influenced by whether or not the desired substrate is commercially available. Second, the depth of the resulting metabolic library can be affected by the initial growth conditions of the plant and by the method by which the label is introduced into the tissue. Third, metabolite extraction, chromatography conditions, and MS settings can enrich for metabolites of different chemical properties and that are of specific interest to the researcher. In the end, adding information about the source of metabolite classes can help identify structural relationships between MS features and reduce the space to be searched to identify a metabolite. Moreover, this data can be applied to retrospectively annotate metabolites in other datasets that were analyzed by similar methods, thus reducing the complexity of subsequent research, such as uncovering relevant gene-metabolic associations in metabolic genome wide association studies or identifying global changes in metabolites in mutants.

Acknowledgments

This work was supported by the US Department of Energy, Office of Science (BER), Grant DE-SC0020368 (CC). J.P.S. was supported by a United States Department of Agriculture National Institute of Food and Agriculture postdoctoral grant 2018-08121/1019231. We thank Dr. Bruce Cooper (Bindley Bioscience Center, Purdue University) for assistance in the acquisition of the LC-MS metabolite profiling data. We also thank Dr. Cole Wunderlich and Dr. Brian Dilkes (Purdue University) for initial work in devising the isotope labeling procedure described here.

References

Adusumilli, R., & Mallick, P. (2017). Data conversion with ProteoWizard msConvert. *Methods in Molecular Biology*, *1550*, 339–368.

Allen, D. K. (2016). Quantifying plant phenotypes with isotopic labeling & metabolic flux analysis. *Current Opinion in Biotechnology*, *37*, 45–52.

Allen, D. K., Bates, P. D., & Tjellstrom, H. (2015). Tracking the metabolic pulse of plant lipid production with isotopic labeling and flux analyses: Past, present and future. *Progress in Lipid Research*, *58*, 97–120.

Allwood, J. W., Williams, A., Uthe, H., van Dam, N. M., Mur, L. A. J., Grant, M. R., et al. (2021). Unravelling plant responses to stress-the importance of targeted and untargeted metabolomics. *Metabolites*, *11*, 558.

Alseekh, S., & Fernie, A. R. (2018). Metabolomics 20 years on: What have we learned and what hurdles remain? *The Plant Journal*, *94*, 933–942.

Benson, A. A., Bassham, J. A., Calvin, M., Goodale, T. C., Haas, V. A., & Stepha, W. (1950). The path of carbon in photosynthesis. V. Paper chromatography and radioautography of the products. *Journal of the American Chemical Society*, *72*, 1710–1718.

Boutet, S., Barreda, L., Perreau, F., Totozafy, J. C., Mauve, C., Gakiere, B., et al. (2022). Untargeted metabolomic analyses reveal the diversity and plasticity of the specialized metabolome in seeds of different *Camelina sativa* genotypes. *The Plant Journal*, *110*, 147–165.

Brown, S. A., & Neish, A. C. (1955). Shikimic acid as a precoursor in lignin biosynthesis. *Nature*, *175*, 688–689.

Brown, S. A., & Neish, A. C. (1956). Studies of lignin biosynthesis using isotopic carbon. V. comparative studies on different plant species. *The Canadian Journal of Biochemistry and Physiology*, *34*, 769–778.

Bueschl, C., Kluger, B., Berthiller, F., Lirk, G., Winkler, S., Krska, R., et al. (2012). Metextract: A new software tool for the automated comprehensive extraction of metabolite-derived LC-MS signals in metabolomics research. *Bioinformatics*, *28*, 736–738.

Bueschl, C., Kluger, B., Neumann, N. K. N., Doppler, M., Maschietto, V., Thallinger, G. G., et al. (2017). Metextract II: A software suite for stable isotope-assisted untargeted metabolomics. *Analytical Chemistry*, *89*, 9518–9526.

Capellades, J., Navarro, M., Samino, S., Garcia-Ramirez, M., Hernandez, C., Simo, R., et al. (2016). geoRge: A computational tool to detect the presence of stable isotope labeling in LC/MS-based untargeted metabolomics. *Analytical Chemistry*, *88*, 621–628.

Chapple, C. C., Vogt, T., Ellis, B. E., & Somerville, C. (1992). An Arabidopsis mutant defective in the general phenylpropanoid pathway. *Plant Cell*, *4*, 1413–1424.

Chokkathukalam, A., Jankevics, A., Creek, D. J., Achcar, F., Barrett, M. P., & Breitling, R. (2013). mzMatch-ISO: An R tool for the annotation and relative quantification of isotope-labelled mass spectrometry data. *Bioinformatics*, *29*, 281–283.

de Oliveira, M. V. V., Jin, X., Chen, X., Griffith, D., Batchu, S., & Maeda, H. A. (2019). Imbalance of tyrosine by modulating TyrA arogenate dehydrogenases impacts growth and development of Arabidopsis thaliana. *The Plant Journal*, *97*, 901–922.

De Vos, R. C., Moco, S., Lommen, A., Keurentjes, J. J., Bino, R. J., & Hall, R. D. (2007). Untargeted large-scale plant metabolomics using liquid chromatography coupled to mass spectrometry. *Nature Protocols*, *2*, 778–791.

Dong, Y., Feldberg, L., & Aharoni, A. (2019). Miso: An R package for multiple isotope labeling assisted metabolomics data analysis. *Bioinformatics*, *35*, 3524–3526.

Feldberg, L., Dong, Y., Heinig, U., Rogachev, I., & Aharoni, A. (2018). DLEMMA-MS-imaging for identification of spatially localized metabolites and metabolic network map reconstruction. *Analytical Chemistry*, *90*, 10231–10238.

Feldberg, L., Venger, I., Malitsky, S., Rogachev, I., & Aharoni, A. (2009). Dual labeling of metabolites for metabolome analysis (DLEMMA): A new approach for the identification and relative quantification of metabolites by means of dual isotope labeling and liquid chromatography-mass spectrometry. *Analytical Chemistry*, *81*, 9257–9266.

Fraser, C. M., & Chapple, C. (2011). The phenylpropanoid pathway in *Arabidopsis*. *Arabidopsis Book*, *9*, e0152.

Hashimoto, T., & Yamada, Y. (2003). New genes in alkaloid metabolism and transport. *Current Opinion in Biotechnology*, *14*, 163–168.

Hemm, M. R., Ruegger, M. O., & Chapple, C. (2003). The Arabidopsis *ref2* mutant is defective in the gene encoding CYP83A1 and shows both phenylpropanoid and glucosinolate phenotypes. *Plant Cell*, *15*, 179–194.

Huang, X., Chen, Y. J., Cho, K., Nikolskiy, I., Crawford, P. A., & Patti, G. J. (2014). X13CMS: Global tracking of isotopic labels in untargeted metabolomics. *Analytical Chemistry*, *86*, 1632–1639.

Mannochio-Russo, H., de Almeida, R. F., Nunes, W. D. G., Bueno, P. C. P., Caraballo-Rodriguez, A. M., Bauermeister, A., et al. (2022). Untargeted metabolomics sheds light on the diversity of major classes of secondary metabolites in the Malpighiaceae botanical family. *Frontiers in Plant Science*, *13*, 1–19.

Murashige, T., & Skoog, F. (1962). A revised medium for rapid growth and bio assays with tobacco tissue cultures. *Physiologia Plantarum*, *15*, 473–497.

Oh, M., Park, S., Kim, H., Choi, G. J., & Kim, S. H. (2021). Application of UPLC-QTOF-MS based untargeted metabolomics in identification of metabolites induced in pathogen-infected rice. *Plants (Basel)*, *10*, 1–13.

Perez de Souza, L., Alseekh, S., Naake, T., & Fernie, A. (2019). Mass spectrometry-based untargeted plant metabolomics. *Current Protocols in Plant Biology*, *4*, e20100.

Simpson, J. P., Olson, J., Dilkes, B., & Chapple, C. (2021). Identification of the tyrosine- and phenylalanine-derived soluble metabolomes of sorghum. *Frontiers in Plant Science*, *12*, 714164.

Simpson, J. P., Wunderlich, C., Li, X., Svedin, E., Dilkes, B., & Chapple, C. (2021). Metabolic source isotopic pair labeling and genome-wide association are complementary tools for the identification of metabolite-gene associations in plants. *Plant Cell*, *33*, 492–510.

Smith, C. A., Want, E. J., O'Maille, G., Abagyan, R., & Siuzdak, G. (2006). XCMS: Processing mass spectrometry data for metaboite profiling using nonlinear peak align ment, matching, and identification. *Analytical Chemistry*, *78*, 779–787.

Tohge, T., Wendenburg, R., Ishihara, H., Nakabayashi, R., Watanabe, M., Sulpice, R., et al. (2016). Characterization of a recently evolved flavonol-phenylacyltransferase gene provides signatures of natural light selection in *Brassicaceae*. *Nature Communications*, *7*, 12399.

Voll, L. M., Allaire, E. E., Fiene, G., & Weber, A. P. (2004). The Arabidopsis *phenylalanine insensitive growth* mutant exhibits a deregulated amino acid metabolism. *Plant Physiology*, *136*, 3058–3069.

Wang, P., Guo, L., Jaini, R., Klempien, A., McCoy, R. M., Morgan, J. A., et al. (2018). A ^{13}C isotope labeling method for the measurement of lignin metabolic flux in *Arabidopsis* stems. *Plant Methods*, *14*, 51.

Weng, J. K., Li, Y., Mo, H., & Chapple, C. (2012). Assembly of an evolutionarily new pathway for α-pyrone biosynthesis in Arabidopsis. *Science*, *337*, 960–964.

Zhou, S., Kremling, K. A., Bandillo, N., Richter, A., Zhang, Y. K., Ahern, K. R., et al. (2019). Metabolome-scale genome-wide association studies reveal chemical diversity and genetic control of maize specialized metabolites. *Plant Cell*, *31*, 937–955.

SECTION III

Environment

CHAPTER ELEVEN

Modification of G-protein biochemistry and its effect on plant/environment interaction

Sona Pandey[a,*], Kate Harline[a], and Swarup Roy Choudhury[a,b]

[a]Donald Danforth Plant Science Center, St. Louis, MO, United States

[b]Department of Biology, Indian Institute of Science Education and Research (IISER) Tirupati, Tirupati, Andhra Pradesh, India

*Corresponding author: e-mail address: spandey@danforthcenter.org

Contents

Abstract

Heterotrimeric GTP-binding proteins comprised of Gα, Gβ and Gγ subunits are key regulators of a multitude of signaling pathways in eukaryotes. In plants, G-proteins are currently a focus of intense research due to their involvement in modulation of many agronomically important traits such as seed yield, organ size, abscisic acid (ABA)-dependent signaling and stress responses, plant defense responses, symbiosis and nitrogen use efficiency. The mechanistic details of G-protein biochemistry in modulating these processes in plants remain largely unknown. Although the core G-protein components and their activation/deactivation chemistries are broadly conserved throughout eukaryotic evolution, their regulation seems to have been rewired in plants to meet specific needs. Plant G-proteins may be spontaneously active and/or are regulated by phosphorylation-dependent changes, by the activity of lipid second messengers such as phospholipases, or may even have nucleotide-exchange independent regulation. Regardless of these deviations from the established norm, the biochemical

Methods in Enzymology, Volume 676
ISSN 0076-6879
https://doi.org/10.1016/bs.mie.2022.06.006

properties of plant G-proteins are key to affecting plant phenotypes and responses. Detailed characterization of such activities, *in vitro* and *in planta*, will pave the way for precise manipulation of these proteins for future agricultural needs.

Abbreviations

GAPs GTPase activity accelerating proteins
GDP guanosine diphosphate
GEF guanine nucleotide exchange factor
GPCR G-protein coupled receptor
G-proteins guanine nucleotide-binding proteins
GTP guanosine-5′-triphosphate
RLK receptor-like kinase

1. Introduction

Heterotrimeric GTP-binding proteins (G-proteins hereafter) comprised of Gα, Gβ and Gγ subunits are key regulators of a multitude of signaling pathways in most eukaryotes. The G-protein complex is localized at the plasma membrane and couples the signal perception by G-protein coupled receptors (GPCRs) to intracellular effectors.

1.1 Classical G-protein signaling cycle

As per the classical paradigm, the G-protein complex switches between the inactive and active states depending on the nucleotide-bound form of Gα. During resting phase, Gα is GDP-bound and remains associated with the Gβγ dimer (Syrovatkina, Alegre, Dey, & Huang, 2016). Activation occurs when a signal-dependent change in the conformation of a GPCR causes exchange of GDP on Gα for GTP. The GPCRs essentially act as guanine nucleotide exchange factors (GEFs) in regulation of G-protein cycle. GTP-bound Gα dissociates from the Gβγ dimer and both freed entities can interact with downstream effectors to transduce the signal, representing the active state of signaling (Fig. 1). Deactivation occurs via the inherent GTPase activity of Gα which causes hydrolysis of bound GTP, regenerating its GDP-bound form. GDP-Gα associates with Gβγ restoring the trimeric complex, ready to be activated for the next round of signaling (Offermanns, 2003). Because of its cyclic nature, both activation and deactivation steps need to be synchronized for effective, continuous G-protein signaling (Ross, 2008; Siderovski & Willard, 2005); however, the inherent GTPase activity of Gα is slow and accessory proteins such as regulator of G-protein signaling (RGS) proteins act as GAPs (GTPase activity accelerating

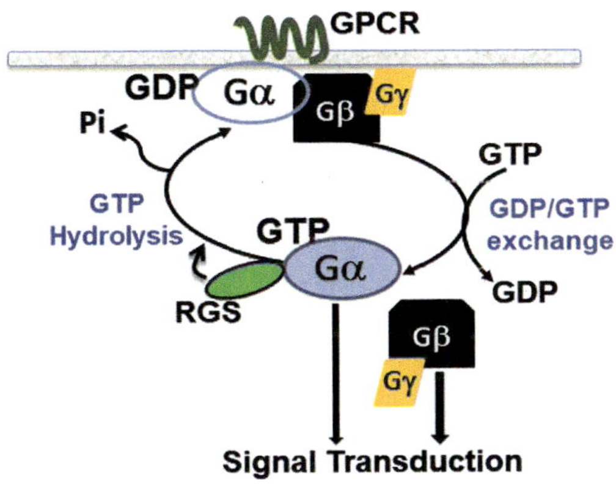

Fig. 1 Classic heterotrimeric G-protein signaling model. GPCR: G-protein coupled, 7 transmembrane receptor, RGS: regulator of G-protein signaling.

proteins) to enhance it. Therefore, both GEFs and GAPs are central to effective functioning of G-protein cycle.

1.2 Unique aspects of G-protein signaling in plants

The core G-protein components and their biochemistries are broadly conserved throughout eukaryotic evolution but several differences exist across plant lineages (Pandey, 2019). In general, plants have relatively small numbers of each of the G-protein components compared to metazoan. The fully sequenced genomes of model plants such as Arabidopsis or rice have only one canonical Gα and one Gβ coding genes, with a few Gγ. In contrast, humans have 23 Gα, 5 Gβ and 12 Gγ proteins, which may form more than 1000 potential heterotrimeric combinations. Higher numbers of G-proteins are present only in plants with recently duplicated genomes such as Camelina or soybean (Bisht, Jez, & Pandey, 2011; Roy Choudhury, Bisht, Thompson, Todorov, & Pandey, 2011; Roy Choudhury, Marlin, & Pandey, 2019).

Plants also have several unique G-protein components, which expand their networks (Mohanasundaram, Dodds, Kukshal, Jez, & Pandey, 2022; Roy Choudhury et al., 2019). The extra-large Gα (XLGα) proteins, present in 3 or more copies in most plants, have a large N-terminal extension with the canonical Gα-like domain. These proteins interact with the Gβγ dimer, although their regulation by classic G-protein cycle or by guanine nucleotide exchange remains to be established (Chakravorty, Gookin, Milner, Yu, & Assmann, 2015; Maruta et al., 2021). Similarly, most plants have multiple copies of a unique cysteine-rich Gγ proteins, which is involved in controlling many yield-related traits (Li et al., 2012; Mohanasundaram et al., 2022; Roy Choudhury et al., 2011).

In addition to the presence of these unique variants of G-proteins, some plant groups have lost the key G-protein components. For example, the entire Bryopsida moss group does not contain a canonical Gα protein and relies specifically on the XLGα proteins-dependent G-protein signaling. Similarly, many plant groups spanning the entire plant evolutionary spectrum exhibit random instances of RGS protein loss (Mohanasundaram et al., 2022).

The regulatory mechanisms of plant G-protein signaling also differ significantly from the established metazoan models. Canonical GPCRs that can facilitate GTP/GDP exchange have not been identified in plants, to date (Roy Choudhury et al., 2019). *In vitro* biochemical and structure-function characterization of plant Gα proteins suggests that these may have an extremely fast inherent rate of GDP/GTP exchange, and are likely GTP-bound without requiring a GPCR (Jones et al., 2011). Plant Gα proteins are also exceptionally slow GTPases, at least an order of magnitude slower than the slowest known human Gα (Hackenberg et al., 2016; Johnston et al., 2007). A fast (or spontaneous) acquisition of GTP combined with its very slow hydrolysis implies that the plant Gα proteins mostly remain in their GTP-bound, active form; unless their GTPase activity is promoted by GAPs such as RGS. This model entails that effective deactivation is the critical, rate-limiting regulatory step of plant G-protein signaling, and that GAPs are the key regulatory proteins (Fig. 2A). However, as was mentioned, many

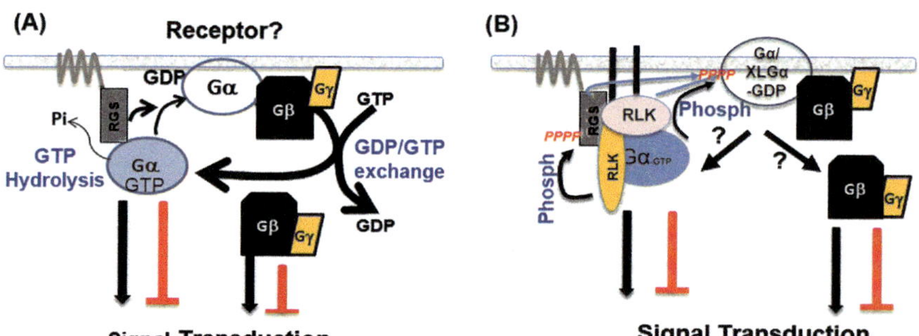

Fig. 2 Two potential models of G-protein signaling in plants. (A) Model based on classic G-protein signaling but the identity of receptors is unknown. The GDP/GTP exchange activity is proposed to be fast and spontaneous, while GTP hydrolysis is very slow. (B) An alternate model of G-protein signaling in plants, which is regulated by receptor-like kinase (RLK)-mediated phosphorylations. Signal can be transduced by either Gα or XLGα. Signal may be independent of the nucleotide exchange reactions. The Gα and Gβγ proteins may act as positive or negative regulators.

plants do not have an RGS protein, and the extent to which additional proteins such as phospholipases are sufficient for deactivation in such cases remains to be established (Mohanasundaram et al., 2022; Pandey, 2016; Roy Choudhury et al., 2019; Roy Choudhury & Pandey, 2016b).

Several plant-specific regulatory mechanisms have also been identified in recent years. These include the interaction and activation/deactivation of Gα proteins with receptor-like kinases, which are prevalent in plants (Pandey, 2020; Roy Choudhury & Pandey, 2016a). Both Gα and RGS proteins are regulated by RLK-dependent phosphorylation mechanisms (Fig. 2B) (Roy Choudhury & Pandey, 2015, 2022). It has been hypothesized that the extant diversity of G-protein signaling may arise at the level of receptors (hundreds in plants) and receptor-mediated unique phosphorylation signatures. Furthermore, a nucleotide exchange/hydrolysis-independent mechanism of G-protein signaling has also been proposed (Maruta et al., 2019).

1.3 Specificity of response regulation by G-proteins

The relatively simple repertoire of G-proteins affects almost all aspects of plant life and influence key agronomic traits such as overall yield, seed shape and size, oil content etc. The loss of G-protein function has been shown to affect normal developmental progression, which is shaped by both internal and external inputs, as well as the plants' response to changing environment or interactions with microbes, which depends on external signals. G-proteins exhibit these effects by regulating processes as fundamental and diverse as control of cell division and ion channel activity, modification of cell membrane and cell wall properties, response to multiple hormones, water and nutrient use efficiency and overall plant architecture (Roy Choudhury et al., 2019).

In general, the specificity of signal-response coupling was presumed to occur downstream of the heterotrimer, where active proteins interact with unique effector proteins, depending on specific signal, tissue type or the developmental stage of the plant. Only a few such effectors have been identified and linked to an explicit response to date. However, given the deviation of plant G-protein components and their regulation from the established metazoan paradigm, it is anticipated that novel mechanisms exist that need to be explored, both at the biochemical and the genetic level, to link G-protein-mediated signal transduction with specific responses. In the following sections, we describe biochemical approaches currently in practice to study different aspects of plant G-protein signaling.

2. Methods

2.1 Protein purification

In this part we describe methods for purification of recombinant Gα, XLGα, RGS, PLDα, and the kinase domain of RLKs from plants. A general purification protocol that works for these proteins is included, with any protein-specific modification listed.

2.1.1 Cloning of protein expression constructs

These methods are described for Arabidopsis Gα (AtGPA1, At2g26300), XLGα (AtXLG2, At4g34390), RGS (AtRGS1, At3g26090), PLDα (AtPLDα1, At3g15730) and BAK1 (AtBAK1, At4g33430) proteins but are easily adaptable to similar proteins from additional plants with minor modifications, as needed.

a. Full-length AtGPA1, PLDα1, AtXLG2, C-terminal RGS domains of AtRGS1 (amino acid 251–459), and C-terminal kinase domain of AtBAK1 (amino acid 247–615) were amplified from Arabidopsis seedling cDNA using gene-specific forward and reverse primers with *Nhe*I and *Not*I restriction enzymes (New England Biolabs) sites, respectively.

b. Amplified products were purified using a PCR purification kit (Qiagen) and cloned into the pET28a bacterial expression vector (Novagen) by restriction digestion and ligation. Ligated recombinant plasmids were introduced into *E. coli* competent cells (TOP10) via heat shock-mediated bacterial transformation.

c. Recombinant plasmids were isolated using plasmid isolation kit (Qiagen) from a single colony of transformed cells and verified by sequencing.

d. Recombinant plasmids harboring the gene were transformed into *E. coli* competent cells (BL21, DE3) by bacterial transformation.

2.1.2 Cell culture and protein induction

a. A single colony from the transformed cells was cultured in 3 mL of Luria broth media (Sigma) containing 50 mg/mL kanamycin (Gold Biotech) and incubated at 37 °C for 16 h with constant shaking.

b. Next day, 0.05 mL from the primary culture was added to a fresh 5 mL secondary culture containing 50 mg/mL kanamycin and incubated at 37 °C for 4 h with constant shaking. After 4 h, 0.5 mL from secondary culture was added to a fresh 250 mL tertiary culture containing 50 mg/mL kanamycin and incubated at 37 °C for 3 h with constant shaking. At this point the $O.D_{600}$ of bacterial cultures reached between 0.7 and 0.9.

c. Protein expression was induced by adding isopropyl ß-D-1-thiogalactopyranoside (IPTG, Gold Biotechnology) to the tertiary culture and cells were grown at 25 °C for 16 h. *For AtGPA1 and AtXLG2 purification, 1 mM IPTG was used for induction, while 0.5 mM IPTG was used for other proteins.

2.1.3 Recombinant protein purification

a. All purification steps were performed at 4 °C. Cells were pelleted by centrifugation at 6000 x *g* for 15 min at 4 °C (Beckman Avanti J-25).

b. The pelleted cells were mixed thoroughly using plastic pipettes in 25 mL lysis buffer (50 mM Tris-HCl, pH 8.0; 500 mM NaCl; 5 mM $MgCl_2$; 20 mM imidazole; 10% glycerol (v/v); 1% Tween (v/v), 1 mM phenylmethylsulfonyl fluoride (PMSF), 1 X protease inhibitor cocktail, Promega). The lysate was maintained at 4 °C.

c. The mechanical cell lysis was performed using a sonicator (Sonic Dismembrator 550, Fisher Scientific) using a 6-min cycle of 1 s on 1 s off at power setting 5.

d. The lysed cell mixture was centrifuged for 15 min at 15,000 x *g* at 4 °C, and the resulting supernatant was transferred to a new, pre-chilled tube.

e. The supernatant was passed through a 1 mL Ni-NTA-agarose resin column (Qiagen), maintained at 4 °C and unbound fraction was collected in a separate tube. The column was washed with 30 mL wash buffer (50 mM Tris-HCl, pH 8.0; 500 mM NaCl; 5 mM $MgCl_2$; 20 mM imidazole; 10% glycerol).

f. Proteins bound to the column were eluted in 4 to 5 fractions (1 mL each) by using elution buffer (50 mM Tris-HCl, pH 8.0; 500 mM NaCl; 5 mM $MgCl_2$; 250 mM imidazole; 10% glycerol).

g. Purified recombinant proteins were stored at −80 °C and verified by 10% SDS-PAGE analysis (Fig. 3A).

2.2 Site-directed mutagenesis

Key amino acids critical for the biochemical activities of Gα, XLGα, RGS, PLDα and RLKs are well-described. In the case of Gα proteins a change in a specific amino acid may convert the protein to its inactive form (any change in its GTP-binding region, e.g., Ser53 in AtGPA1) or in an constitutively active form (can no longer hydrolyse bound GTP, e.g., Gln222 in AtGPA1). Similarly, changes in specific amino acids in RGS protein

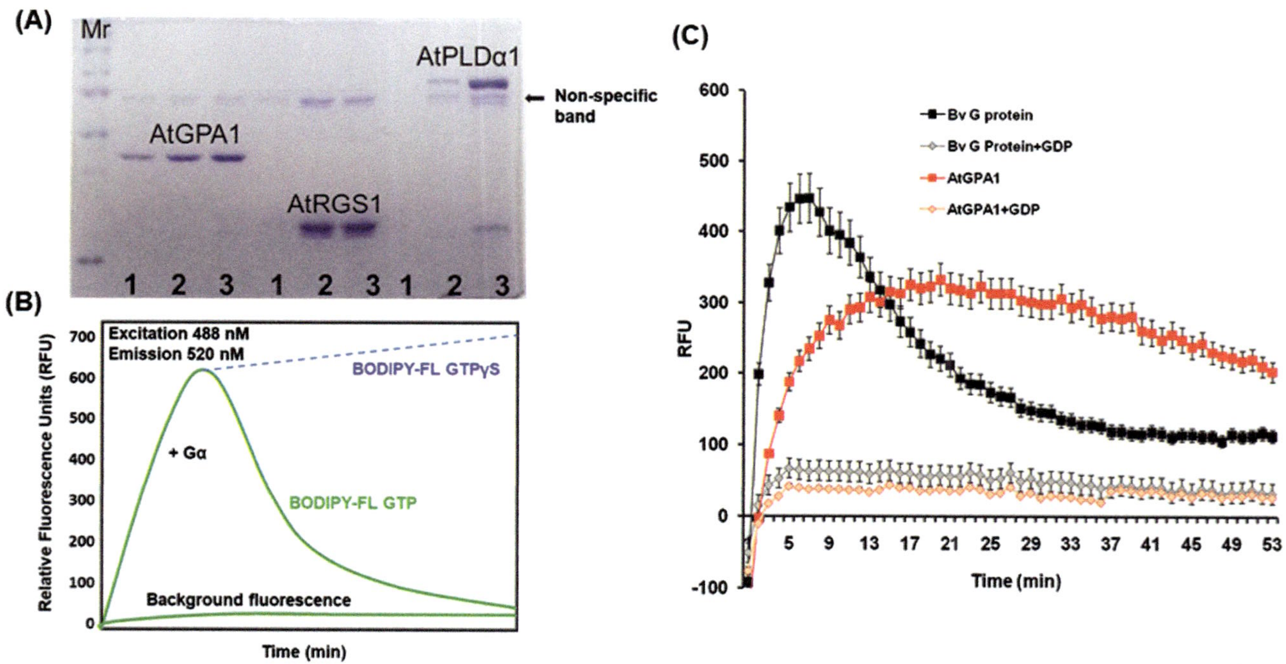

Fig. 3 Protein purification and G-protein activity measurement. (A) Purification of recombinant AtGPA1, C-terminal of AtRGS1, and AtPLDα1 proteins. The numbers denote different elution fractions for each protein. (B) Principle of BODIPY based GTP-binding assay. Upon binding with an active GTP-binding protein, the fluorescent moiety is released, shown as increased fluorescence over time. With BODIPY-FL GTP dye, the fluorescence decreases overtime due to the GTPase activity of Gα proteins. No decrease in fluorescence is observed with the non-hydrolysable GTP-γS version. (C) GTP-binding and hydrolysis activities of AtGPA1 and a bovine Gα (Bv) protein. The bovine protein shows faster rate of GTP-hydrolysis, seen as a fast downward slope compared to AtGPA1. The assay is specific for GTP/GDP, as no increase in fluorescence is seen when non-fluorescent GDP (or GTP, not shown) are included in the assays. Data show average of three experiments, error bars represent ±S.D.

(e.g., Glu320 in AtRGS1), PLDα1 or BAK1 may make them completely inactive. Specifically serine (S), threonine (T) and tyrosine (Y), the amino acids that can be phosphorylated by kinases, can be mutated to aspartate (D) or glutamate (E) to generate phospho-mimetic versions of the proteins, or to alanine (A) to create the phospho-deficient versions. These amino acid variants are critical for the study of protein activity *in vitro* and, the effects of such varied activities *in planta*, by introducing them to plants by transformation. General steps for generating site-directed protein variants are described, which can be applied to create the protein variant of interest, using specific primer sequences.

a. To change a particular amino acid, specific primers were designed by altering nucleotide sequences of genes. For example, to change AtGPA1 to AtGPA1^{Q222L}, the following primers were used:
AtGPA1^{Q222L} FP: TTGTTTGACGTGGGTGGAC(A►T)GAGA AATGAGAGGAGGAAA
AtGPA1^{Q222L} RP: TTTCCTCCTCTCATTTCTC(T►A)GTCC ACCCACGTCAAACAA

b. Cloned plasmids of interest (pET28a:Gα or pET28a:RGS1 or pET28a: PLDα1) were used as template plasmids for PCR using PfuTurbo DNA Polymerase (Stratagene) and modifying primers. Amplification was performed as follows: 95 °C for 2 min, 25 cycles of 95 °C for 30 s, 55 °C for 1 min and 68 °C for 8 min followed by final extension at 68 °C for 15 min. PCR cycles should be standardized for each mutagenesis.

c. Amplified PCR products were digested by restriction enzyme DpnI (New England Biolabs) for 37 °C for 1 h.

d. Digested PCR product (2 μL) was transformed into *E. coli* competent cells (Top10) via heat shock transformation.

e. Multiple single colonies from the transformed cells were cultured in 3 mL of Luria broth at 37 °C for 16 h for plasmid isolation.

f. Plasmids were isolated using a plasmid isolation kit (Qiagen) and sequenced to verify the desired amino acid change.

g. Mutant recombinant plasmids harboring the gene were introduced into *E. coli* competent cells (BL21/Rosetta) by heat shock-mediated bacterial transformation for further recombinant protein purification and activity assays.

2.3 GTP-binding and hydrolysis assays

In this section, we describe real-time GTP-binding and -hydrolysis assays for recombinant Gα and XLGα proteins using fluorescence or absorbance-based methods. These assays provide fast, reliable and high throughput quantification of these activities. The results can be further substantiated using additional stopped flow-based or radio-isotope labeled analysis, if needed.

2.3.1 GTP binding assay using BODIPY-GTP or BODIPY GTPγS

A real-time fluorescence-based assay was used to measure GTP-binding and hydrolysis. BODIPY-GTPγS-FL (4,4-difluoro-4-bora-3a,4a-diaza-s-indacene-GTPγS Fluorophore, Invitrogen) was used to detect GTP-binding, whereas BODIPY-GTP-FL (Invitrogen) was used to detect both GTP-binding and -hydrolysis.

a. BODIPY GTP reagents were diluted in 50 mL assay buffer (10 mM Tris-HCl, pH 8.0 and 10 mM $MgCl_2$) to make 10 nM BODIPY-GTPγS-FL or BODIPY-GTP-FL solutions.

b. Recombinant Gα protein was adjusted to 200 nM in BODIPY assay buffer.

c. Diluted Gα protein (100 μL) was added to 96 well, flat bottom polystyrene microplates (Greiner Bio One) and mixed thoroughly. Reaction mixes containing 100 nM BSA (instead of Gα), and without any protein were included as negative controls with each run. A total three to five replicates for each reaction were performed.

d. Injection tubes were washed with sterile water, followed by BODIPY-assay buffer before starting the reaction.

e. 100 μL of 10 nM BODIPY-GTPγS or 10 nM BODIPY-GTP FL solutions were added to the protein solution to initiate the reaction using the automatic injector of the plate reader. The final protein concentration should be 100 nM.

f. Fluorescence readings were recorded using a Tecan Pro plate reader using excitation at 488 nm, emission at 520 nm, at 25 °C every 15 s for up to 100 cycles (total time per read: ~30 min). These parameters can be adjusted to desired specifications for any compatible fluorescence plate reader.

g. Fluorescence readings for the negative controls were subtracted at each time point and relative fluorescence of the proteins as a factor of time was plotted (any software of your choice). When using BODIPY GTP-FL, the increase in fluorescence with time denotes GTP-binding and subsequent decrease in fluorescence with time denotes GTP hydrolysis by an active Gα protein (Fig. 3B).

h. To ascertain specificity of GTP-binding and hydrolysis, 5 μM of non-labeled nucleotides (GTP/GDP/ATP/ADP) can be added to the assay buffer before starting the reaction. The additives should also be included in the negative controls (Fig. 3C).

i. To quantify the effect of RGS1 or PLDα1 on the rate of GTP hydrolysis, varying concentrations of purified, recombinant RGS1 and/or PLDα1 proteins were added to the reaction mix in the presence or absence of Gα proteins. A reaction mix containing only RGS1 or PLDα1 protein was included as a negative control for these assays. The assays were performed for native as well as variant Gα proteins to study the effects of specific mutations on GTP-binding/hydrolysis activities as well as in the presence of native or mutant RGS1 or PLDα1 proteins.

j. At least five replicates were performed for each reaction and data were averaged. Error bars represent ±SE.

2.3.2 GTPase activity quantification using phosphate release assay

a. The EnzCheck® Phosphate Assay Kit (Invitrogen,) was used to analyze the phosphate released by GTP hydrolysis by Gα proteins in the presence of RGS1 or PLDα1 proteins. This kit used a spectrophotometric method for the fast and sensitive quantitation of inorganic phosphate (P_i) released from enzymatic reactions.

b. The principle is based on the quantitation of P_i consumed during enzymatic conversion of 2-amino-6-mercapto-7-methylpurine riboside (MESG) to ribose 1-phosphate and 2-amino-6-mercapto-7-methylpurine by purine nucleoside phosphorylase (PNP).

c. The standard 200 μL reaction mixture contains 10 μL 20 × reaction buffer, 40 μL MESG substrate solution, 2 μL purine nucleoside phosphorylase (PNP) and H_2O. The entire reaction mix was incubated for 10 min at 22 °C before adding PNP.

d. The Pi production was measured as a change in absorbance at 360 nm up to 30 min at 25 °C using microplate reader (Tecan Pro).

e. A standard curve of phosphate release was prepared by using different concentrations of inorganic phosphate provided in the kit (Fig. 4A).

f. To analyze Pi release from Gα proteins in the presence of RGS1 or PLDα1 proteins, the reaction mix was prepared similar to the standard reaction without the phosphate standard.

g. The prepared reaction mixture was incubated with Gα (5 μg) and varying concentrations of RGS1 (0–2.5 μg) or PLDα1 (0–2.5 μg) along with GTP (1 mM) to quantify the amount of P_i released (Fig. 4B). The amount of P_i released was estimated from the corresponding values of

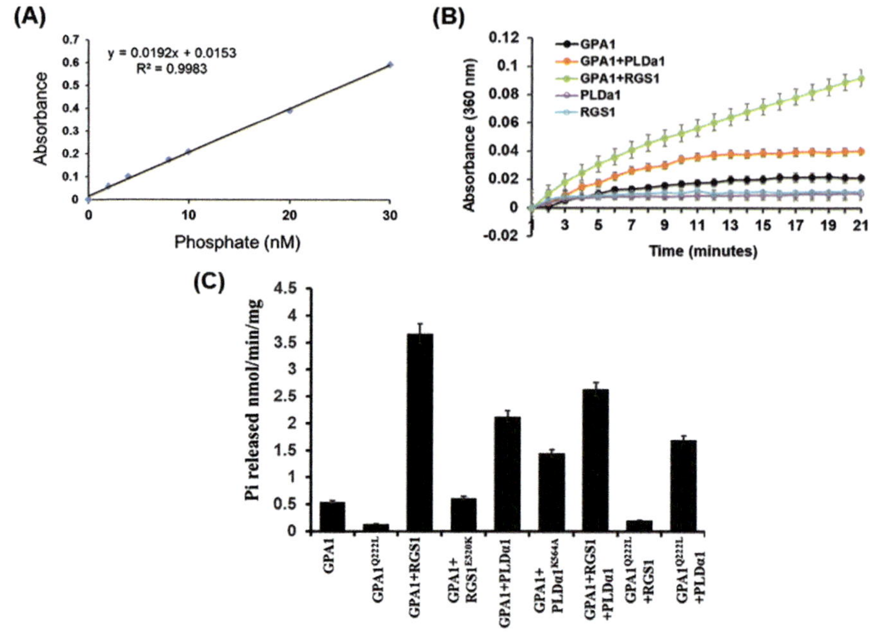

Fig. 4 GTPase activity measurement. (A) Generation of standard curve with known amount of phosphate in the assay mixture (B) Measurement of Pi release by AtGPA1 in the presence and absence of RGS1 and PLDα1 proteins. (C) Quantification of Pi released from AtGPA1 in the presence of RGS1 and PLDα1. GPA1^{Q222L} is a mutant GPA1 with no inherent GTPase activity. RGS^{E320K} is a mutant version of RGS1, which exhibits no GAP activity. PLDα1^{K564A} is a mutant version of PLDα1, which shows reduced GAP activity. Data show average of five experiments, error bars represent ±S.E.

the standard curve. Protein variants of Gα, RGS1 or PLDα1 can be added to the reaction mix to evaluate their effects on GTP-hydrolysis (Fig. 4C).

h. For each assay, three to five biological replicates were performed, and data were averaged. Error bars represent ±SE.

2.4 Phospholipase activity assay

Phospholipases acting as GAPs is well-established in mammalian systems, where phospholipase Cβ (PLCβ) isoforms act both as GAPs and as effectors of Gα proteins (Ross, 2008, 2011). Plants lack classical PLCβ homologs precluding such a possibility. We have shown that phospholipase D proteins (e.g. PLDα1) fulfill this role in plants (Pandey, 2016). Interestingly, these two plant GAPs, RGS1 and PLDα1 interact with each other as well as with core G-proteins to form higher-order protein complexes *in vivo*.

Furthermore, RGS1 and PLDα1 regulate the activity of each other in a double negative regulatory loop. The net outcome of such complex interactions may be an exquisitely controlled level and duration of active Gα, modulating the specificity of response regulation (Hackenberg et al., 2016).

2.4.1 Phospholipase Dα1 activity assay

Phospholipase D activity was quantified using Amplex® Red Phospholipase D Assay Kit (Life Technologies). In this assay a substrate, phosphatidylcholine (lecithin) was cleaved by PLD enzymes (e.g., PLDα1) to generate choline and phosphatidic acid. Choline is oxidized by choline oxidase to betaine and H_2O_2. H_2O_2 reacts with Amplex Red reagent in presence of horseradish peroxidase (HRP) to generate Resorufin, a fluorescent product, which is quantified using a fluorescent plate reader.

a. The standard 100 μL reaction mixture was prepared using 10 μL of Amplex Red reagent stock solution, 10 μL of HRP stock solution, 10 μL of choline oxidase stock solution, 5 μL of the lecithin and 965 μL of 1 × reaction buffer.

b. Resorufin, resulting from PLDα1 activity, was quantified for up to 60 min at 37 °C using excitation at 530 nm and emission at ~590 nm with a microplate reader (Tecan Pro).

c. To detect PLD activity, ~250 ng of recombinant purified PLDα1 was used either in the presence or absence of varying concentrations Gα and RGS1 recombinant proteins. The background fluorescence value was corrected by subtracting the values of no-PLDα1 control. Protein variants of Gα, RGS1 or PLDα1 can be added to the reaction mix to evaluate their effects.

d. For each assay, three to five biological replicates were performed, and data were averaged (Fig. 5). Error bars represent ±SE.

2.5 Protein phosphorylation assays

Phosphorylation of Gα, XLGα and RGS proteins by RLKs has emerged as a key G-protein regulatory mechanism in recent years. The proteins can be phosphorylated, *in vitro*, in the presence of a relevant kinase (e.g., cytoplasmic domain of an RLK) and phosphorylated amino acids can be identified by liquid chromatography mass spectrometry (LC-MS/MS)-based analysis. By generating phospho-mimetic and phospho-deficient versions of these proteins by site-directed mutagenesis, the effects of such phosphorylations can also be detected on their GTP-binding and hydrolysis.

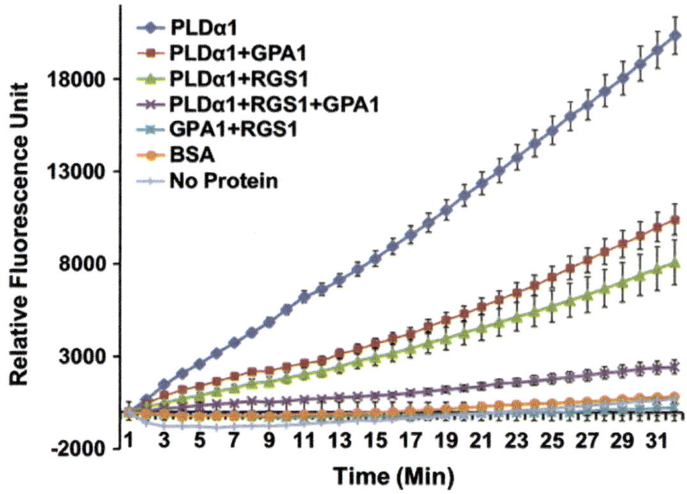

Fig. 5 Phospholipase activity measurement. Measurement of phospholipase activity of PLDα1 and its regulation by AtGPA1 and RGS1 proteins. Data show average of three experiments, error bars represent ±S.E.

2.5.1 In vitro phosphorylation assay

a. Recombinant purified RLK (0.5 μg) was mixed with 1 μg of native or mutant AtGPA1 or AtRGS1 were incubated in 20 μL of 25 mM Tris-HCl (pH 8.0), 15 mM $MgCl_2$, 2 mM $MnCl_2$, 7.5 mM NaCl, 0.05 mM EDTA, 0.05 mM DTT, 5% glycerol, and 75 μM ATP at 27 °C for 30 min.

b. The reaction was stopped by adding 5× SDS sample loading buffer followed by boiling for 5 min. Samples were analyzed via 10% SDS-PAGE.

c. To detect phosphorylated proteins, the gel was stained with Pro-Q® Diamond phosphoprotein gel stain (Invitrogen) following manufacturer's instructions and imaged using the Typhoon 9410 variable mode imager (Molecular Dynamics). For protein staining, this same gel was immersed in SYPRO™ ruby protein gel stain (Invitrogen) followed by imaging using the Typhoon 9410 variable mode imager.

2.5.2 LC-MS/MS analysis to detect phosphorylated amino acids

a. For LC-MS/MS analysis, 20 μL of the kinase-substrate protein solution was reduced using 10 mM tris(2-carboxyethyl)phosphine (TCEP, Thermo Scientific) and alkylated with Iodoacetemide (20 mM) before digestion with trypsin (0.2 μg) overnight. Digest was acidified with 1% trifluoroacetic acid (TFA) and cleaned up with C18 tip.

b. The extracted peptides were dried and each sample was resuspended in 20 μL 5% acetonitrile (AN)/0.1% formic acid (FA). Two μL (~1 μg) peptides were analyzed by LC-MS with a Dionex RSLCnano HPLC coupled to a LTQ-Orbitrap Velos Pro (Thermo Scientific) mass spectrometer using a 2 h gradient. Peptides were resolved using 75 μm × 25 cm PepMap C18 column (Thermo Scientific).

c. All MS/MS samples were analyzed using Mascot (Matrix Science, London, UK; version 2.5.1.0). Mascot was set up to search the provided sequence as well as common contaminants. The digestion enzyme was set as trypsin. Mascot was searched with a fragment ion mass tolerance of 0.60 Da and a parent ion tolerance of 10 ppm. Oxidation of methionine, carbamidomethylation of cysteine, acetylation of N-terminal of protein and phosphorylation of serine, threonine and tyrosine were specified in Mascot as variable modifications.

d. Scaffold (version Scaffold_4.5.3 Proteome Software Inc., Portland, OR) was used to validate MS/MS based peptide and protein identifications. Peptide identifications were accepted if they could be established at greater than 80.0% probability by the Peptide Prophet algorithm (Keller, Nesvizhskii, Kolker, & Aebersold, 2002) with Scaffold delta-mass correction. Protein identifications were accepted if they could be established at greater than 99.0% probability and contained at least 2 identified peptides. Protein probabilities were assigned by the Protein Prophet algorithm (Nesvizhskii, Keller, Kolker, & Aebersold, 2003). Proteins that contained similar peptides and could not be differentiated based on MS/MS analysis alone were grouped to satisfy the principles of parsimony. Proteins sharing significant peptide evidence were grouped into clusters.

3. Conclusions

Changes in G-protein biochemistry is central to regulation of many signaling pathways, in almost all life forms. In metazoan systems, these biochemical mechanisms are exceptionally well-characterized, as these also serve as targets for many pharmaceutical compounds. It has become evident that G-proteins are involved in controlling almost all aspects of plants life cycle, but their detailed biochemical characterization remains limited. This is partly due to the fact that even though the basic biochemical characteristics of specific proteins of the complex are similar to those found in other organisms, their regulation differs significantly from the established norm.

Plant G-proteins exhibit multiple types of regulations, and in many cases these can be connected to specific plant phenotypes. The existence of such flexible regulatory mechanisms implies that the G-protein networks in plants have been rewired to achieve the plasticity needed for plant growth and development. It also suggests that the relatively smaller numbers of G-protein components can be modified in several different ways to achieve the required multiplicity.

Detailed biochemical characterization of G-protein signaling components in plants is still limited to a handful of species, and mostly involves the Gα/XLGα and RGS proteins. Except for PLDα1 and a PLAδIII enzyme, the entire expanded family of phospholipases remains uncharacterized in the context of G-proteins. Similarly, multiple RLKs have been shown to interact with and phosphorylate Gα, detailed characterization of signal-dependent RLK activation and their effect on G-protein signaling is only beginning to be evaluated. Characterization of the biochemical properties and regulation of plant Gβ proteins remains restricted to a handful of studies showing their interactions with certain proteins. The methods discussed in this chapter are a step forward to begin these characterizations, and then link specific biochemical variations of these proteins with distinct phenotypes. With the availability of techniques which can precisely manipulate protein levels or activity *in planta* (e.g., specific base-edited plants), these biochemical characterizations will be pivotal to improve discrete plant traits, with precision breeding or editing, without the need to completely remove a gene or ectopically increase the protein level. Such precise manipulations are expected to be key to our future needs.

Acknowledgments

S.P. conceived methods chapter outline. K.H. and S.R.C. performed experiments. S.P. and S.R.C. wrote the chapter, analyzed data, and prepared figures. All authors have read and edited the manuscript. Work in S.P. lab is supported by the National Science Foundation grants NSF-IOS-1157942 and NSF-MCB-1714693. K.H. was supported by an NSF funded Research Experience for Undergraduate grant NSF-DBI-REU-1156581 to S.P.

References

Bisht, N. C., Jez, J. M., & Pandey, S. (2011). An elaborate heterotrimeric G-protein family from soybean expands the diversity of plant G-protein networks. *New Phytologist*, *190*(1), 35–48. https://doi.org/10.1111/j.1469-8137.2010.03581.x.

Chakravorty, D., Gookin, T. E., Milner, M., Yu, Y., & Assmann, S. M. (2015). Extra-large G proteins (XLGs) expand the repertoire of subunits in Arabidopsis heterotrimeric G protein signaling. *Plant Physiology, 169*, 512–529. Retrieved from http://www.ncbi.nlm.nih.gov/pubmed/26157115. https://doi.org/10.1104/pp.15.00251.

Hackenberg, D., McKain, M. R., Lee, S. G., Roy Choudhury, S., McCann, T., Schreier, S., et al. (2016). Galpha and regulator of G-protein signaling (RGS) protein pairs maintain functional compatibility and conserved interaction interfaces throughout evolution despite frequent loss of RGS proteins in plants. *The New Phytologist, 216*(2), 562–575. Retrieved from http://www.ncbi.nlm.nih.gov/pubmed/27634188. doi:10.1111/nph.14180.

Johnston, C. A., Taylor, J. P., Gao, Y., Kimple, A. J., Grigston, J. C., Chen, J. G., et al. (2007). GTPase acceleration as the rate-limiting step in Arabidopsis G protein-coupled sugar signaling. *Proceedings of the National Academy of Sciences of the United States of America, 104*(44), 17317–17322. Retrieved from http://www.ncbi.nlm.nih.gov/pubmed/17951432. https://doi.org/10.1073/pnas.0704751104.

Jones, J. C., Duffy, J. W., Machius, M., Temple, B. R., Dohlman, H. G., & Jones, A. M. (2011). The crystal structure of a self-activating G protein alpha subunit reveals its distinct mechanism of signal initiation. *Science Signaling, 4*(159), ra8. Retrieved from http://www.ncbi.nlm.nih.gov/pubmed/21304159. https://doi.org/10.1126/scisignal.2001446.

Keller, A., Nesvizhskii, A. I., Kolker, E., & Aebersold, R. (2002). Empirical statistical model to estimate the accuracy of peptide identifications made by MS/MS and database search. *Analytical Chemistry, 74*(20), 5383–5392. Retrieved from http://www.ncbi.nlm.nih.gov/pubmed/12403597.

Li, S., Liu, Y., Zheng, L., Chen, L., Li, N., Corke, F., et al. (2012). The plant-specific G protein gamma subunit AGG3 influences organ size and shape in Arabidopsis thaliana. *The New Phytologist, 194*(3), 690–703. Retrieved from https://www.ncbi.nlm.nih.gov/pubmed/22380792. https://doi.org/10.1111/j.1469-8137.2012.04083.x.

Maruta, N., Trusov, Y., Chakravorty, D., Urano, D., Assmann, S. M., & Botella, J. R. (2019). Nucleotide exchange-dependent and nucleotide exchange-independent functions of plant heterotrimeric GTP-binding proteins. *Science Signaling, 12*(606). Retrieved from https://www.ncbi.nlm.nih.gov/pubmed/31690635. doi:10.1126/scisignal.aav9526.

Maruta, N., Trusov, Y., Urano, D., Chakravorty, D., Assmann, S. M., Jones, A. M., et al. (2021). GTP binding by Arabidopsis extra-large G protein 2 is not essential for its functions. *Plant Physiology, 186*(2), 1240–1253. Retrieved from https://www.ncbi.nlm.nih.gov/pubmed/33729516. https://doi.org/10.1093/plphys/kiab119.

Mohanasundaram, B., Dodds, A., Kukshal, V., Jez, J. M., & Pandey, S. (2022). Distribution and the evolutionary history of G-protein components in plant and algal lineages. *Plant Physiology*. Retrieved from https://www.ncbi.nlm.nih.gov/pubmed/35377452. doi:10.1093/plphys/kiac153.

Nesvizhskii, A. I., Keller, A., Kolker, E., & Aebersold, R. (2003). A statistical model for identifying proteins by tandem mass spectrometry. *Analytical Chemistry, 75*(17), 4646–4658. Retrieved from http://www.ncbi.nlm.nih.gov/pubmed/14632076.

Offermanns, S. (2003). G-proteins as transducers in transmembrane signalling. *Progress in Biophysics and Molecular Biology, 83*(2), 101–130. Retrieved from http://www.ncbi.nlm.nih.gov/entrez/query.fcgi?cmd=Retrieve&db=PubMed&dopt=Citation&list_uids=12865075. https://doi.org/10.1016/s0079-6107(03)00052-x.

Pandey, S. (2016). Phospholipases as GTPase activity accelerating proteins (GAPs) in plants. *Plant Signaling & Behavior, 11*(5), e1176821. Retrieved from http://www.ncbi.nlm.nih.gov/pubmed/27124090. https://doi.org/10.1080/15592324.2016.1176821.

Pandey, S. (2019). Heterotrimeric G-protein signaling in plants: Conserved and novel mechanisms. *Annual Review of Plant Biology, 70*, 213–238. Retrieved from https://www.ncbi.nlm.nih.gov/pubmed/31035831. https://doi.org/10.1146/annurev-arplant-050718-100231.

Pandey, S. (2020). Plant receptor-like kinase signaling through heterotrimeric G-proteins. *Journal of Experimental Botany*, *71*(5), 1742–1751. Retrieved from https://www.ncbi.nlm.nih.gov/pubmed/31930311. https://doi.org/10.1093/jxb/eraa016.

Ross, E. M. (2008). Coordinating speed and amplitude in G-protein signaling. *Current Biology*, *18*(17), R777–R783. Retrieved from http://www.ncbi.nlm.nih.gov/pubmed/18786383. https://doi.org/10.1016/j.cub.2008.07.035.

Ross, E. M. (2011). Galpha(q) and phospholipase C-beta: Turn on, turn off, and do it fast. *Science Signaling*, *4*(159), pe5. Retrieved from http://www.ncbi.nlm.nih.gov/pubmed/21304157. https://doi.org/10.1126/scisignal.2001798.

Roy Choudhury, S., Bisht, N. C., Thompson, R., Todorov, O., & Pandey, S. (2011). Conventional and novel Gg protein families constitute the heterotrimeric G-protein signaling network in soybean. *PLoS One*, *6*(8), e23361. Retrieved from http://www.ncbi.nlm.nih.gov/pubmed/21853116. https://doi.org/10.1371/journal.pone.0023361.

Roy Choudhury, S., Marlin, M. A., & Pandey, S. (2019). The role of Gbeta protein in controlling cell expansion via potential interaction with lipid metabolic pathways. *Plant Physiology*, *179*(3), 1159–1175. Retrieved from https://www.ncbi.nlm.nih.gov/pubmed/30622152. https://doi.org/10.1104/pp.18.01312.

Roy Choudhury, S., & Pandey, S. (2015). Phosphorylation-dependent regulation of G-protein cycle during nodule formation in soybean. *Plant Cell*, *27*(11), 3260–3276. Retrieved from http://www.ncbi.nlm.nih.gov/pubmed/26498905. https://doi.org/10.1105/tpc.15.00517.

Roy Choudhury, S., & Pandey, S. (2016a). Interaction of heterotrimeric G-protein components with receptor-like kinases in plants: An alternative to the established signaling paradigm? *Molecular Plant*, *9*(8), 1093–1095. Retrieved from https://www.ncbi.nlm.nih.gov/pubmed/27250573. https://doi.org/10.1016/j.molp.2016.05.012.

Roy Choudhury, S., & Pandey, S. (2016b). The role of PLDalpha1 in providing specificity to signal-response coupling by heterotrimeric G-protein components in Arabidopsis. *The Plant Journal*, *86*(1), 50–61. Retrieved from http://www.ncbi.nlm.nih.gov/pubmed/26935351. https://doi.org/10.1111/tpj.13151.

Roy Choudhury, S., & Pandey, S. (2022). SymRK-dependent phosphorylation of Galpha protein and its role in signaling during soybean (Glycine max) nodulation. *The Plant Journal*, *110*(1), 277–291. Retrieved from https://www.ncbi.nlm.nih.gov/pubmed/35048428. https://doi.org/10.1111/tpj.15672.

Siderovski, D. P., & Willard, F. S. (2005). The GAPs, GEFs, and GDIs of heterotrimeric G-protein alpha subunits. *International Journal of Biological Sciences*, *1*(2), 51–66. Retrieved from http://www.ncbi.nlm.nih.gov/pubmed/15951850.

Syrovatkina, V., Alegre, K. O., Dey, R., & Huang, X. Y. (2016). Regulation, signaling, and physiological functions of G-proteins. *Journal of Molecular Biology*, *428*(19), 3850–3868. Retrieved from http://www.ncbi.nlm.nih.gov/pubmed/27515397. https://doi.org/10.1016/j.jmb.2016.08.002.

CHAPTER TWELVE

Soybean seed proteomics: Methods for the isolation, detection, and identification of low abundance proteins

Wonseok Kim[a,†], Joel Nott[b,†], Sunhyung Kim[a], and Hari B. Krishnan[a,c,*]
[a]Division of Plant Science and Technology, University of Missouri, Columbia, MO, United States
[b]The Protein Facility of the Iowa State University Office of Biotechnology, Ames, IA, United States
[c]Plant Genetics Research Unit, USDA, Agricultural Research Service, Columbia, MO, United States
*Corresponding author: e-mail address: hari.b.krishnan@usda.gov

Contents

Abstract

The salt-soluble globulins, glycinins (11S globulin), and β-conglycinins (7S globulin), are the most abundant seed proteins of soybean seeds. Together, these two groups of proteins account for 60–70% of total soybean seed proteins. Proteomic assessment of the less abundant soybean seed proteins using general isolation protocols is

† These authors contributed equally.

Methods in Enzymology, Volume 676
ISSN 0076-6879
https://doi.org/10.1016/bs.mie.2022.07.001

challenging due to the overwhelming abundance of storage proteins. Development of a simple, fast, and inexpensive method to remove most storage proteins from a seed extract will significantly enhance the study of the nonabundant proteins within seeds. We have developed two simple methods for the depletion of abundant seed proteins resulting in the enrichment of low abundance proteins from soybean seeds. Here, we provide a detailed procedure for the isolation, separation, identification, and quantification of low abundance seed proteins of soybean.

1. Introduction

Seeds have played a vital role in human diet as a major source of dietary protein. For this reason, seeds of crop plants have been intensively studied. Seeds store large amount of reserve materials in form of proteins, carbohydrates, and lipids that are mobilized during seed germination. Among crop plants, legumes have been extensively used as a major source of protein because they accumulate higher concentration of proteins than cereals. Seeds accumulate various proteins, most of which have either metabolic or structural roles. In addition, the seeds commonly accumulate a group of abundant proteins termed as storage proteins (Shewry, Napier, & Tatham, 1995), which are responsible for determining overall total protein content of the seed. Because of their abundance and economic importance, seed storage proteins were among the earliest proteins that were isolated and crystalized (Shewry et al., 1995). The systematic study of seed proteins was first initiated by Osborne (1924), who classified them into four groups (albumins, globulins, prolamins, and glutelins) based on their solubility properties. From that point onwards, numerous analytical, biochemical, and physiological methods have been extensively used to characterize seed proteins from a wide variety of source plants. The advent of "omics" technology has immensely accelerated the identification and characterization of seed storage proteins.

Proteomics is the large-scale study of proteomes. Seed proteomics has progressed rapidly due to improvements in instrumentation and refinements in protein isolation and separation techniques and technologies (Miernyk, 2014; Miernyk & Hajduch, 2011; Smolikova et al., 2020). The proteome of seeds, which reflects the underlying transcriptome, differs from each other. For example, the seed proteome of maize is quite distinct from that of pea (Bourgeois et al., 2009; Yu et al., 2016). Proteomics is often used for investigating the rates of protein production, steady-state abundance, post-translational modifications, protein-protein interactions, and to identify proteins involved in specific biological process (Low et al., 2021; McWhite et al., 2020; Zecha et al., 2022). Progress in protein separation techniques, in conjunction with technological advances in mass spectrometry, has allowed proteomics to develop quickly, where it is now playing an increasingly key role in functional genomics (Smolikova et al., 2020; Zhang et al., 2022).

Soybean (*Glycine max* [L.] Merr) seeds feature exceptionally high levels of protein (~40%) and is the primary source of protein for animal feed worldwide (Cabanos, Matsuoka, & Maruyama, 2021; Krishnan, 2000; Nielsen, 1996). Soybean seed proteins are composed of two salt-soluble protein classes: glycinins (11S globulin) and β-conglycinins (7S globulin) (Derbyshire, Wright, & Boulter, 1976). Both the glycinins and β-conglycinins are encoded by multigene families (Cho, Davies, & Nielsen, 1989; Li & Zhang, 2011). Six functional genes encoding glycinin (Gy1, Gy2, Gy3, Gy4, Gy5, and Gy7) have been reported in soybean (Beilinson et al., 2002; Nielsen et al., 1989). Each glycinin is synthesized as a precursor protein, which is post-translationally cleaved into acidic and basic polypeptides that are linked by a disulfide bond (Staswick, Hermodson, & Nielsen, 1984). In contrast to other glycinins, the glycinin produced by Gy4 undergoes cleavage at two sites resulting in two acidic (A4 and A5) peptides and one basic (B3) peptide (Momma et al., 1985) with theoretical molecular masses of 11 kDa (A5), 29 kDa (A4), and 21 kDa (B3), respectively (Momma et al., 1985). As many as 15 β-conglycinin genes have been identified in soybean cultivar "Williams 82" genome (Li & Zhang, 2011); however, most of these genes appear to be pseudogenes and only three genes encode the most highly seed-expressed α′-, α-, and β-conglycinins (Kitamura, Takagi, & Shibasaki, 1976; Thanh & Shibasaki, 1978).

Together, the 7S and 11S globulins account for 60–70% of soybean total seed proteins (Krishnan, 2000; Nielsen, 1996). The other seed proteins only account for a minor portion of the total seed protein content. Recent advances in mass spectrometry have led to the identification of numerous soybean seed proteins that have significantly increased the proteome coverage of soybean seed (Agrawal, Hajduch, Graham, & Thelen, 2008; Hajduch, Ganapathy, Stein, & Thelen, 2005; Islam et al., 2019; Islam, Krishnan, & Natarajan, 2020, 2021; Komatsu et al., 2017; Min, Hyeon, et al., 2020; Mooney, Krishnan, & Thelen, 2004; Ohyanagi, Sakata, & Komatsu, 2012). However, analysis of soybean proteome consisting of thousands of proteins has limited dynamic resolution because only the most abundant proteins can be detected (van Wijk, 2001).

Proteomic assessment of the less abundant proteins within soybean seed using general isolation protocols is challenging when the overwhelming majority, sometimes 60–80%, is made up of storage proteins. Development of a simple, fast, and inexpensive method to remove most storage proteins from a seed extract would significantly enhance the study of the nonabundant proteins within seeds. Methods have been reported in the literature for the depletion of abundant seed proteins based upon different physiological and/or biochemical principles (Kim et al., 2015; Miernyk & Hajduch, 2011; Miernyk & Johnston, 2006; Min, Park, et al., 2020). Our laboratory has developed simple, rapid, and inexpensive methods to isolate and enrich low abundant proteins from soybean seeds (Krishnan, Oehrle, & Natarajan, 2009; Natarajan, Krishnan, Lakshman, & Garrett, 2009). Here, we provide a detailed protocol for the isolation, separation, identification, and quantification of low abundance seed proteins of soybean. A schematic outline for the isolation, separation, and proteomic analysis of low abundant soybean seed proteins is shown in Fig. 1.

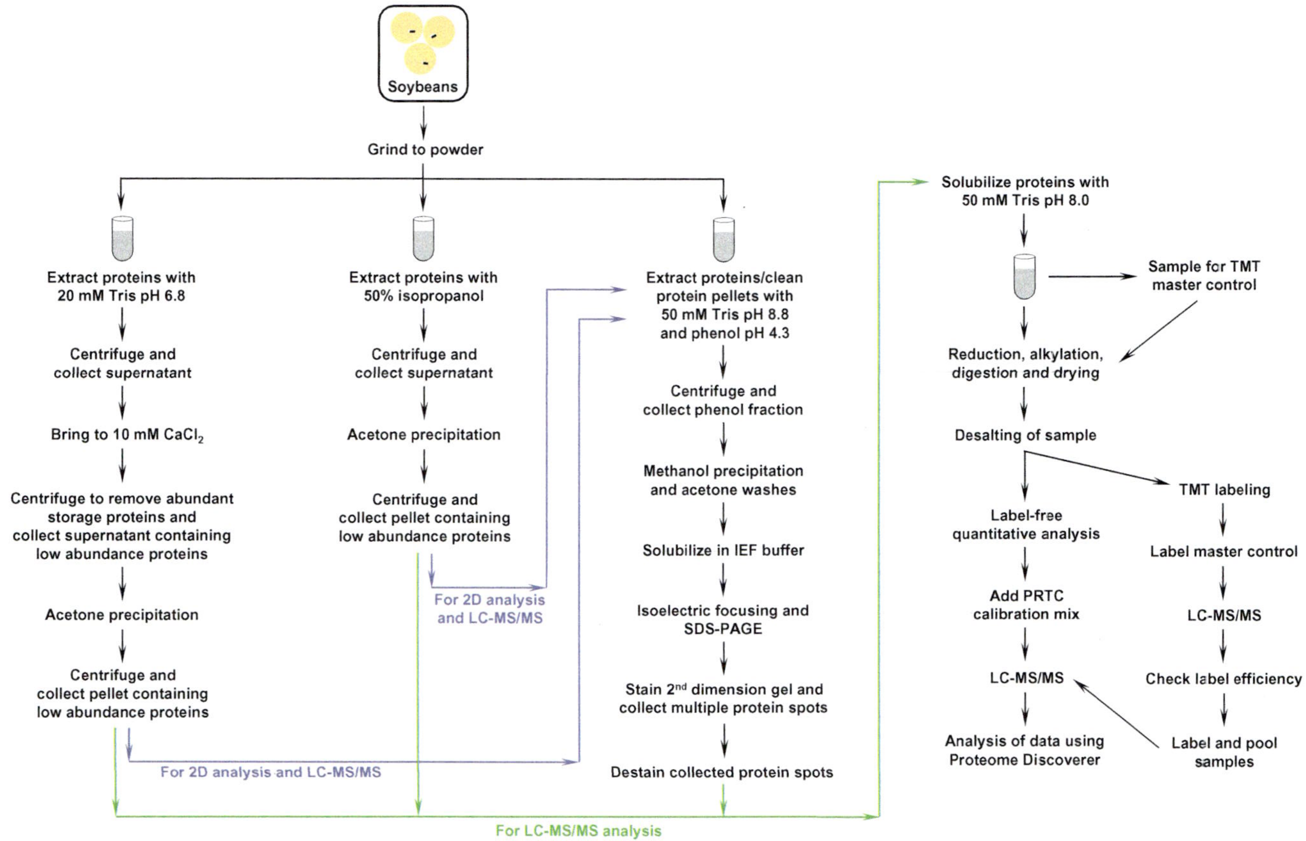

Fig. 1 A schematic outline for the isolation, separation, and proteomic analysis of low abundance soybean seed proteins.

2. $CaCl_2$ fractionation method to obtain low abundance proteins

Depletion of abundant seed storage proteins has been achieved, with different degree of success, through various strategies based on immuno-removal, isoelectric precipitation, salting out, affinity chromatography, random peptide-beads, and fractional solubility (Kim et al., 2015; Miernyk & Hajduch, 2011; Miernyk & Johnston, 2006). Inherent difficulties are encountered with most of these techniques due to low efficiency, cross-reactivity, broad-range affinity, low reproducibility, and expense of technique. Among these methods, the calcium chloride ($CaCl_2$) fractionation protocol developed in our laboratory to selectively remove storage proteins is the easiest, has the greatest capacity and efficiency, and is the least expensive. Our method is quick and can be performed with common laboratory equipment and reagents.

2.1 Materials and equipment

Tris base [Tris(hydroxymethyl)aminomethane]
Centrifuge capable of achieving 16,000 x *g*
Centrifuge tubes capable of withstanding 16,000 x *g*
100 mM $CaCl_2 \cdot 2H_2O$ (calcium chloride dihydrate)
100% Acetone
−20 °C freezer

2.2 Protocol

1. Grind soybean seeds into a fine powder using a mortar and pestle
2. Weigh out 200 mg of seed powder and transfer into a 15 mL centrifuge tube
3. Add 10 mL of 20 mM Tris-HCl, pH 6.8
4. Mix the tube contents for 10 min, with vigorous shaking, at room temperature
5. Clarify the slurry by centrifugation at 16,000 × *g* for 15 min
6. Remove the supernatant and place it into a clean tube
7. Add the required amount of 100 mM $CaCl_2$ stock solution to obtain a final concentration of 10 mM
8. Mix the tube contents for 1 min and leave it at room temperature for 5 min
9. Clarify the slurry by centrifugation at 16,000 × *g* for 15 min

10. Discard the pellet that contains predominantly the abundant seed storage proteins
11. Carefully remove the supernatant, which contains less abundant seed proteins, and place it into a clean tube
12. To this, add 3 volumes of ice-cold 100% acetone, mix the contents, and place it at −20 °C freezer overnight
13. Centrifuge the samples at 16,000 × *g* for 15 min
14. Air-dry the pellet at room temperature for 10 min
15. For one-dimensional (1D) electrophoretic gel analysis (Fig. 2), the pellet can be directly dissolved in SDS-sample buffer. For two-dimensional (2D) electrophoretic analysis gel analysis (Fig. 3), re-isolate the pellet from Step 14 beginning with Step 2 (Section 4.1.2).

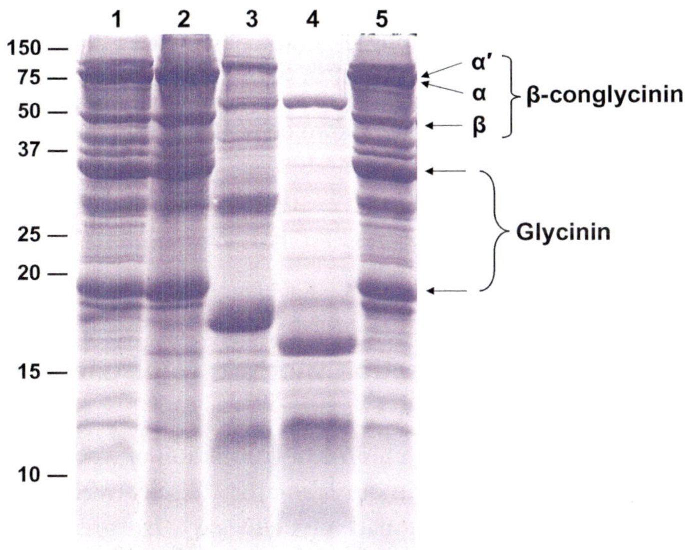

Fig. 2 One-dimensional sodium dodecyl sulfate-polyacrylamide gel electrophoresis (1D SDS-PAGE) analysis of soybean seed proteins. Total soybean seed proteins (lanes 1 and 5), soybean storage proteins specifically removed by calcium precipitation (lane 2), low abundance soybean seed proteins obtained by calcium fractionation (lane 3), and 50% isopropanol extracted soybean seed proteins (lane 4) were analyzed by 15% SDS-PAGE. Separated proteins were visualized by staining the gel with Coomassie Blue. Polypeptides corresponding to α′, α and β-subunits of β-conglycinin and the acidic and basic subunits of glycinin are shown with arrows. The position and sizes of protein markers in kDa are shown on the left side of the figure.

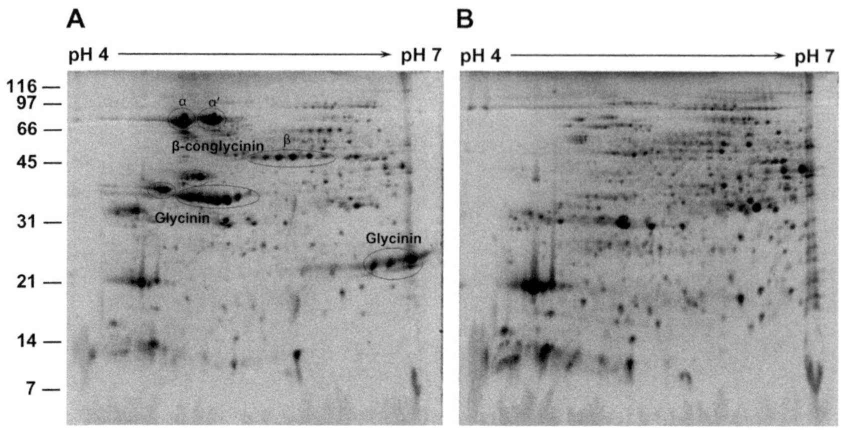

Fig. 3 Two-dimensional (2D) gel electrophoresis separation of soybean seed proteins. Total (A) and calcium chloride fractionated (B) soybean seed proteins were separated by isoelectric focusing on IPG strips (pH 4–7) and then by SDS-PAGE on 13.5% gels. The gels were stained with Colloidal Coomassie Blue G-250. Calcium chloride fractionation removes approximately 90% of highly abundant seed proteins from the protein extract (A), allowing for nearly 300 previously inconspicuous proteins present in soybean seed to be more detectable (B). Protein spots corresponding to α′, α and β-subunits of β-conglycinin and the acidic and basic subunits of glycinin are circled (A). The position and sizes of protein markers in kDa are shown on the left side of the figure.

16. For mass spectrometry analysis, resuspend the pellet from Step 14 in minimal volume of 50 mM Tris-HCl, pH 8.0, and proceed with Step 1 (Section 4.2.2).

3. Isopropanol extraction method to obtain low abundance proteins

In addition to the calcium chloride fractionation protocol in Section 2, we developed another method in our laboratory to selectively remove storage proteins. Isopropanol fractionation is also easy, has great capacity and efficiency, and is inexpensive. The protein pellet obtained from the final protocol step can be further processed for either gel electrophoresis or mass spectrometry.

3.1 Materials and equipment

50% (*v*/v) Isopropanol

100% Acetone

Centrifuge capable of achieving 16,000 × *g*

Centrifuge tubes capable of withstanding 16,000 × *g*
−20 °C freezer

3.2 Protocol

1. Grind soybean seeds into a fine powder using a mortar and pestle
2. Weigh out 500 mg of seed powder and transfer into a 15 mL tube containing 10 mL of 50% isopropanol
3. Mix the tube contents for 15 min, with vigorous shaking, at room temperature
4. Clarify the slurry by centrifugation at 16,000 × *g* for 15 min
5. Remove the clear supernatant, measure the volume, and place it into a clean tube
6. To this, add 3 volumes of ice-cold acetone, mix the contents, and place it at −20 °C freezer overnight
7. Centrifuge the samples at 16,000 × *g* for 15 min
8. Air-dry the pellet at room temperature for 10 min
9. For 2D gel analysis, re-isolate the pellet from Step 8 beginning with Step 2 (Section 4.1.2).
10. For mass spectrometry analysis, resuspend the pellet from step 8 in minimal volume of 50 mM Tris-HCl, pH 8.0, and proceed with Step 1 (Section 4.2.2).

4. Proteomic analysis of low abundance soybean seed proteins

4.1 Two-dimensional gel electrophoresis

For 2D electrophoretic gel analysis, fractionated proteins were separated using the method outlined below. Protein concentration estimate, after solubilized in isoelectric focusing (IEF) resuspension buffer was performed using the Bradford method, and 300 μg of protein sample loaded per strip using overnight in-gel rehydration.

4.1.1 Materials and Equipment

Tris base [Tris(hydroxymethyl)aminomethane]
Glycine
Acrylamide/Bis-acrylamide
Ammonium persulfate
N,N,N′,N′-Tetramethylethylenediamine (TEMED)
Coomassie Brilliant Blue G-250

Urea
Thiourea
Methanol
Ammonium acetate
β-mercaptoethanol
Glycerol
Agarose (low EEO)
Saturated phenol (pH 4.3)
3-[(3-Cholamidopropyl)-dimethylammonio]-1-propane-sulfonate (CHAPS)
Sucrose
Acetic acid
3-(4-Heptyl)phenyl-3-hydroxypropyl)dimethyl-ammoniopropanesulfonate) (C7BzO)
2-hydroxy-ethyldisulfide (2-HED)
Dithiothreitol (DTT)
Iodoacetamide (IAM)
Acetonitrile (ACN)
Ammonium bicarbonate
Protease inhibitor cocktail (Plant ProteaseArrest; G-Biosciences, St. Louis, MO, USA)
Immobilized pH Gradient (IPG) strips (Cytiva, Marlborough, MA, USA)
Centrifuge capable of achieving 14,000 × *g*
Swing-bucket centrifuge capable of achieving 5000 × *g*
Centrifuge tubes capable of withstanding 14,000 × *g* and resistant to phenol
−20 °C freezer and − 80 °C freezer

4.1.2 Protocol

1. For total protein extraction from soybean, grind seed to a fine powder using a mortar and pestle
2. To this powder, or to the protein pellet from prior fractionation, add 5 mL of 0.1 M Tris-Cl, pH 8.8, containing 0.9 M sucrose, 0.2 M NaCl, 0.4% (*w/v*) β-mercaptoethanol, and protease inhibitor cocktail, and continue grinding
3. Add 5 mL of saturated phenol, pH 4.3, and continuously mix for 30 min at room temp. *Note: The volume ratio of 1:1 (buffer:phenol) is critical*

4. Obtain the phenolic phase by centrifugation at 5000 × *g* for 20 min at 23 °C in a swing-bucket rotor
5. The upper phenolic phase was removed and added to ten volumes of 100% methanol with 0.1 M ammonium acetate. Proteins were allowed to precipitate for 2 h at −80 °C
6. Centrifuge at 14,000 × *g* for 15 min at 4 °C. Wash this protein pellet twice more in a solution of 100% methanol, 0.1 M ammonium acetate, with 0.01 M DTT. Repeat resuspension/chilling/centrifugation of the insoluble proteins 3 × additionally using a solution of 100% acetone, with 0.01 M DTT
7. Following the final wash, the protein pellet was allowed to air dry slightly, and then dissolved in a small volume of IEF resuspension buffer (7 M urea, 2 M thiourea, 1% (*w*/*v*) CHAPS, and 2% (w/v) C7BzO). Obtain the protein concentration using Bradford method
8. Rehydrate IPG strips with 250 μL of the IEF resuspension buffer, also containing 300 μg of your solubilized protein, along with a final concentration of the following: 5% (*v*/v) glycerol, 2.2% (v/v) 2-hydroxyethyl disulfide, 60 mM DTT, and 0.3% (v/v) IPG buffer (carrier ampholytes).
9. Rehydrate the strips at 22 °C for 15 h prior to focusing
10. Conduct isoelectric focusing as follows: 50 V active rehydration, 2 h; 250 V, 500 Vhrs, fast ramp; 1000 V, 500 Vhrs, fast ramp; 8000 V, 2 h, linear ramp; 8000 V, 60,000–80,000 Vhrs, fast ramp. *Note: Volt hours (Vhrs) depends on pH gradient*
11. Prior to the second dimension, equilibrate the IPG strips with 5% SDS in a urea-based solution (50 mM Tris-Cl pH 8.8, 6 M urea, 30% (*v*/v) glycerol and 0.1% (*w*/*v*) bromophenol blue) containing 2% (*w*/*v*) DTT for 15 min
12. Following step 10, equilibrate the IPG strips with 5% SDS in a urea-based solution (50 mM Tris-Cl pH 8.8, 6 M urea, 30% (v/v) glycerol and 0.1% (*w*/*v*) bromophenol blue) containing 2.5% (w/v) IAM for 15 min
13. Carefully place the IPG strips onto a 15% (w/v) Tris-glycine vertical second dimension and secure into place with warm 1% (w/v) agarose dissolved in SDS-PAGE running buffer (containing 0.2% (w/v) SDS).
14. Run gels at an initial 10 mA/gel for 1 h followed by 25 mA/gel for the remainder of the run (elimination of dye front; approximately 4 h).
15. Remove gels from the cassette and soak them in solution of 5:4:1 (methanol:water:acetic acid) for 30 min

16. Remove the gels from this solution, rinse twice with ultrapure water for 10 min each wash, and stain them with Coomassie G-250 for overnight
17. Destain the gels with ultrapure H_2O for several hours and scan the gel
18. Excise a small gel piece of each protein spot using a 1–1.5 mm spot picker
19. Wash gel plugs briefly in distilled water and then destain completely in a 50% (*v*/v) solution of acetonitrile/water containing 25 mM ammonium bicarbonate. After a 100% acetonitrile wash, the protein contained in the acrylamide gel can be stored at −80 °C freezer
20. Proceed to Section 4.2.1 for mass spectral analysis

4.2 Mass spectral analysis

All solvents should be LC-MS grade and only ultrapure HPLC grade water should be used.

4.2.1 Materials and equipment

SpeedVac Concentrator (Savant)
Thermomixer C with ThermoTop (Eppendorf)
Acetonitrile (ACN)
Formic acid (FA)
Ultrapure water

4.2.1.1 Digestion

Tris base [Tris(hydroxymethyl)aminomethane]
Bovine serum albumin (BSA)
Trypsin/Lys-C (Promega V5073)
Dithiothreitol (DTT)
Iodoacetamide (IAM)
Trifluoroacetic acid (TFA)
Triethylammonium bicarbonate (TEAB)

4.2.1.2 Desalting

BiopureSPN Mini (Nest Group HUMS18V) desalting column.
Eppendorf Centrifuge 5417C.
Methanol washed 2 mL microcentrifuge tubes (ThermoFisher).

4.2.1.3 Tandem mass tag (TMT) labeling

Pierce Quantitative Colorimetric Peptide Assay

TMT Reagents (ThermoFisher)
DriSolv® Acetonitrile (Millipore AX01347)
50% (*v*/v) hydroxylamine (ThermoFisher 90,115)

4.2.1.4 Label-free

Peptide Retention Time Calibration Mixture (PRTC, Pierce 88,320, 0.5 pmol/μL)

4.2.1.5 LC-MS/MS

Isopropanol
Zorbax Chromatography Packing, SB-C18, 5 μm (Agilent)
UChrom C18 3-μm packing (NanoLCMS Solutions)
μM glass capillary (Agilent)
Model P-2000 Laser Puller (Sutter Instruments Co)
Capillary polishing station CPS-2 (EIS Source Solutions)
Pressure Injection Cell (NextAdance)
Easy-nLC 1200 (ThermoFisher)
Nanospray FlexIon Source (ThermoFisher)
Q Exactive Hybrid Quatrupole-Orbitrap Mass Spectrometer (ThermoFisher)

4.2.1.6 Data analysis

Proteome Discoverer (ThermoFisher)
Mascot search engine (Matrix Science)
Protein database in FASTA format (UniProt)

4.2.2 Protocols

4.2.2.1 Digestion

1. Reduction: Thaw the sample at room temperature. Dilute sample with 50 mM Tris, pH 8.0, to a concentration of 0.5 μg/μL in a volume of 100 μL, 25 μg total. BSA control: 50 μL (1 μg/μL), 50 μL 50 mM Tris, 50 μg total
2. Add 5 μL of 0.1 M DTT (to a final concentration of 5 mM), and mix
3. Incubate at 37 °C for 30 min. *Note: Avoid high temperatures (e.g., 50–60 °C) during the reducing step. High temperatures induce protein carbamylation in the presence of urea in the extract solution. Proteins are efficiently reduced at 37 °C*
4. Alkylation: Add 1.5 μL of 1 M IAM (to a final concentration of 15 mM), and mix

5. Incubate at room temperature for 30 min in the dark
6. Digestion: Add 400 μL of 50 mM Tris-HCl (pH 8). Dilution is needed to reduce extraction buffer concentration for optimal trypsin activity
7. Add enzyme to the desired enzyme to protein ratio. If using the trypsin/Lys-C mix, use 1:25 enzyme:protein ratio (2 μg trypsin/Lys-C:50 μg sample).
8. Incubate overnight at 37 °C in a Thermomixer C
9. Acidify sample by adding FA (5 μL) to a final concentration of 1% and remove particulate material by centrifuging at 14,000 × *g* for 10 min using a benchtop microcentrifuge. *Note: Acidification often results in cloudiness. Be sure to centrifuge the acidified sample*
10. Dry using the SpeedVac

4.2.2.2 Desalting

1. Reconstitute sample in 40 μL sample buffer (5% (*v*/v) acetonitrile/0.1% (v/v) TFA/water).
2. Desalt the sample following the manufacturer's protocol for the BioPure SPN Mini columns
3. SpeedVac to dryness and store at −20 °C until analysis on the LC-MS/MS (either label-free or TMT).

4.2.2.3 Tandem mass tag (TMT) labeling

1. Setup: Remove an aliquot from each sample and pool into a single tube. This will be used as a master control for TMT channel 11. Pool enough from each sample to allow for 10 μg of this control per 10-plex sample in the experiment. You will also need additional amount of the master control for testing digestion and labeling efficiency (i.e., if there are 100 samples, at least 110 μg of the master control will be needed).
2. Digest the master control and 25 μg of each sample (see Section 4.2.2.1).
3. Desalt the master control (see Section 4.2.2.2).
4. Reconstitute the master control to 1 μg/μL with 5% (*v*/v) acetonitrile/0.1% (v/v) FA/water
5. Transfer an aliquot of the master control (typically 1 μg) to a sample vial with insert and run on the LC-MS/MS
6. SpeedVac to dryness and store at −20 °C

   Note: If master control run on the LC-MS/MS looks good (proteins identified, high number of peptide spectrum matches), proceed with labeling.

7. Master control labeling: Reconstitute the master control in 100 mM TEAB to a concentration of 0.25 μg/μL
8. Remove 4 μL of master control to a new tube and add 16 μL distilled water (for protein concentration determination).
9. Determine the concentration using the Pierce Quantitative Colorimetric Peptide Assay kit
10. Label with TMT 131C according to the manufacturer's protocol
11. Transfer an aliquot of the master control (typically 1 μg) to a sample vial with insert and run on the LC-MS/MS. *Note: Analyze the master control labeled run to determine the labeling efficiency. If the labeling efficiency is good, proceed with sample labeling*
12. Sample labeling: Desalt the samples (see Section 4.2.2.2)
13. SpeedVac to dryness
14. Reconstitute the samples in 100 mM TEAB to a concentration of 0.25 μg/μL
15. Remove 4 μL of sample to a new tube and add 16 μL distilled water (for protein concentration determination).
16. Determine the concentration using the Pierce Quantitative Colorimetric Peptide Assay kit
17. Label each sample with TMT label according to the manufacturer's protocol
18. Pool 5 μg of each sample (labels 1–10) plus 5 μg of labeled master control (the volume is dependent upon the concentration of each sample that was determined in steps #9 and #16 above).
19. SpeedVac to dryness
20. Reconstitute TMT sample in 55 μL 5% (*v*/v) acetonitrile/0.1% (*v*/v) FA/water. This will give a final concentration of 1 μg/μL
21. Transfer to a sample vial with insert and run on the LC-MS/MS (typically a 1–5 μg injection).

4.2.2.4 Label-free

For label-free quantitative analysis, a known amount of Pierce Peptide Retention Time Calibration Mixture (PRTC, product #88320, 0.5 pmol/μL) will be spiked into the sample. The PRTC peptides can be used to normalize the data.

1. Digest (see Section 4.2.2.1) 25 μg of sample and desalt (see Section 4.2.2.2).
2. Add 23.75 μL 5% Buffer B (5% (*v*/v) acetonitrile/water/0.1% (v/v) FA) to the dried sample in the microcentrifuge tube. Final concentration of the sample will be 1 μg/μL

3. Add 1.25 μL PRTC (625 fmol) to the microcentrifuge tube. Final concentration of the PRTC will be 25 fmol/μL
4. Vortex and centrifuge
5. Transfer to a sample vial with insert for and run on the LC-MS/MS (typically a 1–5 μg injection).

4.2.2.5 LC-MS/MS

LC conditions (Easy nLC-1200)

- Column: Pulled glass emitter 75 μm × 20 cm capillary. The tip is packed with C18 5 μm material and the remainder of the column is packed with C18 3 μm material
- Buffer A: 0.1% (*v*/v) FA/water
- Buffer B: 80% (v/v) acetonitrile/0.1% (v/v) FA/water
- Wash solution 1 (WS1): isopropanol
- Wash solution 2 (WS2): 50% (v/v) acetonitrile/water
- Wash Solution 3 (WS3): 0.1% (v/v) FA/water
- Maximum pressure for all steps is 300 bar

Column equilibration: 10 μL**

Autosampler wash:

1 × 25 μL wash with WS3

3 × 25 μL washes with WS2

3 × 25 μL washes with WS

3 × 25 μL washes with WS2

3 × 25 μL washes with WS3

Gradient:

Both methods are run at 300 nL/min and incorporate a sawtooth gradient at the end of the run to facilitate column cleaning and reduce carryover.

Label-free gradient:

0–35% B in 80 min

35–70% B in 10 min

70–100% B in 3 min

3 sawtooth gradients from 2% to 100% B

TMT gradient:

0–35% B in 210 min

35–70% B in 10 min

70–100% B in 3 min

3 sawtooth gradients from 2% to 100% B

MS/MS settings (Q Exactive):

MS/MS is performed using a Q Exactive Hybrid Quatrupole-Orbitrap Mass Spectrometer coupled with a NanoSpray FlexIon Source using positive polarity, a spray voltage of 2.65 kV, capillary temperature of 250 °C and an S-Lens RF level of 50.

Label-free:

Full MS from 400 to 2000 *m/z* at 70,000 resolution with an AGC target of 5e5 and maximum inject time of 80 ms.

20 dd-MS2 at 17,500 resolution with an AGC target of 2e5 and maximum inject time of 80 ms. Isolation window of 2.0 m/z and NCE of 27.

Charge exclusion of +1, 6–8, >8, peptide match set to preferred and dynamic exclusion of 6.0 s.

TMT:

Full MS from 400 to 2000 m/z at 70,000 resolution with an AGC target of 1e6 and maximum inject time of 80 ms.

20 dd-MS2 at 35,000 resolution with an AGC target of 1e5 and maximum inject time of 50 ms. Fixed first mass of 110.0 m/z, isolation window of 1.2 m/z and NCE of 32.

Charge exclusion of +1, 6–8, >8, peptide match set to on and dynamic exclusion of 6.0 s.

After each sample run, a wash run is performed (3 sawtooth gradients), followed by 1–2 blank runs using a shortened gradient.

4.2.2.6 Data analysis

Data analysis is performed using Proteome Discoverer (PD; Thermo Fisher Scientific). Different workflows are used for the TMT and the label-free samples.

Label-free:

Mascot search against the database of interest with dynamic modifications of oxidation (Met) and deamidation (Asn, Gln) and static modification of carbamidomethyl (Cys). Sequest HT search against the PRTC peptides with static modification of heavy isotope labeled K and R. Normalization of data using the PRTC peptides. Precursor mass tolerance is set to 10 ppm and fragment mass tolerance is set to 0.02 Da.

After the runs for the sample set have been completed, create a study in PD. Add any study factors that may be used for comparing the samples and assign these factors to the samples. Add the .msf files for each of the samples to the study and create a new analysis comparing all the samples in the sample set. Run a consensus workflow comparing all the samples. Run a consensus workflow with normalization and scaling on the PRTC peptides as well as imputation.

TMT:

Create a new quantitative method and add the reporter ion distributions for the TMT kit used. Mascot and Sequest HT search against the database of interest with dynamic modifications of oxidation (Met) and deamidation (Asn, Gln) and static modification of carbamidomethyl (Cys) and TMT (Lys, N-terminus). Precursor mass tolerance is set to 10 ppm and fragment mass tolerance is set to 0.02 Da.

After the runs for the sample set are complete, create a study in PD. Add any study factors that may be used for comparing the samples and assign these factors to the samples. Assign the samples in channel 131C as controls. Add the .msf files for each of the samples to the study and create a new analysis comparing all the samples in the sample set. Run a consensus workflow comparing all the samples. Run a consensus workflow with normalization and scaling on controls as well as imputation.

5. Conclusions

Here, we have presented two simple, inexpensive, and rapid methods to isolate low abundance seed proteins from soybean seeds. By employing these simple procedures, one can selectively remove most of the abundant seed storage proteins. Our methodology enables the nonabundant proteins in soybean seed to be more accessible for proteomic analysis. We have also provided detailed information on the resolution of low abundance proteins by two-dimensional polyacrylamide gel electrophoresis (2D-PAGE) and the procedure for the identification of individual proteins by mass spectrometry. Procedure for TMT labeling is also detailed for the identification and quantification of proteins in different kinds of samples. The calcium chloride precipitation procedure could also be applied to other legumes to facilitate the enrichment of low abundance seed proteins. The methodology outlined here should help advance proteomic research of low abundance seed proteins of many legumes.

Acknowledgments

The authors would like to thank Nathan Oehrle for his excellent technical assistance. Funding for this research was provided by Agricultural Research Service, USDA (8042-21220-234-00D). Mention of a trademark, vendor, or proprietary product does not constitute a guarantee or warranty of the product by the USDA and does not imply its approval to the exclusion of other products or vendors that may also be suitable. The US Department of Agriculture, Agricultural Research Service, Midwest Area, is an equal opportunity, affirmative action employer and all agency services are available without discrimination.

References

Agrawal, G. K., Hajduch, M., Graham, K., & Thelen, J. J. (2008). In-depth investigation of the soybean seed-filling proteome and comparison with a parallel study of rapeseed. *Plant Physiology*, *148*, 504–518.

Beilinson, V., Chen, Z., Shoemaker, R. C., Fischer, R. L., Goldberg, R. B., & Nielsen, N. C. (2002). Genomic organization of glycinin genes in soybean. *Theoretical and Applied Genetics*, *104*, 1132–1140.

Bourgeois, M., Jacquin, F., Savois, V., Sommerer, N., Labas, V., Henry, C., et al. (2009). Dissecting the proteome of pea mature seeds reveals the phenotypic plasticity of seed protein composition. *Proteomics*, *9*, 254–271.

Cabanos, C., Matsuoka, Y., & Maruyama, N. (2021). Soybean proteins/peptides: A review on their importance, biosynthesis, vacuolar sorting, and accumulation in seeds. *Peptides*, *143*, 170598. https://doi.org/10.1016/j.peptides, 170598. Epub 2021 Jun 18. PMID: 34153351.

Cho, T. J., Davies, C. S., & Nielsen, N. C. (1989). Inheritance and organization of glycinin genes in soybean. *Plant Cell*, *1*, 329–337.

Derbyshire, E., Wright, D., & Boulter, D. (1976). Legumin and vicilin, storage proteins of legume seed. *Phytochemistry*, *15*, 3–24.

Hajduch, M., Ganapathy, A., Stein, J. W., & Thelen, J. J. (2005). A systematic proteomic study of seed filling in soybean. Establishment of high-resolution two-dimensional reference maps, expression profiles, and an interactive proteome database. *Plant Physiology*, *137*, 1397–1419.

Islam, N., Bates, P. D., John, M. K., Krishnan, H. B., Zhang, Z., Luthria, D. L., et al. (2019). Quantitative proteomic analysis of low linolenic acid transgenic soybean reveals perturbations of fatty acid metabolic pathways. *Proteomics*, *19*, 1–11. https://doi.org/10.1002/pmic.201800379.

Islam, N., Krishnan, H. B., & Natarajan, S. S. (2020). Proteomic profiling of fast neutron-induced soybean mutant unveiled pathways associated with increased seed protein content. *Journal of Proteome Research*, *19*, 3936–3944. https://doi.org/10.1021/acs.jproteome.0c00160.

Islam, N., Krishnan, H. B., & Natarajan, S. S. (2021). Quantitative proteomic analyses reveal the dynamics of protein and amino acid accumulation during soybean seed development. *Proteomics*. https://doi.org/10.1002/pmic.202100143. Article e2100143.

Kim, Y. J., Wang, Y., Gupta, R., Kim, S. W., Min, C. W., Kim, Y. C., et al. (2015). Protamine sulfate precipitation method depletes abundant plant seed-storage proteins: A case study on legume plants. *Proteomics*, *15*, 1760–1764.

Kitamura, K., Takagi, T., & Shibasaki, K. (1976). Subunit structure of soybean 11S globulin. *Agricultural and Biological Chemistry*, *40*, 1837–1844.

Komatsu, S., Wang, X., Yin, X., Nanjo, Y., Ohyanagi, H., & Sakata, K. (2017). Integration of gel-based and gel-free proteomic data for functional analysis of proteins through soybean proteome database. *Journal of Proteomics*, *163*, 52–66.

Krishnan, H. B. (2000). Biochemistry and molecular biology of soybean seed storage proteins. *Journal of New Seeds*, *2*, 1–25.

Krishnan, H. B., Oehrle, N. W., & Natarajan, S. S. (2009). A rapid and simple procedure for the depletion of abundant storage proteins from legume seeds to advance proteome analysis: A case study using glycine max. *Proteomics*, *9*, 3174–3188. https://doi.org/10.1002/pmic.200800875.

Li, C., & Zhang, Y. M. (2011). Molecular evolution of glycinin and beta-conglycinin gene families in soybean (Glycine max L. Merr.). *Heredity*, *106*, 633–641.

Low, T. Y., Syafruddin, S. E., Mohtar, M. A., Vellaichamy, A., Rahman, N. S., Pung, Y., et al. (2021). Recent progress in mass spectrometry-based strategies for elucidating protein–protein interactions. *Cellular and Molecular Life Sciences*, *78*, 5325–5339.

McWhite, C. D., Papoulas, O., Drew, K., Cox, R. M., June, V., Dong, X., et al. (2020). A pan-plant protein complex map reveals deep conservation and novel assemblies. *Cell*, *181*, 460–474.

Miernyk, J. A. (2014). Seed proteomics. *Methods in Molecular Biology*, *1072*, 361–377. https://doi.org/10.1007/978-1-62703-631-3_26 (PMID: 24136535).

Miernyk, J. A., & Hajduch, M. (2011). Seed proteomics. *Journal of Proteomics*, *74*, 389–400.

Miernyk, J. A., & Johnston, M. L. (2006). Chemical cross-linking immobilized-concanavalin A for use in proteomic analyses. *Preparative Biochemistry & Biotechnology*, *36*, 203–213.

Min, C. W., Hyeon, H., Gupta, R., Park, J., Cheon, Y. E., Lee, G. H., et al. (2020). Integrated proteomics and metabolomics analysis highlights correlative metabolite-protein networks in soybean seeds subjected to warm-water soaking. *Journal of Agricultural and Food Chemistry*, *68*, 8057–8067.

Min, C. W., Park, J., Bae, J. W., Agrawal, G. K., Rakwal, R., Kim, Y., et al. (2020). In-depth investigation of low-abundance proteins in matured and filling stages seeds of *Glycine max* employing a combination of protamine sulfate precipitation and TMT-based quantitative proteomic analysis. *Cell*, *9*(6), 1517. https://doi.org/10.3390/cells9061517.

Momma, T., Negoro, T., Hirano, H., Matsumoto, A., Udaka, K., & Fukazawa, C. (1985). Glycinin A5A4B3 mRNA: cDNA cloning and nucleotide sequencing of a splitting storage protein subunit of soybean. *European Journal of Biochemistry*, *149*, 491–496.

Mooney, B. P., Krishnan, H. B., & Thelen, J. J. (2004). High-throughput peptide mass fingerprinting of soybean seed proteins: automated workflow and utility of UniGene expressed sequence tag databases for protein identification. *Phytochemistry*, *65*, 1733–1744.

Natarajan, S. S., Krishnan, H. B., Lakshman, S., & Garrett, W. M. (2009). An efficient extraction method to enhance analysis of low abundant proteins from soybean seed. *Analytical Biochemistry*, *394*, 259–268.

Nielsen, N. C. (1996). Soybean seed composition. In D. P. S. Verma, & R. C. Shoemaker (Eds.), *Soybean: Genetics, molecular biology and biotechnology* (pp. 127–163). Tucson, AZ: CAB.

Nielsen, N. C., Dickinson, C. D., Cho, T., Thanh, V. H., Scallon, B. J., Fisher, R. L., et al. (1989). Characterization of glycinin gene family in soybean. *Plant Cell*, *1*, 313–328.

Ohyanagi, H., Sakata, K., & Komatsu, S. (2012). Soybean Proteome Database 2012: Update on the comprehensive data repository for soybean proteomics. *Frontiers in Plant Science*, *3*, 110.

Osborne, T. B. (1924). *The vegetable proteins*. London: Longmans, Green.

Shewry, P. R., Napier, J. A., & Tatham, A. S. (1995). Seed storage proteins: structures and biosynthesis. *Plant Cell*, *7*, 945–956.

Smolikova, G., Gorbach, D., Lukasheva, E., Mavropolo-Stolyarenko, G., Bilova, T., Soboleva, A., et al. (2020). Bringing new methods to the seed proteomics platform: Challenges and perspectives. *International Journal of Molecular Sciences*, *21*(23), 9162. https://doi.org/10.3390/ijms21239162. PMID: 33271881; PMCID: PMC7729594.

Staswick, P. E., Hermodson, M. A., & Nielsen, N. C. (1984). Identification of the cystines which link the acidic and basic components of the glycinin subunits. *Journal of Biological Chemistry*, *259*, 13431–13435.

Thanh, V., & Shibasaki, K. (1978). Major proteins of soybean seeds: subunit structure of β-conglycinin. *Journal of Agricultural and Food Chemistry*, *26*, 692–695.

van Wijk, K. J. (2001). Challenges and prospects of plant proteomics. *Plant Physiology*, *126*, 501–508.
Yu, T., Li, G., Dong, S., Liu, P., Zhang, J., & Zhao, B. (2016). Proteomic analysis of maize grain development using iTRAQ reveals temporal programs of diverse metabolic processes. *BMC Plant Biology*, *16*, 241.
Zecha, J., Gabriel, W., Spallek, R., Chang, Y., Mergner, J., Wilhelm, M., et al. (2022). Linking post-translational modifications and protein turnover by site-resolved protein turnover profiling. *Nature Communications*, *13*, 165. https://doi.org/10.1038/s41467-021-27639-0.
Zhang, M., Liu, S., Wang, Z., Yuan, Y., Zhang, Z., Liang, Q., et al. (2022). Progress in soybean functional genomics over the past decade. *Plant Biotechnology Journal*, *20*(2), 256–282. https://doi.org/10.1111/pbi.13682.

CHAPTER THIRTEEN

Proteomics and phosphoproteomics of C_3 to CAM transition in the common ice plant

Noé Perron[a], Bowen Tan[b], Craig P. Dufresne[c], and Sixue Chen[a,b,d,*]
[a]Plant Molecular and Cellular Biology Program, University of Florida, Gainesville, FL, United States
[b]Department of Biology, Genetics Institute, University of Florida, Gainesville, FL, United States
[c]Thermo Fisher Scientific Training Institute, West Palm Beach, FL, United States
[d]Proteomics and Mass Spectrometry, Interdisciplinary Center for Biotechnology Research (ICBR), University of Florida, Gainesville, FL, United States
*Corresponding author: e-mail address: schen@ufl.edu

Contents

Abstract

Among all post-translational modifications of proteins, phosphorylation is one of the most common and most studied. Since plants are sessile organisms, many physiological processes on which their survival depends are regulated by phosphorylation and dephosphorylation. Understanding the extent to which a plant proteome is phosphorylated at specific developmental stages and/or under certain environmental conditions is essential for identifying molecular switches that regulate physiological processes and responses. While most phosphoproteomic workflows proposed in the literature provide tools to exclusively analyze phosphorylated proteins, it is imperative to examine both the proteome and the phosphoproteome to reveal the true complexity of a biological process. Here we describe a mass spectrometry-based phosphoproteomics workflow to analyze both total and phosphorylated proteins. Our method includes phenol-based protein extraction, as well as techniques to measure the quantity and quality of protein

Methods in Enzymology, Volume 676
ISSN 0076-6879
https://doi.org/10.1016/bs.mie.2022.06.004

extracts. In addition, we compare in detail the efficiency and suitability of in-gel and in-solution trypsin digestion methods. A metal oxide affinity chromatography technique for rapid and efficient enrichment of phosphorylated peptides and an LC-MS/MS method for analysis of the phosphorylated peptides are described. Finally, we present and discuss the results generated by applying this workflow to our study of the C_3 to CAM transition in the common ice plant (*Mesembryanthemum crystallinum*). Overall, our workflow provides robust methods for the identification of phosphoproteins and total proteins. It can be broadly applied to many other organisms and sample types, and the results provide a more accurate picture of the molecular switches that regulate different biological processes.

1. Introduction

Protein post-translational modifications (PTMs) are at the heart of a wide range of cellular processes and drive protein functional and structural diversity by increasing the number of proteoforms (Chou, 2020; Walsh, 2006). Due to their sessile nature, plants rely heavily on PTMs for rapid responses to environmental variations and stresses (Arsova, Watt, & Usadel, 2018). Of the over 400 known types of PTMs, protein phosphorylation is the most studied and one of the most common proteoforms in a proteome (Khoury, Baliban, & Floudas, 2011). It is characterized by the reversible phosphorylation of amino acid residues (usually serine, threonine and tyrosine), which is orchestrated by the opposing activities of kinases (which add phosphate groups) and phosphatases (which remove them) (Arsova et al., 2018). Protein phosphorylation is known to be involved in the regulation of major plant physiological processes such as drought stress signaling and tolerance (Zhang, Zhu, Gong, & Zhu, 2022), growth and development (Van Leene et al., 2019), photosynthesis (Longoni & Goldschmidt-Clermont, 2021), and carbon metabolism (Wu, Sklodowski, Encke, & Schulze, 2014).

Proteome-wide phosphorylation analysis can be accomplished using mass-spectrometry (MS)-based phosphoproteomics methods (Mayya & Han, 2009; Nukarinen et al., 2016; Pi et al., 2018; Silva-Sanchez, Li, & Chen, 2015). Current phosphoproteomics approaches aim to reveal the entire phosphoproteome by detecting phosphoproteins and the individual amino acid residues modified with phosphate groups. Over the past decade, a large number of phosphopeptide enrichment strategies, MS methods, and bioinformatic tools for phosphorylation site mapping, phopshoprotein identification and quantification have been developed (Silva-Sanchez et al., 2015; Urban, 2022). Because phosphoproteins are often present in very low abundance and stoichiometry in biological samples, phosphoproteomics datasets are generated using different phosphoprotein and phosphopeptide enrichment methods. This enrichment step is critical to ensure adequate amounts of phosphoproteins and/or phosphopeptides prior to MS analysis. Some of the most commonly used enrichment methods include immobilized metal affinity chromatography (IMAC), immunoaffinity enrichment, metal oxide affinity chromatography (MOAC), ion-exchange chromatography, and hydrophilic interaction chromatography (Javeed et al., 2020; Silva-Sanchez et al., 2015; Subba & Prasad, 2021; Thingholm & Larsen, 2016). Phosphoproteomics is an important tool for understanding the roles of protein phosphorylation/dephosphorylation switches in vital cellular processes that govern plant growth, development and interaction with the environment.

To date, the vast majority of phosphoproteomics workflows described in the literature focus exclusively on phosphoprotein analysis, while often neglecting the total proteome. Yet, the integration of proteome and phosphoproteome datasets is essential to fully uncover the mechanisms of a given regulatory process, as not only phosphorylated proteins, but also non-phosphorylated proteins provide important information about PTM dynamics, stoichiometry, signaling and metabolic pathways. In this chapter, we describe a MS-based phosphoproteomics workflow to determine not only protein levels but also protein phosphorylation levels in plant tissues in an untargeted fashion. This workflow includes a phenol extraction protocol for the extraction of total proteins from complex samples, a protein quantification step via Bradford assay, and a protein quality and quantity control via sodium dodecyl sulfate polyacrylamide gel electrophoresis (SDS–PAGE). We compare in-gel and in-solution trypsin digestion, two major proteolytic methods of proteomics. In addition, we discuss in great detail a protocol for phosphopeptide enrichment using a MOAC technique. Finally, we describe a liquid chromatography tandem mass spectrometry (LC-MS/MS) method used to analyze the peptide samples and provide guidance on how to process the raw data and interpret them using proteomic data analysis software.

2. Methods

2.1 Protein extraction, quantification, and quality control

In general, proteins are isolated from 1 g of plant tissue using a phenol extraction method, which is a standard and efficient proteomic technique for separating proteins from nucleic acids, starch, metabolites, and other molecules

Fig. 1 A representative gel image of proteins extracted from the ice plant leaves using the phenol extraction method. A total of 20 μg of protein per sample from eight independently extracted protein samples were applied on SDS-PAGE for quality assessment. The clear homogeneous bands between samples suggest good protein quality and consistent amounts across the replicate extractions.

(Faurobert, Pelpoir, & Chaïb, 2007). Because the phenol protein extraction method has been widely used and the protocol has been previously detailed (e.g., Faurobert et al., 2007; Pavokovic, Križnik, & Krsnik-Rasol, 2012; Sheffield, Taylor, Fauquet, & Chen, 2006), in this chapter we simply demonstrate its usefulness for extracting proteins from the common ice plant (*Mesembryanthemum crystallinum*) for total proteomic and phosphoproteomic analyses. Because leaf tissue of the common ice plant largely consists of water (95–98% of the fresh weight of the leaf, Kim et al., 2021), it is recommended to perform the extraction on 2 g of tissue from this specific plant. The quantification step is performed following a Bradford assay (Bradford, 1976), which is a spectrophotometric method allowing an accurate measurement of the protein concentration in a solution, based on the absorbance of Coomassie Brilliant Blue G-250 at 595 nm. The quality and equal quantity of the extracted proteins can be observed using SDS-PAGE by loading 20 μg of the protein extract (Fig. 1).

2.2 In-gel vs. in-solution trypsin digestion

For each leaf protein sample that will go through phosphoenrichment, perform the trypsin digestion protocol on an additional aliquot of the same sample for total protein analysis. In general, 5–10% of the total protein amount is spared for total proteomic analysis. In addition to complex protein samples, it is essential to perform the digestion steps on 10 μg of purified alpha-casein, which is a highly phosphorylated milk protein that can serve as a positive control for phosphopeptide enrichment procedures.

2.2.1 Equipment and consumables

In-gel and in-solution digestion:

- Savant SPD1010 SpeedVac Concentrator, Thermo Fisher Scientific
- Rocking Platform Shaker
- Vortex mixer
- Eppendorf 5417R Refrigerated Centrifuge
- Vortex mixer
- 37 °C incubator
- Thermomixer, Thermo Fisher Scientific
- 1.5 mL Eppendorf tubes
- Laboratory Electrophoresis Power Supply Model 200/2.0, Bio-Rad
- Laboratory Fume Hood

Cleanup and solid phase extraction:

- Eppendorf 5417R Refrigerated Centrifuge
- Vortex mixer
- 1.5 mL Eppendorf tubes
- 2 mL Eppendorf tubes
- BioPureSPN Macro TARGA FastEq Desalting Column (50–200 μL elution volume, 35–350 μg maximum capacity)

2.2.2 Reagents and solvents

In-gel and in-solution digestion:

- Bio-Safe Coomassie stain, Bio-rad
- ddH_2O
- 100 mM ammonium bicarbonate (ABC) in double-distilled water (ddH_2O)
- 100% acetonitrile (ACN)
- 80% ACN with 20% of 1% formic acid (FA) in ddH_2O (*v*/v) (in-gel digestion)
- 60% ACN in ddH_2O (*v*/v) (in-gel digestion)
- 100 mM ABC in ddH_2O (*v*/v)
- 10 mM dithiothreitol DTT in 100 mM ABC. Prepare shortly before use
- 55 mM chloroacetamide in 100 mM ABC (in-gel digestion). Prepare shortly before use and keep in the dark

- 500 mM chloroacetamide in 100 mM ABC (in-solution digestion). Prepare shortly before use and keep in the dark
- 20 mM ABC
- 50 mM ABC
- 10% SDS in ddH_2O (*w/v*) (in-gel digestion)
- 20 mM ABC/ACN (1:1 v/v) (in-gel digestion)
- 100 mM ABC/ACN (1:1 v/v) (in-gel digestion)
- 1% formic acid (FA) in ddH_2O (v/v)
- 100% formic acid (FA)
- Trypsin (V5111, sequencing grade, Promega, Madison). Prepare a solution of 5 ng/μL trypsin in 50 mM ammonium bicarbonate (pH 7.8) shortly before use

Cleanup and solid phase extraction:

- 100% Methanol
- 100% Chloroform
- 0.1% Trifluoroacetic Acid (TFA) in ddH_2O (v/v)
- 80% ACN and 0.1% FA in ddH_2O (v/v)
- ddH_2O

2.2.3 In-gel trypsin digestion

In in-gel digestion protocols, proteins are cleaved into peptides while trapped in the polyacrylamide matrix. This procedure has a number of advantages, the main ones being that SDS-PAGE cleans protein samples, and the resulting workflow has a reduced number of steps (no cleanup or solid phase extraction steps are required). In addition, femtomolar concentrations of proteins remain stable in the gel (Shevchenko et al., 2001).

2.2.3.1 Procedure

Gel destaining and band excision

1. The samples and alpha-casein are loaded in a 12% SDS-PAGE gel, and the gel is run in a laboratory electrophoresis power supply containing 10% SDS solution at 70 V for about 20 min until the bands reach 1 cm below the stacking/separating gel limit
2. Rinse the gel in a plastic tray during 3 periods of 5 min with enough double-distilled water (ddH_2O) to cover the gel. This step aims to remove the SDS and immobilize the proteins in the gel matrix
3. Remove the water and add Bio-Safe Coomassie stain to the plastic tray until the gel is completely covered. Put the tray on a rocking platform shaker for 1 h

4. Rinse the gel in ddH_2O three times, each time for about 20 min to allow for good visualization of the bands. A single band should appear for alpha-casein. Image the gel if needed at this point
5. Carefully excise all the bands with a clean blade and cut them into approximately 2×2 mm cubes to minimize the length of the subsequent destaining step. Transfer the excised bands into a 1.5 mL tube. If some proteins are stuck in the stacking gel (presence of dye in the stacking gel), it is preferable to include it in the 1.5 mL tube

Band destaining, reduction and alkylation

6. Add 300 μL 100% acetonitrile (ACN) to the gel pieces and incubate during 10 min on shaker or vortex mixer. Spin down and remove ACN
7. Add 300 μL 100 mM ABC/ACN (1:1 *V*/V). Incubate on shaker for 30 min. Spin down and remove all liquid from the tube
8. Repeat steps 6 and 7 until the blue stain from the Coomassie Brilliant Blue (CBB) has been completely removed. Shaking the tubes containing the gel pieces on a thermomixer at 50 °C can help to destain the gel faster. It is possible to store the gel pieces in the 100 mM ABC/ACN solution overnight at 4 °C at this stage
9. Add 300 μL 100% ACN. Incubate for 5 min and remove all liquid from the tube
10. Allow for the gel pieces to dry for several minutes. It is also possible to dry the samples in a SpeedVac for 1 min
11. Add enough 10 mM dithiothreitol (DTT) in 100 mM ammonium bicarbonate (ABC) to cover the gel. Incubate on a thermomixer at 55 °C and 700 rpm (rpm) for 30 min, and remove all liquid
12. Add enough 55 mM fresh chloroacetamide in 100 mM ABC to cover the gel. Incubate at 30 min in the dark at room temperature for alkylation. Remove all liquid from the tube
13. Wash the gel pieces with 300 μL 20 mM ABC/ACN (1:1 *V*/V). Incubate at room temperature for 15 min. Remove all liquid from the tube
14. Dehydrate the gel pieces in a SpeedVac for a couple of minutes

Trypsin digestion

15. Add trypsin with an enzyme to protein ratio of 1:100 (*w*/w). Add enough 50 mM ABC to cover the gel pieces. Incubate at 37 °C for 30 min. If the gel pieces absorbed some of the liquid, add more 50 mM ABC to cover the gel pieces
16. Incubate at 37 °C for 16 h

17. Centrifuge and transfer all liquid a new 1.5 mL tube. Add 30 μL 80% ACN with 20% of 1% formic acid (FA) to the gel pieces to stop the reaction. Incubate on a shaker at room temperature for 30 min
18. Centrifuge and transfer all liquid to a new 1.5 mL tube
19. Add 30 μL 60% ACN to the gel pieces. Vortex for 10 min, and combine all liquid with the previous peptide extract (from step 17).
20. Lyophilize the sample using a SpeedVac (typically for 1 h). At this stage, the samples can be stored at −20 °C for a few months until phosphoenrichment

2.2.4 In-solution trypsin digestion

In-solution trypsin digestion is an alternative way to cleave the proteins into peptides. This method does not require separation by SDS-PAGE and thus does not require a destaining step, which can sometimes take several hours; however, it is necessary to make sure the protein solution is compatible for trypsin digestion (e.g., pH should be ∼8, and make sure there is no trypsin inhibitor). For buffer exchange or cleanup, a methanol/chloroform protocol can be followed. In addition, a SPE step is often suggested to desalt the samples and avoid clogging of the LC system. These two additional steps should be performed prior to phosphoenrichment.

2.2.4.1 Procedure

2.2.4.1.1 Methanol and chloroform protein cleanup This protocol is adapted for samples containing 100 μg proteins, and the volumes given fit in a 1.5 mL tube. The volumes of the reagents added need to be adjusted according to the protein amount.

1. Resuspend the proteins in 400 μL methanol and vortex
2. Add 100 μL chloroform and vortex
3. Add 300 μL ddH_2O and vortex
4. Centrifuge during 10 min at 12,000 rpm and discard the upper phase
5. Gently add 300 μL of methanol to the tube without breaking the disc. Mix gently
6. Centrifuge during 10 min at 12,000 rpm and discard the supernatant
7. Lyophilize the pellet using a SpeedVac. At this stage, the samples can be stored at −20 °C

2.2.4.1.2 Trypsin digestion

1. Aliquot the volume of extracted protein solution needed in a 1.5 mL tube (we recommend performing this step using 200 μg of proteins), and adjust to volume to 100 μL with 100 mM ABC

2. Add 3 μL 10 mM DTT to the 100 μL sample solution
3. Vortex and incubate at room temperature for 1 h
4. Add 11 μL 500 mM CAA to obtain a final concentration of CAA of 55 mM
5. Incubate in the dark at room temperature for 1 h
6. Adjust the volume to 500 μL with 100 mM ABC
7. Add trypsin with an enzyme to protein ratio of 1:80 (*w*/w).
8. Incubate at 37 °C for 16 h
9. Add 1.5 μL 100% FA to lower the pH below 4 to stop the reaction
10. Lyophilize the sample in a SpeedVac. At this stage, the samples can be stored at −20 °C

2.2.4.1.3 Solid phase extraction (SPE)

1. Column conditioning
   1.1. Place the BioPureSPN Macro TARGA FastEq Desalting Column (50–200 μL elution volume, 35–350 μg maximum capacity) in a 2 mL collection tube
   1.2. Add 200 μL 100% ACN to the columns and centrifuge at 50 × *g* for 1 min
   1.3. Flush the column with 200 μL ddH_2O and centrifuge at 5000 × *g* for 1 min
2. Column equilibration
   2.1. Add 200 μL 0.1% trifluoroacetic acid (TFA), centrifuge at 700 × *g* for 1 min
   2.2. Repeat step 2.1. one more time
   2.3. Discard the 2 mL collecting tube and blot dry any moisture on the outside of the column
3. Sample loading
   3.1. Place the column in a new 2 mL collection tube and load 50–200 μL sample to the column. Centrifuge at 500 × *g* during 3 min. Repeat this step if you have more than 200 μL of protein sample
4. Column washing
   4.1. Add 200 μL 0.1% TFA to the column and centrifuge at 500 × *g* for 4 min
   4.2. Repeat step 4.1. one more time. You can keep the 2 mL collection tube and label it as "wash."
5. Peptide elution
   5.1. Place the column in a new 2 mL collection tube to collect the eluate

5.2. Add 200 μL of 80% ACN and 0.1% FA in ddH_2O and centrifuge at $500 \times g$ for 1 min

5.3. Repeat step 5.2. one more time. At this stage, the samples can be stored at −20 °C until phosphoenrichment

2.2.5 Comparison

While in-gel digestion has the advantage of skipping the cleanup and solid-phase extraction steps, and thus can be less time-consuming, it may not be suitable for large amounts of protein, because each well in the gel has a relatively small limiting volume. In general, this first described technique is more suitable for small amounts of proteins and simplified samples unless samples can be concentrated. In addition, SDS-gel helps to clean the samples from small molecules and salts. The destaining step at the beginning of the in-gel digestion protocol can be time consuming depending on the amount of protein and Coomassie Brilliant Blue in the gel pieces. In-solution digestion, on the other hand, is straightforward unless samples require protein cleanup with chloroform and methanol; however, it often requires the SPE procedure to clean up and desalt the samples. In general, in-solution digestion is most effective for samples containing large amounts of protein and for PTM analysis where large amounts of starting materials are needed for PTM enrichment.

To effectively compare the two methods, we performed the workflow described in this chapter on six ice plant (*Mesembryanthemum crystallinum*) leaf samples using one of the two trypsin digestion methods presented above: three samples were digested in-gel, and the other three in solution. For each digestion, 100 μg of protein was used, but only 2 μg of the protein digests was used for LC-MS/MS without phosphoenrichment. As shown in the bar graphs in Fig. 2, the LC-MS/MS results showed significant differences between the amounts of total and phosphorylated proteins generated by the two methods. Of the 1157 protein groups identified by the in-gel digestion method on average with high confidence (1% False Discovery Rate (FDR)), 62 proteins (5.4%) carried at least one phosphate group. The in-solution digestion method led to the identification of 743 protein groups with high confidence (1% FDR) 27 phosphorylated proteins (3.6%) were identified using this method. These results suggest that the in-gel digestion technique may lead to the identification of a larger number of proteins. The in-solution samples could have experienced loss in the SPE steps. If a LC-MS system has a precolumn setup for loading and eluting to the analytical column, the SPE steps may be unnecessary.

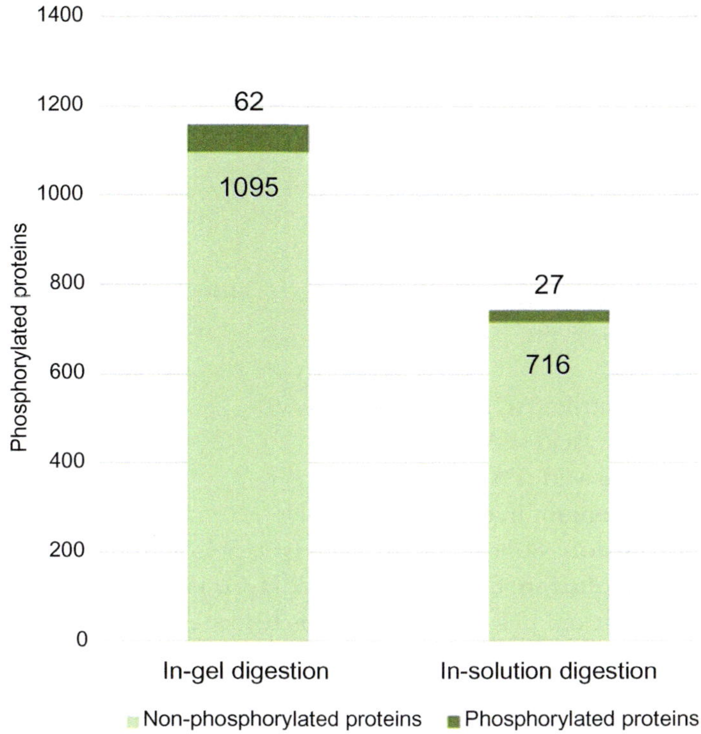

Fig. 2 Comparison between the number of total and phosphorylated proteins obtained by using in-gel and in-solution trypsin digestion methods. Please note that the samples were not phosphoenriched.

2.3 Phosphoenrichment

The phosphopeptide enrichment step is performed using a MOAC technique consisting in passing the digested peptides through mini-columns, in which titanium dioxide (TiO_2) and zirconium dioxide (ZrO_2) chromatographic media are packed. This method relies on the potential of quadrivalent metal ions to bind specifically to negatively-charged phosphopeptides. Thus, using this technique, non-phosphorylated peptides are efficiently separated from phosphorylated peptides which are trapped in the inner surface of the mini-column, and after elution, can be specifically analyzed by LC-MS/MS. The phosphoenrichment step is the most critical in this workflow, and it is essential to perform each step with care.

Equipment and consumables

- Savant SPD1010 SpeedVac Concentrator, Thermo Fisher Scientific
- Eppendorf 5417R Refrigerated Centrifuge

- Vortex mixer
- NuTip™ $TiO_2 + ZrO_2$ 10–200 μL Part No. NT2TIZR, Glygen
- Branson 5510R-MT Ultrasonic Cleaner, Marshall Scientific
- pH meter
- Vortex mixer
- Waste beaker
- 1.5 mL Eppendorf tubes
- Laboratory Fume Hood
- Multichannel micropipette

Reagents and solvents

- ddH_2O
- 80% acetonitrile (ACN) in ddH_2O (*v*/v)
- 100% formic acid (FA)
- 0.1% formic acid (FA) in ddH_2O (v/v)
- 100% ammonium hydroxide (NH_4OH)
- **Load solution**: 80% ACN, 15% ddH_2O, 5% of 100% FA (v/v)
- **Washing solution**: 80% ACN, 19% ddH_2O, 1% of 100% FA (v/v). It is essential that the pH of the solution is below 3. Add few drops of FA if needed
- **Elution solution**: 3% ammonium hydroxide (NH_4OH) in ddH_2O (v/v). It is critical that the pH of the buffer is equal to 11. Add few drops of FA or NH_4OH if needed. Prepare shortly before use

2.3.1 Procedure

The phosphoenrichment protocol should only be performed on samples used for phosphopeptide analysis and on alpha-casein (positive control of the enrichment). For total protein samples, proceed to Section 2.6 of this chapter. Once the column inside the NuTip™ has been in contact with the loading solution, it is essential to keep it wet at all times. Therefore, this protocol should be performed in one session. If you are working with a large quantity of samples, we recommend using a multichannel micropipette. Otherwise, we recommend using a P100 pipette for every step of this protocol. We also recommend collecting the flow-through of the enrichment step in a 1.5 mL tube, lyophilizing to dryness and storing it at −20 °C. If the number of phosphopeptides found after LC-MS/MS analysis is lower than expected, the flow-through can be used for troubleshooting. In addition, the flow-through should have mostly unphosphorylated peptides, which can be used to determine stoichiometry of protein phosphorylation.

1. Resuspend the digested peptides in 12.5 μL 0.1% FA water and 12.5 μL load buffer
2. Sonicate for 3 min, and vortex during 30 min
3. Equilibrate the TiO_2 + ZrO_2 NuTip™ by pipetting 20 μL of load buffer and releasing into a waste beaker. Repeat this step 20 times
4. Take a new 1.5 mL tube and label it "flow-through".
5. Load 25 μL of the peptide mixture into the TiO_2 + ZrO_2 NuTip™. Slowly pipet up and down for 10 min into the "flow-through" tube to allow for specific binding of the phosphopeptides to the embedded chromatographic material. It is crucial to perform this step very carefully, as it will determine the efficiency of the phosphoenrichment process. It is also important to avoid the formation of air bubbles to ensure a constant flow of digested peptide solution through the media. After 10 min, release all liquid into the "flow-through" tube
6. Load 25 μL of the wash buffer into the TiO_2 + ZrO_2 NuTip™ and transfer it to the "flow-through" tube
7. Load 25 more μL of the wash buffer into the TiO_2 + ZrO_2 NuTip™. Pipet up and down 20 times and release all liquid into the "flow-through" tube
8. Take a new 1.5 mL tube and label it "eluate."
9. Load 7 μL of the elution buffer into the TiO_2 + ZrO_2 NuTip™. Pipet up and down 10 times into the "eluate" tube. Repeat this step once by pipetting up and down in the cap of the "eluate" tube. Combine both solutions
10. To avoid loss of the phosphate group under highly basic condition, the eluted peptides were acidified with 5 μL 100% formic acid immediately
11. Lyophilize the enriched samples and the flow-through in a SpeedVac. The dried samples can be kept for few months at −20 °C before use

2.4 Sample processing by LC-MS/MS

For each biological sample to be analyzed, two peptide samples are introduced into the LC-MS/MS: the phosphoenriched sample and the total protein sample for qualitative and quantitative proteomic analyses. In addition, the sample corresponding to the flow-through of the enrichment step can be analyzed for troubleshooting purposes. Before starting the MS analysis, the phosphoenriched and the total protein samples need to be resuspended in 13 μL 0.1% formic acid solution (*v*/v) for injection of 10 μL on the LC.

Sonicate the samples for 5 min, vortex during 30 min, and centrifuge at 14,000 rpm for 15 min. Below is the instrument method used for the samples presented in the next section.

Peptides were directly introduced into a Vanquish Neo ultra-high-performance LC (UHPLC) (Thermo Scientific) connected to an Orbitrap Fusion Lumos™ mass spectrometer (Thermo Scientific). Peptides were injected through direct mode with loop offline at a loading flow of 100 μL/min at a pressure of 1500 bar. Similarly, fast column equilibration was performed at a flow of 100 μL/min and at a pressure of 1500 bar to reduce excessive sample running time. The elution gradient was performed at a nominal flow of 0.35 μL/min through two pump solvent channels, starting at 98% solvent A (100% H_2O) and 2% solvent B (80% acetonitrile), and ending after 75 min at 99% solvent B and 1% solvent A. The tandem MS analysis was performed in positive mode. The default charge state was 2, the ion transfer tube temperature 305 °C, the gas mode static, and the method duration 75 min. The MS spectra scan range was set on 400–2000 *m/z* (mass-to-charge), the Orbitrap™ resolution was 120,000, and the exclusion duration 30 s. The dissociation techniques used were collision-induced dissociation (CID), electron-transfer dissociation (ETD), and higher-energy C-trap dissociation (HCD). The two last methods complement CID, which is a commonly used ion-trap based fragmentation method (Johnson & Carlson, 2015). ETD is particularly useful for labile PTMs (Syka, Coon, Schroeder, Shabanowitz, & Hunt, 2004), while HCD is a method specific to Orbitrap™ mass spectrometers, which yields high resolution spectra (Jedrychowski et al., 2011; Olsen et al., 2007). All the MS/MS ions were detected in the ion-trap detector. A Nano-Spray-Ionization (NSI) source was used as the ion source type, a quadrupole was used as the isolation mode with an isolation window of 1.2 m/z, and the collision energy was 30%.

2.5 MS data processing and analysis

Below are presented the three software packages used for MS data processing and analysis of our ice plant samples. Xcalibur Qual Browser (Thermo Scientific, San Jose, USA) was used for LC chromatogram, MS1 and MS2 spectral inspection.

Mascot Daemon (version 2.4.0.), bundled with Mascot Server (MATRIX SCIENCE), were used for the identification of phosphorylated proteins and peptides. The server needs to be connected before opening the

Mascot Daemon user interface for task submission. Under "task editor" tab, a task name was given and data files with ".mgf" format were added. A pre-saved parameter set (database: green plant (UniProt taxon id 33,090); fixed modification: carbamidomethyl; variable modifications: oxidation (M), phosphorylation (S,T), phosphorylation (Y); peptide tolerance: 10 ppm; MS/MS tolerance: 0.6 Da) was selected under both "task editor" and "parameter editor" tab. After assigning a search title under "parameter editor" tab, the task was submitted to Mascot online server by hitting the "run" button under the "task editor" tab.

Proteome Discoverer 2.5 (Thermo Scientific, San Jose, USA) was used for the identification and quantification of total protein and phosphorylated proteins and peptides. ".raw" files were processed directly by the software without converting them into other file types. Since a tribrid mass spectrometer (using CID, ETD and HCD fragmentation methods) was used to process our samples, a branched processing workflow which separated ETD from CID and HCD was used. In the processing flow, due to the complexity of the protein samples, the SequestHT and Percolator nodes were used in both branches and then merged into an IMP-ptmRS node that better captures post-translationally modified proteins. The same green plant database was used. The precursor mass tolerance was set to 10 ppm and fragment mass tolerance was set to 0.6 Da. "Phospho" was added to the dynamic modifications (S, T and Y). The same processing workflow, without the addition of "Phospho" as a dynamic modification can be used to analyze total proteins.

The example data presented in this section was generated from leaf samples of the common ice plant. The phenol extraction was performed on 2 g of leaf frozen tissue, and the quantity and the quality of the extracted proteins were measured as described in the first section of Section 2 part of this paper. A total of 200 μg of proteins were used for in-gel trypsin digestion, and the phosphoenrichment was performed as described previously. Fig. 3A and B, respectively, illustrate the total number of MS/MS spectra of the phosphopeptides and the total number of phosphorylated residues identified in both ice plant and alpha-casein phosphoenriched samples, highlighting the relatively higher proportion of phosphoserine and phosphothreonine residues in both samples.

Fig. 4 shows a representative MS/MS spectrum generated through HCD and viewed on the Xcalibur software (Thermo Scientific). It illustrates the typical results generated by phosphoproteomics studies. The analysis

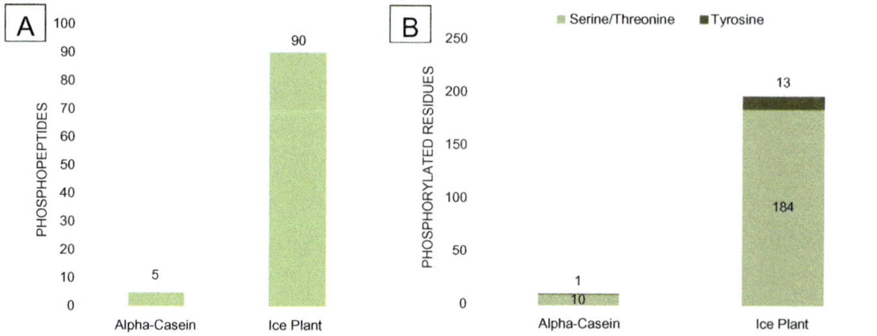

Fig. 3 Total number of phosphopeptide spectra (A) and phosphorylated residues (B) detected in an alpha-casein and an ice plant leaf sample after LC-MS/MS.

performed by Mascot Daemon associated the peptide shown on the figure to amidophosphoribosyltransferase, an enzyme known to catalyze the only committed step of the de novo purine biosynthesis pathway (Hung, Chen, Boldt, Sun, & Li, 2004). The peptide had a retention time of 57.43 min. Knowing the mass-to-charge ratio corresponding to gain or loss of a phosphate group, the phosphorylation status of every identified peptide is calculated by the software based on the relative m/z ratios of the different fragments generated in the MS2 (second phase of tandem mass spectrometry, during which the products ions are generated). The b- and y-ions generated from this peptide in the MS2 are shown in Table 1. b- and y-ions represent the two fragments generated by hydrolysis of a peptide bond. While b-ions are the products when the charge is retained on the N-Terminus of the peptide, y-ions are the products when the charge is retained on its C-Terminus. b7 (m/z 870.3410) and y9 (m/z 1080.4778) had high associated peaks and were identified with very high confidence. The difference between the masses of y9 (m/z 1117.6242) and y8 (m/z 936.6101) is 181.014, which corresponds to the monoisotopic mass of a phosphothreonine. This indicates that the threonine residue in position 7 (T7) is phosphorylated. Similarly, b7 (m/z 870.3410) corresponds to the addition of the monoisotopic masses of two phosphothreonine (181.014), two proline residues (97.05276), one serine (87.03203), one isoleucine (113.08407), and one leucine (113.08407), suggesting the presence of another phosphoresidue in this peptide. The presence of y13 (m/z 1527.8771) and the observation that it only carries one phosphate group excludes the possibility of the serine residue in position 3 being phosphorylated. Taken together, these results thus provide evidence that the peptide shown in Fig. 4 carries two phosphate groups, one on T7 and the other one on T1.

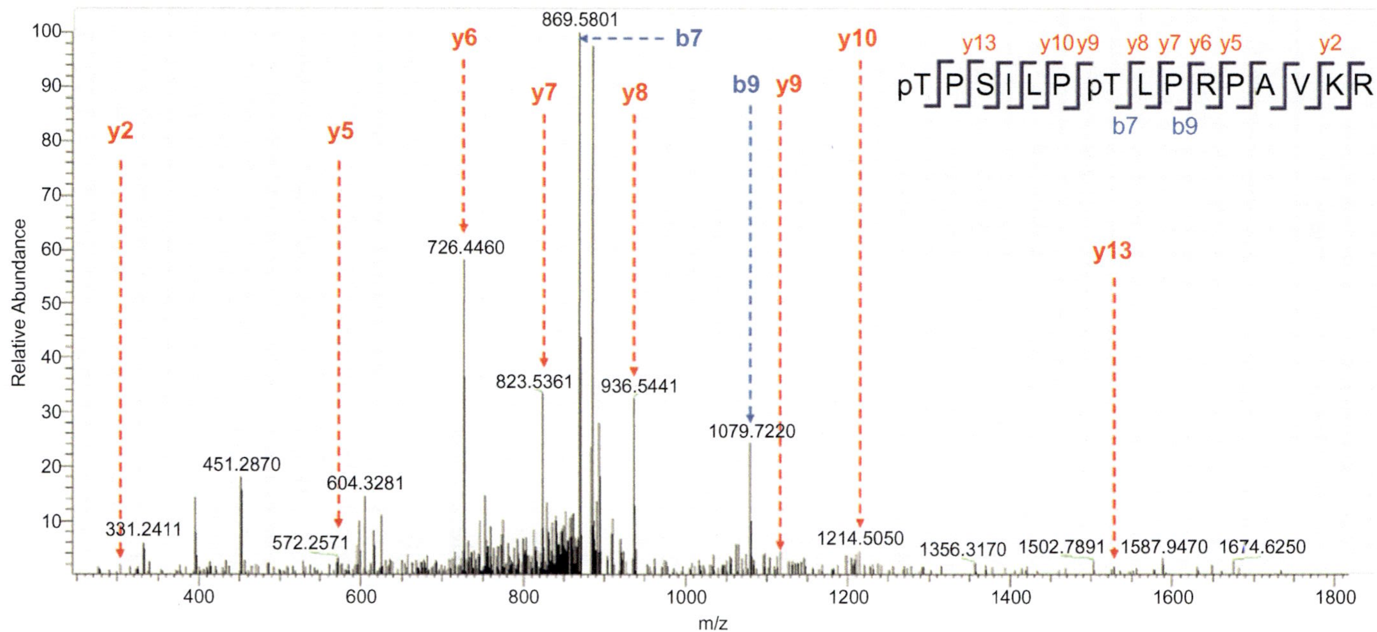

Fig. 4 MS/MS spectrum showing the product ions obtained during MS/MS of a peptide derived from an amidophosphoribosyltransferase . The peptide sequence is shown on the top right corner. The b-ions are shown in blue, and the y-ions in red. (pT) indicate the two phosphothreonine residues identified during the analysis.

Table 1 Product b- and y-ions in bold were generated with high confidence during MS/MS of a peptide associated to amidophosphoribosyltransferase.

#	b-Ions	Residue	y-Ions	#
1	182.0213	pT		15
2	279.0741	P	1624.9298	14
3	366.1061	S	**1527.8771**	13
4	479.1901	I	1440.8450	12
5	592.2742	L	1327.7610	11
6	689.3270	P	**1214.6769**	10
7	**870.3410**	pT	**1117.6242**	9
8	983.4250	L	**936.6101**	8
9	**1080.4778**	P	**823.5261**	7
10	1236.5789	R	**726.4733**	6
11	1333.6317	P	**570.3722**	5
12	1404.6688	A	473.3194	4
13	1503.7372	V	402.2823	3
14	1631.8322	K	**303.2139**	2
15		R	175.1190	1

pT in the residue column indicate the two phosphothreonine residues identified during the analysis. Values in bold represent ions generated with high confidence in the MS2.

2.6 Special considerations

At all times during this workflow, it is crucial to minimize keratin contamination of hair, skin and wool clothing to limit chemical background in samples. In addition, use nitrile gloves instead of latex gloves, and stay away from Western blot stations where milk is used for blotting. Consumables such as tips and tubes should be stored in a dust-free area, and each step of the procedures should be performed on a clean bench.

3. Summary

Most PTM proteomics workflows exclusively focus on PTM analyses, often missing some of the key information provided by non-PTM peptides. Our workflow provides options for generating two complementary datasets by analyzing both the proteome and the phosphoproteome. After phenol extraction, quality and quantity control of the proteins, a trypsin digestion step is performed to cleave the proteins into peptides. In this chapter, we presented two different methods for trypsin digestion. One method requires to run a polyacrylamide gel and perform the in-gel digestion process and the other method is gel-free in-solution protein digestion. While the gel-free method is advantageous when working with large amounts of proteins, more steps may be required than in-gel digestion as it is often necessary to carry out a solid phase extraction step as well as a chloroform/methanol cleanup. Our comparison of the two methods using ice plant protein samples illustrated the utility of in-gel digestion in phosphoproteomics. The most crucial step in almost any phosphoproteomics workflow is the phosphoenrichment step. Here we provide a protocol for rapid and efficient enrichment of phosphorylated peptides using the TiO_2/ZrO_2 mini-column. By using an LC-MS/MS method and following the approach we have described for data processing and analysis, it is possible not only to characterize specific phosphoresidues, phosphopeptides and their corresponding phosphoproteins, but also to combine these results with total proteome analyses. Therefore, the methods presented in this chapter can be used to elucidate complex molecular and physiological processes.

Mesembryanthemum crystallinum, or the common ice plant, has the fascinating ability to switch from C_3 to crassulacean acid metabolism (CAM) photosynthesis when subjected to either salt stress or drought (Guan et al., 2021; Guan, Tan, Kelley, Tian, & Chen, 2020). CAM evolved from the C_3 photosynthetic pathway in plants as an adaptation to low-water environment (Tay, Odang, & Cheung, 2021). It is characterized by nocturnal gas exchange, meaning that the stomata of CAM plants open during the night, when transpiration rate is low, to fix atmospheric carbon dioxide. Therefore, the water use efficiency of plants performing CAM photosynthesis is much higher than that associated with C_3 and C_4 plants (Borland et al., 2014). The majority of crops practice either C_3 or C_4 photosynthesis, which are widely used for food and biofuel production, while none of the major agricultural crops practice CAM photosynthesis. Enhancing crops water use efficiency is fundamental to ensuring global food security as the world's population grows rapidly and our environment changes critically. Protein phosphorylation plays a critical role in CAM. A well-described example is the post-translational activation of phosphoenolpyruvate carboxylase (PEPC) by phosphoenolpyruvate carboxylase kinase at night, allowing nocturnal CO_2 fixation in mesophyll cells (Bakrim, Brulfert, Vidal, & Chollet, 2001). To date, this phosphorylation is the only one that has been fully described. Following the workflow described in this chapter, we are currently working to identify and characterize new major phosphorylation events occurring during the induction of CAM from C_3 by salt stress. Thus, our method may allow a better understanding of the key molecular switches driving this transition. The results may greatly facilitate the introduction of CAM traits into C_3 crops, thereby improving their water use efficiency and stress resistance, and ultimately improving global food security.

Although the C_3 to CAM transition in the ice plant is an ideal molecular and physiological process for phosphoproteomic studies, the methods detailed in this chapter can be applied to answer a wide variety of biological questions, not only in plant biology, but also in major fields of biology and medicine. Because kinases and phosphatases are important components of signaling cascades, our protocol may help to identify new key enzymes in major physiological and biological processes. Our work on the common ice plant has shown that it is possible to conduct both proteomics and phosphoproteomics analysis from simultaneously generated sample sets. Overall, our workflow provides a complete method for characterization of phosphoproteome and total proteome from the same samples.

Acknowledgment

This work was partly supported by the Agriculture and Food Research Initiative (AFRI) (Award No. 2020-67013-31615/accession no. 1022409) from the USDA National Institute of Food and Agriculture, and the University of Florida Plant Molecular and Cellular Biology Program.

References

Arsova, B., Watt, M., & Usadel, B. (2018). Monitoring of plant protein post-translational modifications using targeted proteomics. *Frontiers in Plant Science*, *9*, 1168.

Bakrim, N., Brulfert, J., Vidal, J., & Chollet, R. (2001). Phosphoenolpyruvate carboxylase kinase is controlled by a similar signaling cascade in CAM and C(4) plants. *Biochemical and Biophysical Research Communications*, *286*(5), 1158–1162.

Borland, A. M., Hartwell, J., Weston, D. J., Schlauch, K. A., Tschaplinski, T. J., Tuskan, G. A., et al. (2014). Engineering crassulacean acid metabolism to improve water-use efficiency. *Trends in Plant Science*, *19*(5), 327–338.

Bradford, M. M. (1976). A rapid and sensitive method for the quantitation of microgram quantities of protein utilizing the principle of protein-dye binding. *Analytical Biochemistry*, *72*, 248–254.

Chou, K.-C. (2020). Progresses in predicting post-translational modification. *International Journal of Peptide Research and Therapeutics*, *26*(2), 873–888.

Faurobert, M., Pelpoir, E., & Chaïb, J. (2007). Phenol extraction of proteins for proteomic studies of recalcitrant plant tissues. *Methods in Molecular Biology (Clifton, N.J.)*, *355*, 9–14.

Guan, Q., Kong, W., Zhu, D., Zhu, W., Dufresne, C., Tian, J., et al. (2021). Comparative proteomics of *Mesembryanthemum crystallinum* guard cells and mesophyll cells in transition from C_3 to CAM. *Journal of Proteomics*, *231*, 104019.

Guan, Q., Tan, B., Kelley, T. M., Tian, J., & Chen, S. (2020). Physiological changes in *Mesembryanthemum crystallinum* during the C_3 to CAM transition induced by salt stress. *Frontiers in Plant Science*, *11*.

Hung, W.-F., Chen, L.-J., Boldt, R., Sun, C.-W., & Li, H. (2004). Characterization of Arabidopsis glutamine phosphoribosyl pyrophosphate amidotransferase-deficient mutants. *Plant Physiology*, *135*(3), 1314–1323.

Javeed, R., Hussain, D., Jabeen, F., Saeed, A., Fatima, B., Ashiq, M. N., et al. (2020). Enrichment of HDL proteome and phospholipidome from human serum via IMAC/MOAC affinity. *Biomedical Chromatography*, *34*(1), e4693.

Jedrychowski, M. P., Huttlin, E. L., Haas, W., Sowa, M. E., Rad, R., & Gygi, S. P. (2011). Evaluation of HCD- and CID-type fragmentation within their respective detection platforms for murine phosphoproteomics. *Molecular & Cellular Proteomics: MCP*, *10*(12). M111.009910.

Johnson, A. R., & Carlson, E. E. (2015). Collision-induced dissociation mass spectrometry: A powerful tool for natural product structure elucidation. *Analytical Chemistry*, *87*(21), 10668–10678.

Khoury, G. A., Baliban, R. C., & Floudas, C. A. (2011). Proteome-wide post-translational modification statistics: Frequency analysis and curation of the swiss-prot database. *Scientific Reports*, *1*(1), 90.

Kim, Y. J., Kim, H. M., Kim, H. M., Lee, H. R., Jeong, B. R., Lee, H.-J., et al. (2021). Growth and phytochemicals of ice plant (*Mesembryanthemum crystallinum* L.) as affected by various combined ratios of red and blue LEDs in a closed-type plant production system. *Journal of Applied Research on Medicinal and Aromatic Plants*, *20*, 100267.

Longoni, F. P., & Goldschmidt-Clermont, M. (2021). Thylakoid protein phosphorylation in chloroplasts. *Plant and Cell Physiology*, *62*(7), 1094–1107.

Mayya, V., & Han, D. K. (2009). Phosphoproteomics by mass spectrometry: Insights, implications, applications, and limitations. *Expert Review of Proteomics*, *6*(6), 605–618.

Nukarinen, E., Nägele, T., Pedrotti, L., Wurzinger, B., Mair, A., Landgraf, R., et al. (2016). Quantitative phosphoproteomics reveals the role of the AMPK plant ortholog SnRK1 as a metabolic master regulator under energy deprivation. *Scientific Reports*, *6*(1), 31697.

Olsen, J. V., Macek, B., Lange, O., Makarov, A., Horning, S., & Mann, M. (2007). Higher-energy C-trap dissociation for peptide modification analysis. *Nature Methods*, *4*(9), 709–712.

Pavokovic, D., Križnik, B., & Krsnik-Rasol, M. (2012). Evaluation of protein extraction methods for proteomic analysis of non-model recalcitrant plant tissues. *Croatica Chemica Acta*, *85*, 177–184.

Pi, E., Zhu, C., Fan, W., Huang, Y., Qu, L., Li, Y., et al. (2018). Quantitative phosphoproteomic and metabolomic analyses reveal GmMYB173 optimizes flavonoid metabolism in soybean under salt stress. *Molecular & Cellular Proteomics*, *17*(6), 1209–1224.

Sheffield, J., Taylor, N., Fauquet, C., & Chen, S. (2006). The cassava (*Manihot esculenta* Crantz) root proteome: Protein identification and differential expression. *Proteomics*, *6*(5), 1588–1598.

Shevchenko, A., Loboda, A., Ens, W., Schraven, B., Standing, K. G., & Shevchenko, A. (2001). Archived polyacrylamide gels as a resource for proteome characterization by mass spectrometry. *Electrophoresis*, *22*(6), 1194–1203.

Silva-Sanchez, C., Li, H., & Chen, S. (2015). Recent advances and challenges in plant phosphoproteomics. *Proteomics*, *15*(5–6), 1127–1141.

Subba, P., & Prasad, T. S. K. (2021). Plant phosphoproteomics: Known knowns, known unknowns, and unknown unknowns of an emerging systems science frontier. *OMICS: A Journal of Integrative Biology*, *25*(12), 750–769.

Syka, J. E. P., Coon, J. J., Schroeder, M. J., Shabanowitz, J., & Hunt, D. F. (2004). Peptide and protein sequence analysis by electron transfer dissociation mass spectrometry. *Proceedings of the National Academy of Sciences of the United States of America*, *101*(26), 9528–9533.

Tay, I. Y. Y., Odang, K. B., & Cheung, C. Y. M. (2021). Metabolic modeling of the C_3-CAM continuum revealed the establishment of a starch/sugar-malate cycle in CAM evolution. *Frontiers in Plant Science*, *11*, 573197.

Thingholm, T. E., & Larsen, M. R. (2016). Phosphopeptide enrichment by immobilized metal affinity chromatography. *Methods in Molecular Biology (Clifton N.J.)*, *1355*, 123–133.

Urban, J. (2022). A review on recent trends in the phosphoproteomics workflow. From sample preparation to data analysis. *Analytica Chimica Acta*, *1199*, 338857.

Van Leene, J., Han, C., Gadeyne, A., Eeckhout, D., Matthijs, C., Cannoot, B., et al. (2019). Capturing the phosphorylation and protein interaction landscape of the plant TOR kinase. *Nature Plants*, *5*(3), 316–327.

Walsh, C. (2006). *Posttranslational modification of proteins: Expanding nature's inventory*. Roberts and Company Publishers.

Wu, X., Sklodowski, K., Encke, B., & Schulze, W. X. (2014). A kinase-phosphatase signaling module with BSK8 and BSL2 involved in regulation of sucrose-phosphate synthase. *Journal of Proteome Research*, *13*(7), 3397–3409.

Zhang, H., Zhu, J., Gong, Z., & Zhu, J.-K. (2022). Abiotic stress responses in plants. *Nature Reviews Genetics*, *23*(2), 104–119.

CHAPTER FOURTEEN

Identification of MPK4 kinase interactome using TurboID proximity labeling proteomics in *Arabidopsis thaliana*

Chuwei Lin[a,b], Incheol Yeo[c], Craig P. Dufresne[d], Guimei Zhao[a], Sarah Joe[a], and Sixue Chen[a,b,e,f,*]

[a]Department of Biology, Genetics Institute, University of Florida, Gainesville, FL, United States
[b]University of Florida Genetics Institute (UFGI), University of Florida, Gainesville, FL, United States
[c]Department of Biochemistry & Biophysics, Texas A&M University, College Station, TX, United States
[d]Thermo Fisher Scientific Training Institute, West Palm Beach, FL, United States
[e]Plant Molecular and Cellular Biology Program, University of Florida, Gainesville, FL, United States
[f]Proteomics and Mass Spectrometry, Interdisciplinary Center for Biotechnology Research (ICBR), University of Florida, Gainesville, FL, United States
*Corresponding author: e-mail address: schen@ufl.edu

Contents

Abstract

TurboID is a new and efficient proximity labeling system that was first developed in living mammalian cells. TurboID is a modified bacterial biotin ligase that can be fused to a bait protein, which can then modify proximal interacting proteins with biotin. Prey proteins subsequently labeled with biotin tags will be pulled down with streptavidin-coated beads and identified by mass spectrometry-based proteomics. TurboID has been recently applied to living plant cells and provided promising results in identification of interacting proteins. Mitogen-activated protein kinase 4 (MPK4) is important

Methods in Enzymology, Volume 676
ISSN 0076-6879
https://doi.org/10.1016/bs.mie.2022.06.005

for plant growth, development, and defense; however, the molecular mechanisms underlying the range of MPK4 functions are not completely known. Here we use modern proteomics together with the TurboID in a proof-of-concept study to profile the MPK4 interactome and uncover the functions of MPK4 in plant signaling cascades.

1. Introduction

Proximity labeling (PL) is an emerging approach to map direct protein–protein interactions. The principle is to link a biotin ligase to the protein of interest (bait) so that spatially proximal proteins (prey) will be tagged with biotin, which can be pulled down using streptavidin, an anti-biotin monoclonal antibody, or Tamavidin 2-REV (Chen & Perrimon, 2017; Das, Macharia, Lin, & Wong, 2019; Khan, Youn, Gingras, Subramaniam, & Desveaux, 2018; Motani & Kosako, 2020). PL is always coupled with LC-MS/MS for quantitative proteomic analysis. The advantages of PL over traditional affinity purification-MS include capturing *in vivo* direct interacting proteins and the transient interactions. The first PL developed was BioID, using a bioengineered biotin ligase from *Escherichia coli* (Roux, Kim, Raida, & Burke, 2012). It was developed in mammalian cells, and later gained popularity in plants to map interactome in rice, tobacco, and Arabidopsis (Chen & Perrimon, 2017; Das et al., 2019; Khan et al., 2018). BioID requires a relatively long biotin incubation and labeling time of 18–24 h (Roux et al., 2012). Another PL uses engineered ascorbate peroxidases APEX and APEX2 (Martell et al., 2012). They were derived from monomeric ascorbate peroxidases originally from pea and soybean that catalyze oxidation of biotin-phenol to biotin-phenoxyl radical in the presence of H_2O_2. Comparing with the 35 kDa BioID, APEX and APEX2 have a smaller size of 28 kDa, and a rapid labeling time of 30 s to 1 min (Lobingier et al., 2017). In addition to biotin-phenol, APEX and APEX2 can also catalyze the polymerization and local deposition of diaminobenzidine, which allows it to couple with electron microscopy analysis (Martell et al., 2012). As efficient as they are, both APEX and APEX2 require H_2O_2, which can be toxic to cells, and may induce stresses and introduce noise. Another tool is the next generation of the biotin ligase BioID2, which is a much smaller bioengineered biotin ligase from *Aquifex aeolicus*. It requires less biotin than BioID to label the same amount of proximity proteins because of the smaller tag improves the localization of the bait protein

(Kim et al., 2016). TurboID is a significantly improved version of BioID derived from the same biotin ligase from the *E. coli*. It is the same size as BioID, but contains 15 mutations, enabling higher activity and significantly less labeling time of ~10 min. In addition, it is about 22-fold more efficient in terms of biotinylation of prey proteins (Branon et al., 2018). A MiniTurboID was developed at the same time. It is a truncated version of TurboID lacking the N-terminal DNA binding domain (Branon et al., 2018). It requires the same labeling time as TurboID, and is still better than BioID, but labels fewer prey proteins than TurboID. Because TurboID displays short labeling time and good efficiency in biotinylating prey proteins, it has been widely used, e.g., in plants to identify protein interaction networks (Zhang, Song, et al., 2019). In plants, TurboID labels the highest number of prey proteins in 3 h (Mair, Xu, Branon, Ting, & Bergmann, 2019; Zhang, Song, et al., 2019). Here, we used a stable transgenic Arabidopsis line with a *TurboID-MPK4* construct (Fig. 1) to profile interactome of the MPK4.

Mitogen-activated protein kinase 4 (MPK4) was initially identified as a negative regulator in systemic acquired resistance (SAR) in *Arabidopsis thaliana* (Petersen et al., 2000). It plays important roles in different biological processes, including plant immunity, cytokinesis, reproduction, and photosynthesis (Beck, Komis, Muller, Menzel, & Samaj, 2010; Beck, Komis, Ziemann, Menzel, & Samaj, 2011; Gawroński et al., 2014; Kosetsu et al., 2010; Lin & Chen, 2018; Liu et al., 2011; Zeng, Chen, Ellis, & Laboratories, 2011). MPK4 can be activated by flg22, a 22-amino-acid peptide from bacterial flagellum, which triggers downstream immune responses (Zhang, Schneider, et al., 2019). In this study, TurboID was used as a tool to profile MPK4 interactome. Different from previous TurboID studies in plants, where most often TurboID was transiently expressed in tobacco leaves, the TurboID fused with the MPK4 was stably transformed into *A. thaliana*. The fully expanded leaves of the transgenic *Arabidopsis* were used for the TurboID-proteomics experiments in this chapter.

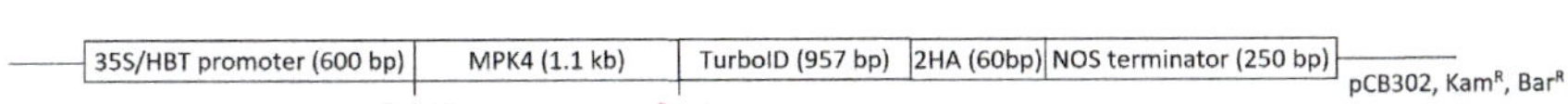

Fig. 1 The *MPK4-TurboID* construct transformed in to *1-1* and *1-4* Arabidopsis lines. The expression of *MPK4-TurboID* is driven by 35S promoter. A 2× HA tag is attached at the C-terminal end of TurboID for Western blot detection. Transgenic plants are resistant to Basta herbicide.

2. Sample preparation

In general, transgenic *A. thaliana* leaves are treated with flg22 and at the same time incubated with biotin for 3h to enable thorough TurboID labeling. After protein extraction, the proteins are desalted to remove excessive free biotin (Zhang, Song, et al., 2019). Because streptavidin has a strong binding affinity with biotin, and biotinylated proteins are difficult to elute off the streptavidin beads, previous studies recommend on-bead trypsin digestion (Branon et al., 2018; Mair et al., 2019; Zhang, Song, et al., 2019); however, for different bait proteins fused with TurboID, it is difficult to estimated biotinylated protein amount remaining on beads and trypsin amount used for digestion, which may lead to over-digestion or incomplete digestion. In this protocol, we recommend performing in-gel trypsin digestion. First, during electrophoresis, biotinylated proteins may be better separated from the streptavidin beads. Second, visualizing proteins on the gel allows for estimation of protein amount and trypsin amount to use for the proteolysis into peptides.

2.1 Plant materials

Four *A. thaliana* lines were used in this experiment, namely, vector control (*VC*) plants, *mpk4 mutant*, and two transgenic lines *MPK4-TurboID/mpk4 1-1* (*1-1*) and *MPK4-TurboID/mpk4 1-4 (1-4)* (all in the *Landsberg erecta*, *Ler* background) were used for experiments. The TurboID constructs have a 2HA tag for Western detection (Fig. 1). Plants were grown in Pro-line C/GP germination mix (Jolly Gardener, FL, USA) at short day conditions: 24 °C in light for 10h and 20 °C in dark for 14h with 70% relative humidity. The light was white light with an intensity of 160 μmol/(m^2s). Five-week-old plants were used for the TurboID-proteomics sample preparation.

2.2 Protein extraction and quantification

Equipment and consumables

- Mortar and pestle
- Rocking Platform Shaker
- Vortex mixer
- Eppendorf 5417R Refrigerated Centrifuge
- Thermomixer
- 1.5 mL Eppendorf tubes
- Laboratory Electrophoresis Power Supply Model 200/2.0, Bio-Rad

Reagents and solvents

- 50 μM biotin
- 10 μM flg22 (Sigma-Aldrich, St. Louis, MO, USA)
- RIPA lysis buffer: 50 mM Tris-HCl pH 7.5, 500 mM NaCl, 1 mM EDTA, 0.1% SDS, 0.5% deoxycholic acid, 1% Triton X-100 and protease and phosphatase inhibitor (Thermo Scientific, Rockford, IL, USA)
- Zeba™ spin desalting column (Thermo Scientific, Rockford, IL, USA)
- Dynabeads™ MyOne™ Streptavidin C1 (Invitrogen, Waltham, MA, USA)
- Buffer I: 2% SDS
- Buffer II: 50 mM HEPES, 500 mM NaCl, 1 mM EDTA, 0.1% deoxycholic acid, and 1% Triton X-100
- Buffer III: 10 mM Tris-HCl pH 7.5, 250 mM LiCl, 1 mM EDTA, 0.1% deoxycholic acid, and 1% Triton X-100
- 50 mM Tris-HCl pH 7.5
- Bio-Safe Coomassie stain (Bio-rad, CA)
- Pierce™ Bradford Protein Assay Kit (Thermo Scientific, Rockford, IL, USA)
- Laemmli protein loading buffer: 1 mg/mL bromphenol blue, 62.5 mM Tris base, 2% SDS, 10% glycerol, and 0.1% 2-mercaptoethanol

Procedure

1. For treatment group, leaves were sprayed with 10 μM flg22. Control group was sprayed with water
2. Plants leaves were incubated with a 50 μM biotin solution for 3 h
3. Leaves were frozen in liquid nitrogen immediately and then ground into fine powder with mortar and pestle
4. For each biological replicate, 100 mg fine powder was used to mix with 500 μL RIPA lysis buffer
5. The mixture was shaken on a shaker at 4 °C for 1 h to ensure extraction of the most amounts of proteins
6. The mixture was centrifuged at 16,500*g* for 10 min, and the supernatant was collected to a new tube
7. The supernatant was passed through a Zeba™ spin desalting column to remove excessive free biotin
8. The concentration of the desalted protein extract was determined by Bradford assay (Thermo Fisher Scientific) according to manufacturer instructions
9. Two portions of 15 μg crude extract were aliquoted for Western blot with an anti-HA antibody and streptavidin to detect MPK4-TurboID expression and biotinylated proteins, respectively (Fig. 2).

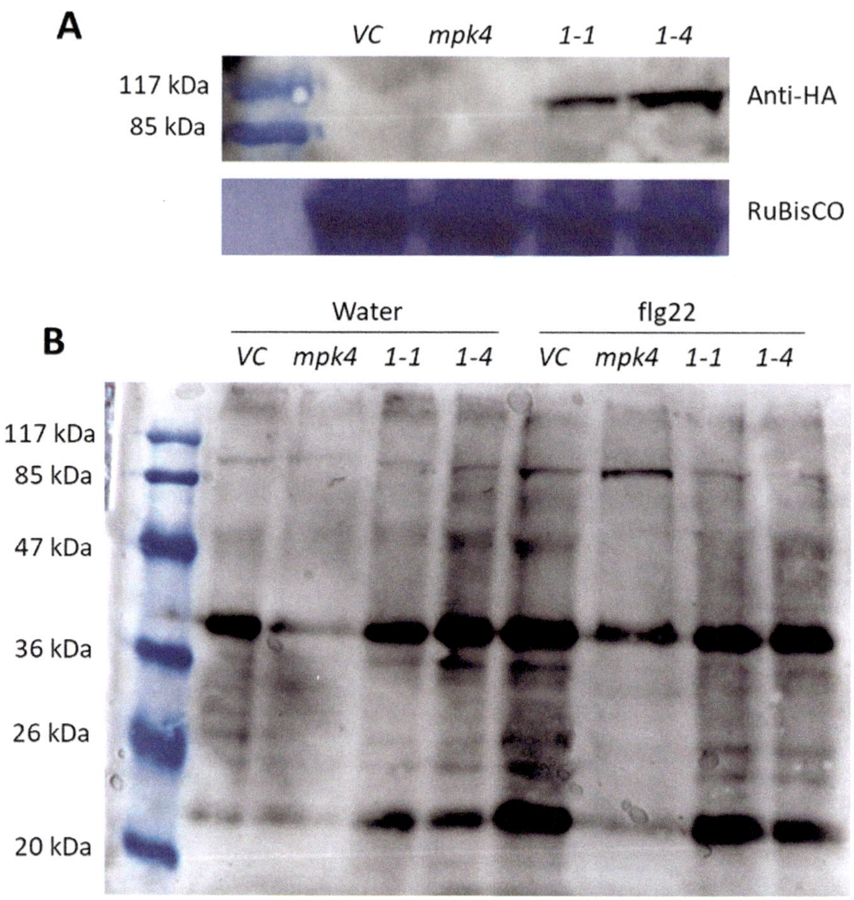

Fig. 2 Western blots of vector control (*VC*), *mpk4*, *1-1* and *1-4* plants. (A) Western blot with anti-HA. Two transgenic lines (*1-1* and *1-4*) successfully expressed the MPK4-TurboID-HA. (B) Western blot with streptavidin to reveal biotinylated proteins in different samples.

10. Every 500 μL crude extract was incubated with 20 μL Dynabeads™ MyOne™ Streptavidin C1 (Invitrogen, Waltham, MA), agitating at 4 °C overnight (16 h).
11. Next day, the beads were pelleted by centrifugation at 5000*g* for 30 s, and the supernatant was discarded
12. The pelleted beads were washed with buffer I. It is important to make sure beads were evenly resuspended by pipetting. Then, the beads were pelleted by centrifugation, and the supernatant was removed
13. The pelleted beads were washed a second time with buffer II in the same manner as above

14. The pelleted beads were washed with buffer III
15. The pelleted beads were washed with 50 mM Tris-HCl multiple times until the supernatant was clear with no green color
16. The beads were resuspended in 30 μL Laemmli protein loading buffer and heated at 90 °C for 10 min
17. Proteins were separated by SDS-PAGE using 12% gels at 100 V for 1 cm in separating gel (Fig. 3).
18. The SDS gels were stained with Coomassie Brilliant Blue for 2 h, followed by destaining for another 2 h

2.3 In-gel trypsin digestion

For each sample, the washed streptavidin beads were loaded into SDS-PAGE gels. Because gel was only run for 1 cm, proteins are concentrated in separating gels, which reduced gel volume for in-gel trypsin digestion. Since some proteins still remained in the stacking gel, so that stacking gels were also included for the in-gel trypsin digestion.

Equipment and consumables

- Clean glass plate
- Scalpel
- 1.5 mL Eppendorf tubes
- Savant SPD1010 SpeedVac Concentrator, Thermo Fisher Scientific
- Rocking Platform Shaker
- Vortex mixer
- Eppendorf 5417R Refrigerated Centrifuge
- 37 °C incubator

Reagents and solvents

- 100% acetonitrile (ACN)
- 100 mM ammonium bicarbonate (ABC)
- 100 mM ABC/ACN (1:1 v/v)

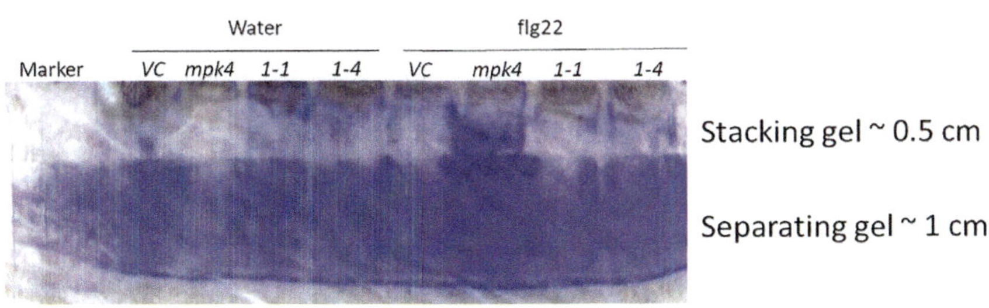

Fig. 3 SDS-PAGE gel separation of the protein extracts for in-gel trypsin digestion. Proteins only run for 1 cm in the separating gel. *VC*, vector control.

- 10 mM dithiothreitol (DTT) in 100 mM ABC. Prepare shortly before use
- 55 mM chloroacetamide (CAA) in 100 mM ABC. Prepare shortly before use and keep in the dark
- 20 mM ABC/ACN (1:1 v/v)
- 6 ng/μL trypsin (V5111, sequencing grade, Promega, Madison, WI, USA) in 50 mM ABC (pH 7.8)
- 80% ACN with 20% of 1% formic acid (FA)
- 60% ACN

Procedure

1. Destained gels (including both stacking and separating gels) were cut into 1 mm^3 small pieces on a clean glass plate using a scalpel and transferred to a 1.5 mL tube
2. The gel pieces were mixed with 300 μL 100% ACN and shaken for 10 min. Then ACN was removed
3. The gel pieces were then mixed with 300 μL 100 mM ABC/ACN and shaken for 30 min. Then supernatant was removed
4. Repeat steps 2 & 3 one or several more times until Coomassie Brilliant Blue stain was gone
5. The gel pieces were incubated with 300 μL 100% ACN for 5 min. Then supernatant was removed
6. The gel pieces were lyophilized in a SpeedVac for 2–3 min
7. The dried gel pieces were covered with 10 mM DTT in 100 mm ABC for 30 min at room temperature
8. Then, the same volume of 55 mM fresh CAA in 100 mM ABC was added and incubated at room temperature for 30 min in dark. The supernatant was removed
9. The gel pieces were washed with 300 μL 20 mM ABC/ACN for 15 min. Then supernatant was removed
10. The gel pieces were dehydrated with 100 μL 100% ACN for 5 min. Then supernatant was removed
11. The gel pieces were dried in a SpeedVac for 2–3 min
12. Then the gel pieces were covered with 6 ng/μL trypsin solution for 30 min
13. After absorption of the trypsin solution, 50 mM ABC was added to cover up all the gel pieces
14. The samples were incubated at 37 °C for 16 h
15. After a quick centrifugation, 30 μL 80% ACN with 20% of 1% FA was added, shaking for 30 min
16. The supernatant was transferred to a new tube, and the remaining gel pieces were mixed with 30 μL 60% ACN, shaking for 10 min

17. The supernatants were combined with that from step 16
18. The combined supernatants were lyophilized to dryness in the SpeedVac

2.4 Ziptip desalting

After in-gel trypsin digestion, peptides were desalted with ZipTip to remove contaminants, such as SDS. Hydrophobic peptides were retained on the C18 resin, while salts were washed off by 0.1% FA. Later, peptides were eluted with high concentration of organic solvent, ACN. Desalted peptides were used for proteomic analysis.

Equipment and consumables

- ZipTip® Pipette Tips with C18 Resin (ZTC18S960, MilliporeSigma, St. Louis, MO, USA)
- 1.5 mL Eppendorf tubes
- Savant SPD1010 SpeedVac Concentrator, Thermo Fisher Scientific

Reagents and solvents

- 100% ACN
- 0.1% FA
- 80% ACN with 0.1% FA

Procedure

1. Dried peptides were dissolved in 20 µL 0.1% FA by sonication for 5 min and shaking for 15 min, and then centrifugation at 2000 rpm for 15 min
2. Pipetteman was set to 10 µL, and a ZipTip was attached. The ZipTip was activated with 100% ACN by pipetting 5 times. Please keep the C18 resin in solution all the time and avoid introducing air
3. The ZipTip was equilibrated with 0.1% FA for 5 times by pipetting up and down
4. The samples were pipetted up and down slowly 20 times through the ZipTip to load the peptides on the resin
5. The ZipTip was washed with 0.1% FA 10 times to wash out unbonded molecules and salts
6. The peptides were eluted with 10 µL 80% ACN with 0.1% FA twice into a new tube
7. The peptide samples were lyophilized dry using the SpeedVac

3. Liquid chromatography tandem mass spectrometry (LC-MS/MS)

Before starting the LC-MS/MS analysis, the peptides were resuspended in 13 µL 0.1% formic acid solution (*v*/v) for injection of

10 μL on the LC column. Peptides were directly introduced into a Vanquish Neo ultra-high-performance liquid chromatography (UHPLC) (Thermo Scientific, San Jose, CA, USA) connected to an Orbitrap Fusion Lumos™ mass spectrometer (Thermo Scientific, San Jose, CA, USA). Peptides were injected through direct mode with loop offline at a loading flow of 100 μL/min and a maximum pressure of 1180 bar. The column used was EASY-Spray column Acclaim PepMap C18 (100 Å, 2 μm particle size, 75 μm i.d. × 25 cm). After valve switching, the loop was put online with the analytical column and the peptides were loaded onto the column. The elution gradient was performed at a nominal flow of 0.35 μL/min through two pump solvent channels with 0.1% formic acid in water as solvent A and 0.1% formic acid in 80% ACN as solvent B. The LC gradient was initially held at 2% solvent B, then a linear gradient was imposed from 2% B to 40% B over 72 min, followed by increasing from 40% B to 100% B over 6 min, and was ended at 100% for 12 min. The tandem MS analysis was performed in positive mode at 1800 V. The default charge state was 2, the ion transfer tube temperature was 305 °C, with static gas mode and a run duration of 90 min. The full MS1 spectra scan range was set on 400 to 2000 m/z (mass-to-charge), the Orbitrap™ resolution was 120,000, the maximum injection time was 50 ms, and automatic gain control (AGC) target was 4e5. Peptide bearing charge +2–6 were selected with a maximum intensity of 1e20 and minimum intensity of 1e4. Dynamic exclusion of 30 s was used to prevent resampling the high abundance peptides. For the MS/MS, the dissociation techniques used were collision-induced dissociation (CID), which was done at 30% of normalized collision energy. All the MS1 ions were selected in quadrupole with an Isolation window of 1.2 m/z. The MS/MS ions were separated in the ion-trap with AGC target as 1e4 and a maximum injection time of 35 ms.

4. Data processing and analysis

Proteome Discoverer™ (version 2.5, Thermo Scientific, Bremen, Germany) was used to search the MS/MS spectra from the protein samples. The SEQUEST algorithm in the Proteome Discoverer™ was used to process raw data files. Spectra were searched using the Arabidopsis TAIR10 database with the following parameters: 10 ppm mass tolerance for MS1 and 0.6 Da as mass tolerance for MS2, two maximum missed tryptic cleavage sites, a fixed modification of carbamidomethylation (+57.021) on cysteine residues and a dynamic modification of methionine oxidation (+15.995).

The Search results were filtered at 1% false discovery rate (FDR) and at least two unique peptides per protein for protein identification. Relative protein abundance in the samples was measured using label-free quantification of peptide peak areas. Proteins identified and quantified in all the four biological samples were used, and no data imputation was performed. Peptides in samples were quantified as the area under the chromatogram peak. FDR cutoffs for both peptide and protein identification were set as 1%.

The data presented in this section was generated from 100 mg fresh weight leaf samples. In water treatment group, there were 108 proteins found in *VC* sample, 49 proteins in *mpk4*, 124 proteins in 1-1 and 158 proteins in 1-4 (Fig. 4A). In flg22 treatment group, there were 148 proteins found in *VC* sample, 58 proteins in *mpk4*, 139 proteins in 1-1 and 163 proteins in *1-4* (Fig. 4B). Overall, flg22 treatment groups had more proteins detected by LC-MS. Among them, 78 proteins were found only in *1-1* and *1-4* with water, 44 only in *1-1* and *1-4* with flg22, but not *VC* and *mpk4*. In total, 93 proteins only appeared in *1-1* and *1-4* samples (Fig. 4C). Among them 49 proteins were unique in water treatment group, 15 proteins were unique in flg22 treatment group, and 29 proteins were shared by both groups. Although overall flg22 treatment groups have more proteins, there were less unique proteins in *1-1* and *1-4* with flg22 treatment, indicating that after activation by flg22, there were fewer proteins interact with MPK4 at the treatment time point. We presented PCA analysis, an unsupervised data-driven approach for data complexity reduction as linear combinations of multivariate protein abundances (Fig. 4D). As a result, *VC*, *mpk4* and two transgenic lines can be clearly separated, but water groups and flg22 treatment groups were not well separated. PC1 captured 52.7% of the total variance within the dataset. In addition, PC2 can explain another 9.8% of overall variance. The next three orthogonal components accounted for 7.2%, 4.9% and 3.6% of the variance, respectively; making the first five components explained over 78.1% of the total variance cumulatively.

Among the 93 proteins appeared in *1--1* or *1-4* samples, a well-known interacting protein, MKK1, was found. MKK1 is an upstream interacting protein, which involves in MEKK1-MKK1/MKK2-MPK4 cascade involved in plant immunity. MKK1 was found in *1-1*, *1-1*_flg22, *1-4* and *1-4*_flg22 samples, but not in the *VC* and *mpk4* samples. Three unique peptides of MKK1, namely KLKPSNLLINHR (Fig. 5A), ISGSLYSNK (Fig. 5B), and ILTSTSSLANSFVGTYPYMSPER (Fig. 5C), were used to identify the MKK1 in samples. The result indicates that MPK4-TurboID

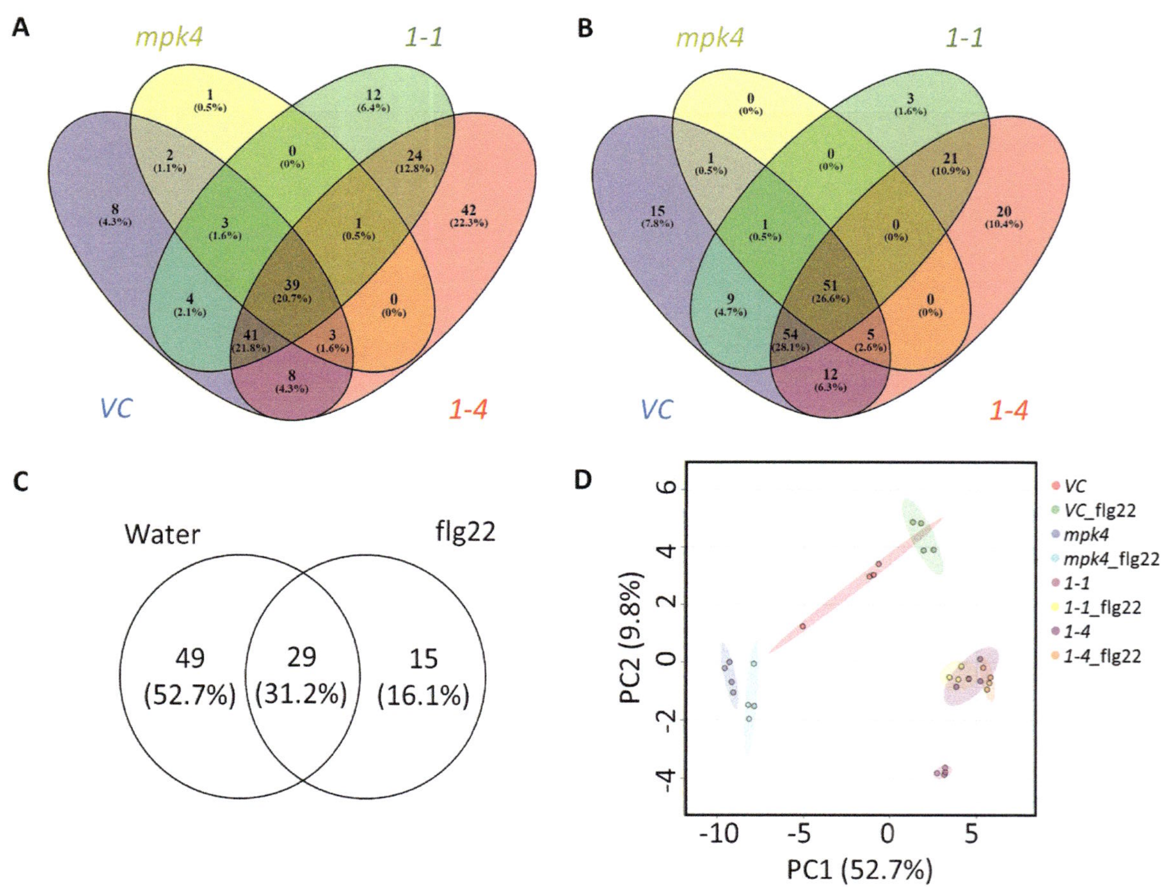

Fig. 4 Analysis of the TurboID-proteomics data. (A) Venn diagram of proteins found in *VC*, *mpk4*, *1-1* and *1-4* samples under water treatment. (B) Venn diagram of proteins found in *VC*, *mpk4*, *1-1* and *1-4* samples after flg22 treatment. (C) Venn diagram of unique proteins from the two transgenic lines under water and flg22 treatments. (D) PCA analysis of the eight groups of samples.

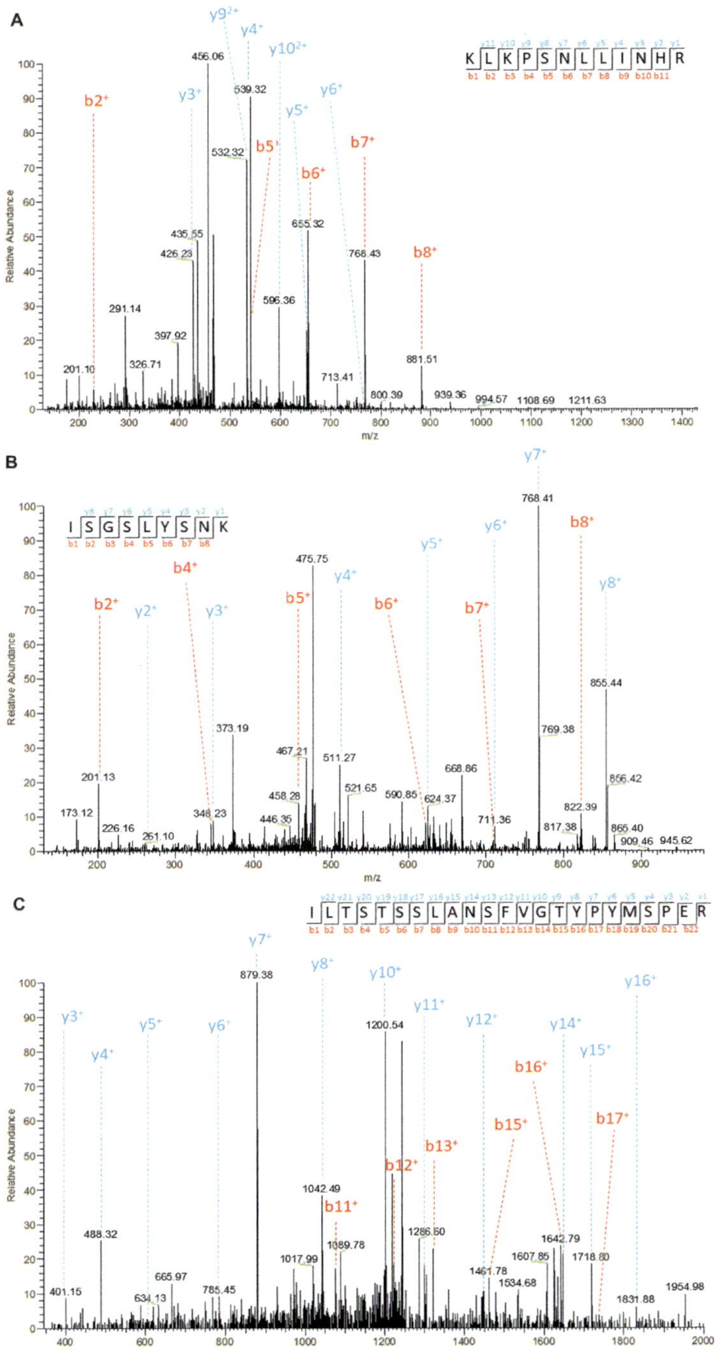

Fig. 5 MS/MS spectra of three unique peptides used to identify the MKK1 in the MPK4-TurboID proteomics experiment. (A) KLKPSNLLINHR; (B) ISGSLYSNK; (C) ILTSTSSLANSFVGTYPYMSPER.

can successfully capture MPK4 interacting proteins. The 93 proteins unique in the *1-1* or *1-4* samples provide an important resource for discovering potential MPK4 interacting proteins.

5. Summary

Many TurboID experiments in plants use a transient expression method. Depending on the transformation efficiency, not all the cells may express the bait-TurboID. Here we used stable *MPK4*-TurboID transgenic lines and 100 mg plant leaves as a proof of concept for the methodology. We were able to identify many proteins that were only present in the two transgenic lines compared to the vector control and the *mpk4* mutant. Since biotin is naturally produced in plants, the proper controls are important for identifying the direct interacting proteins with the bait. In summary, TurboID is a powerful tool to study the MPK4 interactome. Compared to traditional affinity purification-MS, TurboID can label proteins *in vivo* that have transient and weak interactions with MPK4. The TurboID-proteomics method presented here has broad applications and can be applied to many different types of organisms, tissues, and cells. With the stable transgenic line expressing *MPK4*-TurboID, we aim to fully profile the MPK4 interactomes in different cell-types under different conditions in a temporal manner.

Acknowledgment

This work was supported by Department of Biology, the University of Florida, Gainesville, FL, USA and partly by the Agriculture and Food Research Initiative (Award No. 2020-67013-31615/accession no. 1022409) from the USDA National Institute of Food and Agriculture, USA.

Reference

Beck, M., Komis, G., Muller, J., Menzel, D., & Samaj, J. (2010). Arabidopsis homologs of nucleus- and phragmoplast-localized kinase 2 and 3 and mitogen-activated protein kinase 4 are essential for microtubule organization. *Plant Cell*, *22*, 755–771. https://doi.org/10.1105/tpc.109.071746.

Beck, M., Komis, G., Ziemann, A., Menzel, D., & Samaj, J. (2011). Mitogen-activated protein kinase 4 is involved in the regulation of mitotic and cytokinetic microtubule transitions in Arabidopsis thaliana. *The New Phytologist*, *189*, 1069–1083. https://doi.org/10.1111/j.1469-8137.2010.03565.x.

Branon, T. C., Bosch, J. A., Sanchez, A. D., Udeshi, N. D., Svinkina, T., Carr, S. A., et al. (2018). Efficient proximity labeling in living cells and organisms with TurboID. *Nature Biotechnology*, *36*, 880–898. https://doi.org/10.1038/nbt.4201.

Chen, C.-L., & Perrimon, N. (2017). Proximity-dependent labeling methods for proteomic profiling in living cells. *Wiley Interdisciplinary Reviews: Developmental Biology*, *6*, e272. https://doi.org/10.1002/wdev.272.

Das, P. P., Macharia, M. W., Lin, Q., & Wong, S.-M. (2019). In planta proximity-dependent biotin identification (BioID) identifies a TMV replication co-chaperone NbSGT1 in the vicinity of 126 kDa replicase. *Journal of Proteomics*, *204*, 103402. https://doi.org/10.1016/j.jprot.2019.103402.

Gawroński, P., Witoń, D., Vashutina, K., Bederska, M., Betliński, B., Rusaczonek, A., et al. (2014). Mitogen-activated protein kinase 4 is a salicylic acid-independent regulator of growth but not of photosynthesis in Arabidopsis. *Molecular Plant*, *7*, 1151–1166. https://doi.org/10.1093/mp/ssu060.

Khan, M., Youn, J.-Y., Gingras, A.-C., Subramaniam, R., & Desveaux, D. (2018). In planta proximity dependent biotin identification (BioID). *Scientific Reports*, *8*, 9212. https://doi.org/10.1038/s41598-018-27500-3.

Kim, D. I., Jensen, S. C., Noble, K. A., Birendra, K. C., Roux, K. H., Motamedchaboki, K., et al. (2016). An improved smaller biotin ligase for BioID proximity labeling. *Molecular Biology of the Cell*, *27*, 1188–1196. https://doi.org/10.1091/mbc.E15-12-0844.

Kosetsu, K., Matsunaga, S., Nakagami, H., Colcombet, J., Sasabe, M., Soyano, T., et al. (2010). The MAP kinase MPK4 is required for cytokinesis in Arabidopsis thaliana. *The Plant Cell*, *22*, 3778–3790. https://doi.org/10.1105/tpc.110.077164.

Lin, C., & Chen, S. (2018). New functions of an old kinase MPK4 in guard cells. *Plant Signaling & Behavior*, *13*(5), e1477908. https://doi.org/10.1080/15592324.2018.1477908.

Liu, J.-Z., Horstman, H. D., Braun, E., Graham, M. A., Zhang, C., Navarre, D., et al. (2011). Soybean homologs of MPK4 negatively regulate defense responses and positively regulate growth and development. *Plant Physiology*, *157*, 1363–1378. https://doi.org/10.1104/pp.111.185686.

Lobingier, B. T., Hüttenhain, R., Eichel, K., Miller, K. B., Ting, A. Y., von Zastrow, M., et al. (2017). An approach to spatiotemporally resolve protein interaction networks in living cells. *Cell*, *169*, 350–360. https://doi.org/10.1016/j.cell.2017.03.022.

Mair, A., Xu, S.-L., Branon, T. C., Ting, A. Y., & Bergmann, D. C. (2019). Proximity labeling of protein complexes and cell-type-specific organellar proteomes in Arabidopsis enabled by TurboID. *eLife*, *8*, 1–45. https://doi.org/10.7554/eLife.47864.

Martell, J. D., Deerinck, T. J., Sancak, Y., Poulos, T. L., Mootha, V. K., Sosinsky, G. E., et al. (2012). Engineered ascorbate peroxidase as a genetically encoded reporter for electron microscopy. *Nature Biotechnology*, *30*, 1143–1148. https://doi.org/10.1038/nbt.2375.

Motani, K., & Kosako, H. (2020). BioID screening of biotinylation sites using the avidin-like protein Tamavidin 2-REV identifies global interactors of stimulator of interferon genes (STING). *Journal of Biological Chemistry*, *295*(32), 11174–11183. https://doi.org/10.1074/jbc.RA120.014323.

Petersen, M., Brodersen, P., Naested, H., Andreasson, E., Lindhart, U., Johansen, B., et al. (2000). Arabidopsis MAP kinase 4 negatively regulates systemic acquired resistance. *Cell*, *103*, 1111–1120. https://doi.org/10.1016/S0092-8674(00)00213-0.

Roux, K. J., Kim, D. I., Raida, M., & Burke, B. (2012). A promiscuous biotin ligase fusion protein identifies proximal and interacting proteins in mammalian cells. *The Journal of Cell Biology*, *196*, 801–810. https://doi.org/10.1083/jcb.201112098.

Zeng, Q., Chen, J., Ellis, B. E., & Laboratories, M. S. (2011). AtMPK4 is required for male-specific meiotic cytokinesis in Arabidopsis. *The Plant Journal*, *67*, 895–906. https://doi.org/10.1111/j.1365-313X.2011.04642.x.

Zhang, T., Schneider, J. D., Lin, C., Geng, S., Ma, T., Lawrence, S. R., et al. (2019). MPK4 phosphorylation dynamics and interacting proteins in plant immunity. *Journal of Proteome Research*, *18*, 826–840. https://doi.org/10.1021/acs.jproteome.8b00345.

Zhang, Y., Song, G., Lal, N. K., Nagalakshmi, U., Li, Y., Zheng, W., et al. (2019). TurboID-based proximity labeling reveals that UBR7 is a regulator of N NLR immune receptor-mediated immunity. *Nature Communications*, *10*. https://doi.org/10.1038/s41467-019-11202-z.

CHAPTER FIFTEEN

Detection of protein persulfidation in plants by the dimedone switch method

Angeles Aroca[a], Ana Jurado-Flores[a], Milos R. Filipovic[b], Cecilia Gotor[a,*], and Luis C. Romero[a,*]
[a]Institute of Plant Biochemistry and Photosynthesis, CSIC-University of Seville, Seville, Spain
[b]Leibniz Institute for Analytical Sciences, ISAS e.V., Dortmund, Germany
*Corresponding authors: e-mail address: gotor@ibvf.csic.es; lromero@ibvf.csic.es

Contents

Abstract

Hydrogen sulfide (H_2S) is a well-known signaling molecule in both animals and plants, endogenously produced by cells, and involved in a wide variety of biological functions. In plants, H_2S regulates a wide range of essential aspects of plant life, including plant responses to numerous stresses and physiological processes as important as abscisic acid (ABA)-dependent stomatal movement, photosynthesis, and autophagy. The best studied molecular mechanism responsible of sulfide signaling is protein persulfidation, a post-translational modification of cysteine residues, where a thiol group (P-SH) is transformed into a persulfide group (P-SSH). In this way, persulfidation has emerged as a new type of cellular redox mechanism that can regulate protein structure and function and interest in this modification has increased exponentially. However, the identification and the development of detection methods have been challenging. Nevertheless, on the basis of the chemical differences between the thiol and the persulfide groups, different methods have been implemented. In plants, different high-throughput proteomic

Methods in Enzymology, Volume 676
ISSN 0076-6879
https://doi.org/10.1016/bs.mie.2022.07.024

analyzes have been performed using a tag-switch method where in the first step all thiols and persulfides are blocked and then in the second step persulfides are selectively labeled using a specific nucleophile. This chapter outlines a new method, previously described in mammals, that has been applied to detect persulfidation in plants and is based on the same chemical premise but consists of chemoselective persulfide labeling with dimedone-based probes. Here, we provide a detailed workflow of this method that includes procedures for the determination of the persulfidation level of a protein extract visualized and quantified by fluorescence on the gel on one side, and on the other, the labeling and purification of persulfidated proteins for identification by mass spectrometry.

1. Introduction

Hydrogen sulfide (H_2S) is a colorless and flammable gas with a characteristic scent of rotten eggs. It is naturally produced in nature found in petroleum, volcanic emissions, and natural gas. It has been always considered as a pollutant in the environment and a hazardous toxic gas to living organisms because it inhibits complex IV of the electron transport chain in mitochondria. Nevertheless, life on Earth emerged in an atmosphere rich in H_2S and, still, H_2S is necessary for some organisms such as sulfur-oxidizing bacteria, which use H_2S as an electron donor in anoxygenic photosynthesis to produce oxidized sulfur compounds (Johnston, Wolfe-Simon, Pearson, & Knoll, 2009). Moreover, it is known that H_2S is endogenously produced enzymatically by cells throughout the regna, from bacteria to plants and mammals (Aroca & Gotor, 2022). Therefore, all these data were the first premises to think about H_2S as a molecule with additional functions different from toxic.

In mammals, H_2S was found to act as a neuromodulator (Abe & Kimura, 1996), to induce hibernation states in mice (Blackstone, Morrison, & Roth, 2005), and a regulator of inflammation (Szabo, 2007) and aging (Zivanovic et al., 2019). In plants, it is now well known that H_2S regulates a wide range of vital processes. H_2S protects plants against several stresses such as drought, heat, oxidative, and metal stresses (Aroca, Gotor, & Romero, 2018; Shen et al., 2013; Wang, Shi, Li, & Zhang, 2010; Zhang et al., 2008) and regulates important physiological processes, including abscisic acid (ABA)-dependent stomatal movement (Garcia-Mata & Lamattina, 2010; Scuffi et al., 2014), photosynthesis (Chen et al., 2011), and autophagy (Alvarez et al., 2012; Gotor, Garcia, Crespo, & Romero, 2013; Laureano-Marín et al., 2016).

The production of H_2S in plants occurs mainly during the photosynthetic sulfate-assimilation pathway in chloroplast (Gotor et al., 2019). H_2S is also produced in other cell compartments through different enzymes involved in cysteine metabolism, such as L-cysteine desulfhydrase 1 (DES1), which is the main producer of endogenous cytosolic H_2S (Álvarez, Calo, Romero, García, & Gotor, 2010). Several publications in the last decade reveal the interplay of sulfide with plant hormones and other signaling molecules, such as hydrogen peroxide (H_2O_2) and nitric oxide (NO), and how this interaction can regulate differently several physiological processes (Aroca, Gotor, Bassham, & Romero, 2020; Lisjak et al., 2010; Liu, Hou, Liu, Liu, & Wang, 2011; Scuffi, Lamattina, & García-Mata, 2016; Xie et al., 2014; Zhang et al., 2010).

The molecular mechanisms by which sulfide regulates a huge variety of processes were unknown until the post-translational modification of protein thiols (persulfidation, formerly called *S*-sulfhydration) was identified in mice (Mustafa et al., 2009) and Arabidopsis (Aroca, Serna, Gotor, & Romero, 2015). Since then, multiple papers have identified this process in animals and in plants with numerous proteins described as targets for persulfidation (Aroca & Gotor, 2022).

Persulfidation is a post-translational modification of protein cysteine residues where a thiol group (P-SH) is modified to a persulfide group (P-SSH) (Fig. 1). Persulfidation has emerged as a new type of redox regulation of protein structure and function in diverse biological processes, in a similar way to how nitric oxide (NO) regulates cellular processes through S-nitrosylation (P-SNO) or hydrogen peroxide (H_2O_2) through S-sulfenylation (P-SOH).

Persulfidation affects the structure of proteins, the enzymatic activity, and the subcellular localization of proteins (Aroca, Schneider, Scheibe, Gotor, & Romero, 2017; Aroca et al., 2015; Kimura, 2015; Mustafa et al., 2009). In plants, the role of persulfidation was first demonstrated in the activity of several enzymes, glutamine synthetase (GS2), ascorbate peroxidase (APX1), cytosolic glyceraldehyde 3-phosphate dehydrogenase

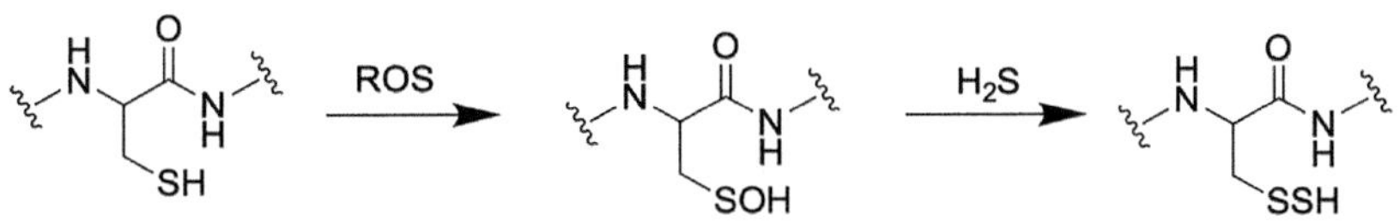

Fig. 1 Proposed mechanism of the reaction of H_2S with protein sulfenic acid to form protein persulfide.

(GapC1) (Aroca et al., 2015). Persulfidation was shown to activate APX1 and GapC1 in Arabidopsis protein extracts and recombinant proteins, while persulfidation inactivated GS2. Furthermore, other studies also confirm the regulation of different enzymes by persulfidation, demonstrate that H_2S-mediated persulfidation regulates the activities of catalase (CAT), APX, and peroxidase (POD) and enhances the antioxidant response of plants to oxidative stress (Li et al., 2020). Persulfidation of 1-aminocyclopropane-1-carboxylic acid (ACC) oxidases 1 and 2 also negatively regulates ethylene biosynthesis in tomato, which identifies another role for persulfidation in plants (Jia et al., 2018). Likewise, persulfidation affects the cytosolic/nuclear subcellular location of GapC1 (Aroca, Schneider, et al., 2017), which could alter its function from its metabolic role in the cytosol to its transcriptional role in mRNA regulation.

More recently, several proteins involved in the autophagy process have been described to be targets of persulfidation, such as autophagy-related protein 2 (ATG2), ATG3, ATG4, ATG5, ATG7, ATG11, ATG13, and ATG18a, as well as other 10 related proteins to autophagy, such as the serine/threonine kinase target of rapamycin (TOR), the regulatory-associated protein of TOR1 (RAPTOR1), the lethal with *sec* thirteen protein 8 (LST8), five subunits of protein phosphatase 2A (PP2A), the regulatory subunit of PP2A (TAP46) and the serine/threonine-protein kinase (VPS15) (Aroca, Benito, Gotor, & Romero, 2017; Jurado-Flores, Romero, & Gotor, 2021; Laureano-Marín et al., 2020). Currently, the role of persulfidation as the underlying mechanism to regulate autophagy has been demonstrated in ATG4 and ATG18a. In both cases, persulfidation of the ATG protein had a direct impact on autophagosome progression to negatively regulate the autophagy process (Aroca, Yruela, Gotor, & Bassham, 2021; Gotor, Aroca, & Romero, 2022; Laureano-Marín et al., 2020).

Various proteins involved in ABA signaling have been also revealed to be susceptible to persulfidation and the role of H_2S in guard cell ABA signaling was demonstrated through persulfidation of specific targets. In this way, several studies showed that ABA triggers the DES1 activity to induce the production of H_2S for persulfidation of DES1 itself, open stomata 1 (OST1), NADPH oxidase RBOHD and the transcription factor ABA insensitive 4 (ABI4), highlighting the complexity of the H_2S mechanism of action to regulate the ABA-dependent stomatal closure process (Chen et al., 2020, 2021; Shen et al., 2020; Zhou et al., 2021).

Furthermore, there is evidence of a crosstalk between reactive oxygen species (ROS), reactive nitrogen species (RNS), and reactive sulfur species

(RSS) regulating biological processes. This provides a link between persulfidation, S-nitrosylation, and S-sulfenylation that allow plants to cope with stress stimuli. More than 600 proteins have been found to be modified by both persulfidation and S-nitrosylation (Aroca et al., 2018), and more than 1700 proteins were modified by persulfidation or S-sulfenylation, which makes 82% of the sulfenylated proteome described in Arabidopsis also undergo persulfidation (Aroca, Zhang, Xie, Romero, & Gotor, 2021). Therefore, numerous studies in plants have demonstrated that persulfidation is a ubiquitous modification of proteins that regulates a wide variety of biological processes and interacts with other redox modifications that might change the outcome of a certain protein, which explains the need for the scientific community to understand the molecular and cellular mechanisms of this modification.

During the last decade, the interest in persulfidation has increased exponentially, but researchers have coped with the intrinsically unstable chemistry of this modification and its similarity to both thiols and disulfides, which made its identification and the development of detection methods very challenging. Due to the asymmetry of the persulfide group, which contains an inner sulfane sulfur susceptible to nucleophilic attack (oxidation state: 0) and an outer sulfur susceptible to electrophilic attack (oxidation state: −1), and its anionic form at physiological pH 7.4 (P-SS-), persulfide groups have a more nucleophilic nature than the thiol groups (Cuevasanta et al., 2015; Filipovic, 2015; Filipovic, Zivanovic, Alvarez, & Banerjee, 2018). Based on these chemical differences, new methods have been described to increase the number of identifications of protein targets.

In plants, three high-throughput proteomic analyses using different labeling methods have been published to reveal more than 3400 proteins in leaves and more than 5200 in roots as modified by persulfidation (Aroca, Benito, et al., 2017; Aroca et al., 2015; Jurado-Flores et al., 2021). Most of these methods were based on chemical differences among the thiol and persulfide groups using a first step to block all thiols and persulfides followed by a second step to specifically label persulfide groups with a reactive nucleophilic compound. In addition to these, a new method has been described in the search for higher specificity and easier detection of persulfidation. Here we report a dimedone switch method, previously described in mammals (Zivanovic et al., 2019), that we now use to detect persulfidation in plants. This method is a chemoselective persulfide labeling approach using dimedone-based probes, which allow visualization of the persulfidation level in the gel or the labeling and purification of persulfidated proteins to be further analyzed by other techniques, such as mass spectrometry. We describe a set of protocols to determine: (a) the level of persulfidation within a protein extract which can be visualized and quantified by fluorescence in the gel (Fig. 2) and (b) to label and purify the persulfidated proteins from a protein extract and identify them by mass spectrometry. The procedures detailed here have been optimized for protein extracts from *Arabidopsis thaliana*, although they can be used for any other species.

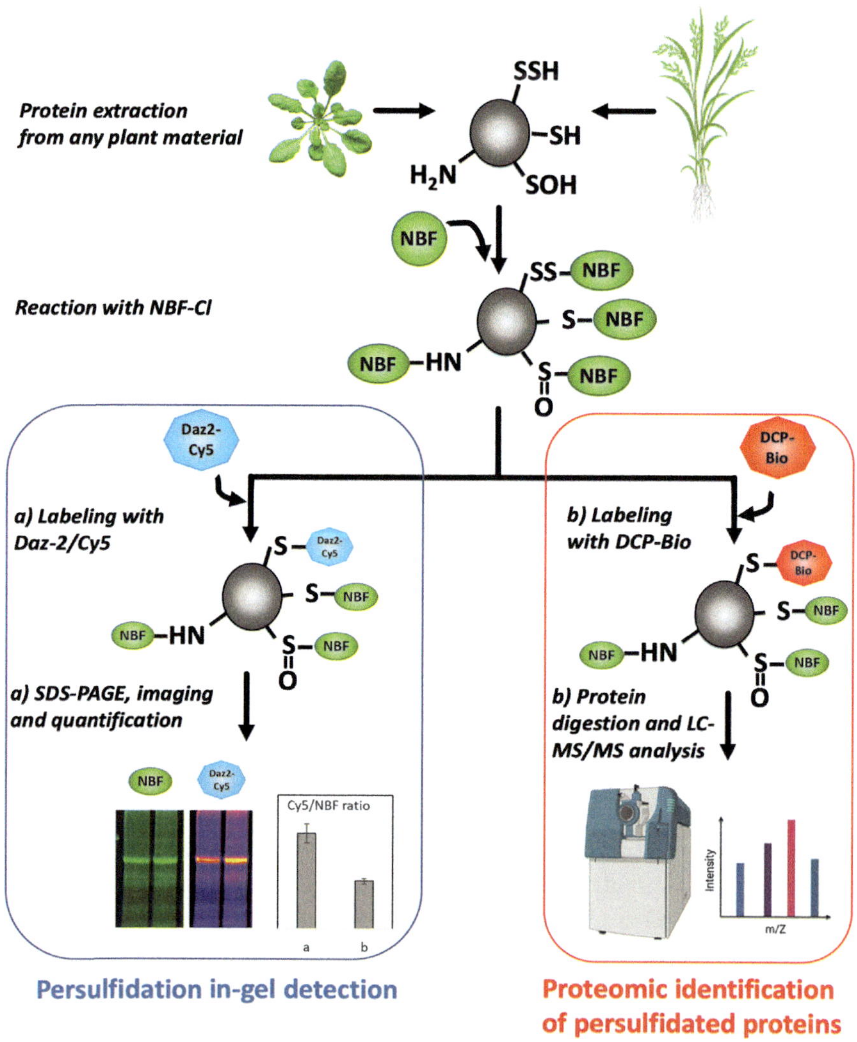

Fig. 2 Graphical scheme of persulfidation in-gel detection protocol, and purification and identification of persulfidated proteins by mass spectrometry.

2. In-gel detection of protein persulfidation of plant proteins

Here we describe the assay to detect the protein persulfidation pattern in a particular plant tissue or under certain conditions, so that the persulfidation levels in different samples can be compared in a semiquantitative way. The dimedone switch detection method consists of two main steps that are performed in protein extracts (Zivanovic et al., 2019). In the first step, all cysteines and amino groups in proteins react with 4-chloro-7-nitrobenzofurazan (NBF-Cl) to yield a variety of adducts with a characteristic green fluorescence signal that could be used to measure the total protein load. In the second step, only the persulfidated cysteine adducts are selectively labeled with a mixture of dimedone azide-2 (Daz2) and Cy5-alkyne, which contains a Cy5-fluorescence moiety for generating a red fluorescence signal. The total labeled protein extracts are separated on SDS-PAGE gels and imaged at 640 nm for the Cy5 signal and 488 nm for the NBF-Cl signal with the fluorescence intensity signal ratio of Cy5/NBF-Cl constituting the total protein persulfidation level (Fig. 2).

2.1 Reaction with NBF-Cl

Materials and equipment

- Liquid nitrogen
- PBS 1 ×, pH 7.4, Gibco™
- EDTA-Na_2-$2H_2O$ Ultrapure, Invitrogen™. Stock solution 0.5 M dissolved in distilled water
- SDS BioChemica, AppliChem. 10% stock solution
- Pierce™ Protease Inhibitor Tablets, EDTA-Free, Thermo Scientific. Stock solution 25 × in distilled water
- 4-Chloro-7-nitrobenzofurazan (NBF-Cl), SigmaAldrich. Stock solution 500 mM in DMSO
- Methanol 99.9%, Fisher Chemical
- Chloroform, Merck
- DC (detergent compatible) Protein Assay, Bio-Rad
- Mortar and pestle
- Eppendorf ThermoMixer™ C
- Eppendorf™ Centrifuge 5424 R
- Vortex VWR® International
- Flowtronic Fume hood
- Bioruptor® Pico minichiller 300, Diagedone SA

Procedure

1. Ground plant tissues (100 mg) in liquid nitrogen using mortar and pestle until a fine powder and resuspend it in 200 μL of cold PBS lysis buffer (PBS 1 × pH 7.4, 1 mM EDTA, 2% SDS) supplemented with 1 × protease inhibitor and 5 mM NBF-Cl
2. Immediately incubate the mixtures at 37 °C for 30 min, protected from light
3. Centrifuge the samples at 15,000 rpm (21,130 × *g*) in a microfuge for 2 min at 4 °C to remove debris, collect the supernatants in a new fresh microtube. Discard the pellets
4. Next, proteins are precipitated by adding to the supernatants (volume should be determined), equal volume of methanol and a quarter of a volume of chloroform and then vortexed (1:1:0.25, protein extract: MeOH:chloroform)
5. Centrifuge the mixtures at 15,000 rpm (21,130 × *g*) for 15 min at 4 °C. The obtained pellet is located between the two layers. Carefully remove the aqueous phase, twist the tube to locate the pellet in the wall of the tube, and carefully remove the chloroform phase (Fig. 3)
6. Wash the pellet with 200 μL methanol, 200 μL distilled water, and 50 μL chloroform, and vortex vigorously
7. Centrifuge again at 15,000 rpm (21,130 × *g*) for 15 min at 4 °C, and remove the liquid as explained before in step 5
8. Perform a last washing adding 500 μL of methanol to the obtained pellet, vortex vigorously and centrifuge at 15,000 rpm (21,130 × *g*) for 15 min at 4 °C

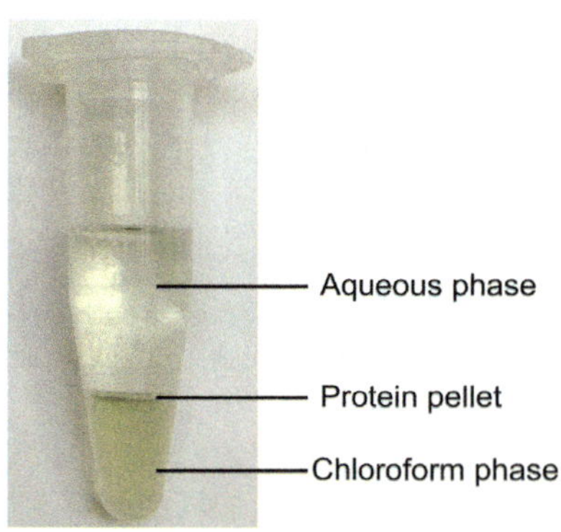

Fig. 3 Separation of phases after centrifugation during methanol/chloroform precipitation.

9. Remove the methanol and air-dry the pellet protected from the light for 10 min at room temperature. At this point, the samples can be stored at −20 °C, and continue the protocol at your convenience
10. Resuspend the samples by first adding 30 μL of 10% SDS directly to the pellet. Add then 120 μL PBS 1 × supplemented with 1 × protease inhibitor. Vortex and made use of the Bioruptor (or sonication bath) if necessary until it is fully dissolved. Program Bioruptor with 10 cycles of 60 s ON and 30 s OFF
11. Determine protein concentration spectrophotometrically using the DC Protein Assay, Bio-Rad
12. Adjust the final protein concentration to 2–3 mg/mL, to continue the second step of labeling

2.2 Labeling with Daz2-Cy5

Materials and equipment

- Acetonitrile, Merck
- PBS 1 ×, pH 7.4, Gibco™
- Dimedone azide-2 (Daz2), Cayman Chemical. Stock solution 50 mM in DMSO
- Cy5-alkyne, Lumiprobe. Stock solution 20 mM in DMSO
- TBTA-Cu(II) complex 10 mM, Lumiprobe
- Dimethyl sulfoxide 99.9% (DMSO), Sigma Aldrich
- L-Ascorbic acid, Sigma Aldrich. Stock solution 100 mM in distilled water (fresh prepared)
- EDTA-Na_2-$2H_2O$ Ultrapure, Invitrogen™. Stock solution 0.5 M dissolved in distilled water
- Methanol 99.9%, Fisher Chemical
- Chloroform, Merck
- SDS BioChemica, AppliChem. 10% (*w*/*v*) stock solution
- Pierce™ Protease Inhibitor Tablets, EDTA-Free, Thermo Scientific. Stock solution 25 × in distilled water
- Eppendorf ThermoMixer™ C
- Eppendorf™ Centrifuge 5424 R
- Vortex VWR® International
- Flowtronic Fume hood

Procedure

1. Prepare the Daz2-Cy5 mix by combining, PBS 1 ×, pH 7.4, 30% (v/v) acetonitrile, 1 mM DAz2, 1 mM Cy5-alkyne, 2 mM TBTA-Cu

complex and 4 mM ascorbic acid, in a final volume of 150 μL. Keep the mixture in a rotator shaker overnight at room temperature to obtain a homogeneous mixture

2. The next day, the reaction is quenched with the final concentration of 20 mM EDTA for 1 h and the final mixture can be aliquoted and stored at −20 °C. Small volume aliquots must be prepared to avoid freezing and thawing (~20 μL). Do not store aliquots for more than 2 months
3. Incubate the protein samples obtained in Section 2.1 (Reaction with NBF-Cl) with 25 μM Daz2-Cy5 mix for 30 min at 37 °C and protected from light
4. Precipitate the proteins following the same procedure described in Section 2.1 (steps 4–9). At this point, the samples can be stored at −20 °C and continue the protocol as desired

2.3 SDS-PAGE of labeled proteins and imaging

Materials and equipment

- PBS 1 ×, pH 7.4, Gibco™
- SDS BioChemica, AppliChem. 10% stock solution
- Pierce™ Protease Inhibitor Tablets, EDTA-Free, Thermo Scientific. Stock solution 25 × in distilled water
- DC (detergent compatible) Protein Assay, Bio-Rad
- 4 × Laemmli buffer, Bio-Rad
- β-Mercaptoethanol, Merck
- Methanol 99.9%, Fisher Chemical
- Acetic acid 99.5% pure, Acros Organics
- Eppendorf ThermoMixer™ C
- Mini-PROTEAN® Tetra Vertical Electrophoresis Cell for Mini Precast Gels and PowerPac™ Basic Power Supply, Bio-Rad
- Precast gels of 10% (w/v) polyacrylamide
- Heidolph Unimax 1010 Orbital Shaker
- Typhoon FLA 9500, GE healthcare
- Varioskan Lux, Thermo scientific
- Multi-RS, Programmable rotator, Biosan
- DC (detergent compatible) Protein Assay, Bio-Rad

Procedure

1. Dissolve the samples obtained in Section 2.2 (Labeling with Daz2-Cy5) in 30 μL of 10% (w/v) SDS and 120 μL of PBS 1 ×, pH 7.4, supplemented with 1 × protease inhibitor

2. Determine protein concentration spectrophotometrically using the DC Protein Assay, Bio-Rad
3. Mix 3 parts of the dissolved sample containing 40–50 μg of protein, with 1 part of 4× Laemmli buffer and supplement with 10% (v/v) β-mercaptoethanol. Boil at 95 °C for 5 min
4. Place the 10% polyacrylamide gel pre-cast on the electrophoresis module
5. Load the mixtures into the wells and run the electrophoresis, first 80 V until the samples go into the separating phase and then 120 V for about 2 h. Cover the electrophoresis module to protect samples from light during the running
6. Prepare the gel fixation solution containing 12.5% (v/v) methanol and 4% (v/v) acetic acid in water
7. Cover the gels with the fixation solution and incubate for 30 min with gentle agitation and protected from light
8. Record the gel at 640 nm for the Cy5 signal and 488 nm for the NBF-Cl signal on the Typhoon 9500 FLA Imager
9. Quantify the persulfidation levels by measuring the Cy5/NBF-Cl fluorescence signal ratio (Fig. 4)

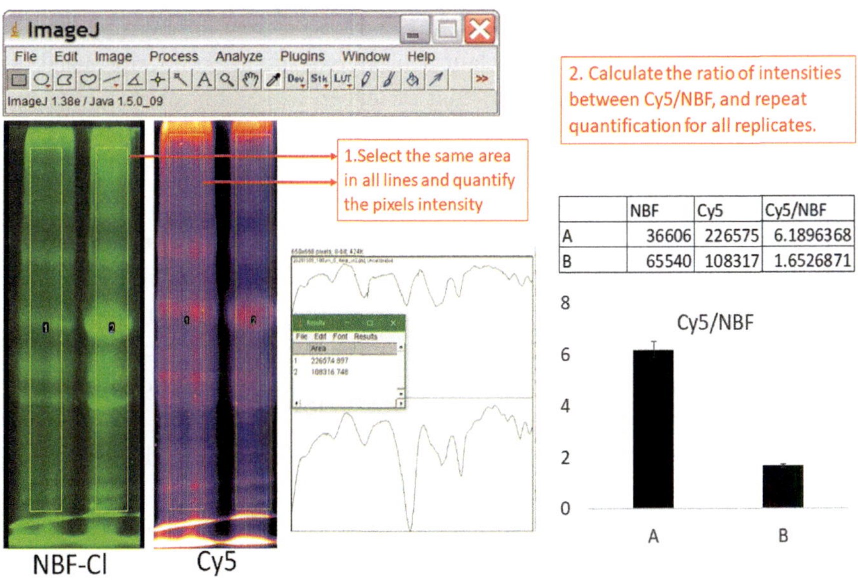

	NBF	Cy5	Cy5/NBF
A	36606	226575	6.1896368
B	65540	108317	1.6526871

Fig. 4 Example of the persulfidation level quantification in protein extracts from Arabidopsis leaves using ImageJ software. The fluorescent signal for NBF-Cl is in green and the Cy5 signal in red color.

2.4 Notes

1. The details of the Daz2-Cy5 mix preparation are as follows: Bring all compounds to room temperature before to start and in an Eppendorf tube
   - First add 60.5 μL of PBS 1 ×, pH 7.4
   - Add 43 μL of acetonitrile and vortex
   - Add 3 μL of 50 mM DAz2 and vortex
   - Add 7.5 μL of 20 mM Cy5-alkyne and vortex
   - Add 30 μL of 10 mM Copper-TBTA complex and vortex
   - Add H_2O to solid ascorbic acid (100 mM) and vortex until well dissolved
   - Add 6 μL of 100 mM ascorbic acid
   - Cover with aluminum foil and leave in a rotator shaker overnight at room temperature
   - Next day add 10 μL of 300 mM EDTA, cover again and leave in a rotator shaker for 1 h
2. Special caution should be taken against the use of reducing agents in the composition of the PBS lysis buffer to avoid reduction of persulfide groups to thiols
3. To have quantitative data with reduced SD, perform the experiment at least with four biological replicates and two experimental replicates for each biological one. When plant protein extracts are used, the variability of the samples is higher than that observed in cell cultures
4. Due to the method detection is based on fluorescent signals, it is very sensitive to small changes during the procedure and the variability can be increased if the replicas are run on SDS-PAGE gels at very different time period. It is highly recommended to run replicates on the same gel and at near time. Furthermore, it is essential to protect the reagents/samples from light when this is indicated in the procedure
5. To avoid erroneous results, it is imperative that the fluorescent signals are not over-exposed. If this is the case, it is necessary to reduce the amount of protein loaded onto the gels
6. Avoid long storage of either the Daz2-Cy5 mixture or labeled proteins because the fluorescence signal can be decreased
7. Instead of a Bioruptor, a sonication bath might be used at 4 °C

3. Identification of persulfidated proteins by mass spectrometry analysis

A similar method based on dimedone probes is also described here to identify persulfidated proteins by proteomic analysis. The method consists first of full blocking with the NBF-Cl reagent as described in the previous procedure and then reaction of the modified persulfidated proteins to specifically tag-switch with DCP-Bio1, a biotinylated form of the dimedone. Finally, biotin-labeled proteins are purified with streptavidin and subjected to trypsin digestion for further mass spectrometry analysis (Fig. 2).

3.1 Labeling with DCP-Bio1

Materials and equipment

- PBS 1×, pH 7.4, Gibco™
- SDS BioChemica, AppliChem. 10% (w/v) stock solution
- Pierce™ Protease Inhibitor Tablets, EDTA-Free, Thermo Scientific. Stock solution 25 × in distilled water
- DCP-Bio1, Merck. Stock solution 25 mM in DMSO
- Sera-Mag Magnetic Streptavidin, Cytiva
- Tween® 20, Fisher Chemical
- H_2O Optima® for LC/MS, Fisher Chemical
- Ammonium hydroxide solution 5 M, Fluka Analytical
- Formic acid for LC/MS, Fisher Chemical
- DC (detergent compatible) Protein Assay, Bio-Rad
- Eppendorf ThermoMixer™ C
- Eppendorf™ Centrifuge 5424 R
- Flowtronic Fume hood
- Vortex VWR® International
- Bioruptor® Pico minichiller 300, Diagedone SA
- Varioskan Lux, Thermo scientific
- Multi-RS, Programmable rotator, Biosan
- Heidolph Unimax 1010 Orbital Shaker
- Invitrogen™DynaMag™-Spin Magnet
- Qubit ™ Protein Assay Kit, Invitrogen
- Qubit 4 Fluorometer, Invitrogen

Procedure

1. In this procedure, grind 500 mg of plant tissue and resuspend in 500 μL of lysis buffer PBS 1 × supplemented with 2% (w/v) SDS, 1 × protease inhibitor, and 5 mM NBF-Cl. Perform the reaction with NBF-Cl previously described in Section 2.1 (Reaction with NBF-Cl), from step 1 to 9
2. Add 200 μM DCP-Bio1 and incubate the mixture at 37 °C for 2 h (Fig. 2)
3. Precipitate the proteins following the same procedure described in Section 2.1 (steps 4–9) (Fig. 3)

4. Resuspend protein samples in 500 μL of PBS 1 × supplemented with 0.1% SDS and 1 × protease inhibitor
5. Add 150 μL of Sera-Mag Magnetic Streptavidin beads and incubate the mixture at 4 °C overnight with gentle agitation on the rotator shaker
6. The microtubes containing the magnetic beads are located in the Invitrogen™DynaMag™-Spin Magnet and after several seconds, the beads are separated from the supernatant, which is discarded
7. The microtubes are removed from the magnet and the magnetic beads are washed by resuspending them in 1 mL of PBS 1 × supplemented with Tween 0.001% (v/v) at room temperature. Afterwards, the microtubes are located again into the Magnet and the supernatant is discarded. This wash is repeated three times
8. The magnetic beads are then washed twice with 1 mL of PBS 1 × as previously described and once with 1 mL of pure H_2O for LC/MS
9. After the six washes, the beads are incubated with 2.25 M ammonium hydroxide at room temperature overnight, with gentle agitation
10. Then, samples are located again in the Invitrogen™DynaMag™-Spin Magnet and supernatants are transferred to fresh microtubes. Magnetic beads are discarded
11. Samples are neutralized with pure formic acid for LC/MS
12. Determine protein concentration spectrophotometrically using the Qubit™ Protein Assay kit

3.2 Liquid chromatography and mass spectrometry analysis (LC-MS/MS)

Protein persulfide labeled proteins with DCP-Bio 1 can be analyzed by a variety of high-resolution LC-MS/MS mass spectrometers after peptide digestion with either trypsin or chymotrypsin following standard protocols. A total of 50 μg of proteins are digested and analyzed by LC-MS/MS. The chromatographic separation and fragmentation of the peptides will depend on the optimal conditions of each commercial brand, recommending analysis in equipment with high sensitivity and resolution.

Regardless of the spectrometer, peptide identification and search setting are following:

- precursor Δm tolerance = 10 ppm
- -fragment Δm tolerance = 0.2 Da
- missed cleavages = 2
- modifications of lysine and arginine: NBF (163.0012)
- modifications of cysteine: NBF (163.0012), hydrolyzed DCP-Bio1 (168.0786) and/or DCP-Bio1 (394.1557)

3.3 Notes

1. As a general rule, all the plasticware used during the treatments of the samples to be analyzed by mass spectrometry is imperative to be of very high quality (Eppendorf-type) to avoid contamination of samples and the obtaining of unusable results
2. The solvents and water used to prepare the buffers should be of HPLC-MS quality
3. It is mandatory to use streptavidin beads with high nominal biotin binding capacities (4500–5500 pmol/mg) to optimize the enrichment of the persulfidated proteins (DCP-Bio1 labeled) and to avoid as much as possible the limitation of this step
4. When it is expected that the protein extract could contain a significant amount of endogenous biotinylated proteins, a previous step prior to incubation with DCP-Bio1 should be performed. The biotinylated proteins are precleared using the streptavidin magnetic beads (low-medium biotin binding capacity can be used) as described before
5. Step 12 for protein quantification might be performed with other protein assay methods, but Qubit is recommended because it is a fluorescent method with high sensitivity

Acknowledgments

The authors acknowledge funding for this work from ERDF A way of making Europe and MCIN/AEI/10.13039/501100011033 (grant No. PID2019-109785GB-I00); and Junta de Andalucía (grant No. P18-RT-3154 and US-1255781). Milos R. Filipovic acknowledges the support of European Research Council (ERC) under the European Union's Horizon 2020 research and innovation program (Grant Agreement No. 864921). A.J.-F. was supported by the Ministerio de Economia y Competitividad through the program of Formación de Personal Investigador.

References

Abe, K., & Kimura, H. (1996). The possible role of hydrogen sulfide as an endogenous neuromodulator. *Journal of Neuroscience*, *16*(3), 1066–1071. https://doi.org/10.1523/JNEUROSCI.16-03-01066.1996.

Álvarez, C., Calo, L., Romero, L. C., García, I., & Gotor, C. (2010). An O-acetylserine (thiol)lyase homolog with l-cysteine desulfhydrase activity regulates cysteine homeostasis in Arabidopsis. *Plant Physiology*, *152*(2), 656–669. https://doi.org/10.1104/pp.109.147975.

Alvarez, C., Garcia, I., Moreno, I., Perez-Perez, M. E., Crespo, J. L., Romero, L. C., et al. (2012). Cysteine-generated sulfide in the cytosol negatively regulates autophagy and modulates the transcriptional profile in Arabidopsis. *Plant Cell*, *24*(11), 4621–4634. https://doi.org/10.1105/tpc.112.105403.

Aroca, A., Benito, J. M., Gotor, C., & Romero, L. C. (2017). Persulfidation proteome reveals the regulation of protein function by hydrogen sulfide in diverse biological processes in Arabidopsis. *Journal of Experimental Botany*, *68*(17), 4915–4927. https://doi.org/10.1093/jxb/erx294.

Aroca, A., & Gotor, C. (2022). Hydrogen sulfide: A key role in autophagy regulation from plants to mammalians. *Antioxidants (Basel)*, *11*(2), 327. https://doi.org/10.3390/antiox11020327.

Aroca, A., Gotor, C., Bassham, D. C., & Romero, L. C. (2020). Hydrogen sulfide: From a toxic molecule to a key molecule of cell life. *Antioxidants*, *9*(7), 621. https://doi.org/10.3390/antiox9070621.

Aroca, A., Gotor, C., & Romero, L. C. (2018). Hydrogen sulfide signaling in plants: Emerging roles of protein persulfidation. *Frontiers in Plant Science*, *9*, 1369. https://doi.org/10.3389/fpls.2018.01369.

Aroca, A., Schneider, M., Scheibe, R., Gotor, C., & Romero, L. C. (2017). Hydrogen sulfide regulates the cytosolic/nuclear partitioning of glyceraldehyde-3-phosphate dehydrogenase by enhancing its nuclear localization. *Plant and Cell Physiology*, *58*(6), 983–992. https://doi.org/10.1093/pcp/pcx056.

Aroca, A., Serna, A., Gotor, C., & Romero, L. C. (2015). S-sulfhydration: A cysteine posttranslational modification in plant systems. *Plant Physiology*, *168*(1), 334–342. https://doi.org/10.1104/pp.15.00009.

Aroca, A., Yruela, I., Gotor, C., & Bassham, D. C. (2021). Persulfidation of ATG18a regulates autophagy under ER stress in Arabidopsis. *Proceedings of the National Academy of Sciences of the United States of America*, *118*(20), e2023604118. https://doi.org/10.1073/pnas.2023604118.

Aroca, A., Zhang, J., Xie, Y., Romero, L. C., & Gotor, C. (2021). Hydrogen sulfide signaling in plant adaptations to adverse conditions: Molecular mechanisms. *Journal of Experimental Botany*, *72*(16), 5893–5904. https://doi.org/10.1093/jxb/erab239%J.

Blackstone, E., Morrison, M., & Roth, M. B. (2005). H2S induces a suspended animation-like state in mice. *Science*, *308*(5721), 518. https://doi.org/10.1126/science.1108581.

Chen, S., Jia, H., Wang, X., Shi, C., Wang, X., Ma, P., et al. (2020). Hydrogen sulfide positively regulates abscisic acid signaling through persulfidation of SnRK2.6 in guard cells. *Molecular Plant*, *13*(5), 732–744. https://doi.org/10.1016/j.molp.2020.01.004.

Chen, S., Wang, X., Jia, H., Li, F., Ma, Y., Liesche, J., et al. (2021). Persulfidation-induced structural change in SnRK2.6 establishes intramolecular interaction between phosphorylation and persulfidation. *Molecular Plant*, *14*(11), 1814–1830. https://doi.org/10.1016/j.molp.2021.07.002.

Chen, J., Wu, F. H., Wang, W. H., Zheng, C. J., Lin, G. H., Dong, X. J., et al. (2011). Hydrogen sulphide enhances photosynthesis through promoting chloroplast biogenesis, photosynthetic enzyme expression, and thiol redox modification in Spinacia oleracea seedlings. *Journal of Experimental Botany*, *62*(13), 4481–4493. https://doi.org/10.1093/jxb/err145.

Cuevasanta, E., Lange, M., Bonanata, J., Coitino, E. L., Ferrer-Sueta, G., Filipovic, M. R., et al. (2015). Reaction of hydrogen sulfide with disulfide and sulfenic acid to form the strongly nucleophilic persulfide. *Journal of Biological Chemistry*, *290*(45), 26866–26880. https://doi.org/10.1074/jbc.M115.672816.

Filipovic, M. R. (2015). Persulfidation (S-sulfhydration) and H2S. *Handbook of Experimental Pharmacology*, *230*, 29–59. https://doi.org/10.1007/978-3-319-18144-8_2.

Filipovic, M. R., Zivanovic, J., Alvarez, B., & Banerjee, R. (2018). Chemical biology of H2S signaling through persulfidation. *Chemical Reviews*, *118*(3), 1253–1337. https://doi.org/10.1021/acs.chemrev.7b00205.

Garcia-Mata, C., & Lamattina, L. (2010). Hydrogen sulphide, a novel gasotransmitter involved in guard cell signalling. *New Phytologist*, *188*(4), 977–984. https://doi.org/10.1111/j.1469-8137.2010.03465.x.

Gotor, C., Aroca, A., & Romero, L. C. (2022). Persulfidation is the mechanism underlying sulfide-signaling of autophagy. *Autophagy*, *18*(3), 695–697. https://doi.org/10.1080/15548627.2021.1936357.

Gotor, C., Garcia, I., Aroca, A., Laureano-Marin, A. M., Arenas-Alfonseca, L., Jurado-Flores, A., et al. (2019). Signaling by hydrogen sulfide and cyanide through post-translational modification. *Journal of Experimental Botany*, *70*(16), 4251–4265. https://doi.org/10.1093/jxb/erz225.

Gotor, C., Garcia, I., Crespo, J. L., & Romero, L. C. (2013). Sulfide as a signaling molecule in autophagy. *Autophagy*, *9*(4), 609–611. https://doi.org/10.4161/auto.23460.

Jia, H., Chen, S., Liu, D., Liesche, J., Shi, C., Wang, J., et al. (2018). Ethylene-induced hydrogen sulfide negatively regulates ethylene biosynthesis by persulfidation of ACO in tomato under osmotic stress. *Frontiers in Plant Science*, *9*(1517). https://doi.org/10.3389/fpls.2018.01517.

Johnston, D. T., Wolfe-Simon, F., Pearson, A., & Knoll, A. H. (2009). Anoxygenic photosynthesis modulated proterozoic oxygen and sustained Earth's middle age. *Proceedings of the National Academy of Sciences of the United States of America*, *106*(40), 16925–16929. https://doi.org/10.1073/pnas.0909248106.

Jurado-Flores, A., Romero, L. C., & Gotor, C. (2021). Label-free quantitative proteomic analysis of nitrogen starvation in Arabidopsis root reveals new aspects of H2S signaling by protein persulfidation. *Antioxidants*, *10*(4), 508. https://doi.org/10.3390/antiox10040508.

Kimura, H. (2015). Physiological roles of hydrogen sulfide and polysulfides. *Handbook of Experimental Pharmacology*, *230*, 61–81. https://doi.org/10.1007/978-3-319-18144-8_3.

Laureano-Marín, A. M., Aroca, A., Perez-Perez, M. E., Yruela, I., Jurado-Flores, A., Moreno, I., et al. (2020). Abscisic acid-triggered persulfidation of the Cys protease ATG4 mediates regulation of autophagy by sulfide. *Plant Cell*, *32*(12), 3902–3920. https://doi.org/10.1105/tpc.20.00766.

Laureano-Marín, A. M., Moreno, I., Aroca, Á., García, I., Romero, L. C., & Gotor, C. (2016). Regulation of autophagy by hydrogen sulfide. In L. Lamattina, & C. García-Mata (Eds.), *Gasotransmitters in plants* (pp. 53–75). Cham: Springer International Publishing.

Li, J., Shi, C., Wang, X., Liu, C., Ding, X., Ma, P., et al. (2020). Hydrogen sulfide regulates the activity of antioxidant enzymes through persulfidation and improves the resistance of tomato seedling to copper oxide nanoparticles (CuO NPs)-induced oxidative stress. *Plant Physiology and Biochemistry*, *156*, 257–266. https://doi.org/10.1016/j.plaphy.2020.09.020.

Lisjak, M., Srivastava, N., Teklic, T., Civale, L., Lewandowski, K., Wilson, I., et al. (2010). A novel hydrogen sulfide donor causes stomatal opening and reduces nitric oxide accumulation. *Plant Physiology and Biochemistry*, *48*(12), 931–935. https://doi.org/10.1016/j.plaphy.2010.09.016.

Liu, J., Hou, L., Liu, G., Liu, X., & Wang, X. (2011). Hydrogen sulfide induced by nitric oxide mediates ethylene-induced stomatal closure of Arabidopsis thaliana. *Chinese Science Bulletin*, *56*(33), 3547–3553. https://doi.org/10.1007/s11434-011-4819-y.

Mustafa, A. K., Gadalla, M. M., Sen, N., Kim, S., Mu, W., Gazi, S. K., et al. (2009). H2S signals through protein S-sulfhydration. *Science Signaling*, *2*(96), ra72. https://doi.org/10.1126/scisignal.2000464.

Scuffi, D., Lamattina, L., & García-Mata, C. (2016). Gasotransmitters and stomatal closure: Is there redundancy, concerted action, or both? *Frontiers in Plant Science*, 7, 277. https://doi.org/10.3389/fpls.2016.00277.

Scuffi, D., Alvarez, C., Laspina, N., Gotor, C., Lamattina, L., & Garcia-Mata, C. (2014). Hydrogen sulfide generated by L-cysteine desulfhydrase acts upstream of nitric oxide to modulate ABA-dependent stomatal closure. *Plant Physiology*, *166*, 2065–2076. https://doi.org/10.1104/pp.114.245373.

Shen, J., Xing, T., Yuan, H., Liu, Z., Jin, Z., Zhang, L., et al. (2013). Hydrogen sulfide improves drought tolerance in Arabidopsis thaliana by MicroRNA expressions. *PLoS One*, *8*(10), e77047. https://doi.org/10.1371/journal.pone.0077047.

Shen, J., Zhang, J., Zhou, M., Zhou, H., Cui, B., Gotor, C., et al. (2020). Persulfidation-based modification of cysteine desulfhydrase and the NADPH oxidase RBOHD controls guard cell abscisic acid signaling. *Plant Cell*, *32*(4), 1000–1017. https://doi.org/10.1105/tpc.19.00826.

Szabo, C. (2007). Hydrogen sulphide and its therapeutic potential. *Nature Reviews. Drug Discovery*, *6*(11), 917–935. https://doi.org/10.1038/nrd2425.

Wang, B. L., Shi, L., Li, Y. X., & Zhang, W. H. (2010). Boron toxicity is alleviated by hydrogen sulfide in cucumber (Cucumis sativus L.) seedlings. *Planta*, *231*(6), 1301–1309. https://doi.org/10.1007/s00425-010-1134-9.

Xie, Y., Zhang, C., Lai, D., Sun, Y., Samma, M. K., Zhang, J., et al. (2014). Hydrogen sulfide delays GA-triggered programmed cell death in wheat aleurone layers by the modulation of glutathione homeostasis and heme oxygenase-1 expression. *Journal of Plant Physiology*, *171*(2), 53–62. https://doi.org/10.1016/j.jplph.2013.09.018.

Zhang, H., Hu, L. Y., Hu, K. D., He, Y. D., Wang, S. H., & Luo, J. P. (2008). Hydrogen sulfide promotes wheat seed germination and alleviates oxidative damage against copper stress. *Journal of Integrative Plant Biology*, *50*(12), 1518–1529. https://doi.org/10.1111/j.1744-7909.2008.00769.x.

Zhang, H., Jiao, H., Jiang, C.-X., Wang, S.-H., Wei, Z.-J., Luo, J.-P., et al. (2010). Hydrogen sulfide protects soybean seedlings against drought-induced oxidative stress. *Acta Physiologiae Plantarum*, *32*(5), 849–857. https://doi.org/10.1007/s11738-010-0469-y.

Zhou, M., Zhang, J., Shen, J., Zhou, H., Zhao, D., Gotor, C., et al. (2021). Hydrogen sulfide-linked persulfidation of ABI4 controls ABA responses through the transactivation of MAPKKK18 in Arabidopsis. *Molecular Plant*, *14*(6), 921–936. https://doi.org/10.1016/j.molp.2021.03.007.

Zivanovic, J., Kouroussis, E., Kohl, J. B., Adhikari, B., Bursac, B., Schott-Roux, S., et al. (2019). Selective persulfide detection reveals evolutionarily conserved antiaging effects of S-sulfhydration. *Cell Metabolism*, *30*(6), 1152–1170 e1113. https://doi.org/10.1016/j.cmet.2019.10.007.

CHAPTER SIXTEEN

CRISPR-Cas mediated genome engineering of cyanobacteria

Annesha Sengupta, Deng Liu, and Himadri B. Pakrasi*
Department of Biology, Washington University, St. Louis, MO, United States
*Corresponding author: e-mail address: pakrasi@wustl.edu

Contents

Abstract

Over the past decade, several cyanobacterial strains have emerged as exciting model systems for unraveling important biological process like photosynthesis and nitrogen fixation. In parallel, novel strains are being developed as platforms for production of various value-added products. To meet either of these goals, synthetic biology tool development has been prioritized, and among many such tools, CRISPR-mediated genome editing tools distinctly hold the potential to revolutionize cyanobacterial research. This chapter reviews our current understanding of the existing and emerging CRISPR-based technologies and their potential application for advanced genome editing in cyanobacterial strains of interest. CRISPR-based tools have gained traction in cyanobacterial research for their ability to target the polyploid genomes in these organisms and generate fully segregated mutants in a remarkably short time. We discuss the native cyanobacterial CRISPR system and the promise they hold for use as precision tools for cyanobacterial genome editing. We elaborate the methodologies for the development of CRISPR-based markerless mutants in cyanobacteria as well as discuss strategies for large scale, regulated genome silencing with CRISPRi. We also highlight some of the emerging CRISPR tools that have shown promise in other prokaryotic and eukaryotic systems but are yet to be adapted for cyanobacterial research.

Methods in Enzymology, Volume 676
ISSN 0076-6879
https://doi.org/10.1016/bs.mie.2022.07.023

1. Introduction

Cyanobacteria are a group of photosynthetic prokaryotes credited for transforming the earth's atmosphere from anoxygenic to oxygen-rich, as we see today. These organisms are considered the ancestor of oxygenic photosynthesis (Heidorn et al., 2011). They have been residing for ages in different ecological niches under unfavorable conditions (Sanchez-Baracaldo, Hayes, & Blank, 2005), thus exhibiting diversity in morphology, physiology and metabolism (Pisciotta, Zou, & Baskakov, 2010; Rippka, Deruelles, Waterbury, Herdman, & Stanier, 1979; Shih, Wu, Latifi, et al., 2013). The geographic diversification has led to cyanobacterial strains with several interesting traits, such as the ability to fix nitrogen, besides oxygenic photosynthesis. Some diazotrophic cyanobacteria can efficiently and effectively partition the two antagonistic processes, atmospheric nitrogen fixing and oxygenic photosynthesis, either temporally or spatially (Bandyopadhyay, Elvitigala, Liberton, & Pakrasi, 2013; Gaudana et al., 2013; Higo & Ehira, 2019; Stockel et al., 2008). The ability of cyanobacteria to temporally segregate metabolic processes led to the clue for the presence of an internal clock (Johnson, Mori, & Xu, 2008). Cyanobacteria are known to be by far the only prokaryote to modulate their metabolism as a function of time (Cohen & Golden, 2015; Dong & Golden, 2008; Jaiswal & Wangikar, 2020). These green bacteria, owing to their faster growth, higher biomass accumulation, efficient photosynthesis, simplified genome and genetic tractability as compared to other higher photosynthetic organisms, have been portrayed as better models to understand various natural phenomena such as photosynthesis, circadian oscillation, ecology, and evolution (Jeong et al., 2020; Sanchez-Baracaldo et al., 2005; Santos-Merino, Singh, & Ducat, 2019; Sengupta, Pakrasi, & Wangikar, 2018). Moreover, the ability of cyanobacteria to harness solar power and use it to fuel the conversion of atmospheric CO_2 into organic molecules has made these photoautotrophs a suitable host for carbon-negative cell factories for the application of biotechnology and bio-remediation (Hagemann & Hess, 2018; Sengupta et al., 2018; Zahra, Choo, Lee, & Parveen, 2020).

The potential of these photosynthetic bacteria has spearheaded the effort to isolate and identify novel cyanobacterial strains. To date, Cyanobase hosts 376 different cyanobacterial strains (Fujisawa, Narikawa, Maeda, et al., 2017; Nakamura, Kaneko, & Tabata, 2000). However, only a limited number of strains are widely used because of easy laboratory cultivation methods, ease of genetic modification, and availability of complete genome information (Heidorn et al., 2011). Table 1 lists some of the widely used cyanobacterial strains. As a proof of concept, cyanobacterial strains have been engineered to channel fixed cellular carbon into various biofuel precursors and value-added products such as sugars (Ducat, Avelar-Rivas, Way, & Silver, 2012; Jacobsen & Frigaard, 2014), isoprene (Gao et al., 2016), ethanol (Choi, Lee, Kim, & Park, 2020; Choi & Park, 2016; Kopka, Schmidt, Dethloff, et al., 2017), 2,3-butanediol (Nozzi & Atsumi, 2015), 1-butanol (Lan & Liao, 2012; Liu, Miao, Lindberg, & Lindblad, 2019), 1,2-propanediol (Li & Liao, 2013), and succinate (Sengupta, Jaiswal, et al., 2020, Sengupta, Pritam, et al., 2020); however, significant improvements are required to translate these lab-scale studies into commercially viable technologies.

One such effort has been isolating faster growing cyanobacterial strains and exploring the metabolic and genetic differences impacting the desirable phenotype (Abernathy et al., 2017; Ungerer et al., 2018). In the past 7 years, four such fast-growing cyanobacteria have been isolated (Table 1) which are phenotypically and physiologically distinct from each other (Jaiswal et al., 2020, 2018; Wlodarczyk et al., 2020; Yu et al., 2015). Genetic engineering of these strains has shown significant improvement in the productivity of products such as sucrose and limonene as compared to the model strains (Ducat et al., 2012; Lin, Saha, Zhang, & Pakrasi, 2017; Lin et al., 2020; Lin, Zhang, & Pakrasi, 2021).

Systematic host engineering by applying synthetic biology tools is one of the strategies to improve strain fitness and heterologous productivity. Synthetic biology involves quick, predictable, and easy assembly/disassembly of synthetic/native modules to develop engineered biological systems. Recently, efforts expand the cyanobacterial synthetic biology toolbox, which comprises a repertoire of native and heterologous transcriptional and translational bio-parts, genome editing and engineering tools, and computational/predictive tools (Santos-Merino et al., 2019; Sengupta, Jaiswal, et al., 2020; Sengupta, Madhu, & Wangikar, 2020; Sengupta et al., 2018; Sengupta, Pritam, et al., 2020). Although reports are available where the synthetic biology approach has been employed for bio-production (Knoot, Khatri, Hohlman, Sherman, & Pakrasi, 2019; Liu et al., 2019; Sengupta, Jaiswal, et al., 2020; Sengupta, Madhu, & Wangikar, 2020; Sengupta, Pritam, et al., 2020), we have not maximized the full potential of these synthetic biology tools.

In this chapter, we focus on one of the most revolutionary advancements in the field of genome engineering, i.e., the use of CRISPR (clustered

Table 1 Cyanobacterial strains widely and mostly used in current days exhibiting unique feature.

Cyanobacterial species	Genome size (Mbp)	Features and application	References
Non-diazotrophic			
Synechocystis sp. PCC 6803	3.6	Model for Photosynthesis studies, one of the earliest strains to be sequenced and thus heavily studied, synthetic biology toolbox	Kondo, Geng, Katayama, and Ikeuchi (2005), Blasi, Peca, Vass, and Kos (2012), Abe et al. (2014), Carrieri, Lombardi, Paddock, et al. (2017), Ferreira et al. (2018), Liu, Liberton, Yu, Pakrasi, and Bhattacharyya-Pakrasi (2018), Liu and Pakrasi (2018), Bartasun et al. (2019), Weisz et al. (2019), Liu, Johnson, and Pakrasi (2020) and Chen, Niedzwiedzki, Bandyopadhyay, Biswas, and Pakrasi (2021)
Synechococcus elongatus PCC 7942	2.7	Ease in genetic modification and model for circadian clock, synthetic biology toolbox	Ditty, Canales, Anderson, Williams, and Golden (2005), Dong and Golden (2008), Dong, Kim, and Golden (2010), Kanno and Atsumi (2017), Sengupta, Sunder, Sohoni, and Wangikar (2019a, 2019b), Sengupta, Jaiswal, et al. (2020), Sengupta, Madhu, and Wangikar (2020), Sengupta, Pritam, et al. (2020) and Li et al. (2022)
Synechococcus sp. PCC 7002	3.0	Euryhaline, fast-growing, metabolic engineering studies, synthetic biology toolbox	Markley, Begemann, Clarke, Gordon, and Pfleger (2015), Gordon et al. (2016), Hendry, Prasannan, Joshi, Dasgupta, and Wangikar (2016), Hendry et al. (2017) and Knoot and Pakrasi (2019)
Synechococcus elongatus UTEX 2973	2.7	Fastest-growing, biotechnological applications, synthetic biology toolbox	Yu et al. (2015), Abernathy, Yu, Ma, et al. (2017), Tan et al. (2018), Ungerer, Wendt, Hendry, Maranas, and Pakrasi (2018), Knoot, Biswas, and Pakrasi (2020), Lin, Zhang, and Pakrasi (2020) and Wendt, Walker, Sengupta, Ungerer, and Pakrasi (2022)

Synechococcus elongatus PCC 11801	2.7	Fast-growing, biotechnological applications	Jaiswal et al. (2018), Sengupta, Jaiswal, et al. (2020), Sengupta, Madhu, and Wangikar (2020), Sengupta, Pritam, et al. (2020) and Srivastava et al. (2021)
Synechococcus elongatus PCC 11802	2.7	Fast-growing	Jaiswal et al. (2020)
Synechococcus sp. PCC 11901	3.0	Fast-growing with high biomass accumulation	Wlodarczyk, Selao, Norling, and Nixon (2020) and Mills, Moreno-Cabezuelo, Włodarczyk, et al. (2022)
Diazotrophic			
Cyanothece sp. ATCC 51142	5.5	Model for nitrogen fixation, understanding circadian rhythm	Stockel et al. (2008), Bandyopadhyay, Stockel, Min, Sherman, and Pakrasi (2010), Bandyopadhyay et al. (2013) and Gaudana et al. (2013)
Anabaena variabilis	7.1	Model for nitrogen fixation	Wealand, Myers, and Hirschberg (1989)
Anabaena (*Nostoc*) sp. PCC 7120	6.4	Model for nitrogen fixation, synthetic biology toolbox	Puerta-Fernandez and Vioque (2011), Ungerer and Pakrasi (2016), Higo, Isu, Fukaya, Ehira, and Hisabori (2018), Higo and Ehira (2019) and Higo, Nishiyama, Nakamura, Hihara, and Ehira (2019)

regularly interspaced short palindromic repeats)-Cas technology in cyanobacteria (Behler, Vijay, Hess, & Akhtar, 2018; Doudna & Charpentier, 2014). This technique has gained popularity due to its ability to target the polyploid genome and facilitate engineering without retaining the antibiotic marker in the genome (Behler et al., 2018; Sengupta et al., 2018). Since the CRISPR-mediated genome engineering uses a Cas nuclease to introduce a double strand break in the genome, homology directed integration of cargo DNA is more efficient than the traditional method of integration. Therefore, this approach has been employed in different model and non-model cyanobacterial strains for the biotechnological purposes (Niu et al., 2019; Sengupta, Jaiswal, et al., 2020; Sengupta, Madhu, & Wangikar, 2020; Sengupta, Pritam, et al., 2020; Ungerer & Pakrasi, 2016). Additionally, the use of deactivated Cas protein (dCas) for gene silencing (CRISPRi) has been a beneficial tool for maximizing productivity by re-distributing and re-wiring intracellular carbon flux (Huang et al., 2016; Knoot et al., 2020; Liu et al., 2020). An advancement of CRISPRi (pooled CRISPRi) was developed which enabled one-pot, high-throughput investigation of gene function of a large array of samples to determine cellular fitness and productivity (Wang et al., 2018; Yao et al., 2020). In this chapter, we review the architecture and function of CRISPR systems native to cyanobacteria and analyze their potential usability in cyanobacteria-based applications. We also provide detailed protocols of plasmid design, DNA transformation, and process of identifying the positive transformants from our lab for CRISPR-mediated genome engineering in cyanobacteria.

2. Native cyanobacterial CRISPR-Cas

CRISPR-Cas loci are RNA-mediated genome editors, naturally found in most bacteria and archaea. It functions as an adaptive immune response to phage invasions and is based on the principle of differentiating the self-DNA from the foreign DNA (Makarova & Koonin, 2015). The CRISPR loci were first identified in *Escherichia coli* by a Japanese group and since then its mechanism and applications are being explored (Doudna & Charpentier, 2014). During a phage infection, CRISPR-Cas immune system integrates a part of the viral genome into its own genome as a signature memory to combat future infections. The process comprises of three steps, integration of the foreign DNA into the host genome as new direct repeats interspaced with the viral DNA, followed by the transcription and processing of the CRISPR array into the crRNA (CRISPR RNA) and finally the Cas protein identifies the foreign DNA with the help of the gRNA and cleaves the foreign DNA, thus preventing any future infection (Bhaya, Davison, & Barrangou, 2011).

Based on the ribonucleoprotein complexity, the CRISPR-Cas system can be categorized into Classes 1 and 2 (Hou et al., 2019). While Class 1 utilizes multi-Cas Cascade protein for driving the defense system, the latter is mediated by a single large modularized Cas effector protein (Tang & Fu, 2018). These Classes could be further sub-divided into types and sub-types. There are so far, eight distinct subtypes (Ecoli, Ypest, Nmeni, Dvulg, Tneap, Hmari, Apern, and Mtube) and are named after the species where it was identified. Some core *cas* genes have a suffix letter such as *cas5a*, the "a" is from the subtype Apern. The *csx* nomenclature was provided for some *cas* genes where no proper link could be established. This classification is prone to change owing to the complex and ever-evolving nature of the CRISPR-Cas system (Makarova & Koonin, 2015).

Cyanobacteria are a complex group of organisms comparatively less explored for their native CRISPR arrangement. In 2013, the Kerfeld group surveyed 126 cyanobacterial genomes for the presence of CRISPR array using a bioinformatics approach and identified the presence of CRISPR-Cas cascade in 86 genomes except for a phylogenetic subclade which comprised of *Prochlorococcus* and marine *Synechococcus* (Cai, Axen, & Kerfeld, 2013). The type, number, and ratio of CRISPR and Cas modules might vary among different organisms (Hou et al., 2019). The largest number of predicted CRISPR loci in a cyanobacterial genome was reported in *Coleofasciculus chthonoplastes* PCC 7420 (Bhaya et al., 2011). Cyanobacterial genome analysis revealed that the Class 1 CRISPR system is more commonly found in cyanobacteria than the Class 2 Cas proteins. In cyanobacteria, two types of Class I CRISPR system (I and III) and only a single type, type V of Class 2 effector Cas protein were identified (Table 2). Generally, the genomes of filamentous cyanobacteria have been observed to be more prone to host the uncommon class of CRISPR, Class 2, as compared to unicellular cyanobacteria (Hou et al., 2019). However, the genome of a diazotrophic unicellular cyanobacterium, *Cyanothece* sp. PCC 8801 has both Class 1 and 2 CRISPR loci. The most explored strains for native CRISPR system are *Synechocystis* sp. PCC 6803, *Synechocystis* sp. PCC 6714, *Microcystis aeruginosa* and *Cyanothece* sp. PCC 8801, *Anabaena* sp. PCC 7120, etc. While *Anabaena* sp. PCC 7120 genome hosts three Class 1, one Class 2 and five orphan spacer-repeat, *Synechocystis* sp. PCC 6803 has three distinct sets of

Table 2 All native Cas proteins from cyanobacteria with its probable characteristics.

CRISPR-Cas protein	Class	Type	Function/characteristics and application
Cas1	1	I	Metal-dependent DNase
Cas2	1	I	RNase specific to U-rich region Double stranded DNase
Cas3′, Cas3″	1	I	Single stranded nuclease and ATP dependent helicase activity Unlike Cas3, encoded as separate gene in the same loci Cas3′: Helicase, Cas3″: Nuclease
Cas4	1	I	Single stranded *exo*-nuclease
Cas5	1	I, III	Binds to the 5′-end of the crRNA
Cas6	1	I	Maturation and pre-processing of crRNA
Cas7	1	I, III	Binds to crRNA
Cas8a1	1	I	Helps in PAM recognition
Cas10	1	III	Multi-domain: helicase, cyclase, polymerase Often associated with nuclease domain different from Type I
Cas11	1	I, III	Additional Cas proteins
Csx19	1	III	Uncharacterized
C2c8 (C2c is class 2 candidate)	2	V	Single effector protein
Cas12k (C2c5; C2c is class 2 candidate)	2	V	Single effector protein

CRISPR-Cascade in one of the endogenous plasmids (Hou et al., 2019; Scholz, Lange, Hein, Hess, & Backofen, 2013). These strains serve as models to understand the CRISPR system in this diverse phylum (Table 1).

The Class 1 CRISPR system has the signature Cas proteins Cas1/Cas2 which functions as a complex, involved in both acquisition and processing of crRNA (Makarova & Koonin, 2015). Mutational studies showed the functional dependency of Cas1 and Cas2 proteins on one another, implying that both Cas proteins are essential. Class 1 CRISPR functions via multi-proteins and Table 2 discusses the function of all the Cas proteins involved in different categories. Most Cas proteins have been observed to be very divergent due to their need for rapid evolution to compete with the phage activity.

However, Cas1 is the most abundant and highly conserved protein, serving as a probe for phylogenetic studies to identify and classify the CRISPR-Cas systems (Makarova & Koonin, 2015). The signature protein for Type I is the Cas3 protein which exhibits both nuclease and helicase activity. Thereby, facilitating large deletion of genomes and this feature has been exploited in heterotrophs to delete large regions of the genome (Csorgo et al., 2020). Cyanobacterial strains host *cas3′* and *cas3″* genes, instead of the *cas3*. While in Cas3 the nuclease domain is fused in the N-terminal domain, if the nuclease and helicase domains are expressed separately by two genes residing in the same locus then the helicase domain is called the Cas3′ and the nuclease is Cas3″ (Makarova & Koonin, 2015). The Class 1 type III arrangement possesses Cas10 in place of Cas3. Cas10 protein is a multi-domain protein with cyclase, polymerase, and helicase activity. Sometimes a nuclease domain, distinctly different from the type I nuclease, is associated with Cas10 (Makarova & Koonin, 2015). Besides these signature and universal proteins, there are other effector Cas proteins involved in various functions such as PAM (Protospacer adjacent motif) recognition, stabilizing the CRISPR complex, or processing of crRNA (Table 2).

The Cas1 and Cas2 proteins were thought to be present in all types of CRISPR arrays, however, the Class 2 CRISPR fails to possess these proteins. Unlike Class 1, the Class 2 CRISPR arrangement involves a single Cas protein in association with other proteins such as Ruv or transposase-like proteins (Pattharaprachayakul, Lee, Incharoensakdi, & Woo, 2020). Due to the simple architecture, the Class 2 Cas proteins such as Cas9, Cas12a and Cas13 have been used for gene editing applications. Though the Cas9 effector proteins have been indicated to be evolved from TnpB-like transposases with HNH nuclease, domain generally found to be abundant in cyanobacteria, there are no reports of the presence of either Cas9 or Cas12a or Cas13-like proteins in cyanobacteria. In cyanobacteria, only one type of Class 2 CRISPR system is found, the types V-U. The C2c8 and C2c5 are the two novel Class 2 Cas proteins identified so far in cyanobacteria and belong to the subtype V-U2 and V-U5 (Hou et al., 2019). The C2c5 protein has been characterized by the inactive RuvC motif and the association with Tn7-like transposase subunits, TnsB, TnsC, and TniQ. Recently, this cyanobacterial native protein, C2c5 protein was designated as effector protein Cas12k belonging to the subtype V-K (Strecker et al., 2019). The CRISPR Cas12k associated Tn7-like transposons (CASTs) from two cyanobacterial strains *Scytonema hofmanni* UTEX B 2349 (ShCAST) and *Anabaena cylindrica* PCC 7122 (AcCAST) has been explored as a synthetic biology tool in heterotrophs.

The CAST systems facilitated specific DNA integration as a function of the sequences of protospacer and PAM (Strecker et al., 2019). In *E. coli*, the use of Cas12k (ShCAST) has led to RNA-guided DNA transposition between 60 and 66 bps downstream of the PAM; however, the overexpression of the system might lead to non-targeted insertion independent of Cas12k (Strecker et al., 2019). Further, optimization of the ShCAST system enabled the integration of large fragments (>30 kb) with improved efficiency of 100% (Cheng et al., 2022), making this editing technology versatile, robust for large scale programmable genome engineering. This budding technology will be very advantageous to perform genome editing in novel cyanobacterial strains or metagenomic samples where the information on recombineering is sparse. Though CRISPR RNA-mediated transposition enables high efficiency integration of large cargo, the system requires future optimization for (i) reducing off-target transposition, (ii) facilitating scarless integration (Cheng et al., 2022) and establish this technology for strain and loci-specific large genome editing.

The CRISPR-Cas arrays native to cyanobacteria showed no evidence of the presence of the potential Cas9 and Cas12a proteins, which are now being predominantly used for synthetic biology purposes (Section 3). Therefore, most CRISPR-mediated studies in cyanobacteria are focused on introducing heterologous CRISPR-Cas9 or CRISPR-Cas12a as a synthetic biology tool for metabolic engineering. Exploring the native systems might provide interesting leads for developing synthetic biology tools. For example, Cas3 from the heterotrophic background has been explored for long deletion because of its dual activity and exploiting the efficiency of Cas3′-Cas3″ and Cas3 might provide new interesting avenues of engineering. Similarly, Cas10 is a multidomain protein exhibiting multiple activities, such as polymerase activity, and understanding this protein might have the potential to be used for synthetic biology purposes. Cas12k is already being examined for its ability to perform precise transposition and this is advantageous for strains that have low homologous recombination efficiency. The next section discusses the methodology and design process for employing the widely used CRISPR-Cas system (Cas9 and Cas12a) for synthetic biology applications.

3. Synthetic biology application of CRISPR technologies in cyanobacteria

Genome editing studies using various CRISPR-Cas systems have been reported in several cyanobacterial species, including *Synechocystis* 6803, *Synechococcus* 7942, *Synechococcus* 2973, and *Anabaena* 7120. Table 3

Table 3 Cyanobacteria that have been genome edited by the CRISPR technology.

Organism	Cas type	Promoter[a]	Purpose	Reference
Synechocystis 6803	Cas12a	P_{Lac} and P_{J23119}	Deletion, insertion, point mutation	Ungerer and Pakrasi (2016)
	Cas9	P_{Tet} and P_{trc}	Deletion	Xiao et al. (2018)
	dCas9	Multiple	Repression	Yao, Cengic, Anfelt, and Hudson (2016)
	dCas9	P_{L22} and P_{L22}	Repression	Shabestary et al. (2018)
	dCas9	P_{L22} and P_{L22}	Repression	Kaczmarzyk, Cengic, Yao, and Hudson (2018)
	dCas9	Native bidirectional	Repression	Kirtania et al. (2019)
	dCas9	P_{cpc560} and P_{trc1O}	Repression	Vasudevan, Gale, Schiavon, et al. (2019)
	dCas9	P_{L22} and P_{L22}	Genome scale repression	Yao et al. (2020)
	dCas12a	P_{rhaBAD} and P_{rhaBAD}	Repression	Liu et al. (2020)
	dCas9	P_{L22} and P_{L22}	Repression	Dietsch, Behle, Westhoff, and Axmann (2021)
	dCas9	P_{psbA2} and P_{L31}	Repression	Santos, Pacheco, Yao, Hudson, and Tamagnini (2021)
	dCas9	P_{L22} and P_{L22}	Repression	Shabestary et al. (2021)
	dCas9	P_{L31} and P_{L22}	Repression	Yunus et al. (2022)
Synechococcus 7942	Cas9	Native	Deletion, insertion	Li et al. (2016)
	dCas9	Multiple	Repression	Huang et al. (2016)
	dCas9	Multiple	Repression	Lee and Woo (2020)
	dCas9	Multiple	Multiple Repression	Li et al. (2022)
	dCas12a	P_{trc} and P_{J23119}	Repression	Choi and Woo (2020)

Continued

Table 3 Cyanobacteria that have been genome edited by the CRISPR technology.—cont'd

Organism	Cas type	Promoter[a]	Purpose	Reference
Synechococcus 2973	Cas12a	P_{Lac} and P_{J23119}	Deletion, insertion, point mutation	Ungerer and Pakrasi (2016)
	Cas9	Native	Deletion	Wendt, Ungerer, Cobb, Zhao, and Pakrasi (2016)
	Cas9	Not available	Insertion	Racharaks, Arnold, and Peccia (2021)
	dCas12a	P_{trc2O} and P_{Lac1O}	Repression	Knoot et al. (2020)
Synechococcus 7002	dCas9	Multiple	Repression	Gordon et al. (2016)
Anabaena 7120	Cas12a	P_{Lac} and P_{J23119}	Deletion, insertion, point mutation	Ungerer and Pakrasi (2016)
	Cas12a	P_{Lac} and P_{J23119}	Large fragment deletion	Niu et al. (2019)
	dCas9	P_{L03} and P_{J23119}	Repression	Higo et al. (2018)
	dCas9	P_{L03} and P_{J23119}	Repression	Higo et al. (2019)
	dCas9	Multiple	Repression	Higo and Ehira (2019)
Synechococcus 11801	Cas12a	P_{Lac} and P_{J23119}	Insertion	Sengupta, Jaiswal, et al. (2020), Sengupta, Madhu, and Wangikar (2020) and Sengupta, Pritam, et al. (2020)

[a]Promoters driving the expression of the Cas coding gene and the guide RNA are displayed.

summarizes studies applying the CRISPR technology for synthetic biology research in cyanobacteria. Note that both CRISPR and CRISPR interference (CRISPRi) technologies have been implemented using two types of Cas proteins: Cas9 and Cas12a.

3.1 Using CRISPR for targeted genome editing in cyanobacteria

Both CRISPR-Cas9 and CRISPR-Cas12a are robust technologies for knocking-out and knocking-in genes to construct cyanobacterial transformants. As RNA-guided nucleases, CRISPR-Cas endonucleases use short CRISPR-derived RNA guides to target complementary DNA. After recognition of a protospacer adjacent motif (PAM) sequence positioned next to the target sequence, the Cas nuclease (Cas9, Cas12a) introduces double-strand DNA breaks. Both Cas9 and Cas12a are class 2 CRISPR systems but are different in their types, of which Cas9 and Cas12a are type II and type V, respectively (Makarova, Zhang, & Koonin, 2017).

The Cas9 effector derived from the bacterium *Streptococcus pyogenesis* was commonly used in the CRISPR-Cas system for genome editing in cyanobacteria. The CRISPR-Cas9 system consists of an RNA-guided Cas9 endonuclease protein, a CRISPR RNA (crRNA) that targets a specific DNA sequence, and a trans-activating RNA (tracrRNA) that recruits Cas9 along with the crRNA (Sander & Joung, 2014). Programming new target sequences for the Cas9 endonuclease activity can be simplified by fusing together the crRNA and tracrRNA to form a single guide RNA (sgRNA). Cas9 recognizes a target with a 3′ G-rich PAM (5′-NGG-3′) and cuts DNA, producing blunt ends (Fig. 1A). Expression of SpCas9 has been linked to toxicity and failure to recover colonies following transformation in cyanobacteria (Wendt et al., 2016). It remains unclear why SpCas9 appears toxic in cyanobacteria, but Cas12a as an alternative and robust CRISPR tool has been widely used in multiple cyanobacteria.

The CRISPR-Cas12a system derived from *Francisella novicida* was first introduced into three strains of cyanobacteria, *Synechocystis* 6803, *Synechococcus* 2973, and *Anabaena* 7120 (Ungerer & Pakrasi, 2016). Compared to Cas9, Cas12a has its distinct features (Bayat, Modarressi, & Rahimpour, 2018). Cas12a uses a single crRNA as gRNA instead of the set of crRNA and tracrRNA used by Cas9. Cas12a catalyzes the maturation of its own gRNA, which allows for efficient multiplex genome editing, whereas maturation of the sgRNA complex of Cas9 relies on processing by the non-Cas ribonuclease RNase III. Cas12a uses a T-rich PAM (5′-TTTN-3′) upstream of the protospacer in contrast to the downstream located G-rich PAM by Cas9. Cas12a has a single nuclease domain that cleaves 17–22 base pairs downstream from the PAM sequence, whereas cleavage by Cas9 occurs within the three base pairs upstream of its PAM. Cas12a generates staggered ends with five nucleotide overhangs in the target DNA, compared to the blunt end cleavage by Cas9 (Fig. 1B).

Genomic double-strand breaks (DSBs) will be repaired mainly by two strategies in cells: non-homologous end-joining repair (NHEJ) and homology-directed repair (HDR) (Phipps & Dubrana, 2022). The error-prone NHEJ repair system is most prevalent in eukaryotes and is predicted to be encoded by 26% of sequenced prokaryotic genomes

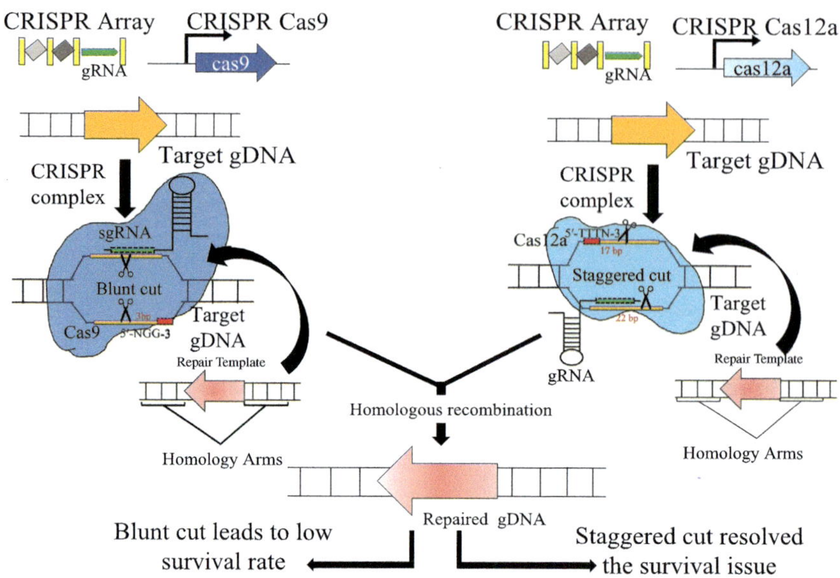

Fig. 1 CRISPR-Cas9/Cas12a-mediated genome editing in cyanobacteria. The CRISPR system consists of a CRISPR array (CRISPR repeats (yellow bars) and spacer (gray box) for gRNA insertion) and Cas protein (blue). The 20 bp gRNA (green arrow) designed for targeting specific genomic region is first cloned between the two CRISPR repeat sequences. The gRNA directs and facilitates the association of the CRISPR-Cas on to the specific position of the genome. While Cas9 introduces a blunt cut 3 bp upstream of the PAM (protospacer adjacent motif), a staggered cut of 5 bp (17–22 bp) is observed for Cas12a downstream of the PAM sequence. The double-strand break introduced need to be repaired either by non-homologous end joining (NHEJ) or homology-directed repair (HDR). For genome engineering, we intend to introduce single or multiple DNA modules of interest (e.g., Promoter-gene-terminator), therefore, besides the CRISPR RNA and Cas protein, a Repair Template with flanking homology arms of 750–900 bp is required to facilitate the repair via HDR. The use of Cas12a have shown higher survival rate than Cas9.

(Sharda, Badrinarayanan, & Seshasayee, 2020). The HDR system allows for precise editing of the genome by adding a repairing template (i.e., donor DNA with the desired genetic property) during the process. Application of CRISPR technology in cyanobacteria has used the HDR strategy in a targeted way to delete small or large DNA fragments, insert DNA fragments into targeted genomic locations, and introduce point mutations. It has been proposed that DSBs generated by the CRISPR system improve the efficiency of HDR in cyanobacteria and the most important benefit of using the CRISPR technology is not introducing any selection marker into the genome. One technical issue that needs to be observant of is to avoid targeting the CRISPR system to the donor DNA, especially when introducing a point mutation to the target gene in the chromosome, simultaneous silent mutations should be introduced for the PAM sequence or the gRNA binding sequence in the donor DNA to avoid the situation (Ungerer & Pakrasi, 2016).

3.2 Using CRISPRi for targeted gene repression in cyanobacteria

Different from the CRISPR system, the CRISPRi system is designed to repress the transcription of target genes. CRISPRi uses deactivated Cas proteins (dCas) rendered unable to cleave target genes but still able to interrupt transcriptional activity (Huang et al., 2016) (Fig. 2). By not cutting DNA, CRISPRi circumvents the disadvantages of CRISPR, which include the genomic instability led by DSBs and the impracticability for investigating the effects of genes that are essential for cell survival. With multiple gRNAs, a single CRISPRi system could simultaneously repress the expression of multiple genes. More importantly, the repressing level of target genes could be tuned through the expression level of dCas proteins and/or

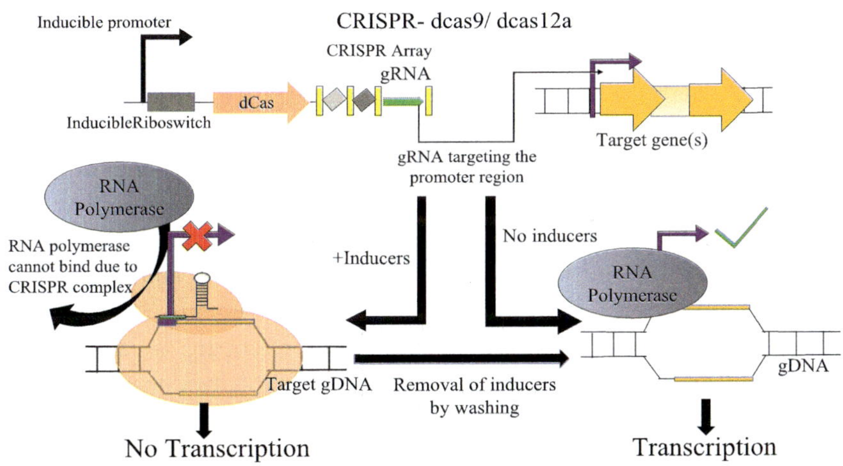

Fig. 2 Detailed description of the CRISPRi system to tune gene expression in cyanobacteria. The mutated Cas protein to abolish the nuclease activity and only retain the DNA binding ability (dCas) is used for CRISPRi where the dcas is expressed under inducible promoter system to regulate the expression level of the dCas proteins. Similar to the CRISPR setup, the gRNA (green arrow) is cloned between two CRISPR repeat (yellow bar) and this gRNA targets the start of the gene of interest (around the transcription start site). With the addition of inducers, the dCas proteins guided by the gRNA associate near the RNA polymerase binding site and hinders the binding the RNA pol to initiate transcription. Therefore knocking down the gene expression. When the induced cells are washed thoroughly to get rid of the inducers, the repression is alleviated leading to expression of the gene of interest.

gRNAs, which is called the dynamic CRISPRi technology. CRISPRi technology has been widely used in cyanobacteria to redirect carbon flux, to knock-down competing carbon pathways for increasing product titers in biotechnological applications and to better understand cyanobacterial physiology and photosynthetic processes.

The nuclease lobe of Cas9 contains an HNH and RuvC domain, which generate nicks on both strands of DNA and gRNA. To remove *endo*-nucleolytic activity from Cas9, two silencing mutations were introduced into each catalytic domain (Qi et al., 2013). Most studies replace His840 and Asp10 with alanines in the HNH domain to specifically abolish the nuclease activity and convert CRISPR-Cas9 into an RNA-guided DNA binding complex. Moreover, the replacement of other catalytic amino acids such as His983, Asp986, or Glu762 can inactivate the RuvC nuclease domain to convert Cas9 into a "nickase" (Xu & Qi, 2019).

As mentioned above, Cas12a is also classified in Class 2 like Cas9 but consists of only a RuvC-like domain, which achieves both DNA/RNA cleavage activity. As with Cas9, mutations of a catalytic residues can deactivate the Cas12a (dCas12a) RuvC nuclease domain (Knoot et al., 2020). Although dCas12a harbors a more desirable repression efficiency than dCas9 and is less toxic to cyanobacteria, there have been fewer successful dCas12a studies than dCas9 reported in cyanobacteria (Table 3).

Since CRISPRi can repress multiple genes concurrently in the cell, an important application of this technology is building a gRNA library that can be prepared in the form of a vector library. This will make it possible to execute CRISPR procedures in a high-throughput manner. The gRNA pool can be prepared as either semi-random sequences or specific sequences. Since the protospacer region of sgRNAs is small enough (~20 bp) to be sequenced by the method of next-generation sequencing, it can serve as a barcode and allows monitoring of the abundance of each clone in the library. This pooled CRISPRi strategy has been successfully used in *Synechocystis* 6803 to investigate the essentiality of genes at the whole genome level and to screen genes responsible for a higher titer of lactate production (Yao et al., 2020).

3.3 Construction of cyanobacterial CRISPR transformants

3.3.1 Design and build vectors containing the CRISPR gene cassette

To construct the cyanobacterial transformants with intended genotypes and projected phenotypes through CRISPR technologies, the vector containing the CRISPR gene cassette needs to be designed and constructed. Both suicide plasmids and self-replicating plasmids have been used to transfer the CRISPR system into cyanobacteria (Li et al., 2016; Wendt et al., 2016). It has been suggested that integrating the gene cassette into the chromosome using the suicide plasmid kept the stable expression of integrated genes (Gu et al., 2015), but it is time consuming for the chromosomal segregation process. The broad host replicating plasmid pRSF1010 and its derivatives have been widely used in cyanobacteria as a vector to express the Cas protein and the corresponding gRNA. Because there is no segregation issue when using the pRSF1010-based vector in cyanobacteria, it significantly shortens the process to build the cyanobacterial transformant. Stability is a concern when the CRISPR target gene is essential or has an important function for cyanobacteria. The phenomenon to eliminate the editing plasmids by cells has been reported in *Synechococcus* 2973 (Knoot et al., 2020).

To design the vector containing the CRISPR system, promoters driving the expression of the CRISPR-Cassette, the binding specificity of gRNA, and the binding loci on the chromosome are important factors. Both constitutive and inducible promoters have been used to drive the expression of (d)Cas protein and gRNA in multiple cyanobacteria (listed in Table 3). All these validated promoters provide a broad choice to regulate the expression level of the CRISPR machinery. Binding specificity of gRNA will affect the on-target efficiency (Wu, Kriz, & Sharp, 2014). Several general rules for gRNA design need to be followed, including 35–65% GC content, no polyT, no secondary structure, and <12bp match by BLAST to the cyanobacterial genome. Multiple tools have been developed to design gRNA in various organisms, such as the Benchling CRISPR tool (Uniyal, Mansotra, Yadav, & Kumar, 2019), but none of them is specific for cyanobacteria, which indicates the necessity to explore in the future.

Another critical factor before building the vector is choosing the loci on the chromosome that the CRISPR protein complex binding, which will determine the final sequence of gRNA. For deletion of DNA fragment, the binding site is preferred to locate in the region to be removed, while for insertion, it is preferred to locate as close as to the site that the sequence will insert. For point mutation of a specific gene, the target nucleotide(s) is preferred to locate in the PAM sequence or the gRNA binding sequence, of which the synonymous mutation in the repairing template will avoid the self-cutting on the editing vector. For gene repression by CRISPRi, the distance of binding loci to the start codon of the gene will affect the repressing activity—the closer to the start codon, the more repression on the target gene (Gordon et al., 2016).

To construct the vector, both Gibson assembly (Gibson et al., 2009) and Golden Gate assembly (Engler, Kandzia, & Marillonnet, 2008) have been validated and widely used as powerful and efficient cloning tools to assemble multiple pieces of DNA fragments in one step of enzymes reaction. Details of manipulating the process of both cloning strategies will not be discussed here. Briefly, multiple DNA fragments need to be prepared by PCR or restriction enzyme digestion from other DNA fragments. The key point is to make sure there is an overlap/overhang between each neighboring DNA fragment. All DNA fragments will be mixed in an equal molar ratio and added to the enzyme mix for running in a thermocycler.

3.3.2 Transfer the vector into the cyanobacterial host

Two methods are widely used to transfer the heterogenous DNA into cyanobacteria: natural transformation and conjugation. The growth phase of the cells is essential for DNA transfer, where the transformation efficiency is significantly reduced when the cells entered the stationary phase.

Some cyanobacteria can take up foreign DNA in the form of plasmids or linear DNA from the environment. The model cyanobacterial strains *Synechocystis* 6803, *Synechococcus* 7942, and *Synechococcus* 7002 are all naturally transformable. The fast-growing strain *Synechococcus* 2973 is not naturally transformable, but a new variant has been developed and genes responsible for natural transformation elucidated in a recent study (Wendt et al., 2022). Here we use *Synechococcus* 2973 as an example to describe the protocol for natural transformation.

Equipment, reagents, and strain used in this procedure are listed here

- Caron plant growth chamber
- Stratos centrifuge
- Nexcelom cell counter
- BG11 medium
- Purified plasmid DNA
- *Synechococcus* 2973

The procedures for natural transformation are described as following:

1. *Synechococcus* 2973 cells were initially streaked onto BG-11 agar plates and incubated at 38 °C and 75 $\mu mol \cdot m^{-2} \cdot s^{-1}$ with 0.6% CO_2
2. Cells were then resuspended in liquid BG-11 medium and transferred to 125-mL Erlenmeyer flasks to be grown overnight under the same conditions
3. Cells were harvested when the cell density reaches about 1×10^8 cells/mL ($OD_{730} = 0.2–0.3$)

4. A volume of 3 mL of cells was spun down for each experimental sample at 2000 × *g* for 10 min to pellet cells
5. After pelleting, 10 mL of sterile BG-11 was added to the cells and the tube was inverted to resuspend the cells. The previous spin was repeated to pellet cells again
6. Cells were then resuspended in 62.5 μL of sterile BG-11 and transferred to 1.5-mL microcentrifuge tubes, where 125 ng donor DNA was added to each experimental sample
7. Tubes were kept in the dark at room temperature overnight, and then transferred to a 38 °C chamber under 40 $\mu mol \cdot m^{-2} \cdot s^{-1}$ of light intensity
8. Cells remained in the light for 4 h prior to plating on the selective solid medium
9. Single colonies can be isolated after days culture and patched to a new plate for genotype validation

The conjugation strategy can be used to transfer DNA into many cyanobacteria. This technique has been widely used for both filamentous and unicellular strains (Ungerer & Pakrasi, 2016). Briefly, the technique is based on the ability of *E. coli* carrying a conjugal plasmid to transfer DNA via conjugation. When a cyanobacterial strain is mixed with such an *E. coli* strain, the cyanobacterial cell will be able to receive a plasmid transferred from the *E. coli* cell. The DNA to be transferred (i.e., the cargo plasmid) must carry the *oriT* region and a *mob*-gene encoding a nickase, which recognizes the *oriT*-site and enables the transfer of the plasmid. The *mob* gene may also be provided by another helper plasmid. Typically, two different *E. coli* strains are used. One is carrying the conjugal plasmid, and the other the helper plasmid and the cargo plasmid to be transferred to the cyanobacterial cell, which is called triparental conjugation. Here we also use *Synechococcus* 2973 as an example to describe the protocol for triparental conjugation, where *E. coli* HB101 with plasmid pRL443 (Elhai, Vepritskiy, Muro-Pastor, Flores, & Wolk, 1997) was the conjugal strain and *E. coli* HB101 with plasmid pRL623 (Wang et al., 1995) and the cargo plasmid was the helper strain.

Equipment, reagents, and strain used in this procedure are listed here

- Caron plant growth chamber
- Shaker (Innova 44, New Brunswick Scientific)
- Bench centrifuge (Eppendorf)
- Nexcelom cell counter
- BG11 medium
- LB medium
- *E. coli* HB101 with plasmid pRL443

- *E. coli* HB101 with plasmid pRL623
- *Synechococcus* 2973

The procedures for conjugation are described as following:

1. Both *E. coli* strains and *Synechococcus* 2973 were grown in LB and BG11 liquid cultures overnight, respectively
2. 100 μL of *E. coli* cultures and 400 μL of *Synechococcus* 2973 culture ($OD_{730}=0.5$) were washed twice by BG11 medium
3. Centrifugation was performed at $4000 \times g$ for 5 min. Both *E. coli* and cyanobacterial cells were gently stirred during the wash step
4. The washed *E. coli* and *Synechococcus* 2973 cells were mixed with 200 μL of BG11 medium and transferred onto a nitrocellulose membrane (0.45 μm pore size) on a BG11 + 5% LB (*v*/v) agar plate
5. The plates were incubated at 38 °C, 100 $\mu mol \cdot m^{-2} \cdot s^{-1}$ light, and ambient CO_2 overnight
6. The membrane was transferred to new BG11 agar plates with appropriate antibiotics at 38 °C, 400 $\mu mol \cdot m^{-2} \cdot s^{-1}$ light, and 0.6% CO_2
7. Single colonies will appear on the membranes after 2–4 days of incubation and be patched to a new plate for genotype validation

3.3.3 Validate the genotype and phenotype of cyanobacterial transformants

To validate that the CRISPR system is successfully transferred into cyanobacterial cells, cells growing on the plate patched from single colonies can be used as the template for colony PCR using primers specific to the Cas protein and gRNA. Another issue that needs to be considered is the chromosomal segregation when deletion or insertion of DNA fragments into the chromosome, which presented multiple copies in cyanobacteria. An extended selection period may be required to fully segregate the modified allele from all wild-type copies, even the CRISPR editing technology should shorten the process. To confirm the genotype and for further analysis, the genomic DNA of transformants can be extracted according to the following procedure.

Equipment, reagents, and buffer used in this procedure are listed here

- Stratos centrifuge
- Heat block
- Water bath
- Bench centrifuge (Eppendorf)
- Liquid nitrogen

- STET buffer (50 mM Tris, 100 mM NaCl, 10 mM EDTA and 0.5% (*v*/v) Triton X-100, pH 8.0)
- Lysozyme (MilliporeSigma)

The procedures for genomic DNA isolation are described as following:

1. Spin down 5–10 mL cyanobacterial cells collected from the liquid culture, re-suspend pellet in 0.5 mL STET buffer, freeze the cells suspension in liquid nitrogen and then incubate cells at 70 °C for 10 min
2. Add 10 mg/mL lysozyme, incubate at 37 °C at least 3 h, add 125 μL 10% SDS, vortex, and incubate at 37 °C for 0.5–1 h
3. Add the equal volume of phenol to the mixture, vortex, centrifuge at 12000 × *g* for 5 min, move the upper phase to a new tube, and repeat this step one more time
4. Add equal volume of chloroform to the mixture form the upper phase, vortex, centrifuge at 12000 × *g* for 5 min, and move the upper phase to a new tube
5. Add 0.6-fold volume of isopropanol and 0.1-fold volume of 3 M Na-acetate (pH 6.2), mix it, incubate it at −20 °C overnight, centrifuge at 12000 × *g* for 10 min, discard the supernatant, add 200 μL of70% (*v*/v) ethanol to gently suspend the pellet, and leave the tube on ice for 5 min
6. Centrifuge at 12000 × *g* for 5 min, remove all the liquid, dry the pellet in the air for 2 min, and suspend the genomic DNA in 100 μL water

When using the CRISPR technology to do the markerless genome editing, such as deletion, insertion, and point mutation, after confirmation of the desired genotype type, the editing vector can be cured from the cyanobacterial transformant. The procedure is simple, transformants will be cultured in a liquid medium without any selection pressure for multiple sequential dilutions (depending on the cyanobacterial species) and cells will be spread to the plate to form single colonies. Colonies will then be checked the antibiotics sensitivity and the ones that lost the resistance ability will be checked by PCR to confirm the curing of the plasmid.

To validate the genotype of transformants containing the CRISPRi system, it is necessary to measure the transcriptional level of targeting genes, which requires the extraction of the total RNA from cells and measurement by quantitative RT-PCR. The following protocol briefly described the whole process.

Equipment, reagents, and buffer used in this procedure are listed here

- Stratos centrifuge
- Heat block
- Colling centrifuge (Eppendorf, 5424R)

- Nano drop spectrophotometer
- PCR Thermocycler machine (Bio-Rad, C1000 Touch)
- Quantitative PCR machine (Bio-Rad, CFX Duet)
- Liquid nitrogen
- RNAWiz (Ambion/Thermo Scientific Science)
- DEPC water (RNase free)
- DNase enzyme (Thermo-fisher Scientific)
- Superscript II Reverse Transcriptase (Thermo-fisher Scientific)
- Random primers for RNA to cDNA (Thermo-fisher Scientific)
- SYBR Green Real-Time PCR Master Mixes (Thermo-fisher Scientific)

The procedures for total RNA isolation and cDNA synthesis are described as following:

1. Harvest cells from liquid culture, wash once by water, freeze in liquid nitrogen, re-suspend cells in 1 mL 70 °C pre-warmed RNAwiz, quickly vortex, and heat at 70 °C from 10 min
2. Add 200 μL chloroform, vortex 20 s, incubate 5 min, centrifuge at 4 °C with 12,000 × *g* for 15 min, move the upper phase to a new tube, and repeat this step one more time
3. Add 0.5-fold volume RNase free water and equal volume isopropanol, mix and incubate on ice or 10 min
4. Centrifuge at 4 °C with 12,000 × *g* for 15 min, wash the pellet by ice-cold 70% ethanol, spin down, air dry the pellet 2 min, dissolve the total RNA with RNAase free water at 60 °C for 5 min, and quantitate the concentration by Nanodrop
5. Remove the contaminated DNA by DNase, repeat the steps 2–4 to collect the pure RNA sample, and synthesize cDNA from 200 ng total RNA using the superscript II reverse transcriptase and random primers
6. Run quantitative PCR (q-PCR) using the QRT-PCR Sybr green dUTP mix on the q-PCR machine and calculate the relative transcriptional levels of target genes using the average threshold cycle (CT) value normalized by the level of house-keeping genes

Another level of phenotype that can be validated is the expression of target proteins in transformants. Cyanobacteria are unique prokaryotes because they possess complex and highly differentiated membrane systems. In addition to having an outer and inner membrane system, inherent to Gram-negative bacteria, cyanobacteria also possess thylakoid membranes, where both electron transport chains of photosynthesis and respiration are located. To extract soluble or membrane proteins from cyanobacteria, the first step is to disrupt the cell. Several approaches have been adopted, including chemically disrupting the cell wall structure (such as lysozyme), mechanically breaking the cells (such as glass beads, French press, and sonication), or a combination of both. Soluble and membrane proteins were separated by using the ultra-speed centrifuge. One more step for the membrane protein after extraction is to solubilize it using a non-ionic detergent, such as Dodecyl-β-D-maltoside (DDM), for further analyze. Many papers published in our lab have described the details of the procedure and we will not cover it here.

4. Summary

CRISPR-Cas is a revolutionary technology that has been borrowed from the bacterial system to be modified and used for editing genome, modulating gene expression and containing genetically modified strains. The antibiotic markerless engineering and the ability to simultaneously target polyploid genome have made CRISPR-Cas technique attractive for multi-genome organisms, like cyanobacteria. Cas12a/Cas9-mediated genome engineering has become widely popular in cyanobacteria. However, these techniques in their current form are limited in cyanobacteria for deleting large genomic regions, genome engineering of novel cyanobacteria where the cellular machinery is not well exploited, gene activation, repression of genes part of the operon, and efficient point mutation.

Cyanobacteria are a complex set of organisms that exhibit various fascinating features and therefore implementing these novel Cas proteins and technologies for genome engineering will truly help understand the complexity of these photoautotrophs and gear toward developing these strains as carbon negative cellular factories; however, there are still several challenges we are facing with this technology in cyanobacteria.

1. Even a long DNA fragment (>110 kb) deletion has been succeeded in *Anabaena* 7120 (Niu et al., 2019), other cyanobacterial species need to be validated. And more importantly, insertion of a long DNA fragment (>50 kb) into the chromosome has never been proved in cyanobacteria. Success to delete and insert long fragment efficiently and robustly will ease the synthetic biology and pathway engineering applications in cyanobacteria
2. We realize that synthetic biology tools are still the key limiting factor for engineering work in cyanobacteria, such as promoters, ribosome binding sites, and terminators. For most of cyanobacteria, a tight, broad

range, and robust inducible promoter is still not available. Establish a library containing various bio-bricks is very useful for further studies
3. For gene repression by CRISPRi, the gene located in the operon is always difficult to design the gRNA. Repressing on the first gene of an operon will generate the polar effect, while if the target gene is not the first gene, the repressing level sometimes is low
4. Only the Cas9 and Cas12a proteins and only CRISPR and CRISPRi technologies have been explored in cyanobacteria. So other Cas proteins with various functions are of great interest to test, which should facilitate the genetic work in cyanobacteria with specific purpose

Acknowledgment

This study was supported by National Science Foundation (NSF) Grant MCB-2037887 to H.B.P. We thank members of the Pakrasi lab for their support and useful discussions.

References

Abe, K., Miyake, K., Nakamura, M., Kojima, K., Ferri, S., Ikebukuro, K., et al. (2014). Engineering of a green-light inducible gene expression system in *Synechocystis* sp. PCC 6803. *Microbial Biotechnology*, 7, 177–183.

Abernathy, M. H., Yu, J., Ma, F., et al. (2017). Deciphering cyanobacterial phenotypes for fast photoautotrophic growth via isotopically nonstationary metabolic flux analysis. *Biotechnology for Biofuels*, *10*, 273.

Bandyopadhyay, A., Elvitigala, T., Liberton, M., & Pakrasi, H. B. (2013). Variations in the rhythms of respiration and nitrogen fixation in members of the unicellular diazotrophic cyanobacterial genus *Cyanothece*. *Plant Physiology*, *161*, 1334–1346.

Bandyopadhyay, A., Stockel, J., Min, H., Sherman, L. A., & Pakrasi, H. B. (2010). High rates of photobiological H_2 production by a cyanobacterium under aerobic conditions. *Nature Communications*, *1*, 139.

Bartasun, P., Prandi, N., Storch, M., Aknin, Y., Bennett, M., Palma, A., et al. (2019). The effect of modulating the quantity of enzymes in a model ethanol pathway on metabolic flux in *Synechocystis* sp. PCC 6803. *PeerJ*, 7, e7529.

Bayat, H., Modarressi, M. H., & Rahimpour, A. (2018). The conspicuity of CRISPR-Cpf1 system as a significant breakthrough in genome editing. *Current Microbiology*, *75*, 107–115.

Behler, J., Vijay, D., Hess, W. R., & Akhtar, M. K. (2018). CRISPR-based technologies for metabolic engineering in cyanobacteria. *Trends in Biotechnology*, *36*, 996–1010.

Bhaya, D., Davison, M., & Barrangou, R. (2011). CRISPR-Cas systems in bacteria and archaea: Versatile small RNAs for adaptive defense and regulation. *Annual Review of Genetics*, *45*, 273–297.

Blasi, B., Peca, L., Vass, I., & Kos, P. B. (2012). Characterization of stress responses of heavy metal and metalloid inducible promoters in *Synechocystis* PCC 6803. *Journal of Microbiology and Biotechnology*, *22*, 166–169.

Cai, F., Axen, S. D., & Kerfeld, C. A. (2013). Evidence for the widespread distribution of CRISPR-Cas system in the Phylum Cyanobacteria. *RNA Biology*, *10*, 687–693.

Carrieri, D., Lombardi, T., Paddock, T., et al. (2017). Transcriptome and proteome analysis of nitrogen starvation responses in *Synechocystis* 6803 ΔglgC, a mutant incapable of glycogen storage. *Algal Research*, *21*, 64–75.

Chen, H. S., Niedzwiedzki, D. M., Bandyopadhyay, A., Biswas, S., & Pakrasi, H. B. (2021). A novel mode of photoprotection mediated by a cysteine residue in the chlorophyll protein IsiA. *mBio*, *12*, e03663-20.

Cheng, Z. H., Wu, J., Liu, J. Q., Min, D., Liu, D. F., Li, W. W., et al. (2022). Repurposing CRISPR RNA-guided integrases system for one-step, efficient genomic integration of ultra-long DNA sequences. *Nucleic Acids Research*, *50*(13), 7739–7750.

Choi, Y.-N., Lee, J. W., Kim, J. W., & Park, J. M. (2020). Acetyl-CoA-derived biofuel and biochemical production in cyanobacteria: A mini review. *Journal of Applied Phycology*, *32*, 1643–1653.

Choi, Y. N., & Park, J. M. (2016). Enhancing biomass and ethanol production by increasing NADPH production in *Synechocystis* sp. PCC 6803. *Bioresource Technology*, *213*, 54–57.

Choi, S. Y., & Woo, H. M. (2020). CRISPRi-dCas12a: A dCas12a-mediated CRISPR interference for repression of multiple genes and metabolic engineering in cyanobacteria. *ACS Synthetic Biology*, *9*, 2351–2361.

Cohen, S. E., & Golden, S. S. (2015). Circadian rhythms in cyanobacteria. *Microbiology and Molecular Biology Reviews*, *79*, 373–385.

Csorgo, B., Leon, L. M., Chau-Ly, I. J., Vasquez-Rifo, A., Berry, J. D., Mahendra, C., et al. (2020). A compact Cascade-Cas3 system for targeted genome engineering. *Nature Methods*, *17*, 1183–1190.

Dietsch, M., Behle, A., Westhoff, P., & Axmann, I. M. (2021). Metabolic engineering of *Synechocystis* sp. PCC 6803 for the photoproduction of the sesquiterpene valencene. *Metabolic Engineering Communications*, *13*, e00178.

Ditty, J. L., Canales, S. R., Anderson, B. E., Williams, S. B., & Golden, S. S. (2005). Stability of the *Synechococcus elongatus* PCC 7942 circadian clock under directed anti-phase expression of the kai genes. *Microbiology*, *151*, 2605–2613.

Dong, G., & Golden, S. S. (2008). How a cyanobacterium tells time. *Current Opinion in Microbiology*, *11*, 541–546.

Dong, G., Kim, Y. I., & Golden, S. S. (2010). Simplicity and complexity in the cyanobacterial circadian clock mechanism. *Current Opinion in Genetics & Development*, *20*, 619–625.

Doudna, J. A., & Charpentier, E. (2014). Genome editing. The new frontier of genome engineering with CRISPR-Cas9. *Science*, *346*, 1258096.

Ducat, D. C., Avelar-Rivas, J. A., Way, J. C., & Silver, P. A. (2012). Rerouting carbon flux to enhance photosynthetic productivity. *Applied and Environmental Microbiology*, *78*, 2660–2668.

Elhai, J., Vepritskiy, A., Muro-Pastor, A. M., Flores, E., & Wolk, C. P. (1997). Reduction of conjugal transfer efficiency by three restriction activities of *Anabaena* sp. strain PCC 7120. *Journal of Bacteriology*, *179*, 1998–2005.

Engler, C., Kandzia, R., & Marillonnet, S. (2008). A one pot, one step, precision cloning method with high throughput capability. *PLoS One*, *3*, e3647.

Ferreira, E. A., Pacheco, C. C., Pinto, F., Pereira, J., Lamosa, P., Oliveira, P., et al. (2018). Expanding the toolbox for *Synechocystis* sp. PCC 6803: Validation of replicative vectors and characterization of a novel set of promoters. *Synthetic Biology*, *3*, ysy014.

Fujisawa, T., Narikawa, R., Maeda, S. I., et al. (2017). CyanoBase: A large-scale update on its 20th anniversary. *Nucleic Acids Research*, *45*, D551–D554.

Gao, X., Gao, F., Liu, D., Zhang, H., Nie, X., & Yang, C. (2016). Engineering the methylerythritol phosphate pathway in cyanobacteria for photosynthetic isoprene production from CO_2. *Energy & Environmental Science*, *9*, 1400–1411.

Gaudana, S. B., Krishnakumar, S., Alagesan, S., Digmurti, M. G., Viswanathan, G. A., Chetty, M., et al. (2013). Rhythmic and sustained oscillations in metabolism and gene expression of *Cyanothece* sp. ATCC 51142 under constant light. *Frontiers in Microbiology*, *4*, 374.

Gibson, D. G., Young, L., Chuang, R. Y., Venter, J. C., Hutchison, C. A., 3rd, & Smith, H. O. (2009). Enzymatic assembly of DNA molecules up to several hundred kilobases. *Nature Methods*, *6*, 343–345.
Gordon, G. C., Korosh, T. C., Cameron, J. C., Markley, A. L., Begemann, M. B., & Pfleger, B. F. (2016). CRISPR interference as a titratable, trans-acting regulatory tool for metabolic engineering in the cyanobacterium *Synechococcus* sp. strain PCC 7002. *Metabolic Engineering*, *38*, 170–179.
Gu, P., Yang, F., Su, T., Wang, Q., Liang, Q., & Qi, Q. (2015). A rapid and reliable strategy for chromosomal integration of gene(s) with multiple copies. *Scientific Reports*, *5*, 9684.
Hagemann, M., & Hess, W. R. (2018). Systems and synthetic biology for the biotechnological application of cyanobacteria. *Current Opinion in Biotechnology*, *49*, 94–99.
Heidorn, T., Camsund, D., Huang, H. H., Lindberg, P., Oliveira, P., Stensjo, K., et al. (2011). Synthetic biology in cyanobacteria engineering and analyzing novel functions. *Methods in Enzymology*, *497*, 539–579.
Hendry, J. I., Prasannan, C. B., Joshi, A., Dasgupta, S., & Wangikar, P. P. (2016). Metabolic model of *Synechococcus* sp. PCC 7002: Prediction of flux distribution and network modification for enhanced biofuel production. *Bioresource Technology*, *213*, 190–197.
Hendry, J. I., Prasannan, C., Ma, F., Mollers, K. B., Jaiswal, D., Digmurti, M., et al. (2017). Rerouting of carbon flux in a glycogen mutant of cyanobacteria assessed via isotopically non-stationary ^{13}C metabolic flux analysis. *Biotechnology and Bioengineering*, *114*, 2298–2308.
Higo, A., & Ehira, S. (2019). Spatiotemporal gene repression system in the heterocyst-forming multicellular cyanobacterium *Anabaena* sp. PCC 7120. *ACS Synthetic Biology*, *8*, 641–646.
Higo, A., Isu, A., Fukaya, Y., Ehira, S., & Hisabori, T. (2018). Application of CRISPR interference for metabolic engineering of the heterocyst-forming multicellular cyanobacterium *Anabaena* sp. PCC 7120. *Plant & Cell Physiology*, *59*, 119–127.
Higo, A., Nishiyama, E., Nakamura, K., Hihara, Y., & Ehira, S. (2019). cyAbrB transcriptional regulators as safety devices to inhibit heterocyst differentiation in *Anabaena* sp. strain PCC 7120. *Journal of Bacteriology*, *201*, e00244-19.
Hou, S., Brenes-Alvarez, M., Reimann, V., Alkhnbashi, O. S., Backofen, R., Muro-Pastor, A. M., et al. (2019). CRISPR-Cas systems in multicellular cyanobacteria. *RNA Biology*, *16*, 518–529.
Huang, C. H., Shen, C. R., Li, H., Sung, L. Y., Wu, M. Y., & Hu, Y. C. (2016). CRISPR interference (CRISPRi) for gene regulation and succinate production in cyanobacterium *S. elongatus* PCC 7942. *Microbial Cell Factories*, *15*, 196.
Jacobsen, J. H., & Frigaard, N. U. (2014). Engineering of photosynthetic mannitol biosynthesis from CO_2 in a cyanobacterium. *Metabolic Engineering*, *21*, 60–70.
Jaiswal, D., Sengupta, A., Sengupta, S., Madhu, S., Pakrasi, H. B., & Wangikar, P. P. (2020). A novel cyanobacterium *Synechococcus elongatus* PCC 11802 has distinct genomic and metabolomic characteristics compared to its neighbor PCC 11801. *Scientific Reports*, *10*, 191.
Jaiswal, D., Sengupta, A., Sohoni, S., Sengupta, S., Phadnavis, A. G., Pakrasi, H. B., et al. (2018). Genome features and biochemical characteristics of a robust, fast growing and naturally transformable cyanobacterium *Synechococcus elongatus* PCC 11801 isolated from India. *Scientific Reports*, *8*, 16632.
Jaiswal, D., & Wangikar, P. P. (2020). Dynamic inventory of intermediate metabolites of cyanobacteria in a diurnal cycle. *iScience*, *23*, 101704.
Jeong, Y., Cho, S. H., Lee, H., Choi, H. K., Kim, D. M., Lee, C. G., et al. (2020). Current status and future strategies to increase secondary metabolite production from cyanobacteria. *Microorganisms*, *8*, 1849.
Johnson, C. H., Mori, T., & Xu, Y. (2008). A cyanobacterial circadian clockwork. *Current Biology*, *18*, R816–R825.

Kaczmarzyk, D., Cengic, I., Yao, L., & Hudson, E. P. (2018). Diversion of the long-chain acyl-ACP pool in *Synechocystis* to fatty alcohols through CRISPRi repression of the essential phosphate acyltransferase PlsX. *Metabolic Engineering*, *45*, 59–66.

Kanno, M., & Atsumi, S. (2017). Engineering an obligate photoautotrophic cyanobacterium to utilize glycerol for growth and chemical production. *ACS Synthetic Biology*, *6*, 69–75.

Kirtania, P., Hódi, B., Mallick, I., Vass, I. Z., Fehér, T., Vass, I., et al. (2019). A single plasmid based CRISPR interference in *Synechocystis* 6803—A proof of concept. *PLoS One*, *14*, e0225375.

Knoot, C. J., Biswas, S., & Pakrasi, H. B. (2020). Tunable repression of key photosynthetic processes using Cas12a CRISPR interference in the fast-growing cyanobacterium *Synechococcus* sp. UTEX 2973. *ACS Synthetic Biology*, *9*, 132–143.

Knoot, C. J., Khatri, Y., Hohlman, R. M., Sherman, D. H., & Pakrasi, H. B. (2019). Engineered production of hapalindole alkaloids in the cyanobacterium *Synechococcus* sp. UTEX 2973. *ACS Synthetic Biology*, *8*, 1941–1951.

Knoot, C. J., & Pakrasi, H. B. (2019). Diverse hydrocarbon biosynthetic enzymes can substitute for olefin synthase in the cyanobacterium *Synechococcus* sp. PCC 7002. *Scientific Reports*, *9*, 1360.

Kondo, K., Geng, X. X., Katayama, M., & Ikeuchi, M. (2005). Distinct roles of CpcG1 and CpcG2 in phycobilisome assembly in the cyanobacterium *Synechocystis* sp. PCC 6803. *Photosynthesis Research*, *84*, 269–273.

Kopka, J., Schmidt, S., Dethloff, F., et al. (2017). Systems analysis of ethanol production in the genetically engineered cyanobacterium *Synechococcus* sp. PCC 7002. *Biotechnology for Biofuels*, *10*, 56.

Lan, E. I., & Liao, J. C. (2012). ATP drives direct photosynthetic production of 1-butanol in cyanobacteria. *Proceedings of the National Academy of Sciences of the United States of America*, *109*, 6018–6023.

Lee, M., & Woo, H. M. (2020). A logic NAND gate for controlling gene expression in a circadian rhythm in cyanobacteria. *ACS Synthetic Biology*, *9*, 3210–3216.

Li, H., & Liao, J. C. (2013). Engineering a cyanobacterium as the catalyst for the photosynthetic conversion of CO_2 to 1,2-propanediol. *Microbial Cell Factories*, *12*, 4.

Li, H., Pham, N. N., Shen, C. R., Chang, C. W., Tu, Y., Chang, Y. H., et al. (2022). Combinatorial CRISPR interference library for enhancing 2,3-BDO production and elucidating key genes in cyanobacteria. *Frontiers in Bioengineering and Biotechnology*, *10*, 913820.

Li, H., Shen, C. R., Huang, C. H., Sung, L. Y., Wu, M. Y., & Hu, Y. C. (2016). CRISPR-Cas9 for the genome engineering of cyanobacteria and succinate production. *Metabolic Engineering*, *38*, 293–302.

Lin, P. C., Saha, R., Zhang, F., & Pakrasi, H. B. (2017). Metabolic engineering of the pentose phosphate pathway for enhanced limonene production in the cyanobacterium *Synechocystis* sp. PCC 6803. *Scientific Reports*, *7*, 17503.

Lin, P. C., Zhang, F., & Pakrasi, H. B. (2020). Enhanced production of sucrose in the fast-growing cyanobacterium *Synechococcus elongatus* UTEX 2973. *Scientific Reports*, *10*, 390.

Lin, P. C., Zhang, F., & Pakrasi, H. B. (2021). Enhanced limonene production in a fast-growing cyanobacterium through combinatorial metabolic engineering. *Metabolic Engineering Communications*, *12*, e00164.

Liu, D., Johnson, V. M., & Pakrasi, H. B. (2020). A reversibly induced CRISPRi system targeting photosystem II in the cyanobacterium *Synechocystis* sp. PCC 6803. *ACS Synthetic Biology*, *9*, 1441–1449.

Liu, D., Liberton, M., Yu, J., Pakrasi, H. B., & Bhattacharyya-Pakrasi, M. (2018). Engineering nitrogen fixation activity in an oxygenic phototroph. *mBio*, *9*, e01029-18.

Liu, X., Miao, R., Lindberg, P., & Lindblad, P. (2019). Modular engineering for efficient photosynthetic biosynthesis of 1-butanol from CO_2 in cyanobacteria. *Energy & Environmental Science*, *12*, 2765–2777.

Liu, D., & Pakrasi, H. B. (2018). Exploring native genetic elements as plug-in tools for synthetic biology in the cyanobacterium *Synechocystis* sp. PCC 6803. *Microbial Cell Factories*, *17*, 48.

Makarova, K. S., & Koonin, E. V. (2015). Annotation and classification of CRISPR-Cas systems. *Methods in Molecular Biology*, *1311*, 47–75.

Makarova, K. S., Zhang, F., & Koonin, E. V. (2017). SnapShot: Class 2 CRISPR-Cas systems. *Cell*, *168*, 328–328.e321.

Markley, A. L., Begemann, M. B., Clarke, R. E., Gordon, G. C., & Pfleger, B. F. (2015). Synthetic biology toolbox for controlling gene expression in the cyanobacterium *Synechococcus* sp. strain PCC 7002. *ACS Synthetic Biology*, *4*, 595–603.

Mills, L. A., Moreno-Cabezuelo, J. A., Włodarczyk, A., et al. (2022). Development of a biotechnology platform for the fast-growing cyanobacterium *Synechococcus* sp. PCC 11901. *Biomolecules*, *12*, 872.

Nakamura, Y., Kaneko, T., & Tabata, S. (2000). CyanoBase, the genome database for *Synechocystis* sp. strain PCC6803: Status for the year 2000. *Nucleic Acids Research*, *28*, 72.

Niu, T. C., Lin, G. M., Xie, L. R., Wang, Z. Q., Xing, W. Y., Zhang, J. Y., et al. (2019). Expanding the potential of CRISPR-Cpf1-based genome editing technology in the cyanobacterium *Anabaena* PCC 7120. *ACS Synthetic Biology*, *8*, 170–180.

Nozzi, N. E., & Atsumi, S. (2015). Genome engineering of the 2,3-butanediol biosynthetic pathway for tight regulation in cyanobacteria. *ACS Synthetic Biology*, *4*, 1197–1204.

Pattharaprachayakul, N., Lee, M., Incharoensakdi, A., & Woo, H. M. (2020). Current understanding of the cyanobacterial CRISPR-Cas systems and development of the synthetic CRISPR-Cas systems for cyanobacteria. *Enzyme and Microbial Technology*, *140*, 109619.

Phipps, J., & Dubrana, K. (2022). DNA repair in space and time: Safeguarding the genome with the cohesin complex. *Genes (Basel)*, *13*, 198.

Pisciotta, J. M., Zou, Y., & Baskakov, I. V. (2010). Light-dependent electrogenic activity of cyanobacteria. *PLoS One*, *5*, e10821.

Puerta-Fernandez, E., & Vioque, A. (2011). Hfq is required for optimal nitrate assimilation in the Cyanobacterium *Anabaena* sp. strain PCC 7120. *Journal of Bacteriology*, *193*, 3546–3555.

Qi, L. S., Larson, M. H., Gilbert, L. A., Doudna, J. A., Weissman, J. S., Arkin, A. P., et al. (2013). Repurposing CRISPR as an RNA-guided platform for sequence-specific control of gene expression. *Cell*, *152*, 1173–1183.

Racharaks, R., Arnold, W., & Peccia, J. (2021). Development of CRISPR-Cas9 knock-in tools for free fatty acid production using the fast-growing cyanobacterial strain *Synechococcus elongatus* UTEX 2973. *Journal of Microbiological Methods*, *189*, 106315.

Rippka, R., Deruelles, J., Waterbury, J. B., Herdman, M., & Stanier, R. Y. (1979). Generic assignments, strain histories and properties of pure cultures of cyanobacteria. *Microbiology*, *111*, 1–61.

Sanchez-Baracaldo, P., Hayes, P. K., & Blank, C. E. (2005). Morphological and habitat evolution in the cyanobacteria using a compartmentalization approach. *Geobiology*, *3*, 145–165.

Sander, J. D., & Joung, J. K. (2014). CRISPR-Cas systems for editing, regulating and targeting genomes. *Nature Biotechnology*, *32*, 347–355.

Santos, M., Pacheco, C. C., Yao, L., Hudson, E. P., & Tamagnini, P. (2021). CRISPRi as a tool to repress multiple copies of extracellular polymeric substances (EPS)-related genes in the cyanobacterium *Synechocystis* sp. PCC 6803. *Life (Basel)*, *11*, 1198.

Santos-Merino, M., Singh, A. K., & Ducat, D. C. (2019). New applications of synthetic biology tools for cyanobacterial metabolic engineering. *Frontiers in Bioengineering and Biotechnology*, *7*, 33.

Scholz, I., Lange, S. J., Hein, S., Hess, W. R., & Backofen, R. (2013). CRISPR-Cas systems in the cyanobacterium *Synechocystis* sp. PCC6803 exhibit distinct processing pathways involving at least two Cas6 and a Cmr2 protein. *PLoS One*, *8*, e56470.

Sengupta, S., Jaiswal, D., Sengupta, A., Shah, S., Gadagkar, S., & Wangikar, P. P. (2020). Metabolic engineering of a fast-growing cyanobacterium *Synechococcus elongatus* PCC 11801 for photoautotrophic production of succinic acid. *Biotechnology for Biofuels*, *13*, 89.

Sengupta, A., Madhu, S., & Wangikar, P. P. (2020). A library of tunable, portable, and inducer-free promoters derived from cyanobacteria. *ACS Synthetic Biology*, *9*, 1790–1801.

Sengupta, A., Pakrasi, H. B., & Wangikar, P. P. (2018). Recent advances in synthetic biology of cyanobacteria. *Applied Microbiology and Biotechnology*, *102*, 5457–5471.

Sengupta, A., Pritam, P., Jaiswal, D., Bandyopadhyay, A., Pakrasi, H. B., & Wangikar, P. P. (2020). Photosynthetic co-production of succinate and ethylene in a fast-growing cyanobacterium, *Synechococcus elongatus* PCC 11801. *Metabolites*, *10*, 250.

Sengupta, A., Sunder, A. V., Sohoni, S. V., & Wangikar, P. P. (2019a). The effect of CO_2 in enhancing photosynthetic cofactor recycling for alcohol dehydrogenase mediated chiral synthesis in cyanobacteria. *Journal of Biotechnology*, *289*, 1–6.

Sengupta, A., Sunder, A. V., Sohoni, S. V., & Wangikar, P. P. (2019b). Fine-tuning native promoters of *Synechococcus elongatus* PCC 7942 to develop a synthetic toolbox for heterologous protein expression. *ACS Synthetic Biology*, *8*, 1219–1223.

Shabestary, K., Anfelt, J., Ljungqvist, E., Jahn, M., Yao, L., & Hudson, E. P. (2018). Targeted repression of essential genes to arrest growth and increase carbon partitioning and biofuel titers in cyanobacteria. *ACS Synthetic Biology*, *7*, 1669–1675.

Shabestary, K., Hernández, H. P., Miao, R., Ljungqvist, E., Hallman, O., Sporre, E., et al. (2021). Cycling between growth and production phases increases cyanobacteria bioproduction of lactate. *Metabolic Engineering*, *68*, 131–141.

Sharda, M., Badrinarayanan, A., & Seshasayee, A. S. N. (2020). Evolutionary and comparative analysis of bacterial nonhomologous end joining repair. *Genome Biology and Evolution*, *12*, 2450–2466.

Shih, P. M., Wu, D., Latifi, A., et al. (2013). Improving the coverage of the cyanobacterial phylum using diversity-driven genome sequencing. *Proceedings of the National Academy of Sciences of the United States of America*, *110*, 1053–1058.

Srivastava, V., Amanna, R., Rowden, S. J. L., Sengupta, S., Madhu, S., Howe, C. J., et al. (2021). Adaptive laboratory evolution of the fast-growing cyanobacterium *Synechococcus elongatus* PCC 11801 for improved solvent tolerance. *Journal of Bioscience and Bioengineering*, *131*, 491–500.

Stockel, J., Welsh, E. A., Liberton, M., Kunnvakkam, R., Aurora, R., & Pakrasi, H. B. (2008). Global transcriptomic analysis of *Cyanothece* 51142 reveals robust diurnal oscillation of central metabolic processes. *Proceedings of the National Academy of Sciences of the United States of America*, *105*, 6156–6161.

Strecker, J., Ladha, A., Gardner, Z., Schmid-Burgk, J. L., Makarova, K. S., Koonin, E. V., et al. (2019). RNA-guided DNA insertion with CRISPR-associated transposases. *Science*, *365*, 48–53.

Tan, X., Hou, S., Song, K., Georg, J., Klahn, S., Lu, X., et al. (2018). The primary transcriptome of the fast-growing cyanobacterium *Synechococcus elongatus* UTEX 2973. *Biotechnology for Biofuels*, *11*, 218.

Tang, Y., & Fu, Y. (2018). Class 2 CRISPR/Cas: An expanding biotechnology toolbox for and beyond genome editing. *Cell & Bioscience*, *8*, 59.

Ungerer, J., & Pakrasi, H. B. (2016). Cpf1 is a versatile tool for CRISPR genome editing across diverse species of cyanobacteria. *Scientific Reports*, *6*, 39681.

Ungerer, J., Wendt, K. E., Hendry, J. I., Maranas, C. D., & Pakrasi, H. B. (2018). Comparative genomics reveals the molecular determinants of rapid growth of the cyanobacterium *Synechococcus elongatus* UTEX 2973. *Proceedings of the National Academy of Sciences of the United States of America, 115*, E11761–E11770.

Uniyal, A. P., Mansotra, K., Yadav, S. K., & Kumar, V. (2019). An overview of designing and selection of sgRNAs for precise genome editing by the CRISPR-Cas9 system in plants. *3 Biotech, 9*, 223.

Vasudevan, R., Gale, G. A. R., Schiavon, A. A., et al. (2019). CyanoGate: A modular cloning suite for engineering cyanobacteria based on the plant MoClo syntax. *Plant Physiology, 180*, 39–55.

Wang, T., Guan, C., Guo, J., Liu, B., Wu, Y., Xie, Z., et al. (2018). Pooled CRISPR interference screening enables genome-scale functional genomics study in bacteria with superior performance. *Nature Communications, 9*, 2475.

Wang, D., Meier, T. I., Chan, C. L., Feng, G., Lee, D. N., & Landick, R. (1995). Discontinuous movements of DNA and RNA in RNA polymerase accompany formation of a paused transcription complex. *Cell, 81*, 341–350.

Wealand, J. L., Myers, J. A., & Hirschberg, R. (1989). Changes in gene expression during nitrogen starvation in *Anabaena variabilis* ATCC 29413. *Journal of Bacteriology, 171*, 1309–1313.

Weisz, D. A., Johnson, V. M., Niedzwiedzki, D. M., Shinn, M. K., Liu, H., Klitzke, C. F., et al. (2019). A novel chlorophyll protein complex in the repair cycle of photosystem II. *Proceedings of the National Academy of Sciences of the United States of America, 116*, 21907–21913.

Wendt, K. E., Ungerer, J., Cobb, R. E., Zhao, H., & Pakrasi, H. B. (2016). CRISPR/Cas9 mediated targeted mutagenesis of the fast growing cyanobacterium *Synechococcus elongatus* UTEX 2973. *Microbial Cell Factories, 15*, 115.

Wendt, K. E., Walker, P., Sengupta, A., Ungerer, J., & Pakrasi, H. B. (2022). Engineering natural competence into the fast-growing cyanobacterium *Synechococcus elongatus* strain UTEX 2973. *Applied and Environmental Microbiology, 88*, e0188221.

Wlodarczyk, A., Selao, T. T., Norling, B., & Nixon, P. J. (2020). Newly discovered *Synechococcus* sp. PCC 11901 is a robust cyanobacterial strain for high biomass production. *Communications Biology, 3*, 215.

Wu, X., Kriz, A. J., & Sharp, P. A. (2014). Target specificity of the CRISPR-Cas9 system. *Quantitative Biology, 2*, 59–70.

Xiao, Y., Wang, S., Rommelfanger, S., Balassy, A., Barba-Ostria, C., Gu, P., et al. (2018). Developing a Cas9-based tool to engineer native plasmids in *Synechocystis* sp. PCC 6803. *Biotechnology and Bioengineering, 115*, 2305–2314.

Xu, X., & Qi, L. S. (2019). A CRISPR-dCas toolbox for genetic engineering and synthetic biology. *Journal of Molecular Biology, 431*, 34–47.

Yao, L., Cengic, I., Anfelt, J., & Hudson, E. P. (2016). Multiple gene repression in cyanobacteria using CRISPRi. *ACS Synthetic Biology, 5*, 207–212.

Yao, L., Shabestary, K., Bjork, S. M., Asplund-Samuelsson, J., Joensson, H. N., Jahn, M., et al. (2020). Pooled CRISPRi screening of the cyanobacterium *Synechocystis* sp PCC 6803 for enhanced industrial phenotypes. *Nature Communications, 11*, 1666.

Yu, J., Liberton, M., Cliften, P. F., Head, R. D., Jacobs, J. M., Smith, R. D., et al. (2015). *Synechococcus elongatus* UTEX 2973, a fast growing cyanobacterial chassis for biosynthesis using light and CO_2. *Scientific Reports, 5*, 8132.

Yunus, I. S., Anfelt, J., Sporre, E., Miao, R., Hudson, E. P., & Jones, P. R. (2022). Synthetic metabolic pathways for conversion of CO_2 into secreted short-to medium-chain hydrocarbons using cyanobacteria. *Metabolic Engineering, 72*, 14–23.

Zahra, Z., Choo, D. H., Lee, H., & Parveen, A. (2020). Cyanobacteria: Review of current potentials and applications. *Environments*, 7, 13.

9780323955614